Infection and Cancer: Bi-Directorial Interactions

Michael R. Shurin • Yasmin Thanavala
Nahed Ismail
Editors

Infection and Cancer: Bi-Directorial Interactions

Springer

Editors
Michael R. Shurin
Department of Pathology
University of Pittsburgh Medical Center
Pittsburgh, Pennsylvania, USA

Yasmin Thanavala
Department of Immunology
Roswell Park Cancer Institute
Buffalo, New York, USA

Nahed Ismail
Department of Pathology
University of Pittsburgh Medical Center
Pittsburgh, Pennsylvania, USA

ISBN 978-3-319-20668-4 ISBN 978-3-319-20669-1 (eBook)
DOI 10.1007/978-3-319-20669-1

Library of Congress Control Number: 2015955137

Springer Cham Heidelberg New York Dordrecht London

Printed on acid-free paper

Springer International Publishing AG Switzerland is part of Springer Science+Business Media (www.springer.com)

Contents

Chapter 1
Infection and Cancer: Multi-directorial Relationship

Michael R. Shurin, Jinbao Zong, and Anton A. Keskinov

Abstract The World Health Organization estimates more than two million cancer cases per year (more than 20 % of the global cancer burden) are attributable to chronic infections, making them the second most preventable cause of cancer. It appears that persistent infections are the leading causes for some of the most important human cancers, such as stomach cancer, cervical cancer and liver cancer. However, fundamental principles regulating the effector functions of immune cells in the tumor environment during systemic infections and host and microbial molecules and pathways that can be targeted for treatment or prevention of cancer progression are not yet well characterized. Identifying these pathways can affect health across populations, creating opportunities to reduce the impact of cancer by preventing or treating infection. The fact that certain chronic infections lie at the root of 20 % of human cancers is expected to render their primary prevention more practicable. On the other side, immunosuppression associated with tumor progression, as well as cancer therapy, predisposes cancer patients to the development of either new infections or reactivations of latent infections. This chapter briefly introduces four main parts of the book and stresses the importance of recognition of different aspects of interactions between infectious agents and neoplastic processes. Specifically, part one provides an overview on the role of chronic infection in cancer development, i.e., infection-associated cancers. Part two focuses on infection diseases induced by cancer treatment, i.e., cancer-associated infections. Part three

M.R. Shurin, M.D., Ph.D., D(ABMLI) (✉)
Departments of Pathology, School of Medicine, University of Pittsburgh,
Scaife Hall S735, 3550 Terrace Street, Pittsburgh, PA 15261, USA

Departments of Immunology, University of Pittsburgh Medical Center, Pittsburgh, PA USA
e-mail: shurinmr@upmc.edu

J. Zong, M.D., Ph.D.
Departments of Pathology, University of Pittsburgh Medical Center,
Scaife Hall S735, 3550 Terrace Street, Pittsburgh, PA 15261, USA

Department of Clinical Laboratory, The Affiliated Hospital of Qingdao University,
Qingdao City, China

A.A. Keskinov, M.D., Ph.D.
Departments of Pathology, University of Pittsburgh Medical Center,
Scaife Hall S735, 3550 Terrace Street, Pittsburgh, PA 15261, USA

M.R. Shurin et al. (eds.), *Infection and Cancer: Bi-Directorial Interactions*, DOI 10.1007/978-3-319-20669-1_1

opens the opportunity to understand the interaction between immune reactions associated with independent development of infection in tumor-bearing hosts, i.e., comorbid development of cancer and infection. And part four discusses viral and bacterial based approaches to cancer therapy.

Keywords Infection-related cancer • Cancer-related infection • Immunosuppression • Infection and cancer • Comorbid diseases

Infection-Associated Cancer

Infectious diseases, which annually claim about 14 million lives, i.e. ~ 25 % of deaths recorded worldwide, are still the primary cause of mortality (Cohen 2000). Regardless of remarkable advances in medical research and treatments during the Twentieth century, almost four billion people are affected by viruses, bacteria, fungi, protozoa, helminthes or prions, and infectious diseases remain among the leading causes of death worldwide (Herrera et al. 2005).

While infectious disease still remains a major problem in many countries, chronic diseases, including such non-communicable conditions as cardiovascular disease, diabetes, respiratory disease and cancer are another major cause of disability and death, not only in developed countries, but also worldwide. According to statistics from the World Health Organization, non-communicable diseases are responsible for 63 % of deaths globally.

Cancer, with more than 12 million people diagnosed annually and more than 8 million people dying from cancer worldwide, is one of the key public health problems. Of the total 59 million global deaths in 2008, ~13 % were attributed to cancer. The projected increase in global cancer burden – from 12.7 million new cases in 2008, to 22.2 million by 2030 (Bray et al. 2012) – not only indicates population growth, but also allied to the frequency and spreading of risk factors. The WHO Global Status Report on non-communicable diseases addresses several risk factors for cancer, including tobacco use, alcohol consumption, little physical activity and unhealthy diet (Vineis and Wild 2014). In fact, it was estimated that a maximum of 60 % of cancer deaths in the United States may be attributed to eight risk factors: tobacco, alcohol, ionizing and solar radiations, occupations, infectious agents, obesity and physical inactivity (Schottenfeld et al. 2013). The increasing burden of cancers in low- and middle-income countries is attributable in part to increasing urbanization, expansion of the adult population at risk and increasing or persistent exposures to tobacco, dietary deficiencies and infectious agents (Schottenfeld et al. 2013).

The proven existence of a causal relationship between infectious diseases and some non-communicable pathophysiological conditions is an important fact. For instance, some infectious agents may cause chronic illness or long-term disability through progressive tissue damage or organ dysfunction via direct effects of persistent infection or immune response to the agent; or by predisposing a person to chronic outcome (O'Connor et al. 2006). Furthermore, infectious agents have emerged as notable determinants, not just complications, of some chronic diseases such as cancer. Directly or indirectly, infectious agents produce longstanding outcomes via several mechanisms including pathways associated with acute infection, persistent active infection, persistent nonreplicating latent infection, immune response to certain infectious agents or their products and malignant transformation. Direct tissue damage or genomic integration explains certain chronic outcomes, but an inflammatory response outlines various established infectious causes of chronic diseases, including some cancers (O'Connor et al. 2006).

A leading example is chronic infections, which cause an estimated at least 16 % of cancers globally with an order of magnitude quite difference in regional contribution (de Martel et al. 2012). As stated by De Martel et al., in 2008, the global population-attributable fraction of cancers associated with infectious agents was 16 %. This portion was greater in less-developed countries (22.9 %) than in more-developed countries (7.4 %), and varied from 3.3 % in Australia and New Zealand to 32.7 % in sub-Saharan Africa. *Helicobacter pylori*, hepatitis B and C viruses, and human papillomaviruses (HPV) caused a major proportion of stomach, liver and cervical cancers. In men, liver and gastric cancers accounted for more than 80 % of infection-related burden of cancer; in women, cervical cancer accounted for about half such cases. Approximately 10–15 % of all human cancers are caused by oncoviruses (Mesri et al. 2014). In some countries the parasitic infection schistosomiasis raises the risk of bladder cancer and the liver fluke increases the risk of cholangiocarcinoma of the bile ducts. Remarkably, almost 30 % of infection-associated cancers occur in people younger than 50 years (de Martel and Franceschi 2009; de Martel et al. 2012). Overall, infection is one of the most important causes of cancer and almost one in every four-five malignancies can be attributed to infectious agents (Shurin 2012).

Thus, it is clearly evident that certain chronic infections can directly support or increase risk of cancer development (Fig. 1.1).

Cancer-Associated Infections

Regardless of the latest approval of new means and technologies for cancer treatment, chemotherapy remains the primary approach to systemic treatment of many common cancers, including lung, breast, colon and ovarian cancer, as well as the hematologic malignancies. Cancer treatment-associated myelosuppression, specifically neutropenia, represents the most common toxicity of antineoplastic therapy (Barreto et al. 2014). Besides common pathogens associated with sinusitis, such as *S. pneumoniae*, *H. influenzae* and *Moraxella catarrhalis*, patients with neutropenia and patients who are immunocompromised are at risk for sinus infections with *P. aeruginosa*, Enterobacteriaceae and moulds. It frequently results in fever and life-threatening infections, provoking hospitalization and sometimes resulting in serious morbidity and even mortality, in spite of available broad-spectrum antibiotics and

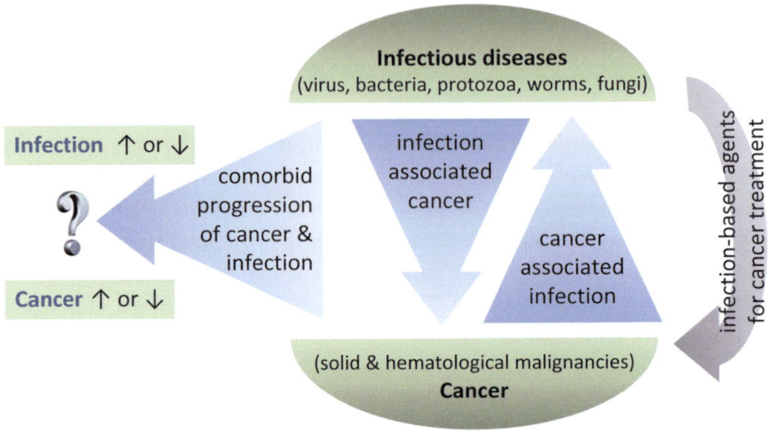

Fig. 1.1 Different pathobiological associations between infectious diseases and cancer. It is well proven that certain viral, bacterial and parasitic infections are associated with cancer development, i.e., can provide the protumorigenic microenvironment, which supports malignant transformation and survival of initial premalignant and malignant cells. This is known as *infection-associated* or *infection-related cancer*. Oppositely, many neoplastic diseases are accompanied by the formation of the local and often systemic immunosuppressive environment, which allows tumor escape from immune recognition. Moreover, conventional cancer chemotherapy may also result in profound immunosuppression. This and other anticancer and related treatments often prompts the development of specific infectious diseases in cancer patients, the phenomenon known as *cancer-related infections*. The phenomenon of independent development of cancer and infection in the same host, i.e., *comorbid cancer-infection progression*, is also a possible scenario, which, however, has not been yet investigated. Inflammatory and antigen-specific immunological reactions associated with the course of acute or chronic infection during tumor progression and development of metastases may strongly affect both disease development in either direction. Finally, based on the success of vaccination, numerous viruses-based and bacteria-based approaches to cancer therapy have been designed and tested in pre-clinical studies and clinical trials

supportive care. Fungal and mycobacterial infections after hematopoietic stem-cell transplantation and viral infections in patient with solid organ transplant are also examples of cancer therapy-related infections.

Chemotherapy-induced myelotoxicity may be a key contributor to fatigue leading to incapacity to complete routine daily activities. Other more severe adverse effects include bleeding events or life-threatening infection manifesting as febrile neutropenia and sepsis. Febrile neutropenia often is a sign of infection in patients with malignancies and bacteremia is documented in about 20 % of cases (Oude Nijhuis et al. 2003). As neutrophils are a major defense against infection, neutropenia may support appearance of a favorable environment for bacterial invasion and multiplication, with the potential for rapidly spreading life-threatening infection. For instance, patients with cancer, particularly those with hematologic malignancies, remain exquisitely vulnerable to infection with gram-negative bacteria as a result of neutropenia, lymphocyte dysfunction, mucositis, and the use of invasive devices (Safdar and Armstrong 2011; Perez et al. 2014). Fifty to 80 % of patients suffering from various hematological malignancies develop infections over the course of their disease and treatment, which contribute to a considerable mortality in these patients (Yadegarynia et al. 2003).

Indwelling central venous catheters for long-term chemotherapy or parenteral nutrition place the cancer patient at significant risk for bloodstream infection, particularly with *Staphylococcal* species. A central catheter can be infected during insertion when bacteria from the skin might infect the insertion tract contaminating the soft tissue and leading to a tunnel or soft tissue infection. In addition, the bacteria can migrate along the line into the bloodstream causing bacteremia. Another possibility is contamination of the port during drug or fluid administration. Some pathogens, such as *Pseudomonas*, *Staphylococcus* and *Candida* sp. may create a biofilm on the catheter preventing immune responses and limiting antibiotic efficacy. In addition to catheter-related infection, cancer surgery may also result in disruption of normal tissue and organs, large incisions and empty tissue spaces that can fill with fluid or blood and become infected. Cancer patients with immune dysfunction and neutropenia after chemoradiotherapy are more prone to developing wound and incision infections or, less commonly, a bacterial infection at the surgical bed site. Other risk factors for infection in patients with cancer include bedrest with subsequent decubiti formation or aspiration pneumonitis. Splenectomy may result in compromised defense mechanisms of the reticuloendothelial system. Mucormycosis of the nasal cavity may result from metabolic acidosis or hyperglycemia. Long-term broad spectrum antibiotics for neutropenic fever may cause candidemia, and patients with breeches of the skin may develop necrotizing fasciitis or Fournier's gangrene (Khayr et al. 2012).

Thus, a significant body of clinical evidence demonstrates that infections pose a major threat to cancer patients as cancer treatment-related immunosuppression and general weakness associated with metastatic disease escalate the risk and severity of different infections (Fig. 1.1). The diminished innate and adaptive immunity in many cancer patients predisposes them to the development of either new infections or reactivations of latent viral infections, such as infections with respiratory syncytial virus, herpes simplex virus, influenza virus, Parainfluenza virus, cytomegalovirus and varicella-zoster virus. Among the major viral infections observed in cancer patients and stem-cell transplantation recipients are hepatitis B and C infections (Torres and Davila 2012).

Concomitant Infections and Cancer

Though many clinical studies describe infection-related cancers and cancer-related infections, very little is known about the interaction between immune response associated with cancer development and immune reactions accompanied appearance and progression of unrelated but concomitant acute or chronic systemic bacterial infections. The incidence of infection and cancer as co-morbid diseases are clinically common cases, but the effect of infection and its therapy on cancer advancement or deterioration remains obscure and not yet investigated. It is still unclear how the host responds to systemic acute or persistent infections during massive immunomodulation induced by tumor progression. Potential alteration of the tumor immunoenvironment and the pre-metastatic niche by immune effector and regulatory cells induced by acute or latent infections has not been considered as a mechanism affecting antitumor immunity. An important question of how independent are the immune responses to infection and unrelated co-morbid cancer occurring in the same or different tissue has not been answered.

Only a few studies evaluated tumor progression during concomitant infections in murine models. For instance, Chen et al. reported that Malaria infection inhibited growth and metastasis and prolonged survival of mice with lung carcinoma due to the activation of IFN-γ-producing NK cells tumor-specific CD8+ T cells (Chen et al. 2011). Kim et al. reported that mice injected with both *Toxoplasma gondii* and tumor cells demonstrated improved survival rates, higher frequency of CD8+ T cells and elevated IFN-γ compared to mice inoculated with tumor cells alone (Kim et al. 2007). However, the impact of bacterial infection on neoplastic growth has not been evaluated. More studies are required to examine whether the immune responses to growing tumor are independent or could be significantly conditioned by the immune responses to concomitant acute or persistent bacterial infections and associated inflammation and immune polarization (Fig. 1.1). Understanding of inflammatory and immune responses induced by new infections in patients with cancer should provide new immunotherapeutic approaches to control cancer development and progression.

Infectious Agents and Cancer Therapy

According to the WHO, at least one-third of all cancer cases are preventable and prevention offers the most cost-effective long-term strategy for the control of cancer. Not infrequently, infection may represent the first misstep along a continuum from pre-malignant lesion to cancer development, and preventing or treating infection or the immune response to infection offers a chance to disrupt the mechanisms supporting appearance and survival of initial premalignant and malignant cells. The use of vaccines against cancer-associated pathogens is one of the most promising areas of ongoing cancer prevention research. Prophylactic vaccines against pathogenic viruses have an excellent record as public health interventions in terms of safety, effectiveness and ability to reach economically disadvantaged populations (De Flora and Bonanni 2011). The development and implementation of human papilloma virus vaccination to prevent cervical cancer is one of the most important advances in cancer prevention in the past decade. Another effective vaccine for cancer prevention is the one against hepatitis B virus and antiviral treatments against chronic hepatitis virus infections resulted in significant reduction in the incidence of hepatocellular carcinoma in treated patients, thus achieving the goal to prevent the occurrence of cancer (Schiller and Lowy 2014). Antiviral treatments for Epstein-Barr virus (EBV), Kaposi's sarcoma-associated herpesvirus (KSHV) and human

T-cell lymphotropic virus type 1 (HTLV-1) had showed some encouraging, although limited results in treating refractory EBV-associated lymphoma and post-transplant lymphoproliferative disorder, KSHV-associated Kaposi's sarcoma in AIDS patients, and HTLV-1-associated acute, chronic and smoldering subtypes of adult T-cell lymphoma, respectively (Shih et al. 2014). It was estimated that by preventing cancer-associated infectious diseases, there would be 26.3 % fewer cases in developing countries (<1.5 million cases/year) and 7.7 % fewer cases in developed countries (<390,000 cases) (Parkin 2006). Attractively, not only prevention but even therapy of an infectious disease and eradication of a pathogen becomes a crucial tool for the primary prevention of certain types of cancer. For instance, analysis of patients with hepatitis C treated with interferon demonstrated a more than 70 % reduction of risk of developing hepatocellular carcinoma (Colombo and Iavarone 2014). Successful efforts to identify effective preventive vaccines may lead to new approaches for therapeutic and prophylactic intervention.

Based on the success of vaccination and in addition to developing preventive vaccines for infection-associated cancer, a significant research effort focuses on microbial-based therapy of cancer, i.e., the use of viruses and bacteria as anticancer agents. It has been shown that viruses and bacteria are naturally capable of homing to the tumor site when systemically administered resulting in high levels of replication locally, either external to (non-invasive species) or within malignant or stromal cells (pathogens) (Cronin et al. 2012). Utilization of the natural ability of disease causing microbes to invade human cells resulted in building a concept of exploiting biological agents for delivery of therapeutic genes to patients to target tumor cells.

Viruses represent an attractive vehicle for cancer gene therapy due to their high efficiency of gene delivery. Many viruses can mediate long-term gene expression, while some are also capable of infecting both dividing and non-dividing cells (Collins et al. 2008). In general, viral therapeutics provides the possibility to express anticancer proteins directly at the tumor site, decreasing exposure to normal tissue during delivery and optimizing therapeutic index. Some viruses are also 'oncolytic', either naturally or by design, and these agents function to eliminate malignant cells selectively before dissemination and infection of adjacent cells and repeating the process (Tedcastle et al. 2012). Oncolytic viruses can be genetically engineered to induce cell lyses through virus replication and expression of cytotoxic proteins. For instance, Herpes simplex virus (HSV) has become one of the most widely clinically used oncolytic agent and different types of HSV have been evaluated in basic or clinical studies (Liu et al. 2013). Numerous gene therapy strategies have been commenced, utilizing a range of viral vectors and demonstrating delivery of immunostimulatory or tumor-modifying molecules directly to tumor or stromal cells, or the direct delivery of tumor antigens or immunomodulators (Lichty et al. 2014). A diverse array of oncolytic viruses and recombinant viral vectors encoding a variety of therapeutic genes or tumor antigens have been tested in pre-clinical studies and provided data which, in some cases, justify their clinical development as potential cancer therapies.

Bacteria present another attractive class of gene vectors, possessing a natural ability to grow specifically within tumors following administration. To date, the genera of bacteria that have been exploited as gene delivery vehicles include *Salmonella*, *Escherichia*, *Listeria*, *Clostridium* and *Bifidobacterium* (Bernardes et al. 2010). These bacteria can be delivered to the tumor site via multiple routes including systemic administration, intratumoral injection or in certain instances orally, and the use of live, attenuated bacteria or their purified products has been investigated. Preclinical studies have demonstrated the capability of different bacterial strains to locally produce therapeutic factors and facilitate specific and effective clinical responses (Cronin et al. 2012). Bacterial-based gene therapy strategies may aim at inducing cancerous cell death directly utilizing 'suicide' genes or indirectly via secreting therapeutic proteins locally within the tumor microenvironment and inducing, for instance, antitumor immune responses. These days, an increasing number of genetically engineered bacteria are emerging in the field, with applications both in therapy and diagnosis. In addition, purified bacterial products are also gaining relevance as new classes of bioactive products to treat and prevent cancer growth and metastasis (Patyar et al. 2010; Bernardes et al. 2013). The example of the US Food and Drug Administration approved, established cancer therapy employing bacteria is Bacillus Calmette-Guérin (BCG), an attenuated *Mycobacterium bovis* strain, which has become the treatment of choice for high risk, superficial bladder cancer in many countries (Askeland et al. 2012). Genetically engineered strains of *Salmonella* have also been proposed for tumor selective therapy. *S. Typhimurium* has been genetically engineered to produce a variety of therapeutic agents, with success in animal models and evaluation in several clinical trials (Roland and Brenneman 2013). Numerous publications provide convincing evidence that genera of bacteria have potential in cancer therapy (Fig. 1.1). This vector class compares favorably with other vectors in terms of safety, practicability and production. While non-pathogenic bacteria have yet to enter clinical trials in the context of tumor targeting vectors, existing knowledge strongly point to their suitability as a vector system for clinical use (Forbes 2010; Patyar et al. 2010; Cronin et al. 2012; Bernardes et al. 2013).

References

Askeland EJ, Newton MR, O'Donnell MA, Luo Y (2012) Bladder cancer immunotherapy: BCG and beyond. Adv Urol 2012:181987

Barreto JN, McCullough KB, Ice LL, Smith JA (2014) Antineoplastic agents and the associated myelosuppressive effects: a review. J Pharm Pract 27(5):440–446

Bernardes N, Seruca R, Chakrabarty AM, Fialho AM (2010) Microbial-based therapy of cancer: current progress and future prospects. Bioeng Bugs 1(3):178–190

Bernardes N, Chakrabarty AM, Fialho AM (2013) Engineering of bacterial strains and their products for cancer therapy. Appl Microbiol Biotechnol 97(12):5189–5199

Bray F, Jemal A, Grey N, Ferlay J, Forman D (2012) Global cancer transitions according to the Human Development Index (2008–2030): a population-based study. Lancet Oncol 13(8):790–801

Chen L, He Z, Qin L, Li Q, Shi X, Zhao S, Chen L, Zhong N, Chen X (2011) Antitumor effect of malaria parasite infection in a murine Lewis lung cancer model through induction of innate and adaptive immunity. PLoS One 6(9), e24407

Cohen ML (2000) Changing patterns of infectious disease. Nature 406(6797):762–767
Collins SA, Guinn BA, Harrison PT, Scallan MF, O'Sullivan GC, Tangney M (2008) Viral vectors in cancer immunotherapy: which vector for which strategy? Curr Gene Ther 8(2):66–78
Colombo M, Iavarone M (2014) Role of antiviral treatment for HCC prevention. Best Pract Res Clin Gastroenterol 28(5):771–781
Cronin M, Stanton RM, Francis KP, Tangney M (2012) Bacterial vectors for imaging and cancer gene therapy: a review. Cancer Gene Ther 19(11):731–740
De Flora S, Bonanni P (2011) The prevention of infection-associated cancers. Carcinogenesis 32(6):787–795
de Martel C, Franceschi S (2009) Infections and cancer: established associations and new hypotheses. Crit Rev Oncol Hematol 70(3):183–194
de Martel C, Ferlay J, Franceschi S, Vignat J, Bray F, Forman D, Plummer M (2012) Global burden of cancers attributable to infections in 2008: a review and synthetic analysis. Lancet Oncol 13(6):607–615
Forbes NS (2010) Engineering the perfect (bacterial) cancer therapy. Nat Rev Cancer 10(11):785–794
Herrera LA, Benitez-Bribiesca L, Mohar A, Ostrosky-Wegman P (2005) Role of infectious diseases in human carcinogenesis. Environ Mol Mutagen 45(2–3):284–303
Khayr W, Haddad RY, Noor SA (2012) Infections in hematological malignancies. Dis Mon 58(4):239–249
Kim JO, Jung SS, Kim SY, Kim TY, Shin DW, Lee JH, Lee YH (2007) Inhibition of Lewis lung carcinoma growth by *Toxoplasma gondii* through induction of Th1 immune responses and inhibition of angiogenesis. J Korean Med Sci 22(Suppl):S38–S46
Lichty BD, Breitbach CJ, Stojdl DF, Bell JC (2014) Going viral with cancer immunotherapy. Nat Rev Cancer 14(8):559–567
Liu S, Dai M, You L, Zhao Y (2013) Advance in herpes simplex viruses for cancer therapy. Sci China Life Sci 56(4):298–305
Mesri EA, Feitelson MA, Munger K (2014) Human viral oncogenesis: a cancer hallmarks analysis. Cell Host Microbe 15(3):266–282
O'Connor SM, Taylor CE, Hughes JM (2006) "Emerging infectious determinants of chronic diseases." Emerg Infect Dis [serial on the Internet] July: http://dx.doi.org/10.3201/eid1207.060037
Oude Nijhuis CS, Gietema JA, Vellenga E, Daenen SM, De Bont ES, Kamps WA, Groen HJ, van der Jagt EJ, van der Graaf WT (2003) Routine radiography does not have a role in the diagnostic evaluation of ambulatory adult febrile neutropenic cancer patients. Eur J Cancer 39(17):2495–2498
Parkin DM (2006) The global health burden of infection-associated cancers in the year 2002. Int J Cancer 118(12):3030–3044
Patyar S, Joshi R, Byrav DS, Prakash A, Medhi B, Das BK (2010) Bacteria in cancer therapy: a novel experimental strategy. J Biomed Sci 17(1):21
Perez F, Adachi J, Bonomo RA (2014) Antibiotic-resistant gram-negative bacterial infections in patients with cancer. Clin Infect Dis 59(Suppl 5):S335–S339
Roland KL, Brenneman KE (2013) Salmonella as a vaccine delivery vehicle. Expert Rev Vaccines 12(9):1033–1045
Safdar A, Armstrong D (2011) Infections in patients with hematologic neoplasms and hematopoietic stem cell transplantation: neutropenia, humoral, and splenic defects. Clin Infect Dis 53(8):798–806
Schiller JT, Lowy DR (2014) Virus infection and human cancer: an overview. Recent Results Cancer Res 193:1–10
Schottenfeld D, Beebe-Dimmer JL, Buffler PA, Omenn GS (2013) Current perspective on the global and United States cancer burden attributable to lifestyle and environmental risk factors. Annu Rev Public Health 34:97–117
Shih WL, Fang CT, Chen PJ (2014) Anti-viral treatment and cancer control. Recent Results Cancer Res 193:269–290
Shurin MR (2012) Cancer as an immune-mediated disease. ImmunoTargets Ther 1:1–6

Tedcastle A, Cawood R, Di Y, Fisher KD, Seymour LW (2012) Virotherapy–cancer targeted pharmacology. Drug Discov Today 17(5–6):215–220

Torres HA, Davila M (2012) Reactivation of hepatitis B virus and hepatitis C virus in patients with cancer. Nat Rev Clin Oncol 9(3):156–166

Vineis P, Wild CP (2014) Global cancer patterns: causes and prevention. Lancet 383(9916):549–557

Yadegarynia D, Tarrand J, Raad I, Rolston K (2003) Current spectrum of bacterial infections in patients with cancer. Clin Infect Dis 37(8):1144–1145

Part I
Infection-Associated Cancers

Chapter 2
Human Tumor Viruses: A Historical Perspective

Joseph S. Pagano

Abstract The fact that infectious diseases and cancer might be often associated either as the complication or cause has been discussed for decades. However, reliable epidemiological and experimental data demonstrating that numerous infectious agents can be etiological factors of human malignancies appeared only in recent years. This chapter overviews data on virus-associated malignancies and discusses potential mechanisms responsible for this association.

Keywords Epstein-Barr virus • Burkitt lymphoma • Nasopharyngeal Carcinoma • Kaposi's sarcoma • Hepatocellular Carcinoma

Introduction

That infection and cancer are so often associated either as complication or cause has been known for many years. However viruses and other infectious agents have emerged only in recent decades as etiological factors of human malignancies. Epstein-Barr virus, discovered 50 years ago (1964) and characterized early as the first human tumor virus, although its exact roles in the malignancies with which it is associated are not straightforward, continues to present the quandaries of ascertaining etiology. Helicobacter pylori was rejected as cause of gastric cancer for years before being accepted, in part because it was not a virus with the then-known mechanisms for transformation at its disposal.

Although avian and rodent viruses were accepted as causes of malignancies in these species as early as 1911, the barrier to viruses in human cancer seemed firm. Means to breach the barrier were being developed unwittingly by John Enders in the 1960s when he devised means to cultivate human cells in culture for the study of poliovirus infection. Not however until 1958 was there an opening to oncogenic studies: Jan Ponten working at the Wistar Institute in Philadelphia showed that a mammalian virus SV 40 could transform normal cells in culture into cells with

J.S. Pagano, M.D. (✉)
Lineberger Comprehensive Cancer Center, University of North Carolina at Chapel Hill, CB#7295, Chapel Hill, NC 27599-7295, USA
e-mail: joseph_pagano@med.unc.edu

M.R. Shurin et al. (eds.), *Infection and Cancer: Bi-Directorial Interactions*, DOI 10.1007/978-3-319-20669-1_2

malignant features. The virus, discovered as a contaminant from the monkey kidneys used to make poliovirus vaccine, was simian not human in origin, but the changes in phenotype that the virus could produce in human diploid fibroblasts were striking: unchecked growth of the cells with loss of contact and mitotic inhibition and stunning transformation of cellular morphology. The field of human tumor virology was launched.

Moreover SV40, a polyomavirus, could produce tumors when injected into hamsters, but not when injected into monkeys, the species from which the virus had been isolated. Nevertheless exploration of SV40 as a possible factor in brain tumors and lymphomas continued for some years, but an association with either remained uncertain. There is, however, evidence that the virus may contribute to pathogenesis of mesothelioma by causing perturbations in cell cycle. In any case SV40 did catalyze the search for a human tumor virus.

The human polyomaviruses, JCV and BKV, which were discovered later, could infect humans, but neither produced tumors. JCV infects the brain where it is the causative agent of Progressive Multifocal Leucoencephalopathy, and it could transform human glial cells in culture but does not cause brain tumors. BKV primarily infects the human urogenital track, is weakly oncogenic in cells from a variety of species of experimental animals, but does not cause tumors.

Only with the discovery of Merkel Cell Virus (MCV) in 2007 was an oncogenic polyomavirus identified (White et al. 2014). It produces a rare tumor of Merkel cells, Merkel cell carcinoma (MCC) in the skin, although intriguingly the virus could be detected in ~80 % but not all MCC. Current thinking is that MCV infection of Merkel cells drives their clonal proliferation. Search for MCV in other skin conditions continues. Intriguingly there are a growing number of new human polyomaviruses, isolated mainly from the respiratory track, but not yet coupled with disease, that should open a new frontier in tumor virology.

The most studied of infectious agents associated with generation of cancer are viruses, & discovery of new tumor viruses continues. That up to 20 % of malignancies are caused by viruses is probably an underestimate in regions of the world such as Africa and Brazil. Not only are additional viruses being identified as oncogenic, but new mechanisms are being uncovered. Examples known for years include inactivation of cellular tumor-suppressor genes, activation of cellular oncogenes and insertional mutagenesis, mostly in latent infection. More recently the importance of inflammatory reactions generated by active viral infection has been appreciated along with the expression of cytokines, both induced and virally encoded. Additionally the importance of cell type as well as viral genotype has emerged, and roles for cofactors identified.

Epstein-Barr Virus

This human herpesvirus epitomizes the levels through which a virus can be linked to malignancy, yet not actually be established as causative (Pagano et al. 2004).

Burkitt Lymphoma

Burkitt lymphoma is named after Dr. Dennis Burkitt, an Irish surgeon who while working in Uganda in the 1950s singled out a hitherto unknown tumor of the jaw in children which was remarkable clinically because of the essentially curative responses to treatment with methotrexate. Although untutored in epidemiology he began to suspect a vector-borne infectious cause because of its patterns of occurrence and endemicity. The report of his observations ignited a search for a virus in the tumor. Efforts by the Epstein laboratory in England to detect a virus directly in tumor tissues by electron microscopy were vexing and not fruitful until Dr. Barr, a cytogeneticist, established cultures of cells derived from BL tissue and only then, 50 years ago in 1964, was able to visualize for the first time herpes-like virus particles. That this was a new human herpesvirus was established by serologic studies by the Henle laboratory in Philadelphia.

DNA-DNA nucleic acid hybridization studies by Harald zur Hausen, and soon thereafter by Nonoyama & Pagano, who used their newly devised quantitative cRNA-DNA hybridization assay (1971) for studies of Burkitt lymphoma tissues obtained by George Klein at the Karolinska Institute from Kenya. The analyses soon revealed that more than 98 % were positive for EBV DNA, usually in lower copy numbers consistent with latently infected tumors. Unexpectedly the genome was retained in novel form in the chromatin, the EBV episome, the first found in eukaryotes save for certain plants (1972). Further the viral episomes were retained in constant copy numbers in successive generations of BL cells in culture. Thus the link of the virus to the lymphoma was firmly established genetically (Pagano 2009).

However the occasional specimen was unexpectedly negative for EBV DNA, and testing later of sporadic BL in the United States disclosed that only 14–20 % were positive. These discordant observations were eventually resolved by findings that disruptions & overexpression of the cMyc oncogene produced by the characteristic chromosomal translocations of BL were the common element in both positive & negative BL. This then was the essential molecular lesion that caused BL, not EBV, which acts as a contributory cofactor through its ability to propel growth of B- lymphocytes (Pagano et al. 2004).

In the meantime the Henle's soon confronted a paradox. EBV antibodies were common in healthy children and adults in the United States, but Burkitt lymphoma was then virtually unknown. Moreover they were able to establish cultures of lymphocytes from the peripheral blood of infected, but not from noninfected, persons. The mystery of the origin of the EBV antibodies was solved thanks to the observant Henle technician, who had been unable to culture cells from her own blood, but tried again—successfully this time–after contracting infectious mononucleosis, and she now had EBV antibodies. Proof that EBV causes infectious mononucleosis came from the landmark prospective sero-epidemiologic study of IM by Niederman and colleagues in Yale college students. During the 4 years they were studied, only EBV-negative, but not students who were already seropositive, contracted infectious mononucleosis; there were no instances of discordance. The study was conclusive.

That EBV, a γ-herpesvirus, causes the benign disease infectious mononucleosis is indisputable, based on the conclusive epidemiologic studies backed by molecular evidence. Infection of B lymphocytes by the virus causes proliferation of these cells, which in immunologically normal hosts is checked by robust reactive T-cell responses manifested in peripheral blood as atypical lymphocytosis and are not tumorigenic, whereas EBV-infected B-cells are potentially tumorigenic in immune-incompetent persons. In normal hosts these responses check the expansion of the infected B-cells.

Initially the virus replicates in epithelial cells in the oropharynx, where it is secreted into the saliva & can be transmitted by oral contact. Replication in these infected cells is cytolytic in contrast to the proliferation the virus produces in B-lymphocytes, which it infects almost simultaneously. After the initial wave of proliferation subsides, the viral genome persists as episomes lifelong in a small fraction of infected germinal-center memory B-cells, which may divide & are perpetuated by ambient proliferative stimuli (White et al. 2014).

EBV also infects and is shed from the human cervix, probably by oral sexual contact, where it does not produce known disease, although there is evidence that infectious mononucleosis can be contracted occasionally. Multiple strains of EBV have been identified in the oropharynx, but none specifically in the vaginal track, probably because not investigated.

Other EBV Lymphomas

Lethal EBV-infected B-cell lymphomas can arise in immunocompromised hosts such as recipients of organ transplants, in patients with AIDS, or in children with rare innate genetic disorders such as Duncan's Syndrome. These lymphomas, which begin as reversible B-cell lymphoproliferation, evolve into polyclonal, then lethal monoclonal lymphomas that are directly caused by EBV.

In contrast although EBV is detected in approximately 40 % of Hodgkin's lymphoma (HL) its pathogenic role is ill-defined, but presumably nontrivial: the virus infects the pathognomonic cell type, the Reed-Sternberg cell--which is of B-cell origin. EBV infects R-S cells & expresses a viral protein, LMP2a, on the outer membrane of the R-S cell; thus it is expressed in the pathologic cell type of HL & it is likely to exert oncogenic function (White et al. 2014).

Another perspective comes from epidemiologic studies that linked EBV and generation of some cases of HL. The studies showed that the incidence of HL was somewhat but significantly greater in persons who had had infectious mononucleosis (but not subclinical EBV infection) earlier in life. Since the syndrome of IM results from a transient disruption of the immune system, and salivary shedding of virus can continue for years, perhaps some type of residual immunodeficiency plays a role in the genesis of HL, although this is speculative.

EBV also rarely infects and NK and T lymphocytes and produces natural killer/T cell lymphomas especially in Korea and Japan.

EBV Infection in Epithelial Malignancies

The first studies of epithelial infection were based on the hypothesis that as with the other herpesviruses EBV was likely to have a primary cell type in which it replicated initially & a different secondary cell type in which the virus persisted lifelong in a latent form. Since EBV replicated in the oropharynx the search centered on epithelial cells shed in the oropharynx. *In situ* nucleic acid cytohybridization assays had been devised independently by the zur Hausen & Pagano laboratories for detection & localization of EBV DNA in tissues and cells. E-S Huang, Pagano and colleagues with the use of cRNA-DNA hybridization then detected EBV DNA directly in oropharyngeal cells obtained from students with IM. They found also that EBV could infect epithelial cells growing in organ cultures although inefficiently. These findings identified not only a primary entry point & possible source for the virus, but also a basis for understanding the pathogenesis of nasopharyngeal carcinoma, an epithelial malignancy.

Finally aside from IM itself Hairy Leukoplakia (HLP) of the lateral tongue is the only instance of lytic EBVinfection in the oropharynx. Striking features of the lesions, which are in the squamous epithelium, are the masses of virions in them. HL occurs mostly in patients with AIDS, responds to treatment with Acyclovir, but may recur. HLP is not thought to be a premalignant lesion, in contrast to lesions of the tongue produced by smoking cigars or pipes.

Nasopharyngeal Carcinoma (NPC)

Carcinoma of the posterior nasopharynx is the prime example of an epithelial malignancy in which EBV plays a causative role. It has many distinctive features. The neoplasm arises in the fossa of Rosenmuller in Waldenstrom's ring, but the malignancy has a proclivity for early spread, & it is most often diagnosed after it has invaded cervical lymph nodes. It is the most common EBV malignancy: it is endemic in Southern China & has high incidence in first generation émigrés to other Asian countries & to the West Coast of the U.S., mainly in men of middle age. NPC occurs sporadically in Western countries with an incidence ~1/100 that of endemic regions. Of the three WHO histopathologic types undifferentiated Type III NPC is the most common (~90 % of cases). All 3 types, whether of sporadic or endemic origins, are latently infected with EBV episomes. It is sometimes argued that WHO type 1 NPC, which are keratinizing carcinomas & are rare, are not infected with EBV. However this Type tends to have low genome copy numbers that may be missed. Moreover the first EBV genome cloned from NPC was from Type I tissue (Nancy Raab-Traub) (Pagano 2009; White et al. 2014; Raab-Traub 2005).

Other fascinating epidemiologic features of NPC are its intermediate high incidence in North Africa & its puzzling bimodal age distribution in the teens & young adults as well as in the middle-aged in the endemic regions. Geographically it is the only region where those affected are Caucasian.

These complexities of incidence of NPC have provoked numerous investigations of possible cofactors that contribute to genesis of the malignancy including dietary, environmental, including exposure to & metabolism of nitrosamines, and genetic. And despite the innumerable strains of EBV that have been detected in saliva, no "oncogenic strain" has been identified for NPC or the other EBV malignancies.

Additionally reasons for the highly invasive phenotype characteristic of NPC has been illuminated by studies showing that the principal EBV oncogene, LMP1, induces a host of cellular factors including MMP9, MUC1, FGF2, VEGF, HIF1alpha, Twist & Snail capable of propelling every facet of the complex processes of invasion, metastasis, angiogenesis and transcription. Further, such factors may be transported in exosomes to the tumor microenvironment & promote tumor progression. Thus EBV likely functions not only as etiologic agent, but also in late stages of oncogenesis in the array of tumors in which LMP1 is expressed (Yoshizaki et al. 2005).

Finally there are the striking elevations in IgA antibodies to EBV antigens that arise before & around the time of detection of NPC. Since the antibodies are to viral lytic proteins they suggest that a period of viral reactivation & active viral replication precedes NPC & propels its onset, perhaps rapidly, presumably by virus entering pre-malignant cells with acquired mutations. In any case, even though their origin is obscure the wave of antibodies is useful diagnostically & may herald onset of tumor. Indeed in endemic regions of China otoscopy of asymptomatic men & blind biopsy of the Rosenmuller Fossa are used for screening (Pagano 2009).

Parotid Tumors

Parotid, but not other, salivary gland tumors have high incidence in North American Inuits and are EBV-infected, & features of their histopathology resemble those of NPC. Interestingly parotid tumors in NPC-endemic regions of China also contain monoclonal EBV episomes.

Kaposi's Sarcoma Herpesvirus

Kaposi's sarcoma (KS) had been described as a distinctive disease of the skin in 1872, but its viral etiology was identified only in 1994, the second of the two human gamma herpes viruses. It is primarily a vascular endothelial lesion found in its classic form in European men. It is more common in regions of Africa where it is endemic and a considerably more aggressive malignancy. However Kaposi's sarcoma (KS) vaulted to prominence in dermatologic patients with AIDS (Friedman-Kien) before that syndrome had been defined, became its herald lesion and, later, its HIV etiology ascertained (Gallo, Montagnier) (Sir and Ou 2010). KSHV genomes are detected in almost all cases of KS regardless of geographic origin or severity of disease. Somewhat unexpectedly the virus is the causative agent of primary effusion lymphoma (PEL), a fatal B cell lymphoma. Cell lines established from PEL are invariably infected with KSHV, but often also co-infected with EBV, which seems to enhance its oncogenic behavior (Bushman et al., 2012). Finally KSHV also causes Multicentric Castleman's Disease, an indolent B-cell lymphoproliferative condition (Raab-Traub 2005; Hayward et al. 2010; Wen and Damania 2009).

The KSHV genome is distinctive because of the homologs to cellular genes it encodes, in contrast to EBV, such as a viral cytokine and an interferon regulatory factor. The latter factor however lacks a DNA-binding domain and thus must partner as a heterodimer with a cellular IRF to exert function. EBV can induce cytokines that enhance oncogenic processes, but not does not encode such proteins. The KSHV K1 gene can immortalize primary endothelial cells and produce angioproliferative KS-like lesions in transgenic mice. Finally both KS and PEL are invasive malignancies, and the KSHV gene product K1 is able to upregulate expression of a matrix metalloproteinase and vascular growth factor (Wen and Damania 2009).

Human Papilloma Viruses

A welter of HPV genotypes began to be recognized in the context of cervical cancer and a loose association began to emerge. It was only when Harold zur Hausen began to discern stereotypic associations of certain genotypes (HPV 16 & 18) with cervical cancer, suggestive of a possible etiologic association, that the subsequent revelatory epidemiologic studies could be designed which would lead to proof that certain types of HPV were causative agents. Since HPV causes warts the laboratory also searched for evidence that HPV might cause skin cancers. Interestingly this suspicion was never verified. However HPV is a cause of anal and vaginal cancer as well as benign genital warts.

HPV has also been implicated as causative agent of a subset of oropharyngeal carcinomas. These cancers of the tongue & tonsil occur in younger patients and have a better prognosis than the cancers associated with smoking cigarettes. Although further definition and proof of etiology awaits conclusive epidemiologic studies HPV is likely causative.

Notably HPVinfection led to creation of type-specific vaccines that could protect against infection with the virus & thus prevent cervical carcinoma. However such prophylactic vaccines have effect on cervical cancer itself or early precancerous changes. Therapeutic vaccines directed against early-stage changes in the cervix are in preclinical trial and have shown considerable promise.

Hepatitis B Virus

Hepatocellular carcinoma is endemic in Asian countries and elsewhere. Taiwanese & American investigators who were at first studying human cytomegalovirus infection, which is highly prevalent in woman in that country and can cause congenital defects, turned their attention to hepatocellular carcinomas, which they suspected might be caused by a virus. Studies of Woodchuck Hepatitis Virus in the United States by Jesse Summers greatly strengthened suspicions. The virus could cause persistent infection of the liver, leading to hepatic fibrosis and ultimately hepatocellular carcinoma in the animals that paralleled the course of hepatic infection in humans. Fortunately even before the virus itself was identified the HBV S-antigen was invariably detectable in the blood of patients with hepatitis, fibrosis and hepatocellular carcinoma itself. Subsequently Palmer Beasley, working with his colleagues in Taiwan designed a sero-epidemiologic study, results of which essentially proved that HBV caused hepatocellular carcinoma. This landmark study was first to prove that a virus could cause a malignancy.

This remarkable journey of discovery led not only to finding the viral culprit, but later to proof produced by a classic sero-epidemiologic study, also designed by Beasley, that hepatitis B virus vaccine could prevent infection with the virus. This indeed was the first vaccine for a human tumor virus, HPV vaccine being the second and only other example.

HBV itself although it is an RNA virus goes through a double-stranded DNA phase and is able to insert into cellular DNA. Although the integration site is not unique it is believed to cause mutational outcomes leading to hepatocellular carcinoma.

The prevalence of the virus is much higher in Asia and other countries than in the United States or other Western countries, because the high prevalence of infection increases the risk to newborns during parturition & leads to persistent productive infection that is more likely to ensue in hepatitis and ultimately hepatocellular carcinoma.

Hepatitis C Virus

Much more prevalent in the United States than HBV, HCV became widespread through blood transfusions before the virus was discovered. It is an RNA virus that produces persistent productive infection in the liver and frequently leads to the sequence of hepatitis, cirrhosis and an increasing incidence of hepatocellular carcinoma. The key to its oncogenic effects is the persistent lytic infection that generates distinctive tissue responses in the liver. The mechanism of oncogenesis is novel in that the basis is the chronic inflammatory response, which has been much studied, that the virus evokes. HCV's RNA genome does not integrate into the cellular genome. Unlike most tumor viruses hepatitis C virus may no longer be detected when the carcinoma is diagnosed, which emphasizes the importance of the distinctive inflammatory responses. The situation while possibly reminiscent of the so-called "hit and run theory" of etiology, is however not the case with HCV--or any other cancer–because the virus is consistently replicating throughout the pro-oncogenic stages if not always at the very end stage of HCC. There are no known instances of this theory (Sir and Ou 2010).

Finally HCV provides the only oncogenic infection that can be eliminated by antiviral therapy. In very recent studies the single antiviral drug (Gilead) completely eliminates infection of the liver in humans thus blocking generation of fibrosis and carcinoma in this ultimately lethal cancer. Another therapy is now available and appears to be effective, but is composed of four different agents including interferon and (blank) with their known side effects.

Human T-Cell Leukemia Virus

Discovered independently by Robert Gallo in the United States and by Hinuma in Japan this is the first known human retrovirus. The Japanese investigators were studying T-cell leukemias that occurred in families in southern Japan. The virus is found in various parts of the world but its familial incidence in Japan indicates that ATL has a genetic basis. The virus does not spread readily. It produces not only subacute T-cell leukemia in adults, but also lymphomas, some of which infiltrate skin (Pagano 2009). This is the only leukemia linked to & caused by a virus. Neither acute myelogenous leukemias of childhood nor chronic B-lymphocytic leukemias in adults have yielded in a search for possible viral etiology.

Conclusion

No single virus is inherently a tumor virus that causes a malignancy, nor does infection of the stomach with Helicobacter pylori necessarily lead to gastric ulcers or cancer. In most instances the infections caused by the viruses may cause illness but are relatively innocuous. Only some of the factors that lead to an oncogenic course of infection are known. However the range of lethal malignancies to which viruses contribute or cause is broad and impressive, and it comprises the largest group of cancers for which causes or substantial contributory factors are known. Both RNA and DNA viruses are represented, and each virus invokes distinct mechanisms of action with some commonalities. Understanding viral mechanisms is in itself a rewarding quest that can also provide insights into central aspects of cell and molecular biology. There is promise that the pace and excitement of discovery is accelerating, attested by the newly isolated polyomaviruses, most not yet linked to disease. Pragmatically the field has on offer many more challenges than accomplishments, starkly witnessed by their paucity so far: two prophylactic vaccines, no therapeutic vaccine, a single antiviral drug. The time is auspicious; we are still on frontiers at every level when the yield on investment both scientifically and materially can grow.

References

Bushman FD, Nabel GJ, Swanstrom R (eds) (2012) HIV: from biology to prevention and treatment cold spring harbor perspectives in medicine. Cold Spring Harbor Laboratory Press, Cold Spring Harbor

Hayward GS, Alcendor DJ, Arav-Boger R (2010) The role of KSHV in pathogenesis of Kaposi's sarcoma. In: Khalili K, Jeang K-T (eds) Viral oncology. Wiley-Blackwell, Hoboken

Pagano JS (2009) EBV diseases, chapter 10. In: Damania B, Pipas JM (eds) DNA tumor viruses. Springer, New York, pp 217–240

Pagano JS, Blaser M, Buendia M-A, Damania B, Khalili K, Raab-Traub N, Roizman B (2004) Infectious agents and cancer: criteria for a causal relation. Semin Cancer Biol 14:453–471

Raab-Traub N (2005) Epstein-Barr virus in the pathogenesis of NPC. In: Robertson ES (ed) Epstein-Barr virus. Caister Academic Press, Norfolk, pp 55–70

Sir D, Ou J-H (2010) Molecular mechanisms of Hepatitis C virus-induced cellular transformation. In: Khalili K, Wang K-T (eds) Viral oncology. Wiley-Blackwell, Hoboken

Wen KW, Damania B (2009) Kaposi sarcoma-associated herpesvirus (KSHV): molecular biology and oncogenesis. Cancer Lett 289(2):140–150

White MK, Pagano JS, Khalili K (2014) Viruses and human cancers: a long road of discovery of molecular paradigms. Clin Microbiol Rev 27(3):463–481. doi:10.1128/CMR.00124-13.PMID: 24982317

Yoshizaki T, Wakisaka N, Pagano JS (2005) Chapter 12: Epstein-Barr virus, invasion and metastasis. In: Robertson E (ed) Epstein-Barr virus. Caister Academic, Norfolk, pp 171–196

Chapter 3
Epidemiology of Virus Infection and Human Cancer

Jessica Liu, Hwai-I Yang, Mei-Hsuan Lee, Wan-Lun Hsu, Hui-Chi Chen, and Chien-Jen Chen

Abstract Based on the comprehensive assessment of epidemiological and mechanistic study findings, the International Agency for Research on Cancer has classified Epstein-Barr virus (EBV), hepatitis B virus (HBV), hepatitis C virus (HCV), Kaposi's sarcoma herpes virus (KSHV), human immunodeficiency virus, type-1 (HIV-1), human T-cell lymphotrophic virus, type-1 (HTLV-1), and human papillomavirus (HPV) as Group 1 human carcinogens. The Merkel cell polyomavirus has recently been documented to cause Merkel cell carcinoma. No causal specificity is observed for many oncogenic viruses. Some of them may cause different cancers, while some cancers may be caused by different viruses. However, only a proportion of infected persons will actually develop cancers, indicating that oncogenic viruses may be necessary but not sufficient to cause specific cancers. Viral, host and environmental cofactors have been assessed for the EBV-associated nasopharyngeal carcinoma, HBV/HCV-associated hepatocellular carcinoma and HPV-associated cervical carcinoma. Persistent infection, high viral load, and genetic/acquired susceptibility factors are important risk predictors for these virus-caused cancers. Risk calculators have been developed for the prediction of the long-term risk of hepatocellular carcinoma among patients affected with chronic hepatitis B and C. Both clinical trials and national programs of immunization or anti-viral therapy have demonstrated a significant reduction in the incidence of cancers caused by HBV, HCV and HPV. Assessing the effects of gene-gene and gene-environment interactions on virus-caused cancers is urgently needed.

J. Liu • H.-I. Yang • W.-L. Hsu • H.-C. Chen
Genomics Research Center, Academia Sinica,
128 Academia Road, Section 2, Taipei 11529, Taiwan

M.-H. Lee
Institute of Clinical Medicine, National Yang-Ming University, Taipei, Taiwan

C.-J. Chen (✉)
Genomics Research Center, Academia Sinica,
128 Academia Road, Section 2, Taipei 11529, Taiwan

Graduate Institute of Epidemiology and Preventative Medicine, National Taiwan University, Taipei, Taiwan
e-mail: chencj@gate.sinica.edu.tw

M.R. Shurin et al. (eds.), *Infection and Cancer: Bi-Directorial Interactions*, DOI 10.1007/978-3-319-20669-1_3

Keywords Cancer • EBV • Epidemiology • HBV • HCV • HIV • HPV • HTLV-I • KSHV

Abbreviations

CLD	chronic liver disease
EBV	Epstein-Barr virus
HBV	hepatitis B virus
HCC	hepatocellular carcinoma
HCV	hepatitis C virus
HIV-1	human immunodeficiency virus, type-1
HPV	human papillomavirus
HTLV-1	human T-cell lymphotrophic virus, type-1
KSHV	Kaposi's sarcoma herpes virus
MCV	Merkel cell polyomavirus

Introduction

Based on the comprehensive assessment of both epidemiological and mechanistic evidence, the International Agency for Research on Cancer (IARC) has assessed the carcinogenicity of biological agents to humans (International Agency for Research on Cancer 2012). Epstein-Barr virus (EBV), hepatitis B virus (HBV), hepatitis C virus (HCV), Kaposi's sarcoma herpes virus (KSHV), human immunodeficiency virus, type-1 (HIV-1), human T-cell lymphotrophic virus, type-1 (HTLV-1), and several types of human papillomavirus (HPV) have been classified as Group 1 human carcinogens. Recently, the Merkel cell polyomavirus (MCV) has been documented as the cause of Merkel Cell carcinoma, although this has not yet been assessed by IARC (Amber et al. 2013; Chang and Moore 2012; Spurgeon and Lambert 2013).

As shown in Table 3.1, EBV causes cancers such as nasopharyngeal carcinoma, Burkitt's lymphoma, immune-suppression-related non-Hodgkin lymphoma, extranodal NK/T-cell lymphoma (nasal type), and Hodgkin's lymphoma in humans with sufficient evidence. Although studies have shown the association between EBV and gastric carcinoma and lympho-epithelioma-like carcinoma, the evidence is considered limited. The evidence showing that HBV and HCV cause hepatocellular carcinoma is sufficient (International Agency for Research on Cancer 2012). There is also sufficient evidence for HCV-caused non-Hodgkin lymphoma, especially B-cell lymphoma, while the evidence for HBV-causeed non-Hodgkin lymphoma and pancreatic cancer is limited (International Agency for Research on Cancer 2012; Yang et al. 2010). There is also limited evidence that HBV and HCV cause cholangiocar-

Table 3.1 Cancers caused by oncogenic viruses with sufficient and limited evidence according to the International Agency for the Research on Cancer

Virus	Cancer sites with sufficient evidence	Cancer sites with limited evidence
Epstein–Barr virus (EBV)	Nasopharyngeal carcinoma, Burkitt's lymphoma, immune suppression-related non-Hodgkin lymphoma, extranodal NK/T-cell lymphoma (nasal type), Hodgkin's lymphoma	Gastric carcinoma, lympho-epithelioma-like carcinoma
Hepatitis B virus (HBV)	Hepatocellular carcinoma	Cholangiocarcinoma, non-Hodgkin's lymphoma, Pancreatic cancer
Hepatitis C virus (HCV)	Hepatocellular carcinoma, non-Hodgkin's lymphoma	Cholangiocarcinoma
Human immunodeficiency virus, type 1 (HIV-1)	Kaposi's sarcoma, non-Hodgkinlymphoma, Hodgkin's lymphoma, cancers of the cervix, anus and conjunctiva	Cancers of the vulva, vagina and penis, non-melanoma skin cancer, hepatocellular carcinoma
Human papillomavirus type 16 (HPV-16)	Cancers of the cervix, vulva, vagina, penis, anus, oral cavity, oropharynx and tonsil	Cancer of the larynx
Human papillomavirus (other types)	Cancer of the cervix[a]	Cancer of the cervix[b]
Human T-cell lymphotrophic virus, type 1 (HTLV-1)	Adult T-cell leukemia and lymphoma	
Kaposi's sarcoma herpes virus (KSHV)	Kaposi's sarcoma, primary effusion lymphoma	Multicentric Castleman's disease
Merkel cell polyomavirus (MCV)	Merkel cell carcinoma[c]	

[a]HPV types other than 16 with sufficient evidence of causing cervical cancer: 18, 31, 33, 35, 39, 45, 51, 52, 56, 58, 59 (HPV-18, 31, 33, 35, 39, 45, 51, 52, 56, 58, 59

[b]HPV types other than 16 with limited evidence of causing cervical cancer: 26, 30, 34, 53, 66, 67, 68, 69, 70, 73, 82, 85, 97 (HPV- 26, 30, 34, 53, 66, 67, 68, 69, 70, 73, 82, 85, 97

[c]Not yet assessed by IARC

cinoma (Fwu et al. 2011). Moreover, the evidence showing that HIV-1 causes Kaposi's sarcoma, non-Hodgkin lymphoma, Hodgkin's lymphoma, as well as cancers of the cervix, anus and conjunctiva is sufficient. However, there is limited evidence to support the hypothesis that HIV-1 causes cancers of the vulva, vagina and penis, non-melanoma skin cancer and hepatocellular carcinoma (International Agency for Research on Cancer 2012).

HPV-16 causes cancers of the cervix, vulva, vagina, penis, anus, oral cavity, oropharynx, and tonsil with sufficient evidence, but the evidence for HPV-16-caused laryngeal cancer is limited. Cervical cancer is also caused by several other oncogenic types of HPV, including HPV-18, 31, 33, 35, 39, 45, 51, 52, 56, 58, and 59 with sufficient evidence. However, the evidence that HPV-26, 30, 34, 53, 66, 67, 68, 69, 70, 73, 82, 85, and 97 also cause cervical cancer is limited. Evidence for HTLV-1-caused adult T-cell leukemia and lymphoma is sufficient, while evidence for KSHV-caused Kaposi's sarcoma and primary effusion lymphoma is also sufficient. However, there is limited evidence proving that KSHV causes multicentric Castleman's disease (International Agency for Research on Cancer2012). Merkel cell polyomavirus (MCV) has been documented to cause Merkel cell carcinoma, but this causal association has not yet been assessed by IARC (Amber et al. 2013; Chang and Moore 2012; Spurgeon and Lambert 2013).

As the proportion of human cancers caused by infectious agents is estimated to be greater than 20 %, the identification of new cancer sites attributable to infectious agents is important because it means that more cancers are potentially preventable (International Agency for Research on Cancer 2012). This chapter will briefly review the epidemiology of oncogenic viruses and their associated cancers.

Transmission Routes and Global Variation in Infection Prevalence of Oncogenic Viruses

The transmission routes and infection prevalences of oncogenic viruses are shown in Table 3.2. EBV is one of the most common viruses in humans, and is highly prevalent throughout the world. Even in remote populations, more than 90 % of adults are infected with EBV (International Agency for Research on Cancer 2012). It is estimated that over 5.5 billion people worldwide are infected with EBV. The virus is primarily transmitted through bodily fluids, particularly saliva, and the age at which primary infection occurs varies significantly. For example, individuals living in overcrowded conditions with poor sanitation are first infected at a younger age than those individuals living in better environments. Two major types of EBV have been identified and differ in their geographical distributions, with EBV-2 more prevalent in Africa and in homosexual men. However, the role of specific EBV types in the development of different cancers remains unclear. As EBV infection is ubiquitous, the specific geographical distributions of EBV-caused malignancies, including endemic Burkitt's lymphoma and nasopharyngeal carcinoma, are more likely attributable to the variation in the distributions of other co-factors, which may activate EBV replication (International Agency for Research on Cancer 2012).

The HBV infects more than 2.0 billion people worldwide, with more than 350 million individuals chronically infected (International Agency for Research on Cancer 2012; Lavanchy 2004). The global distribution of chronic HBV infection varies widely, as shown in Table 3.2. Approximately 45, 43 and 12 % of the world population live in areas where the endemicity of chronic HBV infection is high (seroprevalence of hepatitis B surface antigen >8 %), medium (2–7 %), and low (<2 %) (Centers for Disease Control and Prevention (2014). The prevalence of HBV is highest in sub-Saharan Africa, the Amazon Basin, China, Korea, Taiwan and several countries in Southeast Asia (Lavanchy 2004). In areas of high endemicity, the

Table 3.2 Transmission routes and global variation in infection prevalence of oncogenic viruses

Virus	Transmission routes	Areas of highest and lowest prevalence
EBV	Bodily fluids, especially saliva.	Highly prevalent throughout the world, >90 % of adults
HBV	Highly endemic areas: perinatal or child-to-child	Highest (>8 % seroprevalence of HBsAg): sub-saharan Africa, Amazon basin, China, Korea, Taiwan, and parts of Southeast Asia
	Low endemic areas: occurs in adulthood through injection drug use, among male homosexuals, though medical procedures and transfusions or hemodialysis	Lowest (<2 % seroprevalence of HBsAg): North and Central America, Australia
HCV	Injection drug use	Highest (>2 % seroprevalence of anti-HCV): Egypt, China, Mongolia, northern Africa, Pakistan, southern Italy, parts of Japan
	Iatrogenic exposure	Lowest (<2 % seroprevalence of anti-HCV): all other areas
	Less common: perinatal or sexual transmission	
HIV-1	Sexual activity	High (>15 %): Sub-saharan Africa, medium (6-14 %): the Caribbean, Low (0.5–5 %): Eastern Europe, Central Asia
	Blood contact	Lowest (<0.5 %): Western Europe, parts of East Asia, Australia, Canada, parts of Central America
	Mother to child	
HPV	Mainly by direct contact through sexual intercourse	Highest (20–30 %): Africa, Eastern Europe, Latin America
	Less common: perinatal or iatrogenic transmission	Lowest (6–7 %): Southern and Western Europe, Southeast Asia
HTLV-1	Mother-to-child (such as breastfeeding)	Highest (>1 %): Southwest Japan, Sub-Saharan Africa, the Caribbean, South Africa, parts of Southeast Asia
	Sexual activity	Lowest (<0.01 %): East Asia, China, Russia, parts of Europe, North Africa
	Parenteral transmission (such as transfusions)	
KSHV	Saliva	Highest (>50 %): Sub-Saharan Africa, medium (10–30 %): Mediterranean region, Low (<10 %): Northern Europe, North America, Asia
	Less common: prolonged injection drug use, transfusions, transplantation	
MCV	Unclear	9 % in children younger than 4 years, up to 80 % in 50-year-old adults

lifetime risk of acquiring HBV infection is more than 60 %, with most infections acquired through perinatal and child-to-child transmission, when the risk of chronic infection is greatest. Perinatal (vertical) transmission is predominant in China, Korea, and Taiwan, where the seroprevalence of HBeAg in pregnant women is high, while child-to-child (horizontal) transmission is common in sub-Saharan Africa, where HBeAg seroprevalence in mothers is low. In areas of medium endemicity, mixed HBV transmission patterns occur in infancy, early childhood, adolescence, and adulthood. In areas of low endemicity, most HBV infections occur in adolescents and young adults through injection drug use, male homosexuality, health care practices, and regular transfusions or hemodialysis (International Agency for Research on Cancer 2012; Lavanchy 2004). In addition to the striking geographical variation in the seroprevalence of HBsAg worldwide, eight genotypes of HBV have been discovered, the distribution of which also varies significantly in different countries (International Agency for Research on Cancer 2012).

HCV infects around 130–210 million people worldwide, showing an estimated prevalence of 2.2 % (International Agency for Research on Cancer 2012; Lavanchy 2009; Hajarizadeh et al. 2013). The prevalence of HCV infection (seroprevalence of antibodies against HCV) also varies geographically, and ranges from <0.1 % in the United Kingdom and Scandinavia, to 15–20 % in Egypt (Alter 2007). A high prevalence of HCV infection is also seen in Mongolia, northern Africa, Pakistan, China, southern Italy, and parts of Japan. At least six major HCV genotypes have been identified, the geographical distributions of which also vary widely. Research has shown that responses to anti-viral therapy also vary by HCV genotype; response to therapy is better in patients infected with genotypes 2 or 3 than in those infected with genotypes 1 or 4 (International Agency for Research on Cancer 2012; Fried et al. 2002).

Two major transmission routes for HCV have been identified. They are injection drug use and iatrogenic exposures through transfusion, transplantation, and unsafe therapeutic injection (Table 3.2). While there has been a large reduction in iatrogenic transmission of HCV after 1990 in developed countries such as Japan and Italy, it continues to be a common source of transmission in low-resource countries where disposable needles tend to be re-used. In developed countries, however, injection drug use is the most important transmission route for newly acquired HCV infection. Transmission of HCV through perinatal, sexual and accidental needlestick exposures is less common, as it occurs less efficiently than iatrogenic exposure and injection drug use (International Agency for Research on Cancer 2012; Alter 2007; Lee et al. 2011; Sun et al. 1999).

HIV infects an estimated 34 million people worldwide (UNAIDS 2012; International Agency for Research on Cancer 2012). An estimated 0.8 % of all adults aged 15–49 years worldwide are living with HIV, and the burden varies considerably between countries and regions as shown in Table 3.2. Sub-Saharan Africa remains the most disproportionately affected, with a prevalence of at least 4.9 %. Although the prevalence of HIV infection is nearly 25 times higher in sub-Saharan Africa than it is in Asia, there are still almost 5 million people in South, Southeast, and East Asia living with HIV infection. After sub-Saharan Africa, other regions heavily affected by HIV are the Caribbean, Eastern Europe, and Central Asia, where 1.0 % of adults were living with HIV in 2011. In 2011 alone, 2.5 million people, including 0.39 million children, were newly infected with HIV. Since 2001, the annual incidence of HIV infection has fallen in 33 countries, 22 of them in sub-Saharan Africa. However, after having slowed in the early 2000s, incidence of HIV infection is once again rising in Eastern Europe and Central Asia, and new infections are also on the rise in the Middle East and in North Africa (International Agency for Research on Cancer 2012; UNAIDS 2012).

HIV infection is primarily transmitted through three major routes: sexual intercourse, blood contact, and mother-to-child transmission. HIV infectivity is determined by the interaction between agent, host, and environmental factors. The probability of HIV transmission is highest for blood transfusions, followed by mother-to-child transmission, needle-sharing, man-to-man sexual transmission, and is lowest for heterosexual sexual transmission (UNAIDS 2012; International Agency for Research on Cancer 2012).

HPV infection is highly prevalent throughout the world, and most sexually active individuals will acquire at least one genotype of anogenital HPV infection during their lifetime (International Agency for Research on Cancer 2012). In a meta-analysis of 157,879 women with normal cytology, the estimated oncogenic HPV DNA point prevalence was reported to be as high as 10 %, resulting in an estimate of 600 million infected individuals worldwide (de Sanjose et al. 2007). The point prevalence of HPV infection was highest (20–30 %) in Africa, Eastern Europe, and Latin America; and lowest (6–7 %) in Southern and Western Europe and Southeast Asia, demonstrating a large geographical variation. However, estimated point prevalences are highly dynamic, as both incidence and clearance rates are high.

Among 13 known oncogenic HPV types, the most prevalent types include 16, 18, 31, 33, 35, 45, 52, and 58. HPV type 16 is the most common type across all regions, with a prevalence ranging from 2.3–3.5 %. Anogenital HPV types are spread mainly through sexual transmission through any type of sexual intercourse in teenagers and young adults. Non-sexual routes, including perinatal and iatrogenic transmissions, account for only a minority of HPV infections (Table 3.2).

Globally, HTLV-1 infects an estimated 15–20 million people (International Agency for Research on Cancer 2012). HTLV-1 infection is characterized by micro-epidemic hotspots surrounded by low prevalence areas (Proietti et al. 2005). The prevalence of HTLV-1 infection ranges from <0.1 % in China, Korea and Taiwan, to 20 % in Kyushu and Okinawa of Japan. Regions of high endemicity include south-western Japan, parts of sub-Saharan Africa, the Caribbean Islands, and South Africa. HTLV-1 is primarily transmitted through vertical transmission, sexual transmission and parenteral transmission. While vertical transmission through breast-feeding has a high probability of resulting in mother-to-child infection, in-utero infectivity is low due to the limited trafficking of HTLV-1 infected lymphocytes across the placenta. The efficiency of sexual transmission of HTLV-1 depends on the proviral load and usage of a condom. There has been a significant reduction in parenteral transmission through transfusions due to the sensitive serological examination of blood products. Needle sharing associated with injection drug use is another parenteral route for HTLV-1 transmission (International Agency for Research on Cancer 2012).

Prevalence of KSHV has also been shown to have wide geographical variance (Dukers and Rezza 2003). Seroprevalence ranges from 2 to 3 % in northern Europe, to 82 % in the Congo. Prevalence rates are generally low (<10 %) in northern

Europe, the USA, and Asia, elevated in the Mediterranean region (10–30 %), and high in sub-Saharan Africa (>50 %). KSHV is primarily transmitted via saliva. In countries where KSHV prevalence is high, infection occurs during childhood and increases with age. KSHV may also be transmitted with low efficiency through prolonged injection drug use, blood transfusions, and organ transplantation (International Agency for Research on Cancer 2012).

Lastly, studies have shown that the Merkel cell polyomavirus is a very common virus infecting the human population. Prevalence of infection seems to increase with age; seroprevalence in children younger than 4 years of age was found to be around 9 %, which increased to 35 % by 13 years of age (Chen et al. 2011a). In another study, 80 % of healthy North American adults showed evidence of past MCV exposure (Tolstov et al. 2009).

Global Variation in the Incidence of Virus-Caused Cancers

The incidence rates of some oncogenic virus-caused cancers are shown in Table 3.3. The age-adjusted incidence rates of nasopharyngeal cancer range from <0.1 in low endemic regions to >8.0 per 100,000 in areas of high endemicity. The highest incidence is seen in southern China, Southeast Asia, and sub-Saharan Africa, and the lowest incidence is seen in Europe, western Africa, and Central America. Interestingly, in different cancer registries throughout the world, individuals of

Table 3.3 Global variation in Incidence of virus-caused cancers

Cancer	Incidence rate per 100,000	Area of high and low incidence
Nasopharyngeal cancer	<0.1–8.0+	High: China, Southeast Asia, sub-Saharan Africa
		Low: Western Africa, Central America, parts of Europe
Burkitt's lymphoma	<5–35+	High: Parts of Africa, South America, Papua New Guinea, and the Caribbean
		Low: all other countries
Hepatocellular carcinoma	0.70–94.4	High: East Asia, Southeast Asia, Egypt, sub-Saharan Africa
		Low: Europe, Middle East, Australia, New Zealand, Canada
Cervical cancer	2.14–56.29	High: Latin America, South Asia, sub-Saharan Africa
		Low: Europe, North America, Australia, New Zealand, Middle East
Kaposi's sarcoma	<1.0–30	High: sub-Saharan Africa
		Low: Europe, Australia, North America, East Asia

Chinese ethnicity have the highest incidence of nasopharyngeal cancer. As EBV infection is ubiquitous in humans, the uniquely high incidence of nasopharyngeal carcinoma among individuals of Chinese descent suggests that lifestyles or genetic susceptibility may play an important role in the development of nasopharyngeal cancer (International Agency for Research on Cancer 2012).

The age-adjusted incidence rates of Burkitt's lymphoma are shown in Table 3.3. Central Africa, equatorial South America, Papua New Guinea, and the Caribbean are endemic for Burkitt's lymphoma with an incidence rate of 5–35 per 100,000, but the incidence rate of Burkitt's Lymphoma is relatively low in other countries. As EBV infection is ubiquitous in humans, the extraordinarily high endemicity of Burkitt's lymphoma in Africa suggests that local environments or genetic susceptibility may play an important role in the development of endemic Burkitt's lymphoma (International Agency for Research on Cancer 2012).

The age-adjusted incidence rates of liver cancer range from 0.70 to 94.4 per 100,000. The highest incidence is observed in East Asia, Southeast Asia, Egypt, and sub-Saharan Africa, while the lowest incidence is seen in Europe, Middle East, Australia, New Zealand, and Canada. The geographical variation in liver cancer incidence is consistent with that of the varying seroprevalence of HBV and HCV (International Agency for Research on Cancer 2012).

The age-adjusted incidence rates of cervical cancer range from 2.14 to 56.29 per 100,000. The highest incidence is observed in Latin America, South Asia, and sub-Saharan Africa, while the lowest incidence is seen in Europe, North America, Australia, New Zealand and the Middle East. The geographical variation in cervical cancer incidence is also consistent with the varying seroprevalence of oncogenic HPV (International Agency for Research on Cancer 2012).

The age-adjusted incidence rates of Kaposi's sarcoma range from <1.0 to 30 per 100,000. The highest incidence is observed in sub-Saharan Africa, while the lowest incidence is seen in Europe, Australia, North America and East Asia. The geographical variation in Kaposi's sarcoma incidence is also consistent with that of the varying seroprevalence of KSHV (International Agency for Research on Cancer 2012).

Lifetime Cumulative Incidence of Virus-Caused Cancers

Some viruses may cause more than one cancer, while some cancers can be caused by more than one virus. However, only a proportion of persons infected by these oncogenic viruses will actually develop specific cancers. The cumulative lifetime incidence for virus-caused cancers varies according to different factors. The cumulative lifetime (30–75 years old) risk of developing nasopharyngeal carcinoma was 2.2 % for men seropositive for IgA antibodies against EBV VCA or antibodies against EBV DNase and 0.48 % for those seronegative for both antibodies.

Around one-quarter of patients with patients chronically infected with HBV will develop hepatocellular carcinoma. The cumulative lifetime incidence of HCC shows a striking gender difference, at 27.4 % for men and 8.0 % for women (Huang et al. 2011). The development of HBV-associated hepatocellular carcinoma has been considered as a multistage hepatocarcinogenesis with multifactorial etiology, and involves the interaction between HBV, other viruses such as HCV, chemical carcinogens, host characteristics and genetic susceptibility (Chen et al. 1997; Chen and Chen 2002; Chen and Yang 2011; International Agency for Research on Cancer2012; Huang et al. 2011).

Around one-fifth of patients seropositive for antibodies against HCV (anti-HCV) will develop hepatocellular carcinoma. However, the cumulative lifetime incidence of HCC shows a less significant gender difference, and is approximately 23.7 % for men and 16.7 % for women (Huang et al. 2011). The cumulative lifetime risk of HCC among anti-HCV seropositives with and without detectable serum HCV RNA levels was 24.2 %, and 3.53 %, respectively. In addition, anti-HCV seropositives with detectable serum HCV RNA are at increased risk for liver-related deaths (Lee et al. 2012). Many cofactors are involved in the development of hepatocellular carcinoma in anti-HCV seropositives, which will be discussed below (Lee et al. 2010; International Agency for Research on Cancer 2012; Lee et al. 2014a).

The cumulative lifetime (30–75 years old) risk of cervical cancer for women who are infected with HPV 16, HPV 52, HPV 58, and any Group 1 oncogenic HPV virus was 34.3 %, 23.3 %, 33.4 %, and 20.3 %, respectively (International Agency for Research on Cancer 2012). Women with persistent oncogenic HPV infection have a much higher cumulative risk of cervical cancer than those with only transient infection (Chen et al. 2011b).

Host and Environmental Cofactors of Virus-Caused Cancers

As only a proportion of persons infected by oncogenic viruses will actually go on to develop specific cancers, this strongly suggests the involvement of other cofactors in the carcinogenic process. For example, carcinogenesis would result from the interaction between multiple risk factors, including viral factors, host factors, and environmental factors, as shown in Table 3.4. Typical viral factors include various infection markers such as viral load, genotypes, variants, mutants, and antibody serotiters. Host factors include age, gender, race, anthropometric characteristics, immune status, hormonal levels, personal disease history, and family cancer history. Lastly, environmental factors include chemical carcinogens, nutrients, ionizing radiation, immunosuppressive drugs, and co-infections with other infectious agents. The contribution of several additional factors to the development of virus-associated cancers seems to be substantial, but has not yet been elucidated in detail.

Several cofactors for EBV-caused nasopharyngeal carcinoma have been previously reviewed (Chien and Chen 2003). Viral factors associated with nasopharyngeal carcinoma include elevated serotiters of antibodies against EBV, including anti-EBV VCA IgA, anti-EBV DNase, anti-EBNA1, and elevated serum EBV DNA levels (viral load) (Chien et al. 2001; Hsu et al. 2009). Host factors affecting nasopharyngeal carcinoma include male gender, family history of nasopharyngeal

Table 3.4 Host and environmental co-factors of virus-caused cancers

Virus (cancer)	Viral factors	Host factors	Environmental factors
EBV (nasopharyngeal carcinoma)	Elevated serotiter of antibodies against EBV, EBV viral load	Male gender, family history, genetic polymorphisms (xenobiotic metabolism, DNA repair, human leukocyte antigen)	Cantonese salted fish, Dietary nitrosamine, wood dust, formaldehyde, tobacco, low intake of plant vitamins, fish, green tea, and coffee
HBV (hepatocellular carcinoma)	Persistent infection, viral load, genotype, mutants, serum HBsAg level,	Elder age, male gender, obesity, diabetes, serum androgen and ALT level, family history, parity, genetic polymorphisms (DNA repair, human leukocyte antigen, androgen and xenobiotic metabolism)	Aflatoxins, alcohol, tobacco, carotenoids, selenium, interactions with HCV infection
HCV (hepatocellular carcinoma)	Persistent infection, viral load, genotype, mutants	Elder age, male gender, obesity, diabetes, serum ALT level, family history, genetic polymorphisms	Alcohol, tobacco, betel nut chewing, HBV or HTLV-1 infection, radiation
HPV (cervical carcinoma)	Persistent infection, viral load, genotype	Elder age, number of pregnancies, family history, serum estrogen level, genetic polymorphisms (DNA repair, human leukocyte antigen)	Tobacco, immunosuppression, HIV-1 infection, contraceptives, nutrients, screening frequency

carcinoma (Hsu et al. 2011), and genetic polymorphisms in xenobiotic metabolism enzymes, DNA repair enzymes, and the human leukocyte antigen (Hildesheim et al. 1997, 2002; Cho et al. 2003; Hsu et al. 2012a). Environmental cofactors include consumption of Cantonese salted fish, high dietary intake of nitrite and nitrosamine, occupational exposure to wood dust and formaldehyde, long-term tobacco smoking, and low intake of plant vitamins, fresh fish, green tea, and coffee (Ward et al. 2000; Hildesheim et al. 2001; Hsu et al. 2009, 2012b).

Many studies have examined cofactors of HBV-caused hepatocellular carcinoma. Viral factors associated with HBV-caused hepatocellular carcinoma include positive HBeAg serostatus, elevated serum HBV DNA levels, HBV genotype and mutant types, and elevated serum HBsAg level (Chen et al. 2006; Yang et al. 2002, 2008; Lee et al. 2013). Persistently high HBV DNA levels throughout disease progression also indicate high risk for hepatocellular carcinoma (Chen et al. 2011c). However, reaching HBV DNA undetectability and HBsAg seroclearance have been shown to significantly decrease risk for future hepatocellular carcinoma (Liu et al. 2013a). Host factors include elder age, male gender, persistently elevated serum alanine aminotransferase (ALT) levels, and family history of hepatocellular carcinoma

(Yu et al. 2000a; Yang et al. 2010; Chen et al. 1991, 2011c; Loomba et al. 2013). Interestingly, in a recent study, family history of hepatocellular carcinoma showed a synergistic interaction between family history and HBsAg serostatus, having a multiplicative effect on hepatocellular carcinoma risk (Loomba et al. 2013). In addition, higher parity (Fwu et al. 2009), obesity and diabetes (Chen et al. 2008), elevated serum level of androgen and androgen-related genetic polymorphisms (Yu and Chen 1993; Yu et al. 2000b), and genetic polymorphisms of xenobiotic metabolism enzymes and DNA repair enzymes are also important host factors for HBV-caused hepatocellular carcinoma (Chen et al. 1996a; Yu et al. 1995a, 1999a, 2003). Environmental factors include aflatoxin exposure (Chen et al. 1996b; Wang et al. 1996), alcohol consumption and tobacco smoking (Chen et al. 1991; Wang et al. 2003), inadequate intake of carotenoids and selenium (Yu et al. 1995b, 1999a, b), and co-infection with HCV (Huang et al. 2011). Interestingly, although co-infection with HCV has been shown to result in higher risk for HCC, it was a sub-additive combined effect, and was also associated with later-onset of HCC, suggesting an antagonistic effect between HBV and HCV (Huang et al. 2011).

Recently, studies have been able to further clarify cofactors of HCV-associated hepatocellular carcinoma. Important viral factors associated with HCV-caused hepatocellular carcinoma include elevated serum levels of HCV RNA and HCV genotype 1 (Huang et al. 2011; Lee et al. 2010). Recently, it was also found that patients with genotype 1b HCV infection are at additional risk for hepatocellular carcinoma (Lee et al. 2014b). Host factors include elder age, male gender, obesity, diabetes, elevated serum ALT levels, family history of hepatocellular carcinoma, and genetic polymorphisms (Sun et al. 2003; Chen et al. 2008; Lee et al. 2010; International Agency for Research on Cancer2012). Environmental factors include alcohol consumption, tobacco smoking, betel nut chewing, radiation exposure, and co-infection with HBV or HTLV-1 (International Agency for Research on Cancer 2012; Sun et al. 2003; Huang et al. 2011).

The viral factors associated with oncogenic HPV-associated cervical cancer include persistent infection, elevated viral load, and HPV genotypes and variants (International Agency for Research on Cancer 2012; Chen et al. 2011b, d; Chang et al. 2011). Important host factors include elder age, number of pregnancies, family history of cervical cancer, serum estrogen level, and genetic polymorphisms of DNA repair enzymes and the human leukocyte antigen (Chen et al. 2011b; Chuang et al. 2012; International Agency for Research on Cancer 2012). Environmental factors include tobacco smoking, immunosuppression, HIV-1 co-infection, use of contraceptives, and inadequate intake of micronutrients (International Agency for Research on Cancer 2012). Increased screening frequency has also been shown to indicate lower risk for HPV-caused cervical cancer (Chen et al. 2014).

Prediction of HBV-Caused Hepatocellular Carcinoma

In order to reduce the burden of virus-induced cancers, it is important to understand the natural history of each cancer, including different milestones in its natural history. The most important goal in reducing oncogenic virus-induced cancer incidence is, therefore, determining how to effectively interrupt the progression of the viral infection.

Previous research studies have identified several important serological milestones in the natural history of chronic hepatitis B progression, which are associated with the risk of HCC (Chen and Yang 2011). These milestones, HBeAg seroclearance, HBV DNA undetectability, and HBsAg seroclearance, have each been shown to predict a decreased risk of HCC, while persistence of each is also predictive of an increased risk for HCC (Chen and Yang 2011; Yang et al. 2002; Chen et al. 2006; Liu et al. 2013a). In a recent study, which examined the three milestones together using long-term repeated measurements, HBV DNA undetectability was essential for reducing future HCC risk in chronic carriers of HBV (Liu et al. 2013a). In addition to serological milestones, other viral, host, and environmental factors are important co-factors for HBV-caused hepatocellular carcinoma.

Given that there are many risk factors for each virus-caused cancer, including HBV-caused hepatocellular carcinoma, it would be important to incorporate these risk factors into a risk prediction model or calculator that can predict the cumulative incidence of each cancer. Such risk calculators may provide clinicians with important information for the triage and identification of patients who need intensive treatment, versus those who need only routine follow-up. Moreover, with personalized HCC risk calculations, patients' follow-up intervals, surveillance patterns, and referral strategies can be tailored.

In developing an effective risk calculator for HBV-caused HCC that can be applied to prevention strategies, there are two ways to interrupt the progression of disease; prediction models can be developed for direct long-term prediction of HCC risk, or for the prediction of important milestones in HCC carcinogenesis, including HBeAg seroclearance, HBV DNA undetectability, or HBsAg seroclearance.

To date, several risk prediction models for HBV-caused HCC have been established, each of which incorporates various clinical variables to estimate HCC risk in treatment naïve chronic hepatitis B patients (Table 3.5) (Han and Ahn 2005; Wong et al. 2010; Yang et al. 2010; Yuen et al. 2009). These models incorporated a wide range of clinical parameters, and most were hospital-based cohorts, or did not have satisfactory external validation of the model. In order to perform sufficient external validation of a robust prediction model, the groups behind the IPM model, GAG-HCC model, CUHK Clinical Scoring System, and the REVEAL nomograms jointly established the REACH-B risk score, a 17-point scoring system incorporating gender, age, serum ALT concentration, HBeAg serostatus, and HBV DNA level as its predicting factors (Yang et al. 2011). The REACH-B score was first developed in a large community-based cohort, then externally validated in a composite hospital-based cohort. This model was able to predict the 3-, 5-, and 10-year risk of developing HCC with areas under the receiving operating characteristic curve (AUROC) of 0.811, 0.796, and 0.769, respectively (Yang et al. 2011). The REACH-B score can be widely used in clinical settings, and recently, was also used to examine the efficacy of anti-viral therapy in reducing liver cancer risk in chronic hepatitis B patients.

With the establishment of quantitative HBsAg levels as an important seromarker in the natural history of chronic hepatitis B, the REVEAL nomograms were recently updated to include this novel risk predictor, in addition to the original risk predictors. This upgraded model for HCC risk prediction also provided excellent prediction

Table 3.5 Risk calculators for HBV-caused hepatocellular carcinoma

Name [reference]	Author (Year)	AUROC (95 % confidence interval)	Risk predictors included in the calculator													
			Gender	Age	HCV status	AFP level	ALT level	LC	Alcohol	HBV DNA	CP mutants	Bilirubin	HBV genotype	Family Hx	HBeAg status	HBsAg level
IPM (Han and Ahn 2005)	Han KH, and Ahn SH. (2005)	None reported	X	X	X	X	X	X	X							
GAG-HCC Risk Score (Yuen et al. 2009)	Yuen MF et al. (2009)	5-year: 0.88	X	X				X		X	X					
		10-year: 0.89														
CUHK Clinical Scoring System (Wong et al. 2010)	Wong VW et al. (2010)	(Internal Validation)		X				X		X		X				
		5-year: 0.76 (0.66–0.86)														
		10-year: 0.78 (0.71–0.86)														
REVEAL nomograms (Yang et al. 2010)	Yang HI et al. (2010)	(Internal validation)	X	X			X		X	X			X	X	X	
		5-year: 83.2														
		10-year: 83.0														

REACH-B Score (Yang et al. 2011)	Yang HI et al. (2011)	(External validation) 3-year: 0.902 (0.884–0.918) 5-year: 0.783 (0.759–0·806) 10-year: 0.806 (0.783–0.828)	X	X			X			X					X	
REVEAL nomograms (updated) (Lee et al. 2013)	Lee MH et al. (2013)	(Internal validation) 5-year: 0.84 10-year: 0.86 15-year: 0.87	X	X			X			X			X	X	X	X

LC liver cirrhosis, *CP* core promoter, *Hx* history

accuracy and discriminatory ability, with a 5-year and 10-year AUROC of 0.84 and 0.86, respectively, although it was only validated internally (Lee et al. 2013). Future external validation with clinical patients will confirm its accuracy and reliability.

Prediction HCV-Caused Hepatocellular Carcinoma

Recently, two risk prediction models for HCV-caused hepatocellular carcinoma were also developed and validated (Table 3.6). The first model utilized age, serum ALT levels, AAR (AST/ALT ratio), and a combination of liver cirrhosis, HCV RNA level and HCV genotype to predict HCC among anti-HCV seropositives, while the second model utilized the same variables to predict HCC risk among only those with detectable HCV RNA levels (Lee et al. 2014c). Both models were validated in a high-risk cohort of HCV patients with satisfactory discriminatory ability, showing a 5-year AUROC of 0.73 and 0.70, respectively (Lee et al. 2014c). These risk prediction models have also been validated using the United States Veterans Affairs database, and the 5-year predictability was acceptable with an AUROC of 0.69 (Matsuda et al. 2014).

Prediction of Clinical Milestones of Chronic Hepatitis B Progression

In addition to HCC, several prediction models have also been established for important clinical milestones in HCC progression (Lee et al. 2013; Liu et al. 2013b, 2014). Their predictive factors and calculated AUROCs are shown in Table 3.7. For HBeAg

Table 3.6 Risk calculators for HCV-caused hepatocellular carcinoma

Name [reference]	Author (Year)	AUROC (95 % confidence interval)	Risk predictors included in the calculator					
			Age	ALT level	AAR	LC	HCV RNA level	HCV Genotype
REVEAL-HCV risk calculator (for anti-HCV seropositives) (Lee et al. 2014c)	Lee MH et al. (2014c)	5-year: 0.75 10-year: 0.83 15-year: 0.83 (validation set) 5-year: 0.73	X	X	X	X	X	X
REVEAL-HCV risk calculator (for HCV RNA detectables) (Lee et al. 2014c)	Lee MH et al. (2014c)	5-year: 0.65 10-year: 0.77 15-year: 0.73 (validation set) 5-year: 0.70	X	X	X	X	X	X

LC liver cirrhosis

Table 3.7 Prediction calculators for important clinical milestones in HBV disease Progression

Clinical milestone [reference]	Author (Year)	AUROC (95 % confidence interval)	Predictors included in the calculator								
			Gender	Age	ALT level	Precore mutant	HBV genotype	HBV DNA	HBsAg level	BMI	HBeAg status
HBeAg seroclearance (Liu et al. 2014)	Liu J et al. (2014)	5-year: 0.85 (0.80–0.90)	X		X	X	X	X			
		10-year: 0.78 (0.73–0.83)									
HBV DNA undetectability (Liu et al. 2014)	Liu J et al. (2014)	5-year: 0.77 (0.72–0.82)	X	X					X		
		10-year: 0.73 (0.70–0.76)									
HBsAg seroclearance (Liu et al. 2013b)	Liu et al. (2013a, b)	5-year: 0.89 (0.889–0.891)		X				X	X	X	
		10-year: 0.84 (0.839–0.841)									
Liver cirrhosis (Lee et al. 2013)	Lee MH et al. (2013)	(Internal validation)	X	X	X		X	X	X		X
		3-year: 0.79									
		5-year: 0.80									
		10-year: 0.82									

seroclearance, predictors included gender, ALT level, precore mutation, HBV genotype, and HBV DNA level, with decreased HBV DNA levels being a strong predictor of seroclearance, confirming previous studies which also emphasized the importance of lowering HBV DNA levels for HBeAg seroclearance (Yang et al. 2012). Previous studies had also shown HBV DNA levels to be the strongest predictor of HBV DNA undetectability, however, the establishment of a prediction model incorporating quantitative HBsAg levels showed that HBsAg levels, rather than HBV DNA levels, were the strongest predictor of viral load undetectability (Liu et al. 2014; Yang et al. 2012). Lastly, an accurate prediction model was also established for HBsAg seroclearance. While previous studies emphasized HBV DNA levels as the greatest predictor, this new model showed that, while HBV DNA levels were still important, HBsAg levels were now the most significant predictive factor for HBsAg seroclearance (Liu et al. 2010, 2013b). Finally, an internally validated prediction model was also developed for predicting liver cirrhosis. This model was able to predict the occurrence of HBV-caused liver cirrhosis with moderate accuracy (Lee et al. 2013). Predicting milestones in HBV disease progression is crucial for interrupting the natural progression of disease, and can identify patients who are more likely to reach these important milestones, thus putting them at lower risk for developing HCC. Monitoring these milestones is also an important part of treatment management, and the ability to provide personalized probabilities of reaching important treatment milestones can assist clinicians in appropriate risk communication with their patients. However, as these models have yet to be externally validated, future studies with external cohorts will be needed to validate their clinical applicability. Further studies to develop cost-effective models for use in resource-limited areas are also needed.

Risk calculators for other virus-caused cancers such as nasopharyngeal carcinoma and cervical cancer may also help to improve the triage and clinical management of patients infected with other oncogenic viruses. The development of the risk calculators needs large-scale prospective cohorts, which have been followed for a long period of time with accurate measurements of risk predictors. Demographical characteristics, viral infection biomarkers, family history, and polymorphisms of genetic susceptibility should be incorporated in developing valid and useful cancer risk calculators.

Reduction of Cancer Incidence through Vaccination and Anti-viral Therapy

The most effective strategy for preventing virus-caused cancers is either through vaccination to prevent viral infection all together, or through anti-viral therapy to eliminate the presence of oncogenic viruses in the human host. Currently, vaccines are available for the prevention of HBV-caused hepatocellular carcinoma and HPV-caused cervical cancer, while anti-viral therapies are available for the treatment of chronic infection of HBV, HCV and HIV.

Many clinical trials have demonstrated the efficacy of anti-viral therapy to prevent hepatocellular carcinoma in cirrhotic patients. Moreover, the national HBV immunization program in Taiwan, which was implemented in 1984, has successfully reduced the incidence of hepatocellular carcinoma in vaccinated birth cohorts among youth aged 6–19 years (Chang et al. 1997, 2009; Chien et al. 2006). In a recent study examining 30-year outcomes of the HBV immunization program in Taiwan, results also showed significant declines in incidence and mortality rates of infant fulminant hepatitis, as well as significant declines in mortality rates from chronic liver disease (CLD) and hepatocellular carcinoma (HCC). Between 1977–1980 and 2001–2004, rate ratios of CLD and HCC mortality for individuals aged 5–29 years decreased by more than 90 %, most likely attributed to increasing individual and herd immunity through vaccination (Chiang et al. 2013). A national antiviral therapy program was implemented in 2003 to control chronic hepatitis B or C in Taiwan. It is expected to further reduce the incidence and mortality of hepatocellular carcinoma in treated adult patients. However, its efficacy to prevent hepatocellular carcinoma remains to be assessed.

Many clinical studies have also demonstrated the efficacy of HPV vaccination in preventing cervical neoplasia, a precursor lesion of cervical cancer. The HPV immunization program in Australia has effectively lowered the incidence of cervical neoplasia in vaccinated adolescent cohorts. Studies in the United States have also seen a significant 56 % reduction in the prevalence of vaccine-type HPV infection in teen girls aged 14–19 after the introduction of the HPV vaccine (Markowitz et al. 2013). Another recent study among young women in Denmark showed that vaccination with the HPV vaccine is effective in reducing the risk for cervical cancer precursor lesions, despite having only been available since 2006 in Denmark (Baldur-Felskov et al. 2014).

A multitude of studies have also examined the impact of antiretroviral therapy on the incidence of Kaposi's sarcoma in HIV-infected individuals. In a recent review of studies published between 2009 and 2012, the percent reduction in the population-level incidence of Kaposi's sarcoma attributed to antiretroviral therapy among HIV-infected individuals seen in the period when therapy was available, compared to the period when it was not available, ranged from 78 to 95 % (Semeere et al. 2012). The reduction in Kaposi's sarcoma incidence was similarly striking among multiple clinic-based cohort studies, which saw reductions of 50–90 % since the availability of antiretroviral therapy. However, these results were mostly seen in resource-rich settings. Data from resource-limited settings, where the majority of the burden of Kaposi's sarcoma lies, was limited. More research on the effectiveness of antiretroviral therapy in resource-limited settings is needed (Semeere et al. 2012).

Future Perspectives

In an age of constant technological advancements in proteomic and genomic medicine, an increasing number of biomarkers associated with virus-caused cancers have been identified. These include novel seromarkers, as well as important host genetic susceptibility markers, and host immune markers. Once identified, such markers can be used as markers for the early detection of cancers, or can even be included into risk prediction models that can be applied to clinical practice and used to reduce the global public health burden of virus-caused cancers. For example, multiple microRNAs have been combined for the early diagnosis of hepatocellular carcinoma, although its efficacy and cost-effectiveness as an early diagnosis marker needs to be assessed and further compared with other currently used methods for early diagnosis, including abdominal ultrasonography (Chen and Lee 2011). Repeated measurements of time-dependent biomarkers are also important, as an analysis of their long-term changes or dynamics, rather than a one-time measurement, may improve the accuracy of long-term risk prediction or early detection of virus-caused cancers (Chen 2005). Thus, further longitudinal studies that include repeated and regular measurements of biomarkers are needed to better identify targets for the prevention, diagnosis, or treatment of virus-caused cancers. Doing so will not only improve on current prevention, diagnosis, and treatment strategies, but coupled with assessments of the health economics and cost-effectiveness of such biomarkers, will allow for the development of cost-effective and clinically applicable tools that can be tailored for clinical use, or even for use in resource-limited settings where the burden of virus-caused cancers is often the highest.

References

Alter MJ (2007) Epidemiology of hepatitis C virus infection. World J Gastroenterol 13(17):2436–2441

Amber K, McLeod MP, Nouri K (2013) The Merkel cell polyomavirus and its involvement in Merkel cell carcinoma. Dermatol Surg 39(2):232–238. doi:10.1111/dsu.12079

Baldur-Felskov B, Dehlendorff C, Munk C, Kjaer S (2014) Early impact of human papillomavirus vaccination on cervical neoplasia—nationwide follow-up of young Danish women. J Natl Cancer Inst 106 (3). doi:10.1093/jnci/djt460

Centers for Disease Control and Prevention (2014) Infectious diseases related to travel. In: Francisco A (ed) The yellow book (Updated). Oxford University Press, New York

Chang Y, Moore PS (2012) Merkel cell carcinoma: a virus-induced human cancer. Annu Rev Pathol 7:123–144. doi:10.1146/annurev-pathol-011110-130227

Chang MH, Chen CJ, Lai MS, Hsu HM, Wu TC, Kong MS, Liang DC, Shau WY, Chen DS (1997) Universal hepatitis B vaccination in Taiwan and the incidence of hepatocellular carcinoma in children. Taiwan Childhood Hepatoma Study Group. N Engl J Med 336(26):1855–1859. doi:10.1056/NEJM199706263362602

Chang MH, You SL, Chen CJ, Liu CJ, Lee CM, Lin SM, Chu HC, Wu TC, Yang SS, Kuo HS, Chen DS, Grp THS (2009) Decreased incidence of hepatocellular carcinoma in hepatitis B vaccinees: a 20-year follow-up study. JNCI-J Natl Cancer Inst 101(19):1348–1355. doi:10.1093/Jnci/Djp288

Chang YJ, Chen HC, Lee BH, You SL, Lin CY, Pan MH, Chou YC, Hsieh CY, Chen YMA, Cheng YJ, Chen CJ, Grp C-HS (2011) Unique variants of human papillomavirus genotypes 52 and 58 and risk of cervical neoplasia. Int J Cancer 129(4):965–973. doi:10.1002/Ijc.25724

Chen CJ (2005) Time-dependent events in natural history of occult hepatitis B virus infection: the importance of population-based long-term follow-up study with repeated measurements. J Hepatol 42(4):438–440. doi:10.1016/j.jhep.2005.01.003

Chen CJ, Chen DS (2002) Interaction of hepatitis B virus, chemical carcinogen, and genetic susceptibility: multistage hepatocarcinogenesis with multifactorial etiology. Hepatology 36(5):1046–1049. doi:10.1053/jhep.2002.37084

Chen CJ, Lee MH (2011) Early diagnosis of hepatocellular carcinoma by multiple microRNAs: validity, efficacy, and cost-effectiveness. J Clin Oncol 29(36):4745–4747. doi:10.1200/JCO.2011.39.0054

Chen CJ, Yang HI (2011) Natural history of chronic hepatitis B REVEALed. J Gastroenterol Hepatol 26(4):628–638. doi:10.1111/j.1440-1746.2011.06695.x

Chen CJ, Liang KY, Chang AS, Chang YC, Lu SN, Liaw YF, Chang WY, Sheen MC, Lin TM (1991) Effects of hepatitis B virus, alcohol drinking, cigarette smoking and familial tendency on hepatocellular carcinoma. Hepatology 13(3):398–406

Chen CJ, Yu MW, Liaw YF, Wang LW, Chiamprasert S, Matin F, Hirvonen A, Bell DA, Santella RM (1996a) Chronic hepatitis B carriers with null genotypes of glutathione S-transferase M1 and T1 polymorphisms who are exposed to aflatoxin are at increased risk of hepatocellular carcinoma. Am J Hum Genet 59(1):128–134

Chen CJ, Wang LY, Lu SN, Wu MH, You SL, Zhang YJ, Wang LW, Santella RM (1996b) Elevated aflatoxin exposure and increased risk of hepatocellular carcinoma. Hepatology 24(1):38–42. doi:10.1002/hep.510240108

Chen CJ, Yu MW, Liaw YF (1997) Epidemiological characteristics and risk factors of hepatocellular carcinoma. J Gastroenterol Hepatol 12(9–10):S294–S308

Chen CJ, Yang HI, Su J, Jen CL, You SL, Lu SN, Huang GT, Iloeje UH, Group R-HS (2006) Risk of hepatocellular carcinoma across a biological gradient of serum hepatitis B virus DNA level. JAMA 295(1):65–73. doi:10.1001/jama.295.1.65

Chen CL, Yang HI, Yang WS, Liu CJ, Chen PJ, You SL, Wang LY, Sun CA, Lu SN, Chen DS, Chen CJ (2008) Metabolic factors and risk of hepatocellular carcinoma by chronic hepatitis B/C infection: a follow-up study in Taiwan. Gastroenterology 135(1):111–121. doi:S0016-5085(08)00569-6 [pii] 10.1053/j.gastro.2008.03.073

Chen T, Hedman L, Mattila PS, Jartti T, Ruuskanen O, Soderlund-Venermo M, Hedman K (2011a) Serological evidence of Merkel cell polyomavirus primary infections in childhood. J Clin Virol 50(2):125–129. doi:10.1016/j.jcv.2010.10.015

Chen HC, Schiffman M, Lin CY, Pan MH, You SL, Chuang LC, Hsieh CY, Liaw KL, Hsing AW, Chen CJ, Group C-HS (2011b) Persistence of type-specific human papillomavirus infection and increased long-term risk of cervical cancer. J Natl Cancer Inst 103(18):1387–1396. doi:10.1093/jnci/djr283

Chen CF, Lee WC, Yang HI, Chang HC, Jen CL, Iloeje UH, Su J, Hsiao CHK, Wang LY, You SL, Lu SN, Chen CJ, Elevati REVL (2011c) Changes in serum levels of HBV DNA and alanine aminotransferase determine risk for hepatocellular carcinoma. Gastroenterology 141(4):1240–U1662. doi:10.1053/j.gastro.2011.06.036

Chen HC, You SL, Hsieh CY, Schiffman M, Lin CY, Pan MH, Chou YC, Liaw KL, Hsing AW, Chen CJ, Group C-HS (2011d) Prevalence of genotype-specific human papillomavirus infection and cervical neoplasia in Taiwan: a community-based survey of 10,602 women. Int J Cancer 128(5):1192–1203. doi:10.1002/ijc.25685

Chen YY, You SL, Koong SL, Liu J, Chen CA, Chen CJ (2014) Screening frequency and atypical cells and the prediction of cervical cancer risk. Obstet Gynecol 123(5):1003–1011

Chiang CJ, Yang YW, You SL, Lai MS, Chen CJ (2013) Thirty-year outcomes of the national hepatitis B immunization program in Taiwan. JAMA 310(9):974–976. doi:10.1001/jama.2013.276701

Chien YC, Chen CJ (2003) Epidemiology and etiology of nasopharyngeal carcinoma: gene-environment interaction. Cancer Rev Asia-Pac 1:1–19

Chien YC, Chen JY, Liu MY, Yang HI, Hsu MM, Chen CJ, Yang CS (2001) Serologic markers of Epstein-Barr virus infection and nasopharyngeal carcinoma in Taiwanese men. N Engl J Med 345(26):1877–1882. doi:10.1056/NEJMoa011610

Chien YC, Jan CF, Kuo HS, Chen CJ (2006) Nationwide hepatitis B vaccination program in Taiwan: effectiveness in the 20 years after it was launched. Epidemiol Rev 28:126–135. doi:10.1093/epirev/mxj010

Cho EY, Hildesheim A, Chen CJ, Hsu MM, Chen IH, Mittl BF, Levine PH, Liu MY, Chen JY, Brinton LA, Cheng YJ, Yang CS (2003) Nasopharyngeal carcinoma and genetic polymorphisms of DNA repair enzymes XRCC1 and hOGG1. Cancer Epidemiol Biomarkers Prev 12(10):1100–1104

Chuang LC, Hu CY, Chen HC, Lin PJ, Lee B, Lin CY, Pan MH, You SL, Hsieh CY, Chen CJ, Community-Based Cancer Screening Program-Human Papillomavirus Study G (2012) Associations of human leukocyte antigen class II genotypes with human papillomavirus 18 infection and cervical intraepithelial neoplasia risk. Cancer 118(1):223–231. doi:10.1002/cncr.26227

de Sanjose S, Diaz M, Castellsague X, Clifford G, Bruni L, Munoz N, Bosch FX (2007) Worldwide prevalence and genotype distribution of cervical human papillomavirus DNA in women with normal cytology: a meta-analysis. Lancet Infect Dis 7(7):453–459. doi:10.1016/S1473-3099(07)70158-5

Dukers NH, Rezza G (2003) Human herpesvirus 8 epidemiology: what we do and do not know. AIDS 17:1717–1730

Fried MW, Shiffman ML, Reddy KR, Smith C, Marinos G, Goncales FL Jr, Haussinger D, Diago M, Carosi G, Dhumeaux D, Craxi A, Lin A, Hoffman J, Yu J (2002) Peginterferon alfa-2a plus ribavirin for chronic hepatitis C virus infection. N Engl J Med 347(13):975–982. doi:10.1056/NEJMoa020047

Fwu CW, Chien YC, Kirk GD, Nelson KE, You SL, Kuo HS, Feinleib M, Chen CJ (2009) Hepatitis B virus infection and hepatocellular carcinoma among parous Taiwanese women: nationwide cohort study. J Natl Cancer Inst 101(14):1019–1027. doi:djp146 [pii] 10.1093/jnci/djp146

Fwu CW, Chien YC, You SL, Nelson KE, Kirk GD, Kuo HS, Feinleib M, Chen CJ (2011) Hepatitis B virus infection and risk of intrahepatic cholangiocarcinoma and non-Hodgkin lymphoma: a cohort study of parous women in Taiwan. Hepatology 53(4):1217–1225. doi:10.1002/hep.24150

Hajarizadeh B, Grebely J, Dore GJ (2013) Epidemiology and natural history of HCV infection. Nat Rev Gastroenterol Hepatol 10(9):553–562. doi:10.1038/nrgastro.2013.107

Han KH, Ahn SH (2005) How to predict HCC development in patients with chronic B viral liver disease? Intervirology 48(1):23–28. doi:10.1159/000082091

Hildesheim A, Anderson LM, Chen CJ, Cheng YJ, Brinton LA, Daly AK, Reed CD, Chen IH, Caporaso NE, Hsu MM, Chen JY, Idle JR, Hoover RN, Yang CS, Chhabra SK (1997) CYP2E1 genetic polymorphisms and risk of nasopharyngeal carcinoma in Taiwan. J Natl Cancer Inst 89(16):1207–1212

Hildesheim A, Dosemeci M, Chan CC, Chen CJ, Cheng YJ, Hsu MM, Chen IH, Mittl BF, Sun B, Levine PH, Chen JY, Brinton LA, Yang CS (2001) Occupational exposure to wood, formaldehyde, and solvents and risk of nasopharyngeal carcinoma. Cancer Epidemiol Biomarkers Prev 10(11):1145–1153

Hildesheim A, Apple RJ, Chen CJ, Wang SS, Cheng YJ, Klitz W, Mack SJ, Chen IH, Hsu MM, Yang CS, Brinton LA, Levine PH, Erlich HA (2002) Association of HLA class I and II alleles and extended haplotypes with nasopharyngeal carcinoma in Taiwan. J Natl Cancer Inst 94(23):1780–1789

Hsu WL, Chen JY, Chien YC, Liu MY, You SL, Hsu MM, Yang CS, Chen CJ (2009) Independent effect of EBV and cigarette smoking on nasopharyngeal carcinoma: a 20-year follow-up study on 9,622 males without family history in Taiwan. Cancer Epidemiol Biomarkers Prev 18(4):1218–1226. doi:10.1158/1055-9965.EPI-08-1175

Hsu WL, Yu KJ, Chien YC, Chiang CJ, Cheng YJ, Chen JY, Liu MY, Chou SP, You SL, Hsu MM, Lou PJ, Wang CP, Hong JH, Leu YS, Tsai MH, Su MC, Tsai ST, Chao WY, Ger LP, Chen PR, Yang CS, Hildesheim A, Diehl SR, Chen CJ (2011) Familial tendency and risk of nasopharyngeal carcinoma in Taiwan: effects of covariates on risk. Am J Epidemiol 173(3):292–299. doi:10.1093/aje/kwq358

Hsu WL, Tse KP, Liang S, Chien YC, Su WH, Yu KJ, Cheng YJ, Tsang NM, Hsu MM, Chang KP, Chen IH, Chen TI, Yang CS, Goldstein AM, Chen CJ, Chang YS, Hildesheim A (2012a) Evaluation of human leukocyte antigen-A (HLA-A), other non-HLA markers on chromosome 6p21 and risk of nasopharyngeal carcinoma. PLoS ONE 7(8):e42767. doi:10.1371/journal.pone.0042767

Hsu WL, Pan WH, Chien YC, Yu KJ, Cheng YJ, Chen JY, Liu MY, Hsu MM, Lou PJ, Chen IH, Yang CS, Hildesheim A, Chen CJ (2012b) Lowered risk of nasopharyngeal carcinoma and intake of plant vitamin, fresh fish, green tea and coffee: a case-control study in Taiwan. PLoS One 7(7):e41779. doi:10.1371/journal.pone.0041779

Huang YT, Jen CL, Yang HI, Lee MH, Lu SN, Iloeje UH, Chen CJ, Group R-HHS (2011) Lifetime risk and gender difference of hepatocellular carcinoma among patients affected with chronic hepatitis B and C. J Clin Oncol 29:3643–3650

International Agency for Research on Cancer (2012) A review of human carcinogens: part B – biological agents. IARC monographs on the evaluation of carcinogenic risks to humans, vol 100B. International Agency for Research on Cancer, Lyon

Lavanchy D (2004) Hepatitis B virus epidemiology, disease burden, treatment, and current and emerging prevention and control measures. J Viral Hepat 11(2):97107. doi:487 [pii]

Lavanchy D (2009) The global burden of hepatitis C. Liver Int 29(Suppl 1):74–81. doi:10.1111/j.1478-3231.2008.01934.x

Lee MH, Yang HI, Lu SN, Jen CL, Yeh SH, Liu CJ, Chen PJ, You SL, Wang LY, Chen WJ, Chen CJ (2010) Hepatitis C virus seromarkers and subsequent risk of hepatocellular carcinoma: long-term predictors from a community-based cohort study. J Clin Oncol 28(30):4587–4593. doi:10.1200/JCO.2010.29.1500

Lee MH, Yang HI, Jen CL, Lu SN, Yeh SH, Liu CJ, You SL, Sun CA, Wang LY, Chen WJ, Chen CJ, Group REVEALHS (2011) Community and personal risk factors for hepatitis C virus infection: a survey of 23,820 residents in Taiwan in 1991-2. Gut 60(5):688–694. doi:10.1136/gut.2010.220889

Lee MH, Yang HI, Lu SN, Jen CL, You SL, Wang LY, Wang CH, Chen WJ, Chen CJ, Group REVEALHS (2012) Chronic hepatitis C virus infection increases mortality from hepatic and extrahepatic diseases: a community-based long-term prospective study. J Infect Dis 206(4):469–477. doi:10.1093/infdis/jis385

Lee MH, Yang HI, Liu J, Batrla-Utermann R, Jen CL, Iloeje UH, Lu SN, You SL, Wang LY, Chen CJ, Group REVEALHS (2013) Prediction models of long-term cirrhosis and hepatocellular carcinoma risk in chronic hepatitis B patients: risk scores integrating host and virus profiles. Hepatology 58(2):546–554. doi:10.1002/hep.26385

Lee MH, Yang HI, Yuan Y, L'Italien G, Chen CJ (2014a) Epidemiology and natural history of hepatitis C virus infection. World J Gastroenterol 20(28):9270–9280

Lee MH, Yang HI, Lu SN, Jen CL, You SL, Wang LY, L'Italien G, Chen CJ, Yuan Y, for the R-HCVSG (2014b) Hepatitis C virus genotype 1b increases cumulative lifetime risk of hepatocellular carcinoma. Int J Cancer 135(5):1119–1126. doi:10.1002/ijc.28753

Lee MH, Lu SN, Yuan Y, Yang HI, Jen CL, You SL, Wang LY, L'Italien G, Chen CJ (2014c) Development and validation of a clinical scoring system for predicting risk of HCC in asymptomatic individuals seropositive for anti-HCV antibodies. PLoS One 9(5):e94760. doi:10.1371/journal.pone.0094760

Liu J, Yang HI, Lee MH, Lu SN, Jen CL, Wang LY, You SL, Iloeje UH, Chen CJ, Group R-HS (2010) Incidence and determinants of spontaneous hepatitis B surface antigen seroclearance: a community-based follow-up study. Gastroenterology 139(2):474–482. doi:10.1053/j.gastro.2010.04.048

Liu J, Yang HI, Lee MH, Lu SN, Jen CL, Batrla-Utermann R, Wang LY, You SL, Hsiao CK, Chen PJ, Chen CJ, REVEALHBVSG (2013a) Spontaneous seroclearance of hepatitis B seromarkers and subsequent risk of hepatocellular carcinoma. Gut 63(10):1648–1657. doi:10.1136/gutjnl-2013-305785

Liu J, Lee MH, Batrla-Utermann R, Jen CL, Iloeje UH, Lu SN, Wang LY, You SL, Hsiao CK, Yang HI, Chen CJ (2013b) A predictive scoring system for the seroclearance of HBsAg in HBeAg-seronegative chronic hepatitis B patients with genotype B or C infection. J Hepatol 58(5):853–860. doi:10.1016/j.jhep.2012.12.006

Liu J, Yang HI, Lee MH, Batrla R, Jen CL, Lu SN, Wang LY, You SL, Hsiao CK, Chen CJ (2014) Distinct seromarkers predict different milestones of chronic hepatitis B progression. Hepatology 60(1):77–86. doi:10.1002/hep.27083

Loomba R, Liu J, Yang HI, Lee MH, Lu SN, Wang LY, Iloeje UH, You SL, Brenner D, Chen CJ, Group R-HS (2013) Synergistic effects of family history of hepatocellular carcinoma and hepatitis B virus infection on risk for incident hepatocellular carcinoma. Clin Gastroenterol Hepatol 11(12):1636–1645 e1631–e1633. doi:10.1016/j.cgh.2013.04.043

Markowitz LE, Hariri S, Lin C, Dunne EF, Steinau M, McQuillan G, Unger ER (2013) Reduction in human papillomavirus (HPV) prevalence among young women following HPV vaccine introduction in the United States, National Health and Nutrition Examination Surveys, 2003–2010. J Infect Dis 208(3):385–393. doi:10.1093/infdis/jit192

Matsuda T, McCombs J, Lee MH, Tonnu-Mihara I, L'Italien G, Saab S, Hines P, Yuan Y (2014) External validation of the risk-prediction model for Hepatocellular Carcinoma [HCC] from the REVEAL-HCV study using data from the U.S. Veterans Affairs [VA] Health System. Paper presented at the EASL, London, UK, April 9–13

Proietti FA, Carneiro-Proietti AB, Catalan-Soares BC, Murphy EL (2005) Global epidemiology of HTLV-1 infection and associated diseases. Oncogene 24:6058–6068

Semeere AS, Busakhala N, Martin JN (2012) Impact of antiretroviral therapy on the incidence of Kaposi's sarcoma in resource-rich and resource-limited settings. Curr Opin Oncol 24(5):522–530. doi:10.1097/CCO.0b013e328355e14b

Spurgeon ME, Lambert PF (2013) Merkel cell polyomavirus: a newly discovered human virus with oncogenic potential. Virology 435(1):118–130. doi:10.1016/j.virol.2012.09.029

Sun CA, Chen HC, Lu CF, You SL, Mau YC, Ho MS, Lin SH, Chen CJ (1999) Transmission of hepatitis C virus in Taiwan: prevalence and risk factors based on a nationwide survey. J Med Virol 59(3):290–296

Sun CA, Wu DM, Lin CC, Lu SN, You SL, Wang LY, Wu MH, Chen CJ (2003) Incidence and co-factors of hepatitis C virus-related hepatocellular carcinoma. Am J Epidemiol 157:674–682

Tolstov YL, Pastrana DV, Feng H, Becker JC, Jenkins FJ, Moschos S, Chang Y, Buck CB, Moore PS (2009) Human Merkel cell polyomavirus infection II. MCV is a common human infection that can be detected by conformational capsid epitope immunoassays. Int J Cancer 125(6):1250–1256. doi:10.1002/ijc.24509

UNAIDS (2012) http://data.unaids.org/pub/epislides/2012/2012_epiupdate_en.pdf

Wang LY, Hatch M, Chen CJ, Levin B, You SL, Lu SN, Wu MH, Wu WP, Wang LW, Wang Q, Huang GT, Yang PM, Lee HS, Santella RM (1996) Aflatoxin exposure and risk of hepatocellular carcinoma in Taiwan. Int J Cancer 67(5):620–625. doi:10.1002/(SICI)1097-0215(19960904)67:5<620::AID-IJC5>3.0.CO;2-W

Wang LY, You SL, Lu SN, Ho HC, Wu MH, Sun CA, Yang HI, Chien-Jen C (2003) Risk of hepatocellular carcinoma and habits of alcohol drinking, betel quid chewing and cigarette smoking: a cohort of 2416 HBsAg-seropositive and 9421 HBsAg-seronegative male residents in Taiwan. Cancer Causes Control 14(3):241–250

Ward MH, Pan WH, Cheng YJ, Li FH, Brinton LA, Chen CJ, Hsu MM, Chen IH, Levine PH, Yang CS, Hildesheim A (2000) Dietary exposure to nitrite and nitrosamines and risk of nasopharyngeal cancer in Taiwan. Inter J Cancer 86:603–609

Wong VW, Chan SL, Mo F, Chan TC, Loong HH, Wong GL, Lui YY, Chan AT, Sung JJ, Yeo W, Chan HL, Mok TS (2010) Clinical scoring system to predict hepatocellular carcinoma in chronic hepatitis B carriers. J Clin Oncol Off J Am Soc Clin Oncol 28(10):1660–1665. doi:10.1200/JCO.2009.26.2675

Yang HI, Lu SN, Liaw YF, You SL, Sun CA, Wang LY, Hsiao CK, Chen PJ, Chen DS, Chen CJ, Taiwan Community-Based Cancer Screening Project G (2002) Hepatitis B e antigen and the risk of hepatocellular carcinoma. N Engl J Med 347(3):168–174. doi:10.1056/NEJMoa013215

Yang HI, Yeh SH, Chen PJ, Iloeje UH, Jen CL, Wang LY, Lu SN, You SL, Chen DS, Liaw YF, Chen CJ, Group R-H (2008) Association between Hepatitis B virus genotype and mutants and the risk of hepatocellular carcinoma. J Natl Cancer Inst 100:1134–1143

Yang HI, Sherman M, Su J, Chen PJ, Liaw YF, Iloeje UH, Chen CJ (2010) Nomograms for risk of hepatocellular carcinoma in patients with chronic hepatitis B virus infection. J Clin Oncol 28(14):2437–2444. doi:10.1200/JCO.2009.27.4456

Yang HI, Yuen MF, Chan HL, Han KH, Chen PJ, Kim DY, Ahn SH, Chen CJ, Wong VW, Seto WK (2011) Risk estimation for hepatocellular carcinoma in chronic hepatitis B (REACH-B): development and validation of a predictive score. Lancet Oncol 12:568–574

Yang HI, Hung HL, Lee MH, Liu J, Jen CL, Su J, Wang LY, Lu SN, You SL, Iloeje UH, Chen CJ, Risk Evaluation of Viral Load E, Associated Liver Disease/Cancer HBVSG (2012) Incidence and determinants of spontaneous seroclearance of hepatitis B e antigen and DNA in patients with chronic hepatitis B. Clin Gastroenterol Hepatol: Off Clin Pract J Am Gastroenterol Assoc 10(5):527–534 e521–522. doi:10.1016/j.cgh.2011.12.019

Yu MW, Chen CJ (1993) Elevated serum testosterone levels and risk of hepatocellular carcinoma. Cancer Res 53(4):790–794

Yu MW, Gladekyarborough A, Chiamprasert S, Santella RM, Liaw YF, Chen CJ (1995a) Cytochrome-P450 2e1 and glutathione-S-transferase M1 polymorphisms and susceptibility to hepatocellular-carcinoma. Gastroenterology 109(4):1266–1273. doi:10.1016/0016-5085(95)90587-1

Yu MW, Hsieh HH, Pan WH, Yang CS, Chen CJ (1995b) Vegetable consumption, serum retinol level, and risk of hepatocellular-carcinoma. Cancer Res 55(6):1301–1305

Yu MW, Chiu YH, Chiang YC, Chen CH, Lee TH, Santella RM, Chern HD, Liaw YF, Chen CJ (1999a) Plasma carotenoids, glutathione S-transferase M1 and T1 genetic polymorphisms, and risk of hepatocellular carcinoma: independent and interactive effects. Am J Epidemiol 149(7):621–629

Yu MW, Horng IS, Hsu KH, Chiang YC, Liaw YF, Chen CJ (1999b) Plasma selenium levels and risk of hepatocellular carcinoma among men with chronic hepatitis virus infection. Am J Epidemiol 150(4):367–374

Yu MW, Chang HC, Liaw YF, Lin SM, Lee SD, Liu CJ, Chen PJ, Hsiao TJ, Lee PH, Chen CJ (2000a) Familial risk of hepatocellular carcinoma among chronic hepatitis B carriers and their relatives. J Natl Cancer Inst 92(14):1159–1164. doi:10.1093/jnci/92.14.1159

Yu MW, Cheng SW, Lin MW, Yang SY, Liaw YF, Chang HC, Hsiao TJ, Lin SM, Lee SD, Chen PJ, Liu CJ, Chen CJ (2000b) Androgen-receptor CAG repeat, plasma testosterone levels, and risk of hepatitis B-related hepatocellular carcinoma. J Natl Cancer Inst 92:2023–2028

Yu MW, Yang SY, Pan IJ, Lin CL, Liu CJ, Liaw YF, Lin SM, Chen PJ, Lee SD, Chen CJ (2003) Polymorphisms in XRCC1 and glutathione S-transferase genes and hepatitis B-related hepatocellular carcinoma. J Natl Cancer Inst 95(19):1485–1488

Yuen MF, Tanaka Y, Fong DY, Fung J, Wong DK, Yuen JC, But DY, Chan AO, Wong BC, Mizokami M, Lai CL (2009) Independent risk factors and predictive score for the development of hepatocellular carcinoma in chronic hepatitis B. J Hepatol 50(1):80–88. doi:10.1016/j.jhep.2008.07.023

Chapter 4
Bacterial Infections and Cancer Development

Marianna Agassandian and Galina V. Shurin

Abstract Multiple epidemiological and clinical studies have identified an association between specific bacterial infection and cancers. Several bacterial infections have been proven to serve as risk factors for the initial development and survival of cancerous cells in different tissues and organs. It is estimated that about 20–25 % of all human cancers are caused by chronic infection-derived inflammation. A sustained production of various reactive aldehydes, oxygen and nitrogen species, as well as cytokines, chemokines and growth factors, at the inflammatory microenvironment can perturb normal biological and physiological processes leading to genomic instability and an increased risk of cancer development. Furthermore, toxins produced by bacteria, different bacteria-derived metabolites and some bacterial virulence factors may be also involved in the associated tumorigenic processes via DNA damage response, cell cycle regulation and other mechanisms. Unfortunately, it is still not fully clear which combinations of specific bacteria-associated factors and microenvironmental stimuli contribute to high incidence of cancer during certain infectious diseases. Understanding the mechanisms of how bacterial infections affect cancer development and progression will promote new or more efficient approaches to cancer prevention and therapy.

Keywords Inflammation • Carcinogenesis • Bacterial toxins • DNA damage • Infectious diseases

Abbreviations

AP-1	activator protein-1
BALF	bronchoalveolar lavage fluid
CDT	cytolithal distending toxin
CKD	chronic kidney disease
COX2	cyclooxygenase-2

M. Agassandian (✉) • G.V. Shurin
Departments of Pathology and Immunology, University of Pittsburgh Medical Center,
Scaife Hall S733 3550 Terrace Street, Pittsburgh, PA 15261, USA
e-mail: agassandianm@upmc.edu

M.R. Shurin et al. (eds.), *Infection and Cancer: Bi-Directorial Interactions*, DOI 10.1007/978-3-319-20669-1_4

CRC	colorectal cancer
DC	dendritic cell(s)
EMT	epithelial-to-mesenchymal transition
LPS	lipopolysaccharide
MALT	mucosa-associated lymphoid tissue
MAPK	mitogen-activated protein kinase
NO	nitric oxide
NLR	NOD-like receptors
PI3K	phosphatidylinositol 3-kinase
ROS	reactive oxygen species
PRR	pattern recognition receptor(s)
STAT3	signal transducer and activator of transcription 3
TLR	Toll-like receptor

Introduction

Numerous recent studies have demonstrated that bacterial infections may play a real role in triggering cancer. However, "cancer-associated" bacteria have been observed for the last century, their existence was excluded for a long time, and bacteria research connecting with cancer had never been accepted. The actuality of the bacteria that induced/support cancer has been largely snubbed and even led to deprivation of the medical license like in the case of Swedish physician, Erik Enby. During the last period in history of the cancer-associated bacteria, the potential role of bacteria in carcinogenesis has been largely described by British microbiologist Milton Wainwright. At the present time, scientists investigate the role of bacteria, observed in tumor tissues, as well as in precancerous conditions, and also in locations distant from the primary tumor, in tumorigenesis. However, the cancer-related bacteria have been found in chronic diseases other than cancer, such as lupus, scleroderma, sarcoidosis, and others. It is known that all humans live with many "cancer" bacteria. In a healthy organism, these bacteria are beneficial, but when the immune system is weakened they might induce or support a variety of human diseases, including cancer. These ever-present bacteria were found even within the cell's nucleus, demonstrating the accession of the bacteria to the genetic material of a cell, which might lead to malignant transformation. Recent evidence allows speculating that human disease may be connected with global changes in the bacteria-host symbiotic relationship and is not an attribute of only a single pathogen. Many environmental factors, such as pollutions, radiation, secondary infection, diet or lifestyle, can disturb this relationship and promote carcinogenesis (Schwabe and Jobin 2013).

The cancer development is the complex of multiple steps where the period from infection to appearance of the bacterial-positive tumor can take a long time. The list of infection-associated cancers is growing, however only a small portion of bacteria-infected people develop cancer. Worldwide malignancies can be attributed to bacterial infections (Bouvard et al. 2009) in the range 15–25 % in developing countries and 7–10 % in developed countries (Kuper et al. 2000; Schottenfeld and Beebe-Dimmer 2006a). However, some populations are genetically predisposed to the infections which are associated with cancer and have a higher risk of malignancy (Mager 2006). Cancer-associated bacteria species are diverse as are the significantly varying clinical evidences regarding the association of each pathogen with malignant diseases (Samaras et al. 2010). It was revealed that bacteria can cause or support cancerous process via different and complex mechanisms, which may be organ-specific (Schwabe and Jobin 2013; Samaras et al. 2010). Carcinogenesis could be affected by bacterial toxins, bacterial virulence factors, products of bacterial metabolism, and by the duration and severity of bacteria-associated inflammatory processes (Samaras et al. 2010). It has been demonstrated that chronic bacterial infections, including toxin production, disrupt the cell cycle leading to abnormal cellular growth, inducing DNA damage, apoptosis (Littman et al. 2004a, b; Koyi et al. 2001; Kocazeybek 2003; Nougayrede et al. 2005; Oswald et al. 2005; Lara-Tejero and Galan 2000) and stimulating host immune responses contributing to carcinogenesis via stimulatory and mutagenic effects of cytokines, reactive oxygen species (ROS), cyclooxygenase-2 (COX2), nitric oxide (NO) and other molecules (Schoppmann et al. 2002; Baik et al. 1996; Sheng et al. 2001; Mannick et al. 1996; Biarc et al. 2004). Thus, the key mechanisms by which the bacterial agents can induce carcinogenesis involve chronic infection/inflammation, immune evasion and immune suppression (Kuper et al. 2000).

Bacteria-Host Interactions in Cancer

The symbiotic coexistence between the host and bacteria consists in their separation via multi-level barriers within the host organism, where the majority of bacteria reside in the state of immune tolerance with their host (Schwabe and Jobin 2013). The barrier includes specific cells such as keratinocytes in the skin, and Paneth and goblet cells in the gut that control bacteria number and location, and secrete antibacterial peptides (Salzman et al. 2007; Nestle et al. 2009). In addition, some specific immune cells, such as the gut-associated lymphoid tissue, Langerhans cells in the skin and Th17 cells at mucosal surfaces may also form the protective or immune-triggering barriers (Nestle et al. 2009; Littman and Rudensky 2010). Along with the host, bacteria itself may provide their own barriers through the production of bacteriocins that affect pathobionts and pathogens (Cornforth and Foster 2013).

Any perturbations in these barriers, immune or bacterial, can change innate immune responses. Thus, secondary infection, inflammation or even diet, may alter bacteria-host interactions, can induce bacterial translocation or dysbiosis, may affect inflammatory responses (Schwabe and Jobin 2013), richness, microbial composition, metagenome (Arthur et al. 2012; Holmes et al. 2012; Ley et al. 2006) and lead to bacteria-associated carcinogenesis (Schwabe and Jobin 2013). It is possible that cancer-promoting effects conferred by different classes of bacteria via similar pathways, and perturbations in microbial richness and function affect carcinogenesis in a similar way. Bacterial dysbiosis has also been shown to promote cancer (Couturier-Maillard et al. 2013; Hu et al. 2013). One of the most known conditions that cause dysbiosis is obesity, which may contribute to 20 % of cancer development. Host-derived immune response is also an important provider of dysbiosis, as well as inflammation that mediates the outgrowth of specific bacteria, changes the production of specific metabolites and induces expression of stress-response genes in bacteria (Patwa et al. 2011). However, the mechanisms that contribute to dysbiosis and alterations in microbial richness are still not completely understood.

Inflammation Induced by Bacterial Infection May Promotes Cancer

The connection of inflammation with cancer was first proposed more than 100 years ago by Virchow (Balkwill and Mantovani 2001). The chronic infection leading to tissue inflammation is one of the most known risk factors in human carcinogenesis (Allavena et al. 2008). Thus, the role of inflammatory components in tumor progression is now accepted (Coussens and Werb 2002; Mueller 2006; Mueller et al. 2006). The mechanism of inflammation involves a complex of immune and inflammatory reactions to bacteria and other related agents (Schottenfeld and Beebe-Dimmer 2006a, b). Many years were dedicated to revealing the association between chronic inflammation and cancer. The immune system, which recognizes pathogens or cell damage, activates an influx of neutrophils and macrophages that take up the bacteria, dead cells and debris including proteins, nucleic acids and other molecules released by damaged cells. In response, the cells produce highly reactive chemicals that mediate degradation of the bacteria as well as diffuse into the tissues and may cause its injury. The sustained/chronic inflammation may provide an appropriate microenvironment for the transformation of cells by insertion of oncogenes and inhibition of tumor suppressors leading to initial cancer development (Kuper et al. 2000). The prostaglandins, reactive oxygen and nitrogen species, specific microRNAs, which all are the key mediators of inflammation-induced cancer, may cause changes in cell proliferation, cell death, DNA methylation and DNA mutations that also contribute to carcinogenesis (Ohshima and Bartsch 1994; Shacter and Weitzman 2002; Schetter et al. 2010). The inflammatory responses induced by various bacterial pathogens can accelerate mutagenesis, tissue damage and induce the development of malignancy (Shacter and Weitzman 2002).

Several bacterial infections have been shown to serve as risk factors for the development of cancer at various sites. The extreme scientific interest is focused on the role of the infection/inflammation in the initiation and progression of cancer. The persistent cell activation in the context of chronic infection might promote cell transformation via DNA damage or production of pro-inflammatory factors that sustain chronic inflammation and may support tumor growth (Weitao 2009b). It was revealed that 20–25 % of all human cancers are caused by chronic infection-derived inflammation. Chronic inflammation is a critical component of carcinogenesis by the generation of the pathogenic microenvironment that initiates and promotes cancer development (Jain 2001). A sustained production of various reactive aldehydes, oxygen and nitrogen species as well as cytokines, chemokines and growth factors in the inflammatory microenvironment can perturb normal biological processes leading to genomic instability and the increased risk of cancer development (Balkwill and Mantovani 2001; Coussens and Werb 2002; Gupta and DuBois 2002; Philip et al. 2004; Hussein et al. 2007). Multiple epidemiological and clinical studies have identified an association of specific bacterial infection and cancers. Thus, the persistent inflammation in the context of chronic bacterial infections might promote cell neoplastic transformation via DNA damage or pro-inflammatory factor production.

Mechanisms of Carcinogenesis and Innate Immunity

The innate immune system provides the first line of protection against invading bacteria via phagocytic structures, such as macrophages and dendritic cells (DCs), which are sensitive to pathogen signature through pattern recognition receptors (PRRs) and induce signaling pathways, leading to pro-inflammatory cytokine production. In addition, the Toll-like receptor (TLR) family identifies pathogen-associated molecular patterns on the cell surface or on intracellular vesicles, lysosomes or endosomes. The pathogen-associated molecular patterns activate specific cytosolic PRRs and its activation via secretion of cytokines and chemokines, such as TNF, INF, IL-1, IL-6 and IL-12, leads to inflammatory responses to infection and enhancement of immune system responses (Dallo and Weitao 2010). Multiple pattern recognition receptors that initiate regulatory responses, may not only control the bacteria via antibacterial mediators and thereby suppress cancer, but may also promote resistance to cell death (Pradere et al. 2014). Bacteria may also affect carcinogenesis via the production of tumor-promoting metabolites and the release of carcinogenic molecules.

Bacteria-Induced Activation of TLRs in Carcinogenesis

Bacterial pattern recognition by TLRs is one of the major events in innate immunity, as well as the contributor to carcinogenesis. TLR signaling may promote epithelial tumorigenesis through epithelial cells, stromal fibroblasts and bone-marrow-derived cells. In particular, TLR4, the receptor for Gram-negative bacterial cell wall component LPS, promotes carcinogenesis in the colon, liver, pancreas and skin (Fukata et al. 2011), whereas TLR2 promotes gastric cancer (Tye et al. 2012). The activation of TLR signaling can directly or indirectly affect tumor cell survival pathways via activation of nuclear factor-kB (NF-kB) and the signal transducer activator of transcription 3 (STAT3) (Fukata et al. 2011; Tye et al. 2012). The activation of TLR signaling can directly or indirectly affect tumor cell survival. For example, the pro-survival function of the TLR-myeloid differentiation response 88 (MYD88) pathway was related to the observation that human lymphomas often contain an activating point mutation in MYD88, which triggers NF-kB and STAT3 activation (Ngo et al. 2011). The bacteria-induced TLRs activation on myeloid cells triggered an IL-17 and IL-23 pro-carcinogenic pathway via decreased expression or genetic inactivation of Myd88, Tlr2, Tlr4 or Tlr9 (Grivennikov et al. 2012). TLRs may also promote tumor proliferation through mitogens such as amphyregulin, epiregulin, and hepatocyte growth factor (HGF) (Brandl et al. 2010; Neufert et al. 2013).

Bacteria-Induced Activation of NLRs in Carcinogenesis

Short NOD-like receptors (NLRs) belong to a family of PRRs, which characterized by a central nucleotide-binding oligomerization (NOD) domain (Elinav et al. 2011a, b). Among the PRR family members, NOD2, the pattern recognition receptor that recognizes molecules containing the specific structure called muramyl dipeptide found in bacteria, exerts a key role in the immune system activation (Garaude et al. 2012). Loss of the NOD2 activity is associated with Chron's disease, whereas inactivating polymorphisms in NOD2 associates with increased predisposition to colorectal cancer (CRC) (Khor et al. 2011). Similar to what is seen in patients with NOD2 mutations, *Nod2* deficiency in mice leads to dysbiosis that induces cancer development (Couturier-Maillard et al. 2013; Rehman et al. 2011). The other member of NLRs, NLRP6, which is a component of inflammasomes and contributes to their activation, is also implicated in bacterially promoting carcinogenesis. Similar to *Nod2* deficient mice, *Nlrp6* deficient mice have dysbiosis, which stimulates colitis and colorectal cancer development (Hu et al. 2013). It was also shown that NOD1 deficiency promotes inflammation and genetically-induced colorectal cancer (McGovern et al. 2005). Other NLRs such as NLRP3, NLRP12 and NOD-, LRR- and CARD-containing 4 (NLRC4) promote colitis-associated cancer; however its functional contribution to carcinogenesis remains unclear (Allen et al. 2010, 2012a, b; Hu et al. 2010).

Bacteria-Derived Metabolites and Toxins in Carcinogenesis

In general, bacteria might be linked to cancer by two mechanisms: induction of chronic inflammation (Crowe 2005) and production of carcinogenic metabolites (Salaspuro 2003). Some bacteria induce chronic inflammation associated with increasing (ROS)-mediated genotoxicity in oxygen species, which is contributed to carcinogenesis. Furthermore, toxins produced by bacteria, such as cytolithal distending toxin (CDT), cytotoxic necrotizing factor 1, colibactin and others, can directly modulate tumorigenesis via DNA damage response (Travaglione et al. 2008; Fabbri et al. 2008; Nesic et al. 2004; Cuevas-Ramos et al. 2010). The CDT and colibactin trigger double-strand DNA damage response and genomic instability, which lead to G2/M arrest and cell swelling. Furthermore, different bacteria-derived metabolites, such as hydrogen sulphide and superoxide radicals, can also cause genomic instability (Carbonero et al. 2012; Huycke and Gaskins 2004). Acetaldehyde, produced by bacteria in the digestive tract in humans, is a local carcinogen that enhances gastric cancer risk among gastritis patients (Vakevainen et al. 2002). Bacterial virulence factors are also involved in tumorigenesis. The virulence factors, such as cytotoxin-associated gene A (CagA) or vacuolation cytotoxin A (VacA), can use specific host-derived signaling pathways that lead to the activation of tumor-promoting pathways (Fox and Wang 2007). The other virulence factor of *F. nucleatum*, FadA, interacts with E-cadherin to activate β-catenin signaling pathway and induces colorectal cancer development (Rubinstein et al. 2013).

Effect of Chronic Bacterial Infections and Inflammations on DNA Methylation

DNA methylation is an epigenetic mechanism of gene expression known to be involved in pathogenesis of a variety of diseases. For instance, it was demonstrated that DNA hypermethylation may be involved in increase in the mortality in chronic kidney disease (CKD) (Stenvinkel et al. 2007). In the last two decades, the correlation between cancer development and abnormality in DNA methylation has been reported (Ibrahim et al. 2010, 2011). The IL-6, an inflammatory cytokine, plays an important role in the growth and survival of a variety of tumors. It was recently demonstrated that the raised level of IL-6 expression was associated with the hypermethylation of the p53 gene and inhibition of the expression of tumor suppressors. Furthermore, the link between DNA methylation and mediators of inflammation in the epigenetic control of tumor cell functions provided an explanation of the inflammatory-derived initiation of tumor growth (Hodge et al. 2005a, b). Other studies have demonstrated that hypermethylation of the specific gene promoters contributes to cholangiocarcinoma progression (Stutes et al. 2007). The methylation profiles of six genes such as p16, LOX, HAND1, THBO, p41ARC and APC have revealed their association with multistep gastric carcinogenesis related to *H. pillory* infection (Shin et al. 2010). The methylation of six other genes (DAPK, TWIST, HIN-1, RASSF1A, RARbeta2 and APC) was also associated with inflammatory breast cancer. Thus, chronic bacterial infection and inflammation may be related to the appearance of aberrant genomic DNA methylation, which, at least partly, may explain the mechanism of infection-associated carcinogenesis.

Chronic Infection and Bacteria-Associated Carcinogenesis

Salmonella typhi is one of the most important risk factors leading to carcinogenesis (Lazcano-Ponce et al. 2001; Wistuba and Gazdar 2004). Chronic carriers of this pathogen are at eight-fold increased risk of the development of gallbladder carcinoma, 200-fold inreased risk of hepatobiliary carcinoma and have significantly higher risk to develop cholangiocarcinoma (Caygill et al. 1995; Shukla et al. 2000; Robbins et al. 1988). These bacteria produce β-glucuronidase that provide deconjugation of conjugated toxins and bile acids, which may acquire a carcinogenic action via production of active intermediate substances able to bind DNA with a mutagenic potential (Chipman 1982; Mackowiak 1987; Kinoshita and Gelboin 1978). The discovery of an intracellular-acting cytolethal distending toxin, which directly damages DNA, suggests that toxins might contribute to cancer initiation associated with certwin infections. The *S. typhi* CDT, an intracellular toxin found also in six species of Gram-negative bacteria is only expressed when bacteria are in an intracellular location (Lax 2005; Frisan et al. 2002). This toxin is secreted directly into the cytoplasm of infected cells and catalyzes double-stranded DNA breaks that via repair mechanisms induce G2/M cell cycle arrest associated with the accumulation of the hyperphosphorylated form of cyclin dependent kinase 2 (cdk2) (Lax 2005; Hassane et al. 2003). Thus, CDT of *S. typhi* might be involved in the carcinogenic properties of *S. typhi* carriage. Chronic *S. typhi* infection can contribute to tumorigenesis via products of bile salts degradation by intestinal bacteria as well (Lax and Thomas 2002). It was also shown that typhoid carriage increases the risk of malignancy in the pancreas, lung and colorectum (Caygill et al. 1994, 1995). However, the association of these types of cancers with *S. typhi* infection is weaker than with gallbladder and hepatobilliary carcinomas.

Chlamydia pneumoniae (*C. pneumonae*), a gram-negative compulsory intracellular parasite causing pneumonia, may be associated with an increased risk of lung cancer via an elevation of IgA antibody production (Mager 2006; Littman et al. 2004a; Kocazeybek 2003; Anttila et al. 2003). *C. pneumonae* infections are associated with squamous cell carcinomas, small cell carcinomas and adenocarcinomas of the lung. Smoking assists *C. pneumonae* to invade the lung where the complex of interactions induce the production of oxygen radicals, TNF-α, IL-1 and IL-8 that contribute to lung tissue and DNA damages leading to carcinogenesis (Ohshima and Bartsch 1994; Redecke et al. 1998). *C. pneumonae* can also cause irregular apoptosis in lung tissues by unknown mechanisms.

Other Chlamydia infections, such as *Chlamydia trachomatis*, clinically "silent" species that are often involved in chronic infections of the upper genital tract, can be responsible for significant damage of the reproductive organs leading to epithelial ovarian cancer (Quirk and Kupinski 2001). These infections, as well as the other Chlamydia infections, induce the state of persistent inflammation, which subsequently increases the risk of tumorigenesis in the ovarian surface epithelium (Quirk and Kupinski 2001). Lymphoproliferative disorders, like ocular lymphomas, are linked to *Chlamydia psittaci* infections (Ferreri et al. 2004a) and *Chlamydia trachomatis* infections.

Recently, bacterial and fungal microflora in surgically extracted lung cancer samples has been analyzed using PCR with primers that have been designed to amplify many different strains of microorganisms including Mycoplasma, Streptococcus, Staphylococcus, Bacillus, Haemophilus, Treponema and others (Apostolou et al. 2011). A diversity of pathogens was identified in these samples.

The mycoplasma strains, which increase the ability of cancerous cells to metastasize, have been detected in all tissue samples of patients with lung cancer, whereas *Staphylococcus epidermis* and *Streptococcus mitis* have been observed in one quarter of these samples (Apostolou et al. 2011). Other pathogens were revealed in less than 25 % of tested samples. Staphylococcus (Dancewicz et al. 2009; Szymankiewicz et al. 2006; Rancic et al. 2014; Korona-Glowniak et al. 2003), Bacillus (Szymankiewicz et al. 2006), Listeria (Khardori et al. 1989) and Streptococcus strains (Dancewicz et al. 2009; Szymankiewicz et al. 2006; Rancic et al. 2014), as well as *Haemophilus influenza* (Dancewicz et al. 2009; Szymankiewicz et al. 2006; Rancic et al. 2014; Laroumagne et al. 2011) and *Legionella pneumophila* (Nunnink et al. 1986) were also found in the other studies in patients with respiratory tract and lung malignancies. These results point to a possible etiologic role for chronic infection in lung carcinogenesis.

The number of evidences has demonstrated that *Mycobacterium tuberculosis* may cause the development of cancer. However, the putative link between *M. tuberculosis* and lung cancer is still being under discussion. It was reported that individuals with tuberculosis had a five-fold higher risk of lung carcinomas (Steinitz 1965), bronchogenic carcinomas (Farwell et al. 1978) and Kaposi's sarcoma (Barete et al. 2000; Tamburini et al. 2007). Patients with lung cancer have tuberculosis more frequently than the general population, and bronchogenic carcinomas often appear in the areas of pulmonary scarring from tuberculosis suggesting a possible relationship between *M. tuberculosis* infection and malignancy (Parsonnet 1995a). In fact, similar symptoms and radiological findings in tuberculosis and lung cancer have led to a large number of lung cancer patients being treated for tuberculosis with a delay in the diagnosis of cancer (Tandon et al. 2013). However, the link between active tuberculosis and neoplasm is frequently recognized as the revival of infection in immunocompromised cancer patients rather than a cause-and-effect relationship between infection and malignancy (Flance 1991; Kung et al. 1985; Snyder et al. 1990). The malignancies have been also developed due to an increased level of circulating VEGF related to *Mycobacterium tuberculosis* infection (Samaras et al. 2010; Tamburini et al. 2007). Nevertheless, the high infection rate of *M. tuberculosis* was identified within 300 patients with lung cancer (Wang and Xie 1998). Thus, the results have suggested that there might be a relationship between pulmonary tuberculosis and lung cancer; however, these observations still require direct experimental confirmation.

Schistosoma species are also associated with cancer development. In particular, *Schistosoma haematobium* (*S. haematobium*), a blood fluke, resides in the pelvic organs and systemic venules and capillaries of the human bladder (Nash et al. 1982). *S. haematobium* infections occur in the second decade of life and are connected with many types of malignancy, such as carcinoma of the intestine, liver, uterus and bladder (Schwartz 1984; Koraitim et al. 1995; Mostafa et al. 1999), whereas *Schistosoma mansoni* (*S. mansoni*) as well as *Schistosoma japonicum* (*S. japonicum*) infections are associated with colorectal carcinoma. *Schistosoma japonicum* is also linked with liver carcinoma (Kuper et al. 2000; Mostafa et al. 1999). Cervical schistosomiasis species can also increase development of squamous cell carcinoma of the cervix (Charlewood et al. 1949). Schistosomiasis causes cancer development via chronic inflammatory reaction. The inflammatory infiltrates, consisted of macrophages and neutrophils, are significant sources of endogenous oxygen radicals that relate to the development of carcinogenic N-nitrosamines (Marletta 1988) and are responsible for various mutations, sister chromatid exchanges and DNA damage. Schistosomal infection induces neoplastic progression that has been associated with the activation of H-ras, inactivation of p53 and retinoblastoma genes, which induce uncontrolled cell growth and tumor formation (Mostafa et al. 1999; Sidransky et al. 1991; Knowles and Williamson 1993).

Furthermore, numbers of reports have demonstrated an association of Gram-positive bacterium *Tropheryma whippelii* with carcinogenesis. This pathogen is mainly found in soil infects of the gastrointestinal tract causing rare inflammatory Whipple's disease that then triggers a chronic infection involving nearly every organ (Mohm et al. 1998; Wang et al. 2003; Walter et al. 2001). Whipple's disease can mimic malignant neoplasm related to extraintestinal lymphoma (Gillen et al. 1993), non-Hodgkin-Limphoma (Gruner et al. 2001) and intramucosal gastric adenocarcinoma (Cadenas et al. 1999). The mechanism of lymphoma development might be the result of immunodeficiency status and abnormalities in the humoral response, T cells function or translocation in lymphocytes in patients suffering from the infectious process (Wang et al. 2003; Gillen et al. 1993; Sebok et al. 1997).

Epidemiological studies have revealed that *Clonorchis sinensis* (*C. sinensis*) and *Opisthorchis viverrini* (*O. viverrini*) infections of the bile ducts which have been identified in a great part of the Far East population are predisposing factors for the pathogenesis of cholangiocarcinoma in endemic countries (Flavell 1981; Kim 1984). The carcinogenic processes associated with *C. sinensis* and *O. viverrini* infections are the results of chronic severe infection, immunological disturbances and production of various carcinogenes (Schwartz 1980). The pathologic alterations, such as epithelial hyperplasia, goblet cells metaplasia and adenomatous hyperplasia, might be vulnerable to various exogenous or endogenous carcinogens leading to DNA damage, cytotoxicity and the increase of nitric oxide production (Kim 1984; Sripa et al. 2007; Ohshima et al. 1994; Kirby et al. 1994; Watanapa and Watanapa 2002).

Among the bacterial inflammatory processes, chronic osteomyelitis is the first disease strongly connected with cancer in humans (Song et al. 2002). A long-standing infection leads to neoplasms, mainly located in the tibia but also in the ankle, foot, patella and in a toe (Inglis et al. 1979; Altay et al. 2004; Patel et al. 2002; Patel and Weiner 2002; Ziets et al. 1991). The tumors that grow on chronic osteomyelitis are mostly squamous cell carcinomas (Kirsner et al. 1996; Chang et al. 1998; Sonin et al. 1998; Saglik et al. 2001). However, appearance of basal cell carcinoma, myeloma, lymphoma, fibrosarcoma, angiosarcoma and rhabdomyosarcoma has been also described (Trent and Kirsner 2003; Akbarnia et al. 1976; Baitz and Kyle 1964; Johnston and Miles 1973). The mechanism of carcinogenesis involves the reaction to bacterial pathogens such as *Streptococcus pyogenes*, *Staphylococcus aureus* and *Haemophilus influenza*. Unstable scars and wound tissues around the chronic osteomyelitis zones undergo malignant degeneration

(Mousa 2003; Bauer et al. 2007). The chronic inflammatory processes associated with chronic osteomyelitis increase the rate of DNA turnover, which predisposes cells to malignant transformation by primary carcinogens (Mackowiak 1987).

Hidradenitis suppurativa (HS) is a persistent chronic inflammatory disease of the apocrine glands that involves axillary, perianal, perineal and inflammatory regions (Altunay et al. 2002; Slade et al. 2003; Perez-Diaz et al. 1995). Bacterial pathogens such as *Staphylococcus aureus*, *Streptococcus pyogenes* and *Pseudomonas aeruginosa* can be involved in the cause of HS (Brook and Frazier 1999), which may be associated with squamous cell carcinoma (Yu and Cook 1990; Rosenzweig et al. 2005) and the progression of primary liver and buccal cancer (Lapins et al. 1999). Thus, the development of squamous cell carcinoma associated with HS frequently depends on the duration of the inflammatory processes that produce an inflammatory environment contributing to the processes of tumorigenesis. The secondary bacterial infections and chronic irritation of the skin may also lead to the proliferative epidermal changes, including cancer (Jansen et al. 1998; Jansen and Plewig 1998; Lapins et al. 2001).

Taken together, diverse clinical data suggest that a number of bacterial species can be connected with cancer development. However, clinical evidence regarding the association of each pathogen with specific malignancy varies significantly from well-established to weak-described, making further studies necessary and important (Samaras et al. 2010).

Inflammatory Diseases and Bacteria-Associated Carcinogenesis

According to the estimation of the American Cancer Society, gastric cancer killed nearly 38,500 Americans in 2013. *Helicobacter pylori* (*H. pylori*) is a type of microaerophilic spiral-shaped bacterium founded in the stomachs of almost two-thirds of the world's population. *H. pylori* infection causes colonization of the stomach, which can induce the development of gastric cancer; however, in most infected people, *H. pylori* does not cause illness. In 1994, *H. pylori* was classified as a human carcinogen, or cancer-causing agent, by the International Agency for Research on Cancer. *H. pylori* infection is associated with an increased risk of gastric adenocarcinoma and mucosa-associated lymphoid tissue (MALT), which was supported by the epidemiological studies.

H. pylori is a Gram-negative rod bacterium that lives between the mucus layer of the stomach and the gastric epithelium (Parsonnet 1995a, b). Although immune cells normally recognize and attack invading bacteria located near spots of *H. pylori* infection, they are incapable of reaching the stomach lining. In addition, *H. pylori* interferes with local immune responses that become ineffective at eliminating this bacterium (Atherton 2006; Kusters et al. 2006). *H. pylori* is also involved in many diseases, which have been characterized by immune dysregulation and may lead to autoimmune disorders such as systemic lupus erythematosus, rheumatoid arthritis and Sjorgen's syndrome.

Moreover, *H. pylori* infection causes inflammation and may support the development of gastric malignancy via several signaling pathways acting on gastric epithelia: the indirect pathway inducing inflammation of the gastric epithelia and direct pathway via induction of protein modulation and gene mutations, which both promote gastric carcinogenesis (Crowe 2005; Ponzetto et al. 2000; Sasazuki et al. 2006; Chiba et al. 2008). Many studies have suggested that stem cells play a crucial role in gastric cancer initiation (Houghton et al. 2004). Gastric epithelial cells that make up the gastric gland in mucosa are the progenitor cells. *H. pilory* interacts with and lives within a subgroup of these progenitor cells (Oh et al. 2005) and this *H. pilory* cancer-associated strain regulates the expression of gastric progenitor cells (GER)-associated metabolic and signaling pathways, along with tumor suppressor genes that stimulate the development of gastric cancer in humans (Giannakis et al. 2008). Another study has demonstrated that *H. pilory* infection induces inflammation and lead to stomach malignancy linked to gastric epithelial cell DNA methylation (Niwa et al. 2010). All these results indicate potential interaction of *H. pilory* with gastric progenitor cells (Ding and Zheng 2012). The NF-κB, IL-6, VEGF, HIF, reactive oxygen species and angiogenesis are all involved in stem cell maintenance. It was demonstrated that *H. pilory* infection alters the expression of most of these factors, affects the stem cell differentiation and cause genetic damage, which lead to carcinogenesis (Ding and Zheng 2012).

The cag-pathogenicity island (cagPAI), cytotoxin-associated antigen A (CagA), outer membrane proteins (OMPs) and vacuolating cytotoxin (VacA) are the major virulence factors that have been identified from *H. pilory*. Some of these factors are involved in gastric cancer development via the activation of several intracellular pathways in epithelial cells, such as NF-κb, mitogen-activated protein kinases (MAPK), Wnt/β-catenin, activator protein-1 (AP-1), phosphatidylinositol 3-kinase (PI3K) and signal transducer and activator of transcription 3 (STAT3). However, *H. pilory* CagA itself is the bacterial oncoprotein, which promotes the initiation of gastric cancer. Overexpression of this protein induces multiple malignancies (Ohnishi et al. 2008). Infection with *H. pilory* also disrupts gastric homeostasis and induces the production of cytokines within the local mucosal components, including IL-1, TNF-α and IL-10, which increases the risk of gastric cancer development (Tu et al. 2008, 2011; El-Omar et al. 2000a, b). Thus, *H. pilory* infection triggers inflammation and via interaction with host cells in the local microenvironment affects stem/progenitor cell differentiation, which may potentiate oncogenic transformation (Ding and Zheng 2012). It was also shown that bacterial infection induced epithelial-to-mesenchymal transition (EMT) of colonic crypt cells with acquired characteristics of stem cells promotes spheroid/organoid formation *in vitro* and tumorigenesis *in vivo* (Xia et al. 2012). These studies have demonstrated that *Citrobacter rodentium* induced transmissible colonic hyperplasia and increased stemness promoting cellular transformation in NIH-Swiss mice. The *H. pilory* infection also links to MALT lymphoma where infection chronically stimulates B and T cells which increases the chance of accidental chromosomal changes specific to the development of MALT lymphomas.

Helicobacter hepaticus causes cancer and stomach ulcers in humans. The mice infected with *H. hepaticus* demonstrated conditions similar to inflammatory bowel disease in humans and in 20 weeks developed chronic infections of the liver and colon, with some of the animals developing colon cancer (Mangerich et al. 2012). These studies have revealed intensive tissue damage, different types of damage of DNA, RNA and proteins, and determined the genes that were turned on and off in the infection progression. One of the most important observations was that the liver and colon responded differently to the infection. In the colon, but not in the liver, neutrophils secreted hypochlorous acid which significantly damages proteins, DNA and RNA via addition of a chlorine atom to molecules (Mangerich et al. 2012). This hypochlorous acid kill the bacteria, however, it also leaks into the neighboring tissue and injures the colon epithelial cells (Mangerich et al. 2012). It was also found that DNA repair systems become more active in the liver but less active in the colon, which has demonstrated another difference between the colon and the liver responses to the infection (Mangerich et al. 2012). Furthermore, these studies have identified several unknown types of DNA damage in humans and mice, where one of them involves oxidation of guanine to two new products, spiroiminodihydantoin and guanodinohydanotoin (Mangerich et al. 2012).

Porphyrmomonas gingivalis infection is involved in gum disease and can be associated with pancreatic cancer development. Although pancreatic cancer is associated with a high mortality rate of about 95 %, its etiology is still unknown. *P. gingivalis* infection is the major recognizable risk factor that accounts for almost 40 % of pancreatic cancer cases. One of several theories explaining how these infections can contribute to the pancreatic cancer progression suggests the role of chronic inflammation caused by *P. gingivalis* infection. A second potential mechanism is associated with a weak immune system. The other potential risk factors for pancreatic cancer that suppress immune responses are obesity, diabetes and smoking. *P. gingivalis* infections via stimulation of the new blood cell growth can also directly activate pancreatic tumor cell signaling and growth. Another option is *P. gingivalis-induced* indirect activation of pancreatic cancer pathways, which trigger regulatory immune responses in the tumor microenvironment, but not in the tumor cells.

Streptococcus infantarius might be associated with colon cancer. Cell wall proteins isolated from the bacteria are putatively involved in colorectal inflammation and carcinogenesis (Nguyen and Bellamy 2006; Nguyen et al. 2006). *Streptococcus* infections in humans are often associated with meningitis, spontaneous bacterial peritonitis, neonatal sepsis, septic arthritis and vertebral osteomyelitis.

Streptococcus bovis is normally habits in human gastrointestinal tract and cause bacteremia, endocarditis and urinary infection (Roses et al. 1974). *S. bovis* chronic infection is also frequently associated with hepatic dysfunction, colonic neoplasia, especially carcinoma of the colon and AIDS (Bayliss et al. 1983, 1984a, b; Burns et al. 1985; Gold et al. 2004). It has been revealed that *S. bovis* or its cell-associated proteins (S300) or wall extracted antigens (WEA) promote carcinogenesis in rats (Gold et al. 2004). S300 and WEA can induce the release of chemokines and prostaglandin E2 from the human colonic epithelial cells and overexpression of COX-2 associated with mucosal inflammation, inhibition of apoptosis and enhancement of angiogenesis (Mager 2006; Biarc et al. 2004). *S. bovis* proteins can also stimulate cell proliferation mediated by MAP kinases, which might contribute to cell transformation and genetic mutations (Biarc et al. 2004). Thus, these observations demonstrate that *S. bovis* can trigger cancer development particularly in chronic infection/inflammation diseases, where the bacterial components interfere with cell function (Biarc et al. 2004). Interestingly, *Streptococcus* species, mainly *Streptococcus aureus*, were identified in highly metastatic lung cancer in patients with neutropenia (Lanoix et al. 2011).

Mycoplasmas are self-replicating bacteria that possess very small genomes and lack cell wall components. These bacteria need cholesterol for membrane function and growth, and use UGA codon for tryptophan. Mycoplasmas can cause upper respiratory problems, nerve damage, Crohn's disease, arthritides, and have been linked as a cofactor to AIDS (Baseman and Tully 1997). *Mycoplasma* infections have been revealed to be related to different cancers (Huang et al. 2001a, b). Malignant transformations of the cell lines caused by Mycoplasma infection were associated with cell transformation, chromosomal aberrations and altered morphologies. These transformed cells continued to be abnormal oncogenic cells, even after the elimination of mycoplasmas (Tsai et al. 1995). Other studies have demonstrated that persistent mycoplasma infection leads to cell transformation, karyotypic alterations, and tumorigenicity in the embryo of nude mice (Tsai et al. 1995).

Bacteroides fragilis (*B.fragilis*), which is normal constituent of the gut microflora, can lead to inflammatory bowel disease, which predisposes an increased risk of cancer. *B.fragilis* toxin (BFT) plays a major role in carcinogenesis (Lax 2005; Wu et al. 2003). This toxin is metalloprotease that cleaves E-cadherin that is important for cell-cell interactions. BFT associates with the cytoskeleton through catenins. Degradation of E-cadherin by BFT activates nuclear translocation and c-myc expression, resulting in cell proliferation (Lax 2005; Wu et al. 2003).

Citrobacter rodentium is another carcinogenic bacterium that produces the toxins that cause a colonic disease in mice leading to mucosal hyperplasia and inflammation (Luperchio and Schauer 2001). *C. rodentium* infection promotes adenoma development in mouse models of human colorectal cancer (Newman et al. 2001). This bacteria possess a pathogenicity island (PAI) called the locus of enterocyte effacement (LEE), which encodes a type III secretion system. The mutational analysis of *C. rodentium* has revealed the correlation between the microorganism and hyperplasia associated with bacterial colonization. It is suggested that the mitochondrial associated protein, which perturbs cytoskeletal organization, might play an important role in *C. rodentium* – mediated carcinogenesis (Mundy et al. 2004). It was also demonstrated that *C. rodentium* infection was related to Th17 responses and impairs host defense (Ryz et al. 2012).

Pasturella multocida (*P. multocida*), which possesses a toxic mechanism of carcinogenesis, has been isolated from humans with chronic respiratory infections; however, it is predominantly associated with pig disease whose primary sign is the bone loss. Some strains of this bacterium produce the mitogenic toxin called PMT. PMT is classically acting toxin that enters cells, including fibroblasts, to stimulate signaling pathways linked to the regulation of cell growth. It is suggested that PMT activates one or more heterotrimeric G-proteins, in particular Gq, which relates to stimulation of inositol phosphate accumulation, activation of protein kinase C, MAP kinase ERK1/2 and small G-protein Rho with its downstream pathway. It was demonstrated that even a low concentration of PMT could stimulate the re-entrance of quiescent cells to the cell cycle and promote anchorage-independent cell growth. Injections of recombinant toxin or natural infection induce proliferation of bladder epithelium in a non-inflammatory manner. The majority of signaling proteins such as RhoA, Src, focal adhesion kinase (FAK), EGFR and ERK1/2, which are indirectly stimulated by PMT, a potent growth factor, and are involved in carcinogenesis.

E. coli that cause most of urogenital tract infections form complexes, biofilm-like communities in the bladder, which are resistant to antibiotics, unaffected by the host immune response and promote the re-infections that became chronic (Justice et al. 2004; Mulvey et al. 2001). These bacteria also induce autoimmune cholangitis and anti-mitochondrial antibodies in diabetic (Wang et al. 2013). Many of human prostate infections with *E. coli* show increase in inflammation and tissue damage, while the toxins production contributes to cancer risk (Rippere-Lampe et al. 2001). Development of prostatic intraepithelial neoplasia induced by *E. coli* infection has been directly demonstrated in a mouse model. Mice infected for 5 days have revealed acute inflammation related to neutrophils infiltration and epithelial necrotic debris in the lumen of prostatic glandular. In 12 weeks inflammation became chronic and provided dense inflammatory infiltrates in stroma and prostatic epithelium atypical hyperplasia. At 26 weeks, the prostatic glands exhibited dysplasia associated with DNA damage, increased epithelial cell proliferation and decreased in androgen receptor, p27kip1, PTEN and GSTP1 expression. Thus, *E. coli* infection induces chronic inflammation that leads to prostatic neoplasia (Elkahwaji et al. 2009). It was also demonstrated that many strains of *E. coli* express cycle inhibiting factor (Cif) which is usually injected by the LEE type III secretion system. This toxin is involved in tumorigenesis via induction of G2/M cell cycle arrest by hyperphosphorylation of cdc2, which interacts with cyclin B (Marches et al. 2003). The other toxin expressed by *E. coli* is cytotoxic necrotizing factor (CNF). CNF activate all members of the Rho family of small GTPases and its downstream pathway including FAK, Src and COX (Landraud et al. 2004; Thomas et al. 2001). This toxin also promotes the entrance of quiescent cells to cell cycle, production of multinucleated cells via interactions with normal cytokines and inhibition of apoptosis, which are all involved in tumorigenesis (Lax 2005).

Chronic *Campylobacter jejuni* (*C.jejuni*) infection, as well as *E. coli* infection, can promote intestinal carcinogenesis. CDT toxin has been identified in EHEC and EPEC isolates of *C.jejuni* (Janka et al. 2003). Even short-term exposure to CDT isolated from *C.jejuni* might affect tumorigenesis. This bacterial infection links to small intestinal lymphomas (Lecuit et al. 2004), whereas *Chlamydia psitacci* infection is associated with ocular lymphomas (Ferreri et al. 2004b). It has also been shown that *Campylobacter* species were associated with the development of Gullain-Barre syndrome (GBS), which is an autoimmune disorder of the peripheral nervous system and is characterized by inflammatory demyelinating or motor axonal neuropathy (Nachamkin et al. 1998).

The Role of Bacterial Infection in the Formation of Pre-metastatic Niche

A growing body of evidence suggests that factors such as VEGF-A, TGF-β, TNF-α and other cytokines secreted from the primary tumor sites can activate and systematically alter the microenvironment of secondary organ sites or recruit bone marrow-derived cells promoting establishment of pre-metastatic niches (Kaplan et al. 2005, 2006; Hiratsuka et al. 2006). Furthermore, a systemic effect of acute inflammation on metastasis caused by lipopolysaccharide (LPS) injection has been established: LPS, a bacterial cell wall component, increased the metastatic potential of cancer (Simiantonaki et al. 2007a, b; Harmey et al. 2002). However, recent studies by Yon and colleagues provided a new mechanism by which acute inflammation can raise metastasis in the lung model. The authors used two acute infection models, LPS-induced acute inflammation and the *E. coli* bacterial pneumonia, and revealed the induction of lung metastasis in mice with bacterial lung infections (Yan et al. 2010, 2013). Melanoma, lung, prostate and colorectal cancer cell lines were also evaluated in this study. The bacterial-induced migration of $CXCR4^+$ tumor cells toward the bronchoalveolar lavage fluid (BALF) in bacteria-infected mice was revealed (Yan et al. 2013). It was also found that differential BALF cytokine expression patterns from infected and control mice could affect tumor cell migration. The CXCR4/SDF-1 signaling is responsible for recruitment of normal hematopoietic stem cells to the bone marrow and metastatic tumor cells to distant organs (Muller et al. 2001). However, the investigators have found that extracellular ubiquitin (Ub), alternative CXCR4 ligand, instead of SDF-1, the most likely CXCR4 candidate for tumor cell recruitment, was responsible for chemotaxis. The CXCR4 can be controlled by multiple components of the inflammatory environment, including stress and tissue damage (Jourdan et al. 2000). Interestingly, the CXCR4/Ub axis activates AKT signaling, which may mediate tumor cell migration (Yan et al. 2013), and the use of AKT inhibitors can reduce the risk of metastasis during acute inflammation. Thus, these studies clearly recognized that infection-induced acute inflammation, rather than tumor-induced inflammation, is involved in the formation of a metastatic niche-like environment (Weitao 2009a), and a novel CXCR4/ubiquitin/AKT signaling pathway may be a key mediator of tumor cell recruitment.

Conclusions

The relationship between the human organism and the microbiota is critically important in human life because of the persistent contacts of a human with internal and external microorganisms. These symbiotic relationships are usually beneficial, but certain environmental or microenvironment changes can lead to pathogenic misbalance and even malignancy. A growing body of data suggests a potential contribution of bacteria to the development of cancer through different pathways, including inflammation, mutagenic toxins and metabolites. Importantly, many bacteria-induced chronic infections do not cause or support malignant transformation and only accumulation of specific changes in bacteria-host interactions leading to irreversible mutation of pre-malignant cells or supporting survival of transformed cells promote cancer formation. At present, it is still not completely clear what kind of combination of specific factors and environmental stimuli contributes to high incidence of cancer during certain infectious diseases. Understanding the role of bacterial infection in cancer development requires identification of these factors, their combinations and specific mechanisms. More experimental studies are required to reveal how the diverse bacterial infections may promote neoplastic transformation in order to provide new and effective approaches to cancer prevention.

References

Akbarnia BA, Wirth CR, Colman N (1976) Fibrosarcoma arising from chronic osteomyelitis. Case report and review of the literature. J Bone Joint Surg Am 58:123–125

Allavena P, Garlanda C, Borrello MG, Sica A, Mantovani A (2008) Pathways connecting inflammation and cancer. Curr Opin Genet Dev 18:3–10

Allen IC, TeKippe EM, Woodford RM, Uronis JM, Holl EK et al (2010) The NLRP3 inflammasome functions as a negative regulator of tumorigenesis during colitis-associated cancer. J Exp Med 207:1045–1056

Allen IC, Wilson JE, Schneider M, Lich JD, Roberts RA et al (2012a) NLRP12 suppresses colon inflammation and tumorigenesis through the negative regulation of noncanonical NF-kappaB signaling. Immunity 36:742–754

Allen IC, Lich JD, Arthur JC, Jania CM, Roberts RA et al (2012b) Characterization of NLRP12 during the development of allergic airway disease in mice. PLoS One 7:e30612

Altay M, Arikan M, Yildiz Y, Saglik Y (2004) Squamous cell carcinoma arising in chronic osteomyelitis in foot and ankle. Foot Ankle Int 25:805–809

Altunay IK, Gokdemir G, Kurt A, Kayaoglu S (2002) Hidradenitis suppurativa and squamous cell carcinoma. Dermatol Surg 28:88–90

Anttila T, Koskela P, Leinonen M, Laukkanen P, Hakulinen T et al (2003) Chlamydia pneumoniae infection and the risk of female early-onset lung cancer. Int J Cancer 107:681–682

Apostolou P, Tsantsaridou A, Papasotiriou I, Toloudi M, Chatziioannou M et al (2011) Bacterial and fungal microflora in surgically removed lung cancer samples. J Cardiothorac Surg 6:137

Arthur JC, Perez-Chanona E, Muhlbauer M, Tomkovich S, Uronis JM et al (2012) Intestinal inflammation targets cancer-inducing activity of the microbiota. Science 338:120–123

Atherton JC (2006) The pathogenesis of Helicobacter pylori-induced gastro-duodenal diseases. Annu Rev Pathol 1:63–96

Baik SC, Youn HS, Chung MH, Lee WK, Cho MJ et al (1996) Increased oxidative DNA damage in Helicobacter pylori-infected human gastric mucosa. Cancer Res 56:1279–1282

Baitz T, Kyle RA (1964) Solitary myeloma in chronic osteomyelitis. Report of case. Arch Intern Med 113:872–876

Balkwill F, Mantovani A (2001) Inflammation and cancer: back to Virchow? Lancet 357:539–545

Barete S, Calvez V, Mouquet C, Barrou B, Kreis H et al (2000) Clinical features and contribution of virological findings to the management of Kaposi sarcoma in organ-allograft recipients. Arch Dermatol 136:1452–1458

Baseman JB, Tully JG (1997) Mycoplasmas: sophisticated, reemerging, and burdened by their notoriety. Emerg Infect Dis 3:21–32
Bauer T, David T, Rimareix F, Lortat-Jacob A (2007) Marjolin's ulcer in chronic osteomyelitis: seven cases and a review of the literature. Rev Chir Orthop Reparatrice Appar Mot 93:63–71
Bayliss R, Clarke C, Oakley CM, Somerville W, Whitfield AG et al (1983) The microbiology and pathogenesis of infective endocarditis. Br Heart J 50:513–519
Bayliss R, Clarke C, Oakley CM, Somerville W, Whitfield AG et al (1984a) The bowel, the genitourinary tract, and infective endocarditis. Br Heart J 51:339–345
Bayliss R, Clarke C, Whitehead TP, Whitfield AG (1984b) The management of hyperuricaemia. J R Coll Physicians Lond 18:144–146
Biarc J, Nguyen IS, Pini A, Gosse F, Richert S et al (2004) Carcinogenic properties of proteins with pro-inflammatory activity from Streptococcus infantarius (formerly S.bovis). Carcinogenesis 25:1477–1484
Bouvard V, Baan R, Straif K, Grosse Y, Secretan B et al (2009) A review of human carcinogens–Part B: biological agents. Lancet Oncol 10:321–322
Brandl K, Sun L, Neppl C, Siggs OM, Le Gall SM et al (2010) MyD88 signaling in nonhematopoietic cells protects mice against induced colitis by regulating specific EGF receptor ligands. Proc Natl Acad Sci U S A 107:19967–19972
Brook I, Frazier EH (1999) Aerobic and anaerobic microbiology of axillary hidradenitis suppurativa. J Med Microbiol 48:103–105
Burns CA, McCaughey R, Lauter CB (1985) The association of Streptococcus bovis fecal carriage and colon neoplasia: possible relationship with polyps and their premalignant potential. Am J Gastroenterol 80:42–46
Cadenas F, Sanchez-Lombrana JL, Perez R, Lomo FJ, Madrigal Rubiales B et al (1999) Persistent leucocytosis as initial manifestation of Whipple's disease and development of gastric cancer in the follow up. Rev Esp Enferm Dig 91:785–788
Carbonero F, Benefiel AC, Alizadeh-Ghamsari AH, Gaskins HR (2012) Microbial pathways in colonic sulfur metabolism and links with health and disease. Front Physiol 3:448
Caygill CP, Hill MJ, Braddick M, Sharp JC (1994) Cancer mortality in chronic typhoid and paratyphoid carriers. Lancet 343:83–84
Caygill CP, Braddick M, Hill MJ, Knowles RL, Sharp JC (1995) The association between typhoid carriage, typhoid infection and subsequent cancer at a number of sites. Eur J Cancer Prev 4:187–193
Chang A, Spencer JM, Kirsner RS (1998) Squamous cell carcinoma arising from a nonhealing wound and osteomyelitis treated with Mohs micrographic surgery: a case study. Ostomy Wound Manage 44:26–30
Charlewood GP, Shippel S, Renton H (1949) Schistosomiasis in gynaecology. J Obstet Gynaecol Br Emp 56:367–385
Chiba T, Marusawa H, Seno H, Watanabe N (2008) Mechanism for gastric cancer development by Helicobacter pylori infection. J Gastroenterol Hepatol 23:1175–1181
Chipman JK (1982) Bile as a source of potential reactive metabolites. Toxicology 25:99–111
Cornforth DM, Foster KR (2013) Competition sensing: the social side of bacterial stress responses. Nat Rev Microbiol 11:285–293
Coussens LM, Werb Z (2002) Inflammation and cancer. Nature 420:860–867
Couturier-Maillard A, Secher T, Rehman A, Normand S, De Arcangelis A et al (2013) NOD2-mediated dysbiosis predisposes mice to transmissible colitis and colorectal cancer. J Clin Invest 123:700–711
Crowe SE (2005) Helicobacter infection, chronic inflammation, and the development of malignancy. Curr Opin Gastroenterol 21:32–38
Cuevas-Ramos G, Petit CR, Marcq I, Boury M, Oswald E et al (2010) Escherichia coli induces DNA damage in vivo and triggers genomic instability in mammalian cells. Proc Natl Acad Sci U S A 107:11537–11542
Dallo SF, Weitao T (2010) Bacteria under SOS evolve anticancer phenotypes. Infect Agent Cancer 5:3

Dancewicz M, Szymankiewicz M, Bella M, Swiniarska J, Kowalewski J (2009) Bronchial bacterial colonization in patients with lung cancer. Pneumonol Alergol Pol 77:242–247

Ding SZ, Zheng PY (2012) Helicobacter pylori infection induced gastric cancer; advance in gastric stem cell research and the remaining challenges. Gut Pathog 4:18

Elinav E, Strowig T, Henao-Mejia J, Flavell RA (2011a) Regulation of the antimicrobial response by NLR proteins. Immunity 34:665–679

Elinav E, Strowig T, Kau AL, Henao-Mejia J, Thaiss CA et al (2011b) NLRP6 inflammasome regulates colonic microbial ecology and risk for colitis. Cell 145:745–757

Elkahwaji JE, Hauke RJ, Brawner CM (2009) Chronic bacterial inflammation induces prostatic intraepithelial neoplasia in mouse prostate. Br J Cancer 101:1740–1748

El-Omar EM, Oien K, Murray LS, El-Nujumi A, Wirz A et al (2000a) Increased prevalence of precancerous changes in relatives of gastric cancer patients: critical role of H. pylori. Gastroenterology 118:22–30

El-Omar EM, Carrington M, Chow WH, McColl KE, Bream JH et al (2000b) Interleukin-1 polymorphisms associated with increased risk of gastric cancer. Nature 404:398–402

Fabbri A, Travaglione S, Falzano L, Fiorentini C (2008) Bacterial protein toxins: current and potential clinical use. Curr Med Chem 15:1116–1125

Farwell DJ, Rutledge LJ, Bryant LR, Schechter FG (1978) Localization of bronchogenic carcinoma in tuberculous lobes. South Med J 71:377–379

Ferreri AJ, Campo E, Seymour JF, Willemze R, Ilariucci F et al (2004a) Intravascular lymphoma: clinical presentation, natural history, management and prognostic factors in a series of 38 cases, with special emphasis on the 'cutaneous variant'. Br J Haematol 127:173–183

Ferreri AJ, Guidoboni M, Ponzoni M, De Conciliis C, Dell'Oro S et al (2004b) Evidence for an association between *Chlamydia psittaci* and ocular adnexal lymphomas. J Natl Cancer Inst 96:586–594

Flance IJ (1991) Scar cancer of the lung. JAMA 266:2003–2004

Flavell DJ (1981) Liver-fluke infection as an aetiological factor in bile-duct carcinoma of man. Trans R Soc Trop Med Hyg 75:814–824

Fox JG, Wang TC (2007) Inflammation, atrophy, and gastric cancer. J Clin Invest 117:60–69

Frisan T, Cortes-Bratti X, Thelestam M (2002) Cytolethal distending toxins and activation of DNA damage-dependent checkpoint responses. Int J Med Microbiol 291:495–499

Fukata M, Shang L, Santaolalla R, Sotolongo J, Pastorini C et al (2011) Constitutive activation of epithelial TLR4 augments inflammatory responses to mucosal injury and drives colitis-associated tumorigenesis. Inflamm Bowel Dis 17:1464–1473

Garaude J, Kent A, van Rooijen N, Blander JM (2012) Simultaneous targeting of toll- and nod-like receptors induces effective tumor-specific immune responses. Sci Transl Med 4(120):120ra16

Giannakis M, Chen SL, Karam SM, Engstrand L, Gordon JI (2008) Helicobacter pylori evolution during progression from chronic atrophic gastritis to gastric cancer and its impact on gastric stem cells. Proc Natl Acad Sci U S A 105:4358–4363

Gillen CD, Coddington R, Monteith PG, Taylor RH (1993) Extraintestinal lymphoma in association with Whipple's disease. Gut 34:1627–1629

Gold JS, Bayar S, Salem RR (2004) Association of Streptococcus bovis bacteremia with colonic neoplasia and extracolonic malignancy. Arch Surg 139:760–765

Grivennikov SI, Wang K, Mucida D, Stewart CA, Schnabl B et al (2012) Adenoma-linked barrier defects and microbial products drive IL-23/IL-17-mediated tumour growth. Nature 491:254–258

Gruner U, Goesch P, Donner A, Peters U (2001) Whipple disease and non-Hodgkin lymphoma. Z Gastroenterol 39:305–309

Gupta RA, DuBois RN (2002) Cyclooxygenase-2 inhibitor therapy for the prevention of esophageal adenocarcinoma in Barrett's esophagus. J Natl Cancer Inst 94:406–407

Harmey JH, Bucana CD, Lu W, Byrne AM, McDonnell S et al (2002) Lipopolysaccharide-induced metastatic growth is associated with increased angiogenesis, vascular permeability and tumor cell invasion. Int J Cancer 101:415–422

Hassane DC, Lee RB, Pickett CL (2003) Campylobacter jejuni cytolethal distending toxin promotes DNA repair responses in normal human cells. Infect Immun 71:541–545
Hiratsuka S, Watanabe A, Aburatani H, Maru Y (2006) Tumour-mediated upregulation of chemoattractants and recruitment of myeloid cells predetermines lung metastasis. Nat Cell Biol 8:1369–1375
Hodge DR, Peng B, Cherry JC, Hurt EM, Fox SD et al (2005a) Interleukin 6 supports the maintenance of p53 tumor suppressor gene promoter methylation. Cancer Res 65:4673–4682
Hodge DR, Peng B, Pompeia C, Thomas S, Cho E et al (2005b) Epigenetic silencing of manganese superoxide dismutase (SOD-2) in KAS 6/1 human multiple myeloma cells increases cell proliferation. Cancer Biol Ther 4:585–592
Holmes E, Li JV, Marchesi JR, Nicholson JK (2012) Gut microbiota composition and activity in relation to host metabolic phenotype and disease risk. Cell Metab 16:559–564
Houghton J, Stoicov C, Nomura S, Rogers AB, Carlson J et al (2004) Gastric cancer originating from bone marrow-derived cells. Science 306:1568–1571
Hu B, Elinav E, Huber S, Booth CJ, Strowig T et al (2010) Inflammation-induced tumorigenesis in the colon is regulated by caspase-1 and NLRC4. Proc Natl Acad Sci U S A 107:21635–21640
Hu B, Elinav E, Huber S, Strowig T, Hao L et al (2013) Microbiota-induced activation of epithelial IL-6 signaling links inflammasome-driven inflammation with transmissible cancer. Proc Natl Acad Sci U S A 110:9862–9867
Huang S, Li JY, Wu J, Meng L, Shou CC (2001a) Mycoplasma infections and different human carcinomas. World J Gastroenterol 7:266–269
Huang S, Shou C, Wu J (2001b) Mycoplasma infection in human gastriointestinal carcinoma tissues. Zhonghua Yi Xue Za Zhi 81:601–604
Hussein MA, Coats DK, Harris LD, Sanchez CR, Paysse EA (2007) Ultrasound biomicroscopy (UBM) characteristics of scleral tunnels created with suture needles commonly used during strabismus surgery. Binocul Vis Strabismus Q 22:102–108
Huycke MM, Gaskins HR (2004) Commensal bacteria, redox stress, and colorectal cancer: mechanisms and models. Exp Biol Med (Maywood) 229:586–597
Ibrahim N, He L, Leong CO, Xing D, Karlan BY et al (2010) BRCA1-associated epigenetic regulation of p73 mediates an effector pathway for chemosensitivity in ovarian carcinoma. Cancer Res 70:7155–7165
Ibrahim AE, Arends MJ, Silva AL, Wyllie AH, Greger L et al (2011) Sequential DNA methylation changes are associated with DNMT3B overexpression in colorectal neoplastic progression. Gut 60:499–508
Inglis AM, Morton KS, Lehmann EC (1979) Squamous cell carcinoma arising in chronic osteomyelitis. Can J Surg 22:271–273
Jain RK (2001) Normalizing tumor vasculature with anti-angiogenic therapy: a new paradigm for combination therapy. Nat Med 7:987–989
Janka A, Bielaszewska M, Dobrindt U, Greune L, Schmidt MA et al (2003) Cytolethal distending toxin gene cluster in enterohemorrhagic Escherichia coli O157:H- and O157:H7: characterization and evolutionary considerations. Infect Immun 71:3634–3638
Jansen T, Plewig G (1998) Follicular cysts. Fortschr Med 116:46–49
Jansen T, Muller S, Kaudewitz P, Plewig G (1998) Spinocellular carcinoma. Fortschr Med 116:49–50
Johnston RM, Miles JS (1973) Sarcomas arising from chronic osteomyelitic sinuses. A report of two cases. J Bone Joint Surg Am 55:162–168
Jourdan P, Vendrell JP, Huguet MF, Segondy M, Bousquet J et al (2000) Cytokines and cell surface molecules independently induce CXCR4 expression on CD4+ CCR7+ human memory T cells. J Immunol 165:716–724
Justice SS, Hung C, Theriot JA, Fletcher DA, Anderson GG et al (2004) Differentiation and developmental pathways of uropathogenic Escherichia coli in urinary tract pathogenesis. Proc Natl Acad Sci U S A 101:1333–1338

Kaplan RN, Riba RD, Zacharoulis S, Bramley AH, Vincent L et al (2005) VEGFR1-positive haematopoietic bone marrow progenitors initiate the pre-metastatic niche. Nature 438:820–827
Kaplan RN, Psaila B, Lyden D (2006) Bone marrow cells in the 'pre-metastatic niche': within bone and beyond. Cancer Metastasis Rev 25:521–529
Khardori N, Berkey P, Hayat S, Rosenbaum B, Bodey GP (1989) Spectrum and outcome of microbiologically documented listeria monocytogenes infections in cancer patients. Cancer 64:1968–1970
Khor B, Gardet A, Xavier RJ (2011) Genetics and pathogenesis of inflammatory bowel disease. Nature 474:307–317
Kim YI (1984) Liver carcinoma and liver fluke infection. Arzneimittelforschung 34:1121–1126
Kinoshita N, Gelboin HV (1978) beta-Glucuronidase catalyzed hydrolysis of benzo(a)pyrene-3-glucuronide and binding to DNA. Science 199:307–309
Kirby GM, Pelkonen P, Vatanasapt V, Camus AM, Wild CP et al (1994) Association of liver fluke (Opisthorchis viverrini) infestation with increased expression of cytochrome P450 and carcinogen metabolism in male hamster liver. Mol Carcinog 11:81–89
Kirsner RS, Spencer J, Falanga V, Garland LE, Kerdel FA (1996) Squamous cell carcinoma arising in osteomyelitis and chronic wounds. Treatment with Mohs micrographic surgery vs amputation. Dermatol Surg 22:1015–1018
Knowles MA, Williamson M (1993) Mutation of H-ras is infrequent in bladder cancer: confirmation by single-strand conformation polymorphism analysis, designed restriction fragment length polymorphisms, and direct sequencing. Cancer Res 53:133–139
Kocazeybek B (2003) Chronic Chlamydophila pneumoniae infection in lung cancer, a risk factor: a case-control study. J Med Microbiol 52:721–726
Koraitim M, Kamal B, Metwalli N, Zaky Y (1995) Transurethral ultrasonographic assessment of bladder carcinoma: its value and limitation. J Urol 154:375–378
Korona-Glowniak I, Rybojad P, Malm A, Furmanik F (2003) Microflora colonizing pleural drains after thoracic surgery. Pneumonol Alergol Pol 71:5–11
Koyi H, Branden E, Gnarpe J, Gnarpe H, Steen B (2001) An association between chronic infection with Chlamydia pneumoniae and lung cancer. A prospective 2-year study. APMIS 109:572–580
Kung IT, Lui IO, Loke SL, Khin MA, Mok CK et al (1985) Pulmonary scar cancer. A pathologic reappraisal. Am J Surg Pathol 9:391–400
Kuper H, Adami HO, Trichopoulos D (2000) Infections as a major preventable cause of human cancer. J Intern Med 248:171–183
Kusters JG, van Vliet AH, Kuipers EJ (2006) Pathogenesis of Helicobacter pylori infection. Clin Microbiol Rev 19:449–490
Landraud L, Pulcini C, Gounon P, Flatau G, Boquet P et al (2004) E. coli CNF1 toxin: a two-in-one system for host-cell invasion. Int J Med Microbiol 293:513–518
Lanoix JP, Pluquet E, Lescure FX, Bentayeb H, Lecuyer E et al (2011) Bacterial infection profiles in lung cancer patients with febrile neutropenia. BMC Infect Dis 11:183
Lapins J, Jarstrand C, Emtestam L (1999) Coagulase-negative staphylococci are the most common bacteria found in cultures from the deep portions of hidradenitis suppurativa lesions, as obtained by carbon dioxide laser surgery. Br J Dermatol 140:90–95
Lapins J, Ye W, Nyren O, Emtestam L (2001) Incidence of cancer among patients with hidradenitis suppurativa. Arch Dermatol 137:730–734
Lara-Tejero M, Galan JE (2000) A bacterial toxin that controls cell cycle progression as a deoxyribonuclease I-like protein. Science 290:354–357
Laroumagne S, Salinas-Pineda A, Hermant C, Murris M, Gourraud PA et al (2011) Incidence and characteristics of bronchial colonisation in patient with lung cancer: a retrospective study of 388 cases. Rev Mal Respir 28:328–335
Lax AJ (2005) Opinion: bacterial toxins and cancer–a case to answer? Nat Rev Microbiol 3:343–349

Lax AJ, Thomas W (2002) How bacteria could cause cancer: one step at a time. Trends Microbiol 10:293–299
Lazcano-Ponce EC, Miquel JF, Munoz N, Herrero R, Ferrecio C et al (2001) Epidemiology and molecular pathology of gallbladder cancer. CA Cancer J Clin 51:349–364
Lecuit M, Abachin E, Martin A, Poyart C, Pochart P et al (2004) Immunoproliferative small intestinal disease associated with Campylobacter jejuni. N Engl J Med 350:239–248
Ley RE, Turnbaugh PJ, Klein S, Gordon JI (2006) Microbial ecology: human gut microbes associated with obesity. Nature 444:1022–1023
Littman DR, Rudensky AY (2010) Th17 and regulatory T cells in mediating and restraining inflammation. Cell 140:845–858
Littman AJ, White E, Jackson LA, Thornquist MD, Gaydos CA et al (2004a) Chlamydia pneumoniae infection and risk of lung cancer. Cancer Epidemiol Biomarkers Prev 13:1624–1630
Littman AJ, Thornquist MD, White E, Jackson LA, Goodman GE et al (2004b) Prior lung disease and risk of lung cancer in a large prospective study. Cancer Causes Control 15:819–827
Luperchio SA, Schauer DB (2001) Molecular pathogenesis of Citrobacter rodentium and transmissible murine colonic hyperplasia. Microbes Infect 3:333–340
Mackowiak PA (1987) Microbial oncogenesis. Am J Med 82:79–97
Mager DL (2006) Bacteria and cancer: cause, coincidence or cure? A review. J Transl Med 4:14
Mangerich A, Knutson CG, Parry NM, Muthupalani S, Ye W et al (2012) Infection-induced colitis in mice causes dynamic and tissue-specific changes in stress response and DNA damage leading to colon cancer. Proc Natl Acad Sci U S A 109:E1820–E1829
Mannick EE, Bravo LE, Zarama G, Realpe JL, Zhang XJ et al (1996) Inducible nitric oxide synthase, nitrotyrosine, and apoptosis in Helicobacter pylori gastritis: effect of antibiotics and antioxidants. Cancer Res 56:3238–3243
Marches O, Ledger TN, Boury M, Ohara M, Tu X et al (2003) Enteropathogenic and enterohaemorrhagic Escherichia coli deliver a novel effector called Cif, which blocks cell cycle G2/M transition. Mol Microbiol 50:1553–1567
Marletta MA (1988) Mammalian synthesis of nitrite, nitrate, nitric oxide, and N-nitrosating agents. Chem Res Toxicol 1:249–257
McGovern DP, Hysi P, Ahmad T, van Heel DA, Moffatt MF et al (2005) Association between a complex insertion/deletion polymorphism in NOD1 (CARD4) and susceptibility to inflammatory bowel disease. Hum Mol Genet 14:1245–1250
Mohm J, Naumann R, Schuler U, Ehninger G (1998) Abdominal lymphomas, convulsive seizure and coma: a case of successfully treated, advanced Whipple's disease with cerebral involvement. Eur J Gastroenterol Hepatol 10:893–895
Mostafa MH, Sheweita SA, O'Connor PJ (1999) Relationship between schistosomiasis and bladder cancer. Clin Microbiol Rev 12:97–111
Mousa HA (2003) Bone infection. East Mediterr Health J 9:208–214
Mueller MM (2006) Inflammation in epithelial skin tumours: old stories and new ideas. Eur J Cancer 42:735–744
Mueller C, Laule-Kilian K, Christ A, Brunner-La Rocca HP, Perruchoud AP (2006) Inflammation and long-term mortality in acute congestive heart failure. Am Heart J 151:845–850
Muller A, Homey B, Soto H, Ge N, Catron D et al (2001) Involvement of chemokine receptors in breast cancer metastasis. Nature 410:50–56
Mulvey MA, Schilling JD, Hultgren SJ (2001) Establishment of a persistent Escherichia coli reservoir during the acute phase of a bladder infection. Infect Immun 69:4572–4579
Mundy R, Petrovska L, Smollett K, Simpson N, Wilson RK et al (2004) Identification of a novel Citrobacter rodentium type III secreted protein, EspI, and roles of this and other secreted proteins in infection. Infect Immun 72:2288–2302
Nachamkin I, Allos BM, Ho T (1998) Campylobacter species and Guillain-Barre syndrome. Clin Microbiol Rev 11:555–567
Nash TE, Cheever AW, Ottesen EA, Cook JA (1982) Schistosome infections in humans: perspectives and recent findings. NIH conference. Ann Intern Med 97:740–754

Nesic D, Hsu Y, Stebbins CE (2004) Assembly and function of a bacterial genotoxin. Nature 429:429–433
Nestle FO, Di Meglio P, Qin JZ, Nickoloff BJ (2009) Skin immune sentinels in health and disease. Nat Rev Immunol 9:679–691
Neufert C, Becker C, Tureci O, Waldner MJ, Backert I et al (2013) Tumor fibroblast-derived epiregulin promotes growth of colitis-associated neoplasms through ERK. J Clin Invest 123:1428–1443
Newman JV, Kosaka T, Sheppard BJ, Fox JG, Schauer DB (2001) Bacterial infection promotes colon tumorigenesis in Apc(Min/+) mice. J Infect Dis 184:227–230
Ngo VN, Young RM, Schmitz R, Jhavar S, Xiao W et al (2011) Oncogenically active MYD88 mutations in human lymphoma. Nature 470:115–119
Nguyen GT, Bellamy SL (2006) Cancer information seeking preferences and experiences: disparities between Asian Americans and Whites in the Health Information National Trends Survey (HINTS). J Health Commun 11(Suppl 1):173–180
Nguyen IS, Biarc J, Pini A, Gosse F, Richert S et al (2006) Streptococcus infantarius and colonic cancer: Identification and purification of cell wall proteins putatively involved in colorectal inflammation and carcinogenesis in rats. Int Congr Ser 1289:257–261
Niwa T, Tsukamoto T, Toyoda T, Mori A, Tanaka H et al (2010) Inflammatory processes triggered by Helicobacter pylori infection cause aberrant DNA methylation in gastric epithelial cells. Cancer Res 70:1430–1440
Nougayrede JP, Taieb F, De Rycke J, Oswald E (2005) Cyclomodulins: bacterial effectors that modulate the eukaryotic cell cycle. Trends Microbiol 13:103–110
Nunnink JC, Gallagher JG, Yates JW (1986) Legionnaires' disease in patients with cancer. Med Pediatr Oncol 14:81–85
Oh JD, Karam SM, Gordon JI (2005) Intracellular Helicobacter pylori in gastric epithelial progenitors. Proc Natl Acad Sci U S A 102:5186–5191
Ohnishi N, Yuasa H, Tanaka S, Sawa H, Miura M et al (2008) Transgenic expression of Helicobacter pylori CagA induces gastrointestinal and hematopoietic neoplasms in mouse. Proc Natl Acad Sci U S A 105:1003–1008
Ohshima H, Bartsch H (1994) Chronic infections and inflammatory processes as cancer risk factors: possible role of nitric oxide in carcinogenesis. Mutat Res 305:253–264
Ohshima H, Bandaletova TY, Brouet I, Bartsch H, Kirby G et al (1994) Increased nitrosamine and nitrate biosynthesis mediated by nitric oxide synthase induced in hamsters infected with liver fluke (Opisthorchis viverrini). Carcinogenesis 15:271–275
Oswald E, Nougayrede JP, Taieb F, Sugai M (2005) Bacterial toxins that modulate host cell-cycle progression. Curr Opin Microbiol 8:83–91
Parsonnet J (1995a) Bacterial infection as a cause of cancer. Environ Health Perspect 103(Suppl 8):263–268
Parsonnet J (1995b) The incidence of Helicobacter pylori infection. Aliment Pharmacol Ther 9(Suppl 2):45–51
Patel N, Weiner SD (2002) Osteochondritis dissecans involving the trochlea: report of two patients (three elbows) and review of the literature. J Pediatr Orthop 22:48–51
Patel NM, Weiner SD, Senior M (2002) Squamous cell carcinoma arising from chronic osteomyelitis of the patella. Orthopedics 25:334–336
Patwa LG, Fan TJ, Tchaptchet S, Liu Y, Lussier YA et al (2011) Chronic intestinal inflammation induces stress-response genes in commensal Escherichia coli. Gastroenterology 141(1842–1851):e1841–1810
Perez-Diaz D, Calvo-Serrano M, Martinez-Hijosa E, Fuenmayor-Valera L, Munoz-Jimenez F et al (1995) Squamous cell carcinoma complicating perianal hidradenitis suppurativa. Int J Colorectal Dis 10:225–228
Philip M, Rowley DA, Schreiber H (2004) Inflammation as a tumor promoter in cancer induction. Semin Cancer Biol 14:433–439

Ponzetto A, Pellicano R, Leone N, Berrutti M, Turrini F et al (2000) Helicobacter pylori seroprevalence in cirrhotic patients with hepatitis B virus infection. Neth J Med 56:206–210
Pradere GP, Dapito DH, Schwabe RF (2014) The Yin and Yan of Toll-like receptors in cancer. Oncogen 33(27):3485–3495
Quirk JT, Kupinski JM (2001) Chronic infection, inflammation, and epithelial ovarian cancer. Med Hypotheses 57:426–428
Rancic M, Ristic L, Rancic S, Radovic M, Ciric Z (2014) Pulmonary function parameters as prognostic factors in advanced non-small cell lung cancer. Med Glas (Zenica) 11:58–65
Redecke V, Dalhoff K, Bohnet S, Braun J, Maass M (1998) Interaction of Chlamydia pneumoniae and human alveolar macrophages: infection and inflammatory response. Am J Respir Cell Mol Biol 19:721–727
Rehman A, Sina C, Gavrilova O, Hasler R, Ott S et al (2011) Nod2 is essential for temporal development of intestinal microbial communities. Gut 60:1354–1362
Rippere-Lampe KE, Lang M, Ceri H, Olson M, Lockman HA et al (2001) Cytotoxic necrotizing factor type 1-positive Escherichia coli causes increased inflammation and tissue damage to the prostate in a rat prostatitis model. Infect Immun 69:6515–6519
Robbins S, Chuang VP, Hersh T (1988) The development of hepatobiliary cancer in a carrier of Salmonella typhus. Am J Gastroenterol 83:675–678
Rosenzweig LB, Brett AS, Lefaivre JF, Vandersteenhoven JJ (2005) Hidradenitis suppurativa complicated by squamous cell carcinoma and paraneoplastic neuropathy. Am J Med Sci 329:150–152
Roses DF, Richman H, Localio SA (1974) Bacterial endocarditis associated with colorectal carcinoma. Ann Surg 179:190–191
Rubinstein MR, Wang X, Liu W, Hao Y, Cai G et al (2013) Fusobacterium nucleatum promotes colorectal carcinogenesis by modulating E-cadherin/beta-catenin signaling via its FadA adhesin. Cell Host Microbe 14:195–206
Ryz NR, Patterson SJ, Zhang Y, Ma C, Huang T et al (2012) Active vitamin D (1,25-dihydroxyvitamin D3) increases host susceptibility to Citrobacter rodentium by suppressing mucosal Th17 responses. Am J Physiol Gastrointest Liver Physiol 303:G1299–G1311
Saglik Y, Arikan M, Altay M, Yildiz Y (2001) Squamous cell carcinoma arising in chronic osteomyelitis. Int Orthop 25:389–391
Salaspuro MP (2003) Acetaldehyde, microbes, and cancer of the digestive tract. Crit Rev Clin Lab Sci 40:183–208
Salzman NH, Underwood MA, Bevins CL (2007) Paneth cells, defensins, and the commensal microbiota: a hypothesis on intimate interplay at the intestinal mucosa. Semin Immunol 19:70–83
Samaras V, Rafailidis PI, Mourtzoukou EG, Peppas G, Falagas ME (2010) Chronic bacterial and parasitic infections and cancer: a review. J Infect Dev Ctries 4:267–281
Sasazuki S, Inoue M, Iwasaki M, Otani T, Yamamoto S et al (2006) Effect of Helicobacter pylori infection combined with CagA and pepsinogen status on gastric cancer development among Japanese men and women: a nested case-control study. Cancer Epidemiol Biomarkers Prev 15:1341–1347
Schetter AJ, Heegaard NH, Harris CC (2010) Inflammation and cancer: interweaving microRNA, free radical, cytokine and p53 pathways. Carcinogenesis 31:37–49
Schoppmann SF, Birner P, Stockl J, Kalt R, Ullrich R et al (2002) Tumor-associated macrophages express lymphatic endothelial growth factors and are related to peritumoral lymphangiogenesis. Am J Pathol 161:947–956
Schottenfeld D, Beebe-Dimmer J (2006a) Alleviating the burden of cancer: a perspective on advances, challenges, and future directions. Cancer Epidemiol Biomarkers Prev 15:2049–2055
Schottenfeld D, Beebe-Dimmer J (2006b) Chronic inflammation: a common and important factor in the pathogenesis of neoplasia. CA Cancer J Clin 56:69–83
Schwabe RF, Jobin C (2013) The microbiome and cancer. Nat Rev Cancer 13:800–812

Schwartz DA (1980) Helminths in the induction of cancer: Opisthorchis viverrini, Clonorchis sinensis and cholangiocarcinoma. Trop Geogr Med 32:95–100
Schwartz DA (1984) Carcinoma of the uterine cervix and schistosomiasis in West Africa. Gynecol Oncol 19:365–370
Sebok P, Takacs I, Szabo G, Zeher M, Matolcsy A et al (1997) The presence of t(14;18) chromosome translocation in various types of diseases. Orv Hetil 138:3301–3305
Shacter E, Weitzman SA (2002) Chronic inflammation and cancer. Oncol (Williston Park) 16:217–226, 229; discussion 230–212
Sheng H, Shao J, Washington MK, DuBois RN (2001) Prostaglandin E2 increases growth and motility of colorectal carcinoma cells. J Biol Chem 276:18075–18081
Shin HR, Oh JK, Masuyer E, Curado MP, Bouvard V et al (2010) Epidemiology of cholangiocarcinoma: an update focusing on risk factors. Cancer Sci 101:579–585
Shukla Y, Singh A, Mehrotra NK (2000) Evaluation of carcinogenic and co-carcinogenic potential Quinalphos in mouse skin. Cancer Lett 148:1–7
Sidransky D, Von Eschenbach A, Tsai YC, Jones P, Summerhayes I et al (1991) Identification of p53 gene mutations in bladder cancers and urine samples. Science 252:706–709
Simiantonaki N, Kurzik-Dumke U, Karyofylli G, Jayasinghe C, Kirkpatrick CJ (2007a) Loss of E-cadherin in the vicinity of necrosis in colorectal carcinomas: association with NFkappaB expression. Int J Oncol 31:269–275
Simiantonaki N, Kurzik-Dumke U, Karyofylli G, Jayasinghe C, Michel-Schmidt R et al (2007b) Reduced expression of TLR4 is associated with the metastatic status of human colorectal cancer. Int J Mol Med 20:21–29
Slade DE, Powell BW, Mortimer PS (2003) Hidradenitis suppurativa: pathogenesis and management. Br J Plast Surg 56:451–461
Snyder DL, Pollard M, Wostmann BS, Luckert P (1990) Life span, morphology, and pathology of diet-restricted germ-free and conventional Lobund-Wistar rats. J Gerontol 45:B52–B58
Song LY, Yan WS, Zhao T (2002) Detection of Mycobacterium tuberculosis in lung cancer tissue by indirect in situ nested PCR. Di Yi Jun Yi Da Xue Xue Bao 22:992–993
Sonin AH, Resnik CS, Mulligan ME, Murphey MD (1998) General case of the day. Squamous cell carcinoma arising in a chronic draining sinus tract as a complication of chronic osteomyelitis. Radiographics 18:530–532
Sripa B, Kaewkes S, Sithithaworn P, Mairiang E, Laha T et al (2007) Liver fluke induces cholangiocarcinoma. PLoS Med 4:e201
Steinitz R (1965) Pulmonary tuberculosis and carcinoma of the lung. A survey from two population-based disease registers. Am Rev Respir Dis 92:758–766
Stenvinkel P, Karimi M, Johansson S, Axelsson J, Suliman M et al (2007) Impact of inflammation on epigenetic DNA methylation – a novel risk factor for cardiovascular disease? J Intern Med 261:488–499
Stutes M, Tran S, DeMorrow S (2007) Genetic and epigenetic changes associated with cholangiocarcinoma: from DNA methylation to microRNAs. World J Gastroenterol 13:6465–6469
Szymankiewicz M, Kowalewski J, Dancewicz M (2006) Bacteriological and mycological analysis of material taken from lower respiratory tract in patients with malignancy. Pol Merkur Lekarski 21:218–222
Tamburini J, Grimaldi D, Chiche JD, Bricaire F, Bossi P (2007) Cytokine pattern in Kaposi's sarcoma associated with immune restoration disease in HIV and tuberculosis co-infected patients. AIDS 21:1980–1983
Tandon R, Misra P, Soni VK, Bano N, Misra-Bhattacharya S et al (2013) Unresponsiveness of Mycobacterium w vaccine in managing acute and chronic Leishmania donovani infections in mouse and hamster. Parasitology 140:435–444
Thomas W, Ascott ZK, Harmey D, Slice LW, Rozengurt E et al (2001) Cytotoxic necrotizing factor from Escherichia coli induces RhoA-dependent expression of the cyclooxygenase-2 Gene. Infect Immun 69:6839–6845

Travaglione S, Fabbri A, Fiorentini C (2008) The Rho-activating CNF1 toxin from pathogenic E. coli: a risk factor for human cancer development? Infect Agent Cancer 3:4
Trent JT, Kirsner RS (2003) Wounds and malignancy. Adv Skin Wound Care 16:31–34
Tsai S, Wear DJ, Shih JW, Lo SC (1995) Mycoplasmas and oncogenesis: persistent infection and multistage malignant transformation. Proc Natl Acad Sci U S A 92:10197–10201
Tu S, Bhagat G, Cui G, Takaishi S, Kurt-Jones EA et al (2008) Overexpression of interleukin-1beta induces gastric inflammation and cancer and mobilizes myeloid-derived suppressor cells in mice. Cancer Cell 14:408–419
Tu SP, Quante M, Bhagat G, Takaishi S, Cui G et al (2011) IFN-gamma inhibits gastric carcinogenesis by inducing epithelial cell autophagy and T-cell apoptosis. Cancer Res 71:4247–4259
Tye H, Kennedy CL, Najdovska M, McLeod L, McCormack W et al (2012) STAT3-driven upregulation of TLR2 promotes gastric tumorigenesis independent of tumor inflammation. Cancer Cell 22:466–478
Vakevainen S, Mentula S, Nuutinen H, Salmela KS, Jousimies-Somer H et al (2002) Ethanol-derived microbial production of carcinogenic acetaldehyde in achlorhydric atrophic gastritis. Scand J Gastroenterol 37:648–655
Walter R, Bachmann SP, Schaffner A, Ruegg R, Schoedon G (2001) Bone marrow involvement in Whipple's disease: rarely reported, but really rare? Br J Haematol 112:677–679
Wang A, Xie J (1998) Infection of mycobacterium tuberculosis in lung cancer. Zhongguo Fei Ai Za Zhi 1:92–94
Wang S, Ernst LM, Smith BR, Tallini G, Howe JG et al (2003) Systemic Tropheryma whippleii infection associated with monoclonal B-cell proliferation: a Helicobacter pylori-type pathogenesis? Arch Pathol Lab Med 127:1619–1622
Wang J, Budamagunta MS, Voss JC, Kurth MJ, Lam KS et al (2013) Antimitochondrial antibody recognition and structural integrity of the inner lipoyl domain of the E2 subunit of pyruvate dehydrogenase complex. J Immunol 191:2126–2133
Watanapa P, Watanapa WB (2002) Liver fluke-associated cholangiocarcinoma. Br J Surg 89:962–970
Weitao T (2009a) Bacteria form biofilms against cancer metastasis. Med Hypothesis 72:477–478
Weitao T (2009b) Multicellularity of a unicellular organism in response to DNA replication stress. Res Microbiol 160:87–88
Wistuba II, Gazdar AF (2004) Gallbladder cancer: lessons from a rare tumour. Nat Rev Cancer 4:695–706
Wu S, Morin PJ, Maouyo D, Sears CL (2003) Bacteroides fragilis enterotoxin induces c-Myc expression and cellular proliferation. Gastroenterology 124:392–400
Xia L, Ahmed I, Tawfik O, Anant S, Chandrakesan P (2012) Critical role of notch and Wnt/?-cathenin pathways in the regulation of hyperplasia and/or colitis in response to bacterial infection. Infect Immun 80:3107–3121
Yan HH, Pickup M, Pang Y, Gorska AE, Li Z et al (2010) Gr-1+CD11b+ myeloid cells tip the balance of immune protection to tumor promotion in the premetastatic lung. Cancer Res 70:6139–6149
Yan L, Cai Q, Xu Y (2013) The ubiquitin-CXCR4 axis plays an important role in acute lung infection-enhanced lung tumor metastasis. Clin Cancer Res 19:4706–4716
Yu CC, Cook MG (1990) Hidradenitis suppurativa: a disease of follicular epithelium, rather than apocrine glands. Br J Dermatol 122:763–769
Ziets RJ, Evanski PM, Lusskin R, Lee M (1991) Squamous cell carcinoma complicating chronic osteomyelitis in a toe: a case report and review of the literature. Foot Ankle 12:178–181

Chapter 5
Human Protozoal Infections and Their Potential for Causing Neoplasms

Raavi Gupta, Maja Nowakowski, and M.A. Haseeb

Abstract Protists causing human infections generally produce local or systemic disease through direct cellular injury, and inflammatory response. Some protists have been found associated with human neoplasms, and their carcinogenic potential has received much attention in recent years. Here we outline the epidemiologic and experimental evidence linking *Cryptosporidium* sp., malaria and *Trichomonas vaginalis* to neoplastic changes in humans. Experimental studies in mammalian cells have unraveled the disruptive alterations in many of the normal signaling pathways that are critical in innate and adaptive immunity. The immune deficient states, induced by protists or other concurrent infections, most likely increase susceptibility to infection and contribute to tumorigenesis. The neoplasms that occur in malaria and trichomoniasis often contain Epstein-Barr virus (EBV) and human papilloma virus (HPV), respectively. The relative contribution of protists and different viruses, including HIV, requires delineation. The role of local microbiota in determining susceptibility to *Cryptosporidium* sp. or *T. vaginalis* infection and in carcinogenesis also requires additional investigation. The elucidation of precise mechanisms of tumorigenesis in mono- and polymicrobial infections is expected to identify targets for intervention and treatment.

R. Gupta
Department of Pathology, State University of New York, Downstate Medical Center, Brooklyn, NY, USA

M. Nowakowski
Department of Pathology, State University of New York, Downstate Medical Center, Brooklyn, NY, USA

Department of Medicine, State University of New York, Downstate Medical Center, Brooklyn, NY, USA

M.A. Haseeb (✉)
Department of Pathology, State University of New York, Downstate Medical Center, Brooklyn, NY, USA

Department of Medicine, State University of New York, Downstate Medical Center, Brooklyn, NY, USA

Department of Cell Biology, State University of New York, Downstate Medical Center, MSC 44, 450 Clarkson Avenue, Brooklyn 11203, NY, USA
e-mail: haseeb.siddiqi@downstate.edu

M.R. Shurin et al. (eds.), *Infection and Cancer: Bi-Directorial Interactions*, DOI 10.1007/978-3-319-20669-1_5

Keywords Protists • Carcinogenesis • *Cryptosporidium* • Malaria • *Trichomonas* sp • Proto-oncogene • Cyclooxygenase-2

Abbreviations

BL	Burkitt's lymphoma
COX-2	Cyclooxygenase-2
EBV	Epstein-Barr virus
NF-κB	nuclear factor κB
SCID	severe combined immunodeficiency

Introduction

Host-parasite co-evolution has not rendered all parasites altogether innocuous. The parasites across all taxonomic groups have retained or evolved mechanisms that facilitate their survival in adverse host microenvironments. Thus, it is not surprising to find dysequilibrium in a given host-parasite interaction with contribution from both the parasite factors and the host defense mechanisms. The spectrum of outcomes of such relationships ranges from inapparent to fatal. Detection of any effect of an inapparent infection on the host depends largely on our ability to visualize or measure any change. The longitudinal relationship may also determine the outcome: the harm may occur or become apparent only after the host microenvironment has been sufficiently modified. HIV and other infections causing immunosuppression, and increasing use of immunomodulators in organ transplantation and treatment of autoimmune and chronic diseases, are rendering humans susceptible to opportunistic infections. In addition, self-limiting infections may be running a chronic course with increased likelihood of neoplastic changes to take place. Induction of neoplasms by protists, by themselves or in concert with other pathogens, should be considered part of their pathogenesis.

Despite enormous strides, as evidenced by the sizable literature, there are still more unanswered questions than answers. The preponderance of evidence, despite its origins in a handful of laboratories, supports the potential of *Cryptosporidium* development forms to induce neoplastic changes in gastrointestinal epithelial cells of the host. However, these studies do not address the role of gut microbiome in carcinogenesis as has been done in some studies on trichomoniasis (see below). Whether a related apicomplexan, *Toxoplasma gondii*, which undergoes a similar developmental cycle in feline enterocytes, causes host cell transformation is not known. Neoplastic changes in biliary and respiratory epithelial cells infected with *Cryptosporidium parvum* have not been investigated. Are neoplastic changes restricted to the infected cells or include the adjacent cells or additional cell types that also undergo such changes?

The role of protists in carcinogenesis has been ably reviewed in recent years (Blaser 2008; Benamrouz et al. 2012; Kutikhin et al. 2013). Here we bring that information up-do-date and document some of the evidence from experimental systems, in addition to epidemiologic associations, that may potentially help to devise screening methods for detection of early changes in animals or humans. The criteria for the etiologic association of protists with carcinogenesis are not defined, however, of the three infections discussed in this chapter, falciparum malaria in holoendemic areas is classified as group 2A carcinogen (IARC 2014).

Cryptosporidium sp.

Cryptosporidium sp. is an apicomplexan protist that causes acute gastroenteritis and diarrhea in humans and a wide range of other vertebrates worldwide. The infection is acquired by ingestion of oöcysts (4–6 μm) and the entire life cycle is completed in mucosal epithelial cells in a single host. The organism is an obligate intracellular parasite that resides in an apical parasitophorous vacuole which obtains nourishment from the host cell via a feeder organelle (Valigurová et al. 2008). The intracellular localization was initially described as intracellular-extracytoplasmic (Goebel and Braendler 1982), and subsequently as epicellular (Valigurová et al. 2008) and extra-cytoplasmic (Chalmers and Davis 2010). Most *Cryptosporidium* sp. infect the gut epithelium but dissemination can occur to extra-intestinal sites in severe infections (Lopez-Velez et al. 1995). It is most often found in the small intestine; however, it has been demonstrated in colonic epithelia of patients with AIDS (Orenstein and Dieterich 2001).

In the early 1900s, the organism was found causing diarrhea in avian and mammalian hosts, including livestock. Its clinical significance was recognized in recent decades when it was identified as the cause of life-threatening protracted diarrhea in patients with immune deficiencies and those with AIDS (Meisel et al. 1976; Lasser et al. 1979; Weisburger et al. 1979; Current et al. 1983). Though most human infections are probably caused by *C. parvum* and *C. hominis*, the genus contains many species that have been described on the basis of morphology of developmental forms, biological features (including host specificity) and molecular characterization (Ramirez et al. 2004; Snelling et al. 2007). Genome sequences of the two species have been elucidated (Abrahamsen et al. 2004; Xu et al. 2004), contributing greatly to our understanding of the pathogenesis of cryptosporidiosis.

Possible association of cryptosporidial infection with neoplasia, especially of the gastrointestinal tract, has been suspected for some time (Izquierdo et al. 1988). High proportions (18 %) of patients with colorectal cancer in Poland were found to have cryptosporidiosis (Sulzyc-Bielicka et al. 2007, 2012). This prevalence of cryptosporidiosis was higher than that in the general European population (Semenza and Nichols 2007). The incidence of colorectal cancer in HIV-infected patients, who are known to be highly susceptible to *Cryptosporidium* infection, was 2.3-fold higher than that in the United States general population (Patel et al. 2008). The risk of colorectal cancer in AIDS patients with cryptosporidiosis was found to be higher than in those without it. These tumors were of uncommon types (small cell carcinoma, leiomyosarcoma, carcinoma NOS) in contrast to colonic adenocarcinomas most often found in the general population and in those with AIDS (Shebl et al. 2012).

Tumorigenesis of *C. parvum* in the gastrointestinal tracts of severe combined immunodeficient (SCID) mice has been extensively studied (Certad et al. 2007, 2010a, b, 2012). SCID mice infected with *C. parvum* and treated with oral dexamethasone developed adenomas with low or high grade dysplasia with progression to invasive carcinoma (Certad et al. 2007). Ki-67 positivity of the colonic mucosa supported the neoplastic nature of epithelial transformation and suggested that potential neoplastic alterations begin before the histopathological lesions are observed (Certad et al. 2010b). Another strain of *C. parvum* (TUM1), that causes a more fulminant cryptosporidiosis, was also demonstrated to cause intramuscosal adenocarcinoma in the dexamethasone-treated SCID mice. Higher burden of parasite involvement was associated with an earlier onset and rapid evolution of neoplastic lesion (Certad et al. 2010a, b). Abnormal upregulation of cyclin D1 in low and high grade ileaocecal dysplasia in dexamethasone-treated mice infected with *C. parvum* was an early event in intestinal carcinogenesis. In addition, extraintestinal dissemination of the organisms in these mice led to large cell dysplasia in the liver (Abdou et al. 2013) as previously demonstrated in murine bile ducts (Certad et al. 2010a).

Dysplastic changes have also been reported in bile ducts in a model of IFN-γ knockout mice infected with *C. parvum* (Stephens et al. 1999). These changes may reflect the initial steps toward development of neoplastic changes preceded by chronic infection and inflammation and progressing to malignant transformation in immunosuppressed hosts.

Inhibition of apoptosis in host cells has been observed in this infection. The cells infected with *Crytptosporidum* have evolved mechanisms to keep the host cells in a survival mode, by acquiring resistance to various chemical agents that trigger apoptosis (Chen et al. 2001; Liu et al. 2008). Apoptosis prevention benefits the parasite by stabilizing the host to permit the completion of its life cycle (Heussler et al. 2001). One of the mechanisms for the downregulation of apoptosis in the infected cells is the activation of nuclear factor κB (NF-κB) pathway (Chen et al. 2001). These are a family of transcription factors that regulate the activation of a wide variety of genes that respond to immune or inflammatory signals via a number of intracellular survival signals such as the c-Myc protooncogene, inhibitor of apoptosis and Bcl-xl (Chen et al. 2001). Activation of NF-κB pathway has been observed in many cancers including colon carcinoma (Naugler and Karin 2008); thus, resistance to apoptosis could be an essential step in progression to malignancy (Lowe and Lin 2000). The neighboring cells undergo apoptosis by Fas/Fas ligand-dependent apoptotic mechanisms (Chen et al. 1999). Microarray analysis in an *in vitro* model using *C. parvum* infected human ileocecal HCT8 cells revealed a genome wide alteration of apoptotic genes (Liu et al. 2009) in a biphasic pattern with an early anti-apoptotic stage and late pro-apoptotic stage in infection. Apoptosis may also be enhanced during co-infection with human immunodeficiency virus type 1 which releases a peptide HIV-1 tat from infected T cells and macrophages. Tat inhibits cholangiocyte TLR4 protein expression through translational inhibition thus diminishing the ability of cholangiocytes to initiate an innate immune response to *C. parvum* (O'Hara et al. 2009). It is noteworthy that apoptosis is also inhibited by other apicomplexans through multiple mechanisms: in murine hepatocytes infected with *Plasmodium berghei*, and in several mammalian cell types infected with *Toxoplasma gondii* (Nash et al. 1998; Sturm et al. 2013; Cowman and Kappe 2013).

Another possible pathway towards progression of carcinogenesis in *Cryptosporidium*-infected SCID rodents was shown to be via the Wnt signaling pathway. Decrease in cytoplasmic APC was recorded in addition to abnormal juxtamembranous localization of β-catenin; both of these processes are considered early events in the development of colorectal neoplasia (Takahashi et al. 2000). Abnormalities in the E-cadherin – β-catenin complex have been identified which could result in reduced cell-cell adhesion and conversion to a migratory phenotype (Benamrouz et al. 2014).

Cryptosporidium infection causes a direct cytoskeletal modification of the host cell during sporozoite attachment and invasion of epithelial cells. Attachment of the sporozoite on the apical surface of the epithelial cells induces reorganization of the host-cell membrane around the sporozoite to form a vacuole in which the organism remains intracellular but extra-cytoplasmic. It induces host cell cytoskeletal changes by modulating a localized actin reorganization and channel/transport insertion, causing whole cell and perhaps, tissue-level changes in the cytoskeletal architecture via phosphatidylinositol 3-kinase pathway and other pathways (O'Hara and Chen 2011). Thus, *C. parvum* is able to modulate host-cell cytoskeleton activities and several host-cell biological processes.

The specific mechanisms of cell transformation induced by *Cryptosporidium* remain poorly understood. It is likely that a combination of several pathways is needed to transform infected cells; exploration of other signaling pathways will be useful to elucidate and define a clearer route to neoplasia (Benamrouz et al. 2014).

Malaria

Malaria is a mosquito-borne disease caused by five species of the protozoal genus *Plasmodium*. Following inoculation of the infective form (sporozoite) by an infected anopheline mosquito, the organism undergoes asexual multiplication first in hepatocytes and then in RBCs. Successive developmental cycles in RBCs, involving infection and rupture of RBCs, result in malaise, fever, and other nonspecific symptoms often accompanied by pronounced anemia. Severe malaria may lead to multi-organ dysfunction, coma and death. Approximately 3.4 billion people are at risk of infection in 97 countries (WHO 2013). In 2012, there were an estimated 207 million cases of malaria and 627,000 deaths, including 482,000 children younger than 5 years of age; 90 % of all malaria deaths occur in sub-Saharan Africa.

Burkitt's lymphoma (BL), a high-grade Non-Hodgkin's lymphoma, is endemic in the "lymphoma belt" of equatorial Africa where Burkitt's lymphoma comprises about half of all childhood cancers and up to 90 % of lymphoma diagnoses. The annual incidence of BL has been estimated at 40–50 per million children younger than 18 years, peaking at age 6 years; the disease is twice as common in boys as in girls (Orem et al. 2007; Molyneux et al. 2012). The geographic distribution of endemic BL across Africa and Papua New Guinea overlaps those of holoendemic malaria and early acquisition of Epstein-Barr virus (EBV).

Since the first description of BL (Burkitt 1958), malaria has been implicated in its etiology early and often (Burkitt and Wright 1966; Burkitt 1969; Kafuko 1969). Serologic, molecular, and epidemiologic evidence for the causal role of the other important etiologic agent in BL, EBV, is well established (Epstein et al. 1964; Henle et al. 1969; Zur Hausen et al. 1970; Henle and Henle 1974; de The et al. 1978; de The 1979). These associations were widely accepted for the African Burkitt's lymphoma, but the discovery of cases of American BL in the absence of EBV and malarial infection were interpreted to mean that these infections are neither necessary nor sufficient causes of all BL tumors (Evans 1985). Still, a review of more than 25 years of research on the etiology of BL identified three factors operating in a stepwise fashion: EBV infection in early childhood, malaria and its effect on polyclonal proliferation of EBV-infected B lymphocytes, and chromosomal translocations activating the c-Myc oncogene in these B cells (de The 1993). The translocation in B cells moves Myc on chromosome 8 into the vicinity of promoter elements of immunoglobulin genes on chromosome 14, 2, or 22 (Manolov and Manolova 1972; Dalla-Favera et al. 1982). Thus, BL is often described as a model disease to understand the polymicrobial and genetic basis of cancer. More recently, the general understanding of BL development has included early infection with EBV and malaria acting in concert to increase the absolute number (load) of translocation positive B cells, which would increase the number of cells capable of progressing to BL. The second element and rate-limiting step would be the enhanced survival of the translocation-positive B cells due to circumvention of apoptosis feedback loops mediated by cytokines (Mbulaiteye 2013).

The incidence of BL is higher among HIV-infected immunosuppressed patients in non-endemic areas (Molyneux et al. 2012). In a recent study of BL among children (<15 years) in Malawi, it was found that BL cases were more likely than controls to be HIV-positive (Odds Ratio (OR))=12.4, 95 % Confidence Interval (CI) 1.3–116.2, p=0.03). ORs for BL increased with increasing antibody titers against EBV (p=0.001) and malaria (p=0.01). Among HIV-negative participants, cases were thirteen times more likely than controls to have increased levels of both EBV and malaria antibodies (OR=13.2; 95 % CI 3.8–46.6; p=0.001). Reported use of mosquito nets was associated with a lower risk of BL (OR=0.2, 95 % CI, 0.03–0.9, p=0.04) (Mutalima et al. 2008). Other studies, however, did not find such clear associations between HIV infection and endemic BL (Newton et al. 2001).

The contribution of malaria to increased risk of developing BL is not entirely without controversy: a recent study of children in The Gambia concluded that a single, primary *Plasmodium falciparum* infection is not sufficient to impair immunological control of EBV infection. The authors stated that their data support the idea that chronic repeated exposure to malaria throughout a child's early years is required to alter the EBV-host balance, leading to high EBV loads and a greater likelihood of further increases linked to acute malarial episodes and possibly BL (Jayasooriya et al. 2012).

The precise molecular mechanisms responsible for the effects of EBV, malaria, and HIV on development of BL still remain incompletely understood. However, it is known since the 1970s that malaria causes polyclonal B cell activation (Greenwood 1974), leading to a higher frequency of EBV-infected B cells and a higher EBV load (Lam et al. 1991; Moormann et al. 2005). Alternative mechanisms include an immunosuppressive effect of malaria on EBV-specific immune response, presumably via reduced T cell activity, resulting in higher EBV loads as observed in transplant patients and subjects with HIV infection (Moormann et al. 2005). Additional studies have provided further support for malaria causing altered immunity to EBV, specifically diminishing T cell immunity to EBV lytic but not latent antigens (Snider et al. 2012).

Additional factors contributing to the lymphomagenesis in BL have recently been recognized. A plant, *Euphorbia tirucalli*, with a geographical distribution overlapping that of endemic BL is used as a hedge, herbal remedy and toy in the endemic regions in Africa. It has been demonstrated that exposure to extracts of this plant reactivated EBV from its latent phase as measured by the expression of the EBV Zebra antigen, augmented the expression of EBV early antigens and LMP1, EBNA1, and EBNA2, and was associated with polysomies involving chromosome 8 (Mannucci et al. 2012). The association between *E. tirucalli* and BL is supported by the observation that the incidence of BL has fallen in northern Zambia following the eradication of thickets of *E. tirucalli* (Osato 1998).

It has been clearly demonstrated that endemic BL arises in germinal centers when deregulated expression of activation-induced cytidine deaminase causes a c-Myc chromosomal translocation in a cell that is latently infected with EBV (Klein and Dalla-Favera 2008). Still, the mechanism of malaria contribution to the tumorigenic process leading to BL development remains elusive. Recent studies have proposed that *P. falciparum* targets B cells via multiple pathways to increase the risk of BL. Specifically, *P. falciparum* causes deregulated expression of activation-induced cytidine deaminase, thus increasing the chance of c-Myc translocation. Further, *P. falciparum* increases the number of B cells trafficking through the germinal centers, and increases the frequency of the EBV-infected B cells that are protected from c-Myc induced apoptosis (Torgbor et al. 2014). These new findings provide a better understanding of the interactions between *P. falciparum*, EBV, and the tumorigenic process leading to development of BL, and define possible targets for intervention. Given the extremely heavy disease burden of holoendemic malaria and the prevalence of early EBV, efforts to reduce the incidence of BL must include indirect means of reducing all contributing factors.

Trichomonas vaginalis

Trichomonas vaginalis causes the most prevalent curable sexually transmitted infection globally. It is a flagellate protozoan that exists only as a trophozoite (7–23 μm) and lives in the human vagina and prostate gland. Coitus is the primary mode of transmission. Symptoms may develop in 5–20 days, however, untreated infections may last months to years (Bachmann et al. 2011).

The World Health Organization estimates that globally there are ≈248 million new infections/year of which approximately half are in men (WHO 2011). In the United States *T. vaginalis* infects an estimated 3.7 million people which is more than chlamydial and gonococcal infections combined (Satterwhite et al. 2013), and more women are believed to be infected than men (Miller et al. 2005).

In females, the infection may cause vaginitis, vulvitis, uretheritis, and possibly postpartum endometritis. In men, the infection may be asymptomatic or cause prostatitis and recurrent uretheritis. Most infections, however, are believed to be asymptomatic and recent studies have linked the asymptomatic infections to other health problems. Patients with *T. vaginalis* infections are at an increased risk of HIV acquisition and transmission (Hughes et al. 2012), acquisition of other sexually transmitted infections (Allsworth et al. 2009), preterm labor (Cotch et al. 1997) and pelvic inflammatory disease (Moodley et al. 2002). Clinically apparent infections present as local itching, irritation, erythema, dysuria and vaginal or urethral discharge.

Although culture is the gold standard, laboratory diagnosis is most often made by microscopic examination of vaginal or urethral discharge, urine sediment and prostatic fluid to demonstrate *T. vaginalis*. Other tests including dipstick and nucleic acid amplification assays are also available; when used more widely, the latter may bring to light many more infections that remain otherwise undetected (Muzny et al. 2014).

In recent years human papilloma virus (HPV) has been recognized as the most important etiologic agent for cervical dysplasia and cancer. After controlling for HPV infection, a significant association between *T. vaginalis* infection and cervical neoplasia has been reported in 2–5 % of the patients, and the presence of *T. vaginalis* doubles the risk of cervical neoplasia (Gram et al. 1992; Zhang and Begg 1994; Zhang et al. 1995). In addition, the patients with cervical carcinoma had a threefold increase in the prevalence of *T. vaginalis* antibodies than in age-matched controls (Sayed el-Ahl et al. 2002).

T. vaginalis has been found in the prostatic fluid from asymptomatic men suggesting that the prostate gland might serve as a reservoir for trichomoniasis in men (Mitteregger et al. 2012). This organism has been demonstrated in the prostatic urethra, submucosa, glandular lumina and stroma (Gardner et al. 1986), and in benign prostatic hyperplasia (Mitteregger et al. 2012).

There is a positive correlation between the presence of antibodies against *T. vaginalis* alpha-actinic protein and prostate carcinoma risk (Sutcliffe et al. 2006). Another investigation also found a positive association of *T. vaginalis* infection with extra-prostatic and fatal prostate cancer in one of the studied populations (Stark et al. 2009a). Additional epidemiologic evidence suggests association of *T. vaginalis* infection with a more aggressive prostate carcinoma and death in African American patients (Siegel et al. 2012).

Despite the observed relationship between *T. vaginalis* infection and cervical and prostate carcinoma, only a few studies have evaluated the molecular mechanisms of carcinogenesis. In addition to direct carcinogenesis, the epidemiological association of HIV with *T. vaginalis* may suggest a synergistic effect of the two infections in causing cervical cancers.

Inflammation at the site of infection is believed to be important for carcinogenesis. *T. vaginalis* lipoglycans (*Tv*LGs) mediate the parasite adherence to the host epithelial cells by binding to galectin-1 which is the only identified human receptor for *T. vaginalis*. *Tv*LGs, the most abundant surface molecules of the parasite, modulate inflammatory cellular responses (Ryan et al. 2011). Parasite adherence to vaginal epithelial cells induces expression of monocyte chemoattractant protein-1 and IL-8, the pro-inflammatory cytokines involved in neutrophil recruitment (Kucknoor et al. 2007). Vaginal secretions containing this organism have also been shown to be high in IL-8, leukotriene B4 and neutrophils (Ryu et al. 2004). Neutrophils may contribute to carcinogenesis by secreting a variety of oxygen- and nitrogen-based reactive molecules capable of damaging DNA and nearby cells (Dhanasekeran et al. 2001). Other studies have demonstrated that *T. vaginalis* attachment to vaginal and prostate epithelial cells leads to elevated levels of IL-6, a key mediator of acute inflammatory responses (Han et al. 2012; Twu et al. 2013). In addition, IL-6 down-regulates IL-8 response which is involved in the recruitment of neutrophils to the site of infection and persists in its active form within the immediate environment far longer than other chemoattractants (Shaio et al. 1994; Fichorova et al. 2006). Thus, reduced IL-8 secretion favors chronic infection. Furthermore, IL-6 has been associated with increased prostate cancer incidence, progression and mortality among healthy men (Stark et al. 2009b), and with its worse outcomes (Azevedo et al. 2011). Thus, regulation of IL-6 and IL-8 secretion determines successful colonization of urogenital tracts with *T. vaginalis*, chronicity of this infection, and its carcinogenesis.

Many proto-oncogenes have been found to be up-regulated in both trichomonas infection and in carcinogenesis. One of these, *PIM-1*, is a proto-oncogene of a family of serine/threonine kinases. Its overexpression leads to genomic instability and preservation of potentially cancer-producing genomic alterations by promoting cell survival (Roh et al. 2008; Magnuson et al. 2010). Gene expression studies have demonstrated an altered expression of *PIM-1* in malignant prostate cancer (Valdman et al. 2004). Overexpression of *PIM-1*, most likely induced by IL-6 via JAK/STAT pathway, has been demonstrated in prostate epithelial cells (Sutcliffe et al. 2012). Thus, *PIM-1* expression provides a link between induction of IL-6 secretion in *T. vaginalis* infection and prostate carcinogenesis (Sansone and Bromberg 2012).

Another proto-oncogene, HMGA1, which encodes a chromatin "architectural transcription factor", acts downstream of PIM-1 in an HMGA1-mediated prostate cancer induction pathway (Reeves and Beckerbauer 2001; Sutcliffe et al. 2012). PIM-1/PIM-2 is a synergistic partner with c-Myc and stabilizes it in tumorigenesis

(Zhang et al. 2008). Both c-Myc and HMGA1 have been found to be overexpressed in prostate cancer. Thus, PIM-1/c-Myc/HMGA1 signaling cascade plays an important role in prostate carcinogenesis. Upregulation of HMGA1 has been demonstrated in prostate epithelial cells in the presence of *T. vaginalis* infection most likely through IL-6 secretion, providing further mechanism by which *T. vaginalis* may contribute to prostate carcinogenesis. HMGA1 induces prostate cell chromosomal instability and rearrangement, and its high levels have been found in rapidly proliferating prostate cancer cells and metastasis (Sutcliffe et al. 2012).

Cyclooxygenase-2 (COX-2) is a component of the cellular response to inflammation and is induced by several extra- and intracellular stimuli, including proinflammatory cytokines, infectious agents, mitogens, hormones and growth factors (Dubois et al. 1998; Allaj et al. 2013). Overexpression of COX-2 has been described in many cancers including colon, stomach, breast, lung, urinary bladder and prostate. It plays a role in prostate cancer initiation and progression by affecting cell proliferation, mitosis, cell adhesion, apoptosis and immune surveillance and angiogenesis. *T. vaginalis* has been shown to induce the expression of COX-2 in primary human vaginal epithelial cells (Kucknoor et al. 2007; Sutcliffe et al. 2012). In addition, HMGA1 also up-regulates expression of COX-2 (Tesfaye et al. 2007). *T. vaginalis* secretes large amounts of polyamines which regulate COX-2 levels and participate in altering cell cycle regulation resulting in a proliferative phenotype (Garcia et al. 2005; Cowan et al. 2006).

T. vaginalis inhabits in the genitourinary microenvironment with bacteria, fungi and viruses. It is actively phagocytic and ingests various microbiota and human cells to induce dysbiosis and contributes to increased host susceptibility to viruses, specifically HIV (Brotman et al. 2012; Clemente et al. 2012). In the presence of *T. vaginalis*, vaginal microbiota exhibits low abundance of lactobacilli and higher proportions of *Mycoplasma* and *Prevotella* and other bacteria typically observed in bacterial vaginosis (Brotman et al. 2012). Co-infection with *Mycoplasma*, carried by *T. vaginalis* contributes to reduced nitric oxide production by the macrophages through depletion of arginine in the vagina, and thus potentially interfering with an important host defense mechanism (Morada et al. 2010). Another potential consequence of *T. vaginalis* actively feeding on host microbiota is the acquisition of a number of bacterial genes which encode for enzymes capable of degrading glycans. These *T. vaginalis*-secreted enzymes of bacterial origin degrade mucins and other human glycans which represent the initial barrier to organisms for the colonization of host epithelial cells (Rughooputh and Greenwell 2005; Alsmark et al. 2013).

T. vaginalis infection causes tissue damage and inflammation which is thought to facilitate HIV entry and in *T. vaginalis* – HIV dually infected patients to stimulate HIV production (Hobbs et al. 2008). In addition, *T. vaginalis* can also internalize human viruses through endocytosis which are either released upon parasite death or secreted through the recycling route of the endocytic pathway (Hirt et al. 2011; Hirt 2013). Patients with *T. vaginalis* are 6.5 times more likely to be infected with HPV16 than those without it (Lazenby et al. 2014). HIV is implicated in inhibiting p53 tumor suppressor gene, altering cell cycle regulation, and activating proto-oncogenes that lead to cellular transformation (Amini et al. 2004). In addition, it causes profound immunosuppression, rendering the host susceptible to other viral infections likes HPV and EBV which induce neoplasia.

Conclusions

We have discussed the associations of three of the protists with human neoplasms, and have summarized key findings from the epidemiologic and experimental work. Some of these findings offer new avenues toward understanding the mechanisms of carcinogenesis in these infections. Certain tumors arise when multiple infectious agents are involved but relative contribution of each need to be ascertained with greater accuracy. In experimental models, some degree of immunosuppression is required for neoplastic changes to occur. With increasing understanding of the association or the role of protists in carcinogenesis comes the task of revising guidelines for early diagnosis of infections associated with carcinogenesis and treatment of resultant tumors as has already happened in the case of *Helicobacter pylori*, for example. Much additional work is required to document associations with greater certainty since association alone does not prove causality. A better understanding of the pathogenic mechanisms may also prove useful to devise effective means to interrupt carcinogenesis. Host genetics would also need to be examined in parallel with other investigations since not all infected individuals develop tumors.

References

Introduction

Benamrouz S, Conseil V, Creusy C et al (2012) Parasites and malignancies, a review, with emphasis on digestive cancer induced by *Cryptosporidium parvum* (Alveolata: Apicomplexa). Parasite 19:101–115

Blaser MJ (2008) Understanding microbe-induced cancers. Cancer Prev Res (Phila) 1:15–20

IARC (2014) International agency for research on cancer. Agents classified by the IARC monographs, vol 1–110. World Health Organization, Geneva

Kutikhin AG, Yuzhalin AE, Brusina EB (2013) The role of protozoa in cancer development. In: Kutikhin AG et al (eds) Infectious agents and cancer. Springer, Berlin

Cryptosporidium sp.

Abdou AG, Harba NM, Afifi AF, Elnaidany NF (2013) Assessment of *Cryptosporidium parvum* infection in immunocompetent and immunocompromised mice and its role in triggering intestinal dysplasia. Int J Infect Dis 17:e593–e600

Abrahamsen MS, Templeton TJ, Enomoto S et al (2004) Complete genome sequence of the apicomplexan, *Cryptosporidium parvum*. Science 304:441–445

Benamrouz S, Conseil V, Chabé M et al (2014) *Cryptosporidium parvum*-induced ileo-caecal adenocarcinoma and Wnt signaling in a mouse model. Dis Model Mech 7:693–700

Certad G, Ngouanesavanh T, Guyot K et al (2007) *Cryptosporidium parvum*, a potential cause of colic adenocarcinoma. Infect Agent Cancer 2:22

Certad G, Creusy C, Guyot K et al (2010a) Fulminant cryptosporidiosis associated with digestive adenocarcinoma in SCID mice infected with *Cryptosporidium parvum* TUM1 strain. Int J Parasitol 40:1469–1475

Certad G, Creusy C, Ngouanesavanh T et al (2010b) Development of *Cryptosporidium parvum-induced* gastrointestinal neoplasia in severe combined immunodeficiency (SCID) mice: severity of lesions is correlated with infection intensity. Am J Trop Med Hyg 82:257–265

Certad G, Benamrouz S, Guyot K et al (2012) Fulminant cryptosporidiosis after near-drowning: a human *Cryptosporidium parvum* strain implicated in invasive gastrointestinal adenocarcinoma and cholangiocarcinoma in an experimental model. Appl Environ Microbiol 78:1746–1751

Chalmers RM, Davis AP (2010) Minireview: clinical cryptosporidiosis. Exp Parasitol 124:138–146

Chen X-M, Levine SA, Tietz P et al (1998) *Cryptosporidium parvum* is cytopathic for cultured human biliary epithelia via an apoptotic mechanism. Hepatology 28:906–913

Chen X-M, Gores GJ, Paya CV, LaRusso NF (1999) *Cryptosporidium parvum* induces apoptosis in biliary epithelia by a Fas/Fas ligand-dependent mechanism. Am J Physiol 277:G599–G608

Chen X-M, Levine SA, Splinter PL et al (2001) *Cryptosporidium parvum* activates nuclear factor κB in biliary epithelia preventing epithelial cell apoptosis.Gastroenterology 120: 1774-1783

Cowman AF, Kappe SHI (2013) Malaria's stealth shuttle. Science 313:1287–1290

Current WL, Reese NC, Ernst JV et al (1983) Human cryptosporidiosis in immunocompetent and immunodeficient persons. Studies of an outbreak and experimental transmission. N Engl J Med 308:1252–1257

Goebel E, Braendler U (1982) Ultrastructure of microgametogenesis, microgametes and gametogamy of *Cryptosporidium* sp. in the small intestine of mice. Protistologica 18:331–334

Heussler VT, Küenzi P, Rottenberg S (2001) Inhibition of apoptosis by intracellular protozoan parasites. Int J Parasitol 31:1166–1176

Izquierdo J, Antúnez I, Calderón MT et al (1988) Diarrhea caused by *Cryptosporidium* and colonic neoplasia. Rev Clin Esp 182:393–394

Lasser KH, Lewin KJ, Ryning FW (1979) Cryptosporidial enteritis in a patient with congenital hypogammaglobulinemia. Hum Pathol 10:234–240

Liu J, Enomoto S, Lancto CA et al (2008) Inhibition of apoptosis in *Cryptosporidium parvum-infected* intestinal epithelial cells is dependent on survivin. Infect Immun 76:3784–3792

Liu J, Deng M, Lancto CA et al (2009) Biphasic modulation of apoptotic pathways in *Cryptosporidium parvum*-infected human intestinal epithelial cells. Infect Immun 77(2):837–849

López-Vélez R, Tarazona R, Garcia Camacho A et al (1995) Intestinal and extraintestinal cryptosporidiosis in AIDS patients. Eur J Clin Microbiol Infect Dis 14:677–681

Lowe SW, Lin AW (2000) Apoptosis in cancer. Carcinogenesis 21:485–495

Meisel JL, Perera DR, Meligro C, Rubin CE (1976) Overwhelming watery diarrhea associated with *Cryptosporidium* in an immunosuppressed patient. Gastroenterology 70:1156–1160

Nash PB, Purner MB, Leon RP et al (1998) *Toxoplasma gondii*-infected cells are resistant to multiple inducers of apoptosis. J Immunol 160:1824–1830

Naugler WE, Karin M (2008) NF-κB and cancer – identifying targets and mechanisms. Curr Opin Genet Dev 18:19–26

O'Hara SP, Chen X-M (2011) The cell biology of *cryptosporidium* [sic] infection. Microbes Infect 13:721–730

O'Hara SP, Small AJ, Gajdos GB et al (2009) HIV-1 Tat protein suppresses cholangiocyte toll-like receptor 4 expression and defense against *Cryptosporidium parvum*. J Infect Dis 199:1195–1204

Orenstein JM, Dieterich DT (2001) The histopathology of 103 consecutive colonoscopy biopsies from 82 symptomatic patients with acquired immunodeficiency syndrome: original and look-back diagnoses. Arch Pathol Lab Med 125:1042–1046

Patel P, Hanson DL, Sullivan PS et al (2008) Incidence of types of cancer among HIV-infected persons compared with the general population in the United States, 1992–2003. Ann Intern Med 148:728–736

Ramirez NE, Ward LA, Sreevatsan S (2004) A review of the biology and epidemiology of cryptosporidiosis in humans and animals. Microbes Infect 6:773–785

Semenza JC, Nichols G (2007) Cryptosporidiosis surveillance and water-borne outbreaks in Europe. Euro Surveill 12:E13–E14

Shebl FM, Engels EA, Goedert JJ (2012) Opportunistic intestinal infections and risk of colorectal cancer among people with AIDS. AIDS Res Hum Retroviruses 28:994–999

Snelling WJ, Lin Q, Moore JE et al (2007) Proteomics analysis and protein expression during sporozoite excystation of *Cryptosporidium parvum* (Coccidia, Apicomplexa). Mol Cell Proteomics 6:346–355

Stephens J, Cosyns M, Jones M, Hayward A (1999) Liver and bile duct pathology following *Cryptosporidium parvum* infection of immunodeficient mice. Hepatology 30:27–35

Sturm A, Amino R, van de Sand C et al (2013) Manipulation of host hepatocytes by the malaria parasite for delivery into liver sinusoids. Science 313:1287–1290

Sulzyc-Bielicka V, Kuźna-Grygiel W, Kolodziejczyk L et al (2007) Cryptosporidiosis in patients with colorectal cancer. J Parasitol 93:722–724

Sulźyc-Bielicka V, Kolodziejczyk L, Jaczewska S et al (2012) Prevalence of *Cryptosporidium* sp. in patients with colorectal cancer. Pol Przegl Chir 84:348–351

Takahashi M, Mutoh M, Kawamori T et al (2000) Altered expression of β-catenin, inducible nitric oxide synthase and cyclooxygenase-2 in azoxymethane-induced rat colon carcinogenesis. Carcinogenesis 21:1319–1327

Valigurová A, Jirku M, Koudela B et al (2008) Cryptosporidia: epicellular parasites embraced by the host cell membrane. Int J Parasitol 38:913–922

Weisburger WR, Hutcheon DF, Yardley JH et al (1979) Cryptosporidiosis in an immunosuppressed renal-transplant recipient with IgA deficiency. Am J Clin Pathol 72:473–478

Xu P, Widmer G, Wang Y et al (2004) The genome of *Cryptosporidium hominis*. Nature 431:1107–1112

Malaria

Burkitt D (1958) A sarcoma involving the jaws in African children. Br J Surg 46:218–223

Burkitt DP (1969) Etiology of Burkitt's lymphoma – an alternative hypothesis to a vectored virus. J Natl Cancer Inst 42:19–28

Burkitt D, Wright D (1966) Geographical and tribal distribution of the African Lymphoma in Uganda. Br Med J 1:569–573

Dalla-Favera R, Bregni M, Erikson J et al (1982) Human c-myc onc gene is located on the region of chromosome 8 that is translocated in Burkitt lymphoma cells. Proc Natl Acad Sci U S A 79:7824–7827

de The G (1979) The epidemiology of Burkitt's lymphoma: evidence for a causal association with Epstein-Barr virus. Epidemiol Rev 1:32–54

de The G (1993) The etiology of Burkitt's lymphoma and the history of the shaken dogmas. Blood Cells 19:667–673

de The G, Geser A, Day NE et al (1978) Epidemiological evidence for causal relationship between Epstein-Barr virus and Burkitt's lymphoma from Ugandan prospective study. Nature 274:756–761
Epstein MA, Achong BG, Barr YM (1964) Virus particles in cultured lymphoblasts from Burkitt's lymphoma. Lancet 1:702–703
Evans AS (1985) Epidemiology of Burkitt's lymphoma: other risk factors. IARC Sci Publ 60:197–204
Greenwood BM (1974) Possible role of a B-cell mitogen in hypergammaglobulinaemia in malaria and trypanosomiasis. Lancet 1:435–436
Henle W, Henle G (1974) The Epstein-Barr virus in Burkitt's lymphoma and nasopharyngeal carcinoma. Ann Clin Lab Sci 4:109–114
Henle G, Henle W, Clifford P et al (1969) Antibodies to Epstein-Barr virus in Burkitt's lymphoma and control groups. J Natl Cancer Inst 43:1147–1157
Jayasooriya S, Hislop A, Peng Y et al (2012) Revisiting the effect of acute *P. falciparum* malaria on Epstein-Barr virus: host balance in the setting of reduced malaria endemicity. PLoS One 7:e31142
Kafuko GW (1969) Association of Burkitt's tumour and holoendemic malaria in the West Nile district, Uganda: malaria as a possible etiologic factor. E Afr Med J 46:414–436
Klein U, Dalla-Favera R (2008) Germinal centres: role in B-cell physiology and malignancy. Nat Rev Immunol 8:22–33
Lam KM, Syed N, Whittle H, Crawford DH (1991) Circulating Epstein-Barr virus–carrying B cells in acute malaria. Lancet 337:876–878
Mannucci S, Luzzi A, Carugi A et al (2012) EBV reactivation and chromosomal polysomies: euphorbia tirucalli as a possible cofactor in endemic Burkitt lymphoma. Adv Hematol 2012:149780
Manolov G, Manolova Y (1972) Marker band in one chromosome 14 from Burkitt lymphomas. Nature 237:33–34
Mbulaiteye SM (2013) Burkitt lymphoma: beyond discoveries. Infect Agent Cancer 8:35–38
Molyneux EM, Rochford R, Griffin B et al (2012) Burkitt's lymphoma. Lancet 379:1234–1244
Moormann AM, Chelimo K, Sumba OP et al (2005) Exposure to holoendemic malaria results in elevated Epstein-Barr virus loads in children. J Infect Dis 191:1233–1238
Mutalima N, Molyneux E, Jaffe H et al (2008) Associations between Burkitt lymphoma among children in Malawi and infection with HIV, EBV and malaria: results from a case-control study. PLoS One 3:e2505
Newton R, Ziegler J, Beral V et al (2001) A case–control study of human immunodeficiency virus infection and cancer in adults and children residing in Kampala, Uganda. Int J Cancer 92:622–627
Orem J, Mbidde EK, Lambert B et al (2007) Burkitt's lymphoma in Africa, a review of the epidemiology and etiology. Afr Health Sci 7:166–175
Osato T (1998) Epstein-Barr virus infection and oncogenesis. In: Epstein-Barr virus and human cancer. Gann Monograph on Cancer Research no. 45. Japanese Cancer Association, Japan Scientific Societies Press, Tokyo, Karger, Basel, Switzerland, pp 7–11
Snider CJ, Cole SR, Chelimo K et al (2012) Recurrent *Plasmodium falciparum* malaria infections in Kenyan children diminish T-cell immunity to Epstein Barr virus lytic but not latent antigens. PLoS One 7:e31753
Torgbor C, Awuah P, Deitsch K et al (2014) A Multifactorial role for *P. falciparum* malaria in endemic Burkitt's lymphoma pathogenesis. PLoS Pathog 10:e1004170
WHO (2013) http://www.who.int/malaria/media/world_malaria_report_2013/en/. World Health Organization, Geneva
Zur Hausen H, Schulte-Holthausen H, Klein G et al (1970) EBV-DNA in biopsies of Burkitt's tumours and anaplastic carcinomas of the nasopharynx. Nature 228:1056–1058

Trichomonas vaginalis

Allaj V, Guo C, Nie D (2013) Non-steroid anti-inflammatory drugs, prostaglandins, and cancer. Cell Biosci 3:8

Allsworth JE, Ratner JA, Peipert JF (2009) Trichomoniasis and other sexually transmitted infections: results from the 2001–2004 National Health and Nutrition Examination Surveys. Sex Transm Dis 36:738–744

Alsmark C, Foster PG, Sicheritz-Ponten T et al (2013) Effect of HIV-1 Vpr on cell cycle regulators. Patterns of prokaryotic lateral gene transfers affecting parasitic microbial eukaryotes. Genome Biol 14:R19

Amini S, Khalili K, Sawaya BE (2004) Effect of HIV-1 Vpr on cell cycle regulators. DNA Cell Biol 23:249–260

Azevedo A, Cunha V, Teixeira AL, Medeiros R (2011) IL-6/IL-6R as a potential key signaling pathway in prostate cancer development. World J Clin Oncol 2:384–396

Bachmann LH, Hobbs MM, Seña AC et al (2011) *Trichomonas vaginalis* genital infections: progress and challenges. Clin Infect Dis 53(Suppl 3):S160–S172

Brotman RM, Bradford LL, Conrad M et al (2012) Association between *Trichomonas vaginalis* and vaginal bacterial community composition among reproductive-age women. Sex Transm Dis 39:807–812

Clemente JC, Ursell LK, Parfrey LW, Knight R (2012) The impact of the gut microbiota on human health: an integrative view. Cell 148:1258–1270

Cotch MF, Pastorek JG 2nd, Nugent RP et al (1997) *Trichomonas vaginalis* associated with low birth weight and preterm delivery. The Vaginal Infections and Prematurity Study Group. Sex Transm Dis 24:353–360

Cowan MJ, Coll T, Shelhamer JH (2006) Polyamine-mediated reduction in human airway epithelial migration in response to wounding is PGE2 dependent through decreases in COX-2 and cPLA2 protein levels. J Appl Physiol 101:1127–1135

Dhanasekaran SM, Barrette TR, Ghosh D et al (2001) Delineation of prognostic biomarkers in prostate cancer. Nature 412:822–826

Dubois RN, Abramson SB, Crofford L et al (1998) Cyclooxygenase in biology and disease. FASEB J 12:1063–1073

Fichorova RN, Trifonova RT, Gilbert RO et al (2006) *Trichomonas vaginalis* lipophosphoglycan triggers a selective upregulation of cytokines by human female reproductive tract epithelial cells. Infect Immun 74:5773–5779

Garcia AF, Benchimol M, Alderete JF (2005) *Trichomonas vaginalis* polyamine metabolism is linked to host cell adherence and cytotoxicity. Infect Immun 73:2602–2610

Gardner WA Jr, Culberson DE, Bennett BD (1986) *Trichomonas vaginalis* in the prostate gland. Arch Pathol Lab Med 110:430–432

Gram IT, Macaluso M, Churchill J, Stalsberg H (1992) *Trichomonas vaginalis* (TV) and human papillomavirus (HPV) infection and the incidence of cervical intraepithelial neoplasia (CIN) grade III. Cancer Causes Control 3:231–236

Han IH, Park SJ, Ahn MH, Ryu JS (2012) Involvement of mast cells in inflammation induced by *Trichomonas vaginalis* via crosstalk with vaginal epithelial cells. Parasite Immunol 34:8–14

Hirt RP (2013) *Trichomonas vaginalis* virulence factors: an integrative overview. Sex Transm Infect 89:439–443

Hirt RP, de Miguel N, Nakjang S et al (2011) *Trichomonas vaginalis* pathobiology: new insights from the genome sequence. Adv Parasitol 77:87–140

Hobbs MM, Sena AC, Swygard H et al (2008) *Trichomonas vaginalis* and trichomoniasis. In: Holmes K et al (eds) Sexually transmitted diseases, 4th edn. McGraw-Hill, New York, pp 771–793

Hughes JP, Baeten JM, Lingappa JR et al (2012) Determinants of per-coital-act HIV-1 infectivity among African HIV-1-serodiscordant couples. J Infect Dis 205:358–365

Kucknoor AS, Mundodi V, Alderete JF (2007) The proteins secreted by *Trichomonas vaginalis* and vaginal epithelial cell response to secreted and episomally expressed AP65. Cell Microbiol 9:2586–2597

Lazenby GB, Taylor PT, Badman BS et al (2014) An association between *Trichomonas vaginalis* and high-risk human papillomavirus in rural Tanzanian women undergoing cervical cancer screening. Clin Ther 36:38–45

Magnuson NS, Wang Z, Ding G, Reeves R (2010) Why target PIM1 for cancer diagnosis and treatment? Future Oncol 6:1461–1478

Miller WC, Swygard H, Hobbs MM et al (2005) The prevalence of trichomoniasis in young adults in the United States. Sex Transm Dis 32:593–598

Mitteregger D, Aberle SW, Makristathis A et al (2012) High detection rate of *Trichomonas vaginalis* in benign hyperplastic prostatic tissue. Med Microbiol Immunol 201:113–116

Moodley P, Wilkinson D, Connolly C et al (2002) *Trichomonas vaginalis* is associated with pelvic inflammatory disease in women infected with human immunodeficiency virus. Clin Infect Dis 34:519–522

Morada M, Manzur M, Lam B et al (2010) Arginine metabolism in *Trichomonas vaginalis* infected with *Mycoplasma hominis*. Microbiology 156:3734–3743

Muzny CA, Blackburn RJ, Sinsky RJ et al (2014) Added benefit of nucleic acid amplification testing for the diagnosis of *Trichomonas vaginalis* among men and women attending a sexually transmitted diseases clinic. Clin Infect Dis 59:834–841

Reeves R, Beckerbauer L (2001) HMGI/Y proteins: flexible regulators of transcription and chromatin structure. Biochim Biophys Acta 1519:13–29

Roh M, Franco OE, Hayward SW et al (2008) A role for polyploidy in the tumorigenicity of Pim-1-expressing human prostate and mammary epithelial cells. PLoS One 3:e2572

Rughooputh S, Greenwell P (2005) *Trichomonas vaginalis*: paradigm of a successful sexually transmitted organism. Br J Biomed Sci 62:193–200

Ryan CM, de Miguel N, Johnson PJ (2011) *Trichomonas vaginalis*: current understanding of host-parasite interactions. Essays Biochem 51:161–175

Ryu JS, Kang JH, Jung SY et al (2004) Production of interleukin-8 by human neutrophils stimulated with *Trichomonas vaginalis*. Infect Immun 72:1326–1332

Sansone P, Bromberg J (2012) Targeting the interleukin-6/Jak/stat pathway in human malignancies. J Clin Oncol 30:1005–1014

Satterwhite CL, Torrone E, Meites E et al (2013) Sexually transmitted infections among US women and men: prevalence and incidence estimates, 2008. Sex Transm Dis 40:187–193

Sayed el-Ahl SA, el-Wakil HS, Kamel NM, Mahmoud MS (2002) A preliminary study on the relationship between *Trichomonas vaginalis* and cervical cancer in Egyptian women. J Egypt Soc Parasitol 32:167–178

Shaio MF, Lin PR, Liu JY, Tang KD (1994) Monocyte-derived interleukin-8 involved in the recruitment of neutrophils induced by *Trichomonas vaginalis* infection. J Infect Dis 170:1638–1640

Siegel R, Naishadham D, Jemal A (2012) Cancer statistics, 2012. CA Cancer J Clin 62:10–29

Stark JR, Judson G, Alderete JF et al (2009a) Prospective study of *Trichomonas vaginalis* infection and prostate cancer incidence and mortality: physicians' health study. J Natl Cancer Inst 101:1406–1411

Stark JR, Li H, Kraft P et al (2009b) Circulating prediagnostic interleukin-6 and C-reactive protein and prostate cancer incidence and mortality. Int J Cancer 124:2683–2689

Sutcliffe S, Giovannucci E, Alderete JF et al (2006) Plasma antibodies against *Trichomonas vaginalis* and subsequent risk of prostate cancer. Cancer Epidemiol Biomarkers Prev 15:939–945

Sutcliffe S, Neace C, Magnuson NS et al (2012) Trichomonosis, a common curable STI, and prostate carcinogenesis – a proposed molecular mechanism. PLoS Pathog 8:e1002801

Tesfaye A, Di Cello F, Hillion J et al (2007) The high-mobility group A1 gene up-regulates cyclooxygenase 2 expression in uterine tumorigenesis. Cancer Res 67:3998–4004

Twu O, de Miguel N, Lustig G et al (2013) *Trichomonas vaginalis* exosomes deliver cargo to host cells and mediate host: parasite interactions. PLoS Pathog 9:e1003482
Valdman A, Fang X, Pang ST et al (2004) Pim-1 expression in prostatic intraepithelial neoplasia and human prostate cancer. Prostate 60:367–371
WHO (2011) Prevalence and incidence of selected sexually transmitted infections: *Chlamydia trachomatis*, *Neisseria gonorrhoea*e, syphilis, and *Trichomonas vaginalis*. Methods and results used by the WHO to generate 2005 estimates. World Health Organization, Geneva
Zhang ZF, Begg CB (1994) Is *Trichomonas vaginalis* a cause of cervical neoplasia? Results from a combined analysis of 24 studies. Int J Epidemiol 23:682–690
Zhang ZF, Graham S, Yu SZ et al (1995) *Trichomonas vaginalis* and cervical cancer. A prospective study in China. Ann Epidemiol 5:325–332
Zhang Y, Wang Z, Li X, Magnuson NS (2008) Pim kinase-dependent inhibition of c-Myc degradation. Oncogene 27:4809–4819

Chapter 6
An Update on Helminths in Human Carcinogenesis

Aditya Reddy and Bernard Fried

Abstract This review examines the salient literature on selected helminths involved in carcinogenicity in humans and updates information in an earlier review on this topic by Fried et al. (Cancer Lett 305:239–249, 2011). Herein, we examine the salient cancer-helminth literature from 2011 to 2014 on *Opisthorchis viverreni* and *Clonorchis sinensis* (associated with liver and bile duct cancer), and *Schistosoma haematobium* (associated with urinary bladder cancer). These three trematodes are the only helminths recognized by International Agency for Research on Cancer (IARC) as being human carcinogens. Our review places particular emphasis on the molecular mechanisms involved in helminth induced carcinogenesis.

Keywords Helminths • Cancer • Gene • Trematodes • *Schistosoma haematobium* • *Opisthorchis viverrini* • *Clonorchis sinensis* • Carcinogenesis

Abbreviations

AP	acid phosphatase
CCA	cholangiocarcinoma
CP	Cysteine protease(s)
CsCP	cysteine protease of *Clonorchis sinensis*
ESP	excretory secretory products
HCC	hepatocellular carcinoma
ICC	Intrahepatic cholangiocarcinoma
MAP-2	methionine aminopeptidase
MMP	mitochondrial membrane potential
NDMA	N-nitrosodimethylamine

A. Reddy (✉) • B. Fried
Department of Biology, Lafayette College, Easton, PA 18042, USA
e-mail: adireddymd@gmail.com

M.R. Shurin et al. (eds.), *Infection and Cancer: Bi-Directorial Interactions*, DOI 10.1007/978-3-319-20669-1_6

Introduction

An earlier review by Fried et al. (2011) considered the salient literature on the three trematode parasites, *Opisthorchis viverreni*, *Clonorchis sinensis*, and *Schistosoma haematobium* that have been implicated by the International Agency for Research on Cancer (IARC) as human carcinogens. Since that review covered all aspects of the biology, life history, and epidemiology of these medically important trematodes, we do not consider in detail aspects of the biology of these trematodes in this review. We recommend that the reader refer to our earlier review for information on the biology of these organisms. The purpose of this review is to update the literature on helminths in human carcinogenesis from 2011 to 2014.

Clonorchis sinensis

Clonorchis sinensis is a food-borne zoonotic trematode and the most common human liver fluke in parts of East Asia. Several studies have demonstrated its carcinogenesis in humans and it was reclassified as a group 1 biological carcinogen in 2009. It is still actively transmitted in endemic areas of the Korean peninsula, China, Russia, and Vietnam. Currently it is estimated that more than 200 million people are at risk of infection, with 15–20 million people infected and 1.5–2 million additional individuals showing clinical symptoms or complications. Multiple molecules and genes of the fluke have been identified and characterized.

Clonorchis sinensis infection may result in cholelithiasis, cholecystitis, hepatic fibrosis, and liver tumors. Although total excretory secretory products (ESP) of *C. sinensis* adults induce hepatic fibrosis in vivo in rats, the causative mechanism is not well understood. To study components of the ESP, *C. sinensis* culture medium was collected and analyzed in a study by Zheng et al. (2011). The authors identified a total of 110 proteins, including glycometabolic enzymes (such as fructose-1,6-bisphosphatase (FBPase) and enolase), detoxification enzymes (such as glutamate dehydrogenase, dihydrolipoamide dehydrogenase and cathepsin B endopeptidase), and a number of RAB family proteins. To identify a potential causative agent for hepatic fibrosis, Zheng et al. (2011) expressed and purified a recombinant FBPase, a 1,041-bp gene product that encodes a 41.7-kDa protein. In addition, the authors found that FBPase is an antigen present in the ESP and in circulation. Immunofluorescence showed that FBPase localizes to the intestinal cecum and vitellarium in *C. sinensis* adults. Zheng et al. (2011) conclude that FBPase may be an important antigen present in the ESP of *C. sinensis* and may lay the foundation for additional studies on the development of clonorchiasis-associated hepatic fibrosis.

Hong and Fang (2012) have provided an important update on this human liver fluke. Actual diagnosis of infection depends mainly on detection of eggs in feces but other methods have been developed. ELISA using crude extract antigen is now commonly utilized for diagnosis. Diagnosis by detecting DNAs from eggs in feces has also been developed using PCR, real-time PCR, and LAMP, which have been found to be both sensitive and specific. The authors mention that imaging diagnosis have been studied in depth and are widely used clinically. Evidence of clonorchiasis, such as eggs, DNAs, or images, may lead to recommendations of chemotherapy in endemic areas. Praziquantel is the major chemotherapeutic agent for clonorchiasis and recently tribendimidine was found effective and is now under investigation as a promising chemotherapeutic alternative. Sustainable control programs such as mass chemotherapy with praziquantel and education for prevention of re-infection may reduce its morbidity and eliminate infections in endemic areas.

Cysteine proteases (CP) were associated with the pathogenicity and excystment of *Clonorchis sinensis* in a study by Li et al. (2012). In the study, a full-length sequence encoding cathepsin L from *C. sinensis* (CsCL41.5) was identified from an adult cDNA library. Bioinformatic analysis was performed and showed that CsCL41.5 included typical motifs of cathepsin L and conserved amino acid positions which constituted the active center of the enzyme. Recombinant CsCL41.5 (rCsCL41.5) was highly expressed in the inclusion bodies of *Escherichia coli*, and soluble rCsCL41.5 was obtained after purification and renaturation. Western blotting analysis indicated that CsCL41.5 is an excretory-secretory antigen of *C. sinensis* adult. Immunolocalization demonstrated that CsCL41.5 is distributed in the intestine and eggs in the uterus of adult worm, tegument of metacercaria, oral sucker, and in the cercarial tail. ELISA assays showed that IgG4 was the predominant IgG isotype responding to rCsCL41.5 in sera from actual clonorchiasis patients. Li et al. (2012) concluded that, as in previous studies, there were differences in biological function, efficiency of serodiagnosis, and characterization of immune reactivity between CsCL41.5 and other CPs derived from *C. sinensis*.

A full-length sequence encoding cysteine protease of *Clonorchis sinensis* (CsCP) was isolated from an adult cDNA library by Lv et al. (2012). Cysteine proteases play critical roles in parasite physiology as well as in host-parasite interactions through their modulation of various biological and pathobiological events. Bioinformatics analysis showed that conserved domains and characteristic amino acid residues of cysteine proteases were observed in this sequence. Real-time PCR experiments revealed that CsCP was consecutively transcribed in various developmental stages of the parasite, including adult worm, excysted juvenile, metacercaria and the egg. Recombinant CsCP (rCsCP) was probed by rat anti-CsCP serum, rabbit anti-excretory-secretory products serum and serum from human infected with *Clonorchis sinensis* in Western blot analyses. The results of immunolocalization by Lv et al. (2012) showed that CsCP was mainly located in the oral sucker, excretory bladder and tegument of cercariae and metacercariae, as well as the intestine of adult worm. The rCsCP-based IgG and its isotypes were all detected in sera from human infected with *C. sinensis* by enzyme-linked immunosorbent assay, with IgG1 levels as the highest of the studied isotypes. The authors conclude that CsCP may play an important role in the biology of *C. sinensis* and could be a key diagnostic marker.

Chronic clonorchiasis, caused by direct and continuous contact with *Clonorchis sinensis* worms and their excretory-secretory products, is associated with hepatobiliary damage, inflammation and periductal fibrosis. In the present study, Nam et al. (2012) found that treatment of human cholangiocarcinoma cells with excretory-secretory products triggered increases in free radicals via a time-dependent activation of NADPH oxidase, xanthine oxidase and inducible nitric oxide synthase. In the study, increases in free radicals substantially promoted the degradation of cytosolic IκB-α, nuclear translocation of nuclear factor-κB subunits (RelA and p50), and increased κB consensus DNA-binding activity. Excretory-secretory product-induced nuclear factor-κB activation was markedly attenuated by preincubation with specific inhibitors of each free radical-producing enzyme or the antioxidant, N-acetylcysteine. In addition, excretory-secretory products induced an increase in the mRNA and protein expression of the proinflammatory cytokines, IL-1β and IL-6, in a nuclear factor-κB-dependent manner. The authors concluded that enzymatic production of free radicals in ESP-treated cells is a key component in nuclear factor-κB-mediated inflammation. These findings provide new insights into the pathophysiological role of *C. sinensis* excretory-secretory products in host chronic inflammatory processes, which are initial events in hepatobiliary disease pathogenesis.

The excretory secretory products of *Clonorchis sinensis* are the causative agents of clonorchiasis and biliary diseases. The parasites' ESP play key roles in host-parasite interactions. The protein compositions of ESP at different secretory times are diverse and have not been systemically investigated. A study by Zheng et al. (2013) collected ESP from six different periods (0–3 h, 3–6 h, 6–12 h, 12–24 h, 24–36 h, and 36–48 h) from *C. sinensis* adults. Their findings presented the compositions of different period excretory secretary products from *C. sinensis* adults. Using a shotgun LC-MS/MS analysis, the authors found 187, 80, 103, 58, 248, and 383 proteins, respectively. Among these proteins, the authors selected methionine aminopeptidase 2 (MAP-2, presented in 24–36 h and 36–48 h ESP) and acid phosphatase (AP, presented in 3–6 h, 12–24 h, 24–36 h, and 36–48 h ESP) for further study. Bioinformatics analysis showed that CsMAP-2 has metallopeptidase family M24, unique lysine residue-rich and acidic residue-rich domain, SGTS motif, and an auto-cleavage point. MAP-2 and AP were identified as antigens present in the ESP and circulating antigens by immunoblot analysis, which were also found expressed in the eggs, metacercariae, and adult stages of *C. sinensis*. Immunofluorescence analysis showed that they were located in tegument and intestinal cecum of adult worms. MTT assays showed that they could inhibit hepatic stellate cell line (LX-2) proliferation.

As noted in Chen et al. (2013), *Clonorchis sinensis* afflicts more than 35 million people in the world and approximately 15 million in China, creating a socio-economic burden in epidemic regions. The infection of *C. sinensis* is highly related to CCA and hepatocellular carcinoma (HCC) in humans. It has been documented that excretory/secretory products of *C. sinensis* (*Cs*ESPs) are involved in the pathogenesis of HCC. Severin, expressed at the life stage of egg, metacercaria and adult worm, was a component of *Cs*ESPs. The properties of severin such as sequence signature, actin and calcium ion binding activity have recently been described. Severin causes apoptotic inhibition in spontaneously apoptotic human HCC cell line primary liver cancer cells as determined by use of morphological analysis, detection of apoptosis-associated changes of mitochondrial membrane potential (MMP) as well as Annexin V/PI apoptosis assay. The study by Chen et al. (2013) provided an introductory view of mechanisms involved in the progress of carcinoma associated with the infection of *C. sinensis* and suggested that severin might exacerbate the process of *C. sinensis* infected HCC patients.

To improve the rate of detection of *Clonorchis sinensis* infection, Qiao et al. (2013) compared different specimens from patients with cholecystolithiasis. Feces, gallbladder bile, and gallbladder stones collected from 179 consecutive patients with cholecystolithiasis underwent microscopic examination. Depending upon the results, 30 egg-positive and 30 egg-negative fecal, gallbladder bile, and gallbladder stone specimens, respectively, underwent real-time fluorescent PCR. The authors found detection rates of eggs in feces, bile, and gallbladder stones of 30.7 %, 44.7 %, and 69.8 %, respectively, and the differences were statistically significant ($P<0.01$). The PCR results confirmed that the eggs in the specimens were *C. sinensis* eggs. Eggs obtained from the feces were "fresh" and those obtained from gallbladder stones were "old." Qiao et al. (2013) concluded that microscopic examination of gallbladder stones may improve the detection rates of *C. sinensis* infection, which is important both for developing individualized treatments to prevent the recurrence of gallbladder stones and to prevent the occurrence of severe liver damage and potential cholangiocarcinoma.

In a study by Zhang et al. (2013), the role of *Clonorchis sinensis* lysophospholipase (CslysoPLA) in hepatic fibrosis induced by *C. sinensis* was explored for the first time. Lysophospholipase plays a vital role in virulence and pathogenesis of parasites and fungi. In the livers of cats infected with *C. sinensis*, CslysoPLA was recognized in the lumen between adult worms and surrounding bile duct epithelia with some also detected inside the cells using immunolocalization. Cell Counting Kit-8 assays and cell cycle analysis of human hepatic stellate cell line LX-2 showed that a higher percentage of cells were in the proliferative phase after incubation. Quantitative real-time polymerase chain reaction (RT-PCR) demonstrated an upregulation in fibrogenic genes of smooth muscle α-actin, collagen III, matrix metalloproteinase 2 and tissue inhibitors of metalloproteinase II in LX-2 treated with rCslysoPLA. In addition, quantitative RT-PCR demonstrated that CslysoPLA was differentially expressed at the developmental stages of *C. sinensis* (metacercariae, adult worms and eggs), with the highest levels at the metacercariae stage. Immunolocalization analysis showed that CslysoPLA was distributed in the intestine, vitelline gland, tegument and eggs in the adult worms and in the tegument and vitelline gland in the metacercariae, respectively. Zhang et al. (2013) concluded that CslysoPLA might be involved in the initiation and promotion of *C. sinensis*-related human hepatic fibrosis and advance future studies on its promotion to *C. sinensis-induced* cholangiocarcinogenesis.

Ubiquitin is a functionally important protein expressed in eukaryotic cells usually encoded by multigenic families containing two types of genes which include ubiquitin extension and polyubiquitin genes. The nucleotide and amino acid sequence of *C. sinensis* polyubiquitin, especially that with five tandem ubiquitin repeats (CsPUB5), were analyzed by Huang et al. (2013). One independent monomeric locus and two independent polyubiquitin loci were firstly identified by the authors from the genome of *C. sinensis*. Huang et al. (2013) obtained recombinant CsPUB5 (rCsPUB5) and anti-rCsPUB5 IgG and investigated the ubiquitin transcripts in life cycle of *C. sinensis*. Ubiquitination was common in adult worm of *C. sinensis* and was also observed in the content of biliary tract and intrahepatic biliary epithelium of liver from *C. sinensis* infected rats. The authors confirmed that rCsPUB5 could bind to human intrahepatic biliary epithelial cell by immunofluorescence in vitro. Huang et al. (2013) concluded that the ubiquitin family is constitutively expressed in *C. sinensis* for variety of cellular processes and might be implicated in the genesis and progression of cholangiocarcinoma.

Accumulating evidences indicate that nitric oxide (NO) is a potent mediator with diverse roles in regulating cellular functions, signaling pathways, and variety of pathological processes. Bian et al. (2014) used data from the published genomic for *C. sinensis* to investigate a gene encoding nitric oxide synthase-interacting protein (NOSIP) of *C. sinensis*. Recombinant CsNOSIP (rCsNOSIP) was expressed and purified from *Escherichia coli* BL21. Quantitative RT-PCR indicated that CsNOSIP differentially transcribed throughout the adult worms, metacercariae, and egg stages of *C. sinensis*, and were highly expressed in the adult worms. Moreover, western blot analysis showed that the rCsNOSIP could be detected by the serum from BALB/c mice infected with *C. sinensis* and the serum from BALB/c mice immunized with excretory/secretory products. Furthermore, immunolocalization assay showed that CsNOSIP was specifically localized in the intestine, vitellarium, and eggs of adult worm. Both immunoblot and immunolocalization results demonstrated that CsNOSIP was one component of ESPs of *C. sinensis*, which was supported by SignalP analysis. Bian et al. (2014) concluded that CsNOSIP is an important antigen exposed to host immune system and probably involved in immune regulation of host by inducing a Th2-polarized immune response.

Opisthorchis viverrini

The Southeast Asian liver fluke (*Opisthorchis viverrini*) chronically infects and affects tens of millions of people in regions of Asia, leading to chronic illness and potentially inducing malignant cancer of the bile duct, also known as cholangiocarcinoma (CCA).

Opisthorchiasis is the major public health problem in the endemic areas because *Opisthorchis viverrini* infection causes serious hepatobiliary diseases including CCA. The molecular mechanism of the CCA carcinogenesis induced by the infection remains obscure. To reveal the potential genes and signaling pathways involved in the carcinogenesis, Boonmars et al. (2011) investigated the expression of c-Ski, an oncogene, and two TGF-β signaling pathway relative genes, TGF-β and Smad4, during the development of CCA induced by *O. viverrini* infection in an hamster model, and in human opisthorchiasis associated CCA. Their results showed that the expression of c-Ski gene was greatly up-regulated during the carcinogenesis of CCA in the hamster model. The overexpression of c-Ski was confirmed by immunohistological staining results which indicated the increased expression of c-Ski protein in the cytoplasm of the epithelial lining of hepatic bile ducts. In addition, the immunohistological staining of the specimens of human opisthorchiasis associated CCA revealed the up-regulated expression of c-Ski and Smad4 proteins in the cytoplasm of the epithelial lining of hepatic bile ducts and stomal fibrosis respectively. The expression of TGF-β and Smad4 were up-regulated, which expression kinetics demonstrated as time-dependent upon CCA development. The results from Boonmars et al. (2011) suggest that c-Ski is likely involved in the carcinogenesis of CCA induced by *O. viverrini* infection through regulating TGF-β signaling pathway.

Chronic inflammation induced by the liver fluke *(Opisthorchis viverrini)* infection is the major risk factor of CCA in Northeastern Thailand. Yongvanit et al. (2012) mention multiple factors, all of which appear to be involved in *O. viverrini-associated* inflammatory processes and CCA. Increased levels of proinflammatory cytokines and nuclear factor kappa B that control cyclooxygenase-2 and inducible nitric oxide activities, were found to disturb the homeostasis of oxidants/antioxidants and DNA repair enzymes. Consequently oxidative and nitrative stress-related cellular damage occurs due to the over production of reactive oxygen and nitrogen species in inflamed target cells. In their study, these findings were supported by the detection of high levels of oxidized DNA and DNA bases modified by lipid peroxidation products in both animal and human tissues affected by *O. viverrini-infection.* Treatment of opisthorchiasis patients with praziquantel, an anti-trematode drug, was shown to reduce inflammation-mediated tissue damage and carcinogenesis. The validity of inflammation-related biomolecules and DNA damage products to serve as predictive biomarkers for disease risk evaluation and intervention was also discussed by Yongvanit et al. (2012).

Proteomic analysis was performed by Rucksaken et al. (2012) to search for the diagnostic biomarkers of the early stage of CCA. In the study, CCA was experimentally induced in hamsters by the combination of N-nitrosodimethylamine (NDMA) treatment and *Opisthorchis viverrini* (OV) infection. Pooled plasma of normal control, NDMA-treated, OV-infected and OV + NDMA (ON) treated group was separated by 1-D PAGE, and the trypsin-digested bands were analyzed with LC-MS/MS. Among 82 overexpressed proteins, the study focused on 26 proteins overexpressed in ON group because CCA development was almost exclusively found in this group. The overexpression levels were verified by real-time RT-PCR and western blotting in the liver and plasma. Transcription and translation levels of both candidate molecules increased significantly at 21 days post-treatment before tumor development. Immunohistochemistry revealed KIF18A was expressed in the epithelial cells of newly formed small bile ducts, some inflammatory cells and hepatocytes. These findings suggest that Orm2 and KIF18A could be used as potential biomarkers for early diagnosis of CCA.

Poor prognosis of CCA is primarily due to delayed diagnosis because of the lack of appropriate tumor marker(s) to detect cancer development at an early stage.

Sawanyawisuth et al. (2012) established a S121 monoclonal antibody (mAb) which recognizes an unidentified glycan epitope on MUC5AC, designated as CCA-associated carbohydrate antigen (CCA-CA). This antigen is expressed in human CCA cells but not in normal biliary epithelia. Detection of CCA-CA effectively distinguished CCA patients' sera from normal control sera with high specificity and sensitivity. In their study, Sawanyawisuth et al. (2012) examined a time profile of the expression of CCA-CA by immunohistochemical methods in the liver tissues of *Opisthorchis viverrini* (Ov)-associated CCA in a hamster model. Hamsters were divided into four groups; non-treated, Ov infected, NDMA (N-nitrosodimethamine) treated and Ov+NDMA treated groups, and animals from each group were euthanized at 1, 3 and 6 months post-treatment. CCA-CA was not detected in the normal biliary cells of non-treated hamsters throughout the course of experiment. CCA-CA became detectable in the cytoplasm and apical surface of biliary cells of the NDMA and Ov+NDMA groups at early stage (1 month) of tumor development and increased with tumor progression. In contrast, CCA-CA was detected as nuclear staining at the 1 month post Ov infection and declined thereafter. The findings suggest the possibility that CCA-CA be used as an early marker for CCA.

Opisthorchiasis caused by *Opisthorchis viverrini* induces periductal fibrosis via host immune/inflammatory responses. Plasma protein alteration during host-parasite interaction-mediated inflammation may provide potential diagnostic and/or prognostic biomarkers. To search for target protein changes in *O. viverrini*-infected hamsters, Khoontawad et al. (2012) utilized a 1-D PAGE gel band that was trypsin-digested and analyzed by a LC-MS/MS-based proteomics approach in the plasma profile of infected hamsters, and applied to humans. Sixty seven proteins were selected for further analysis based on at least two unique tryptic peptides with protein ID score >10 and increased expression at least two times across time points. These proteins have not been previously identified in *O. viverrini*-associated infection. Among those, proteins involved in structural (19 %), immune response (13 %), cell cycle (10 %) and transcription (10 %) were highly expressed. Western blots revealed an expression level of protein tyrosine phosphatase alpha (PTPα) which reached a peak at 1 month and subsequently tended to decrease. Fibronectin significantly increased at 1 month and tended to increase with time, supporting proteomic analysis. PTPα was expressed in the cytoplasm of inflammatory cells, while fibronectin was observed mainly in the cytoplasm of fibroblasts and the extracellular matrix at periductal fibrosis areas. In addition, these protein levels significantly increased in the plasma of *O. viverrini*-infected patients compared to healthy individuals, and significantly decreased at 2-months post-treatment, indicating their potential as disease markers. Khoontawad et al. (2012) concluded that plasma PTPα and fibronectin may be associated with opisthorchiasis and the hamster model and could provide the basis for development of novel diagnostic markers in the future.

CCA is a crucial health problem in the northeastern part of Thailand, which is caused by a combination of *Opisthorchis viverrini* infection and nitrosamine. A better understanding of its molecular mechanism is an important step to discover and develop the new diagnostics and therapies for CCA. To reveal the involvement of potential genes in the development of CCA, Boonjaraspinyo et al. (2012) investigated the expression kinetics of platelet-derived growth factor alpha (Pdgfa) and its receptor (Pdgfra) during the tumorigenesis of CCA induced by *O. viverrini* infection with quantitative RT-PCR, and confirmed the expression with immunohistological staining. The results showed that in the hamster model of opisthorchiasis-associated CCA, the expression of Pdgfa was increased after infection plus NDMA administration, reached its peak at 2 months post infection, and remained at the high level until 6 months. Similarly, the expression of Pdgfra increased in a time-dependent manner. The positive immunostaining for PDGFA proteins was observed in the cytoplasm of epithelial tumor cells of hamster CCA. Moreover, the analysis of the expression of these genes in 10 cases of human opisthorchiasis-associated CCA showed that Pdgfa was overexpressed in 80 %, and Pdgfra was overexpressed in 40 % cases (>3.0-fold, compared with the expressions of adjacent normal tissues). This result suggests that PDGFA is likely involved in the tumorigenesis of opisthorchiasis-associated CCA, and may be a promising candidate biomarker for diagnosis and treatment strategies of CCA.

Jex et al. (2012) generated extensive RNA-Seq data (Illumina) representing adult and juvenile stages of *O. viverrini*, and combined these sequences with previously published transcriptomic data, yielding a combined assembly of significantly increased quality and allowing quantitative assessment of transcription in both the juvenile and adult stages. The authors found that this enhanced assembly reveals that, despite the substantial biological similarities between the human liver flukes, *O. viverinni* and *Clonorchis sinensis*, there are previously unrecognized differences in major aspects of their molecular biology. Most notable are differences among the C13 and cathepsin L-like cysteine peptidases, which play key roles in tissue migration, immune evasion and feeding, and, thus, represent potential drug and/or vaccine targets. Furthermore, these data indicate that major lineages of cysteine peptidases of socioeconomically important trematodes have evolved through a process of gene loss rather than independent radiation, contrasting previous proposals.

Chronic inflammation induced by biological, chemical, and physical factors has been found to be associated with the increased risk of cancer in various organs. Murata et al. (2012) revealed that infectious agents including the liver fluke, *Helicobacter pylori*, and human papilloma virus and noninfectious agents such as asbestos fiber induced iNOS-dependent formation of 8-nitroguanine and 8-oxo-7, 8-dihydro-2′-deoxyguanosine (8-oxodG) in cancer tissues and precancerous regions. Results of colocalization of phosphorylated ATM and γ-H2AX with 8-oxodG and 8-nitroguanine in inflammation-related cancer tissues suggest that DNA base damage leads to double-stranded breaks. This is interesting from a genetic instability point of view. The authors also demonstrated that IL-6-modulated iNOS expression via STAT3 and EGFR in Epstein-Barr-virus-associated nasopharyngeal carcinoma and found promoter hypermethylation in several tumor suppressor genes. Such epigenetic alteration likely occurs by controlling the DNA methylation through IL-6-mediated JAK/STAT3 pathways. Collectively, 8-nitroguanine could be a useful biomarker for predicting the risk of inflammation-related cancers.

Kaewkong et al. (2012) performed the first karyotype analysis of *O. viverrini* with a complete and defined nomenclature. Though the taxonomy, morphology, epidemiology and molecular study of O. viverrini have been previously described, a precise karyotypic description is still incomplete. In their study, the chromosomes of *O. viverrini* were prepared from the testes of adult worms retrieved from metacercariae infected-hamsters. The chromosomes of *O. viverrini* were identified in haploid (n=6) meiotic metaphase and in diploid (2n=12) mitotic metaphase by light microscopy. The chromosome number, length and nomenclature of each of the six chromosomes were determined by scanning electron microscopy.

Concurrent infection with the human liver fluke *Opisthorchis viverrini* and NDMA administration induce CCA and liver injury in hamsters. Laothong et al. (2013) found that melatonin protects against liver injury and reduces the alteration of mitochondrial structure, mitochondrial membrane potential, and mitochondrial pro- and anti-apoptotic pathways in various cancer types. To investigate the chemopreventive effect of melatonin on CCA genesis and liver injury, hamsters were treated with a combination of *O. viverrini* infection and NDMA concurrently administered with melatonin (10 mg/kg and 50 mg/kg) for 120 days. Melatonin treatment at 50 mg/kg caused a significant reduction in liver/body weight ratios and decreased tumor volumes leading to an increase in the survival of animals. In the tumor containing tissues, the high-dose melatonin reduced DNA fragmentation and mitochondrial apoptosis by inducing anti-apoptotic protein Bcl-2 in the mitochondrial fraction and down-regulating cytochrome c, pro-apoptotic protein Bax, and caspase-3 in tumor cytosol. Moreover, Laothong et al. (2013) found that melatonin has potent chemopreventive effects in inhibiting CCA genesis and also reduces liver injury in hamster CCA, which, in part, might involve the suppression of CCA by reducing tumor mitochondria alteration.

Intrahepatic cholangiocarcinoma (ICC) is ranked as one of the top five causes of cancer-related deaths. Subrungruang et al. (2013) employed a microarray approach to compare gene expression profiles of ICCs and normal liver tissues from the same patients residing in Northeast Thailand, a region with a high prevalence of liver fluke infection. In ICC samples, 2,821 and 1,361 genes were found to be significantly up- and down-regulated respectively (unpaired t-test, $p<0.05$; fold-change>2.0). For validation of the microarray results, the authors selected 7 up-regulated genes (FXYD3, GPRC5A, CEACAM5, MUC13, EPCAM, TMC5, and EHF) and 3 down- regulated genes (CPS1, TAT, and ITIH1) for confirmation using quantitative RT-PCR; the selected genes resulted in 100 % agreement. The metallothionine heavy metal pathway contained the highest percentage of genes with statistically significant changes in expression. The study by Subrungruang et al. (2013) provides exon-level expression profiles in ICC that should be fruitful in identifying novel genetic markers for possible early diagnosis.

Although *Opisthorchis viverrini* is a risk factor for cholangiocarcinoma, not all infected individuals develop cholangiocarcinoma. Zeng et al. (2013) investigated whether the base excision repair enzyme gene polymorphisms with differentiated repair capacities of inflammation-related deoxyribonucleic acid damage may play a key role and such possible effects from those genes may be increased or diminished. This was investigated in co-existence of polymorphisms of metabolic enzymes. Five non-synonymous single-nucleotide polymorphisms of three genes were examined and relations between those polymorphisms and the risk of cholangiocarcinoma were documented. It was found that any single polymorphism did not have a measurable association with the risk of cholangiocarcinoma. Findings by Zeng et al. (2013) suggested that decreased capacity of the deoxyribonucleic acid-repair gene, a human homolog of the 8-oxoguanine glycosylase 1, may be related to decreased risk of cholangiocarcinoma if damaged cells die prior to malignant transformation.

MicroRNA, an endogenous noncoding RNA modulating gene expression, is a key molecule that by its dysregulation plays roles in inflammatory-driven carcinogenesis. A study by Chusorn et al. (2013) aimed to investigate the role of oncomiR miR-21 and its target, the programmed cell death 4 (PDCD4) in tumor growth and metastasis of the liver fluke *Opisthorchis viverrini*-associated CCA. The expression levels of miR-21 and PDCD4 were analyzed using the TaqMan miRNA expression assay and immunohistochemistry in liver tissues of both *O. viverrini* plus N-nitrosodimethylamine (NDMA)-treated hamsters and human CCA samples ($n=23$ cases). The peak of miR-21 levels were reached at 2 (hyperplastic lesions) and 6 months of the *O. viverrini* plus NDMA-induced group and had a reverse response with its target PDCD4 proteins. In human CCA, miR-21 was overexpressed in tumor tissues when compared with nontumor tissues ($P=0.0034$) and had a negative correlation with PDCD4 protein ($P=0.026$). It was also found that high expression of miR-21 was significantly correlated with shorter survival ($P<0.05$) and lymph node metastasis ($P=0.037$) of CCA patients. In the study, Chusorn et al. (2013) found that transient transfection of pre-miR-21 reduced the PDCD4 level and resulted in an increase of M213 CCA cell growth and wound-induced migration ability. These results indicate that miR-21 plays a role in the carcinogenesis and metastasis of *O. viverrini*-associated CCA by suppressing the function of PDCD4. The authors summarize that modulation of aberrantly expressed miR-21 may be a useful strategy to inhibit tumor cell phenotypes or improve response to chemotherapy.

Liquid chromatography in tandem mass spectrometry (LC-MS/MS) has emerged as an informative tool to investigate oxysterols (oxidized derivatives of cholesterol) in helminth parasite associated cancers. Vale et al. (2013) used LC-MS/MS to investigate soluble extracts of the adult developmental stage of *Opisthorchis viverrini* in experimentally infected hamsters. Using comparisons with known bile acids and the metabolites of estrogens, the LC-MS data indicated the existence of novel oxysterol derivatives in *O. viverrini*. Most of these derivatives were ramified at C-17, in similar fashion to bile acids and their conjugated salts. Several were compatible with the presence of an estrogen core, and/or hydroxylation of the steroid aromatic ring A, hydroxylation of both C-2 and C-3 of the steroid ring and further oxidation into an estradiol-2,3-quinone. Since oxysterols can traverse cell membranes more quickly than cholesterol, Vale et al. (2013) suggested that these compounds might enter biliary epithelia and contribute to CCA.

Schistosoma haematobium

Since 1911 epidemiological evidence has indicated that *S. haematobium* is associated with squamous cell carcinoma of the urinary bladder. However, the mechanisms of this interaction are not clearly defined. Using normal epithelial cells, *S. haematobium* parasite extracts induced cancer-like phenotypes such as proliferation, apoptosis, migration, invasion and tumorigenesis. The parasite extracts on normal urothelium also presented carcinogenic and mutagenic ability. To further elucidate the biological effects of this parasite, new estrogenic molecules were identified in its extracts. These estrogens are also present in the sera of *Schistosoma-infected* patients, and they have the ability to repress ER transcriptional activity both in estrogen-responsive MCF7 cells and normal urothelial HCV29 cells. A review by Botelho et al. (2011) presents some of the recent studies of mass spectrometry of *S. haematobium* extracts and sequence analysis of bladder tissue treated with the same extracts.

To investigate whether mutant stem cells participate in inflammation-related carcinogenesis, Ma et al. (2011) performed an immunohistochemical analysis to examine nitrative and oxidative DNA lesions (8-nitroguanine and 8-oxodG) and a stem cell marker Oct3/4 in bladder tissues obtained from cystitis and bladder cancer patients infected with *S. haematobium*. The authors detected the expression of nuclear factor-κB (NF-κB) and inducible nitric oxide synthase (iNOS), which lead to 8-nitroguanine formation. The staining intensity of 8-nitroguanine and 8-oxodG was significantly higher in bladder cancer and cystitis tissues than in normal tissues. In the study, iNOS expression was co-localized with NF-κB in 8-nitroguanine-positive tumor cells from bladder cancer patients. Oct3/4 expression was significantly increased in cells from *S. haematobium*-associated bladder cancer tissues in comparison to normal bladder and cancer tissues without infection. Oct3/4 was also expressed in epithelial cells of cystitis patients. Moreover, 8-nitroguanine was formed in Oct3/4-positive stem cells in *S. haematobium*-associated cystitis and cancer tissues. In conclusion, inflammation by *S.haematobium* infection may increase the number of mutant stem cells, in which iNOS-dependent DNA damage occurs via NF-κB activation, leading to tumor development.

Urogenital schistosomiasis, chronic infection by *Schistosoma haematobium*, affects 112 million people worldwide. *S. haematobium* worm oviposition in the urinary bladder wall leads to granulomatous inflammation, fibrosis, and egg expulsion into the urine. Ray et al. (2012) used a mouse model of urogenital schistosomiasis to perform the first-ever profiling of the early molecular events that occur in the bladder in response to the introduction of *S. haematobium* eggs. Microarray analysis of bladders revealed rapid, differential transcription of large numbers of genes, peaking 3 weeks post-egg administration. Ray et al. (2012) found that many differentially transcribed genes were related to the canonical Type 2 anti-schistosomal immune response, as reflected by the development of egg-based bladder granulomata. Numerous collagen and metalloproteinase genes were differentially transcribed over time, revealing complex remodeling and fibrosis of the bladder that was confirmed by study authors by the use of Masson's Trichrome staining. Multiple genes implicated in carcinogenesis pathways, including vascular endothelial growth factor-, oncogene-, and mammary tumor-related genes, were differentially transcribed in egg-injected bladders. Interestingly, junctional adhesion molecule, claudin and uroplakin genes, key components for maintaining the urothelial barrier, were globally suppressed after bladder exposure to eggs. This occurred in the context of urothelial hyperplasia and egg shedding in urine. Thus, *S. haematobium* egg expulsion is associated with intricate modulation of the urothelial barrier on both the cellular and molecular levels.

Thanan et al. (2012) explored inflammation mediated activation of stem cells via prostaglandin E2 (PGE2) production mediated by cyclooxygenase-2 (COX-2) expression. COX-2 activation is involved in inflammation-mediated stem cell proliferation/differentiation in urinary bladder carcinogenesis. The authors performed an immunohistochemical analysis of the expression of stem cell markers (Oct3/4 and CD44v6) and COX-2 in urinary bladder tissues obtained from cystitis and cancer patients with and without *Schistosoma haematobium* infections. Immunoreactivity to Oct3/4 was significantly higher in *S. haematobium*-associated cystitis and cancer tissues than in normal tissues. In addition, CD44v6 expression was significantly higher in urinary bladder cancer without *S. haematobium* than in normal tissues. COX-2 was located in the cytoplasmic membrane, cytoplasm, and nucleus of the cancer cells. Thanan et al. (2012) found that the nuclear localization of COX-2, which was reported to function as a transcription factor, was significantly associated with the upregulation of Oct3/4 and CD44v6 in bladder cancer tissues with and without *S. haematobium* infection, respectively.

Chronic infection with the blood fluke, *Schistosoma haematobium*, is associated with squamous cell carcinoma of the bladder. Botelho et al. (2013) noted that soluble extracts of mixed sex adult *S. haematobium* worms (SWAP) have previously demonstrated themselves as tumorigenic, both in vitro and in vivo. In addition, estrogen-related molecules in SWAP of S. haematobium down-regulate estrogen receptors (ERs) alpha and beta in estrogen responsive cells. Moreover, schistosome estrogens occur in the sera of infected individuals and repress transcription of ERs in urothelial cells. Given that eggs of *S. haematobium* are the developmental stage directly responsible for urogenital disease during schistosomiasis haematobia, Botelho et al. (2013) suggested that soluble antigens from *S. haematobium* eggs exhibit similar or more potent tumorigenic capacity. The authors also investigated the tumorigenic potential of soluble egg antigens (Sh-SEA) of *S. haematobium* and the endocrine system in favoring parasitism by schistosomes. The findings confirmed that 6.25 μg/ml of Sh-SEA was enough to stimulate cell proliferation, reduce apoptosis and increase oxidative stress of Sh-SEA-exposed urothelial cells. In addition, the genotoxic effects of Sh-SEA on these cells were studied using alkaline single-cell gel electrophoresis. Furthermore, Liquid Chromatography Diode Array Detection Electron Spray Ionisation Mass Spectrometry indicated the presence of catechol-oestrogens in *S. haematobium* SEA. A prospective estrogen-DNA adduct mediated pathway in *S. haematobium* egg induced bladder cancer was also suggested by Botelho et al. (2013).

References

Bian M, Li S, Wang X, Xu Y, Chen W, Zhou C, Chen X, He L, Xu J, Liang C, Wu Z, Huang Y, Li X, Yu X (2014) Identification, immunolocalization, and immunological characterization of nitric oxide synthase-interacting protein from Clonorchis sinensis. Parasitol Res 113(5):1749–1757

Boonjaraspinyo S, Wu Z, Boonmars T, Kaewkes S, Loilome W, Sithithaworn P, Nagano I, Takahashi Y, Yongvanit P, Bhudhisawasdi V (2012) Overexpression of PDGFA and its receptor during carcinogenesis of *Opisthorchis viverrini*-associated cholangiocarcinoma. Parasitol Int 61:145–150

Boonmars T, Wu Z, Boonjaruspinyo S, Puapairoj A, Kaewsamut B, Nagano I, Pinlaor S, Yongvanit P, Wonkchalee O, Juasook A, Sudsarn P, Srisawangwong T (2011) Involvement of c-Ski oncoprotein in carcinogenesis of cholangiocacinoma induced by *Opisthorchis viverrini* and N-nitrosodimethylamine. Pathol Oncol Res 17:219–227

Botelho MC, Machado JC, Brindley PJ, Correia da Costa JM (2011) Targeting molecular signaling pathways of *Schistosoma haemotobium* infection in bladder cancer. Virulence 2:267–279

Botelho MC, Vale N, Gouveia MJ, Rinaldi G, Santos J, Santos LL, Gomes P, Brindley PJ, Correia da Costa JM (2013) Tumour-like phenotypes in urothelial cells after exposure to antigens from eggs of *Schistosoma haematobium*: an oestrogen-DNA adducts mediated pathway? Int J Parasitol 43:17–26

Chen X, Li S, He L, Wang X, Liang P, Chen W, Bian M, Ren M, Lin J, Liang C, Xu J, Wu Z, Li X, Huang Y, Yu X (2013) Molecular characterization of severin from *Clonorchis sinensis* excretory/secretory products and its potential anti-apoptotic role in hepatocarcinoma PLC cells. PLoS Negl Trop Dis 7:1–10

Chusorn P, Namwat N, Loilome W, Techasen A, Pairojkul C, Khuntikeo N, Dechakhamphu A, Talabnin C, Chan-On W, Ong CK, Teh BT, Yongvanit P (2013) Overexpression of microRNA-21 regulating PDCD4 during tumorigenesis of liver fluke-associated cholangiocarcinoma contributes to tumor growth and metastasis. Tumour Biol 34:1579–1588

Fried B, Reddy A, Mayer D (2011) Helminths in human carcinogenesis. Cancer Lett 305:239–249

Hong ST, Fang Y (2012) *Clonorchis sinensis* and clonorchiasis, an update. Parasitol Int 61:17–24

Huang Y, Liao H, Li W, Hu Y, Huang L, Wang X, Sun J, Chen W, Deng C, Liang C, Wu Z, Li X, Xu J, Yu X (2013) Identification, sequence analysis and characterization of *Clonorchis sinensis* ubiquitin. Exp Parasitol 133:62–69

Jex AR, Young ND, Sripa J, Hall RS, Scheerlinck JP, Laha T, Sripa B, Gasser RB (2012) Molecular changes in *Opisthorchis viverrini* (Southeast Asian liver fluke) during the transition from the juvenile to the adult stage. PLoS Negl Trop Dis 6:e1916

Kaewkong W, Choochote W, Kanla P, Maleewong W, Intapan PM, Wongkham S, Wongkham C (2012) Chromosomes and karyotype analysis of a liver fluke, *Opisthorchis viverrini*, by scanning electron microscopy. Parasitol Int 61:504–507

Khoontawad J, Laothong U, Roytrakul S, Pinlaor P, Mulvenna J, Wongkham C, Yongvanit P, Pairojkul C, Mairiang E, Sithithaworn P, Pinlaor S (2012) Proteomic identification of plasma protein tyrosine phosphatase alpha and fibronectin associated with liver fluke, *Opisthorchis viverrini*, infection. PLoS One 7:e45460

Laothong U, Pinlaor P, Boonsiri P, Pairojkul C, Priprem A, Johns NP, Charoensuk L, Intuyod K, Pinlaor S (2013) Melatonin inhibits cholangiocarcinoma and reduces liver injury in *Opisthorchis viverrini*-infected and N-nitrosodimethylamine-treated hamsters. J Pineal Res 55:257–266

Li Y, Huang Y, Hu X, Liu X, Ma C, Zhao J, Wu Z, Xu J, Yu X (2012) 41.5-kDa Cathepsin L protease from *Clonorchis sinensis*: expression, characterization, and serological reactivity of one excretory-secretory antigen. Parasitol Res 111:673–680

Lv X, Chen W, Wang X, Li X, Sun J, Deng C, Men J, Tian Y, Zhou C, Lei H, Liang C, Yu X (2012) Molecular characterization and expression of a cysteine protease from *Clonorchis sinensis* and its application for serodiagnosis of clonorchiasis. Parasitol Res 110:2211–2219

Ma N, Thanan R, Kobayashi H, Hammam O, Wishahi M, El Leithy T, Hiraku Y, Amro E, Oikawa S, Ohnishi S, Murata M, Kawanishi S (2011) Nitrative DNA damage and Oct3/4 expression in urinary bladder cancer with *Schistosoma haematobium* infection. Biochem Biophys Res Commun 414:344–349

Murata M, Thanan R, Ma N, Kawanishi S (2012) Role of nitrative and oxidative DNA damage in inflammation-related carcinogenesis. J Biomed Biotechnol 2012:623019

Nam JH, Moon JH, Kim IK, Lee MR, Hong SJ, Ahn JH, Chung JW, Pak JH (2012) Free radicals enzymatically triggered by Clonorchis sinensis excretory-secretory products cause NF-κB-mediated inflammation in human cholangiocarcinoma cells. Int J Parasitol 42:103–113

Qiao T, Ma RH, Luo XB, Zheng PM, Luo ZL, Yang LQ (2013) Microscopic examination of gallbladder stones improves rate of detection of *Clonorchis sinensis* infection. J Clin Microbiol 51:2551–2555

Ray D, Nelson TA, Fu CL, Patel S, Gong DN, Odegaard JI, Hsieh MH (2012) Transcriptional profiling of the bladder in urogenital schistosomiasis reveals pathways of inflammatory fibrosis and urothelial compromise. PLoS Negl Trop Dis 6:e1912

Rucksaken R, Khoontawad J, Roytrakul S, Pinlaor P, Hiraku Y, Wongkham C, Pairojkul C, Boonmars T, Pinlaor S (2012) Proteomic analysis to identify plasma orosomucoid 2 and kinesin 18A as potential biomarkers of cholangiocarcinoma. Cancer Biomark 12:81–95

Sawanyawisuth K, Silsirivanit A, Kunlabut K, Tantapotinan N, Vaeteewoottacharn K, Wongkham S (2012) A novel carbohydrate antigen expression during development of *Opisthorchis viverrini*- associated cholangiocarcinoma in golden hamster: a potential marker for early diagnosis. Parasitol Int 61:151–154

Subrungruanga I, Thawornkunob C, Chawalitchewinkoon-Petmitrc P, Pairojkul C, Wongkham S, Petmitrb S (2013) Gene expression profiling of intrahepatic cholangiocarcinoma. Asian Pac J Cancer Prev 14:557–563

Thanan R, Murata M, Ma N, Hammam O, Wishahi M, El Leithy T, Hiraku Y, Oikawa S, Kawanishi S (2012) Nuclear localization of COX-2 in relation to the expression of stemness markers in urinary bladder cancer. Mediators Inflamm 2012:165879

Vale N, Gouveia MJ, Botelho M, Sripa B, Suttiprapa S, Rinaldi G, Gomes P, Brindley PJ, Correia da Costa JM (2013) Carcinogenic liver fluke *Opisthorchis viverrini* oxysterols detected by LC-MS/MS survey of soluble fraction parasite extract. Parasitol Int 62:535–542

Yongvanit P, Pinlaor S, Bartsch H (2012) Oxidative and nitrative DNA damage: key events in opisthorchiasis-induced carcinogenesis. Parasitol Int 61:130–135

Zeng L, You G, Tanaka H, Srivatanakul P, Ohta E, Viwatthanasittiphong C, Matharit M, Chenvidhya D, Jedpiyawongse A, Tanaka M, Fujii T, Sripa B, Ohshima K, Miwa M, Honjo S (2013) Combined effects of polymorphisms of DNA-repair protein genes and metabolic enzyme genes on the risk of cholangiocarcinoma. Jpn J Clin Oncol 43:1190–1194

Zhang F, Liang P, Chen W, Wang X, Hu Y, Liang C, Sun J, Huang Y, Li R, Li X, Xu J, Yu X (2013) Stage-specific expression, immunolocalization of *Clonorchis sinensis* lysophospholipase and its potential role in hepatic fibrosis. Parasitol Res 112:737–749

Zheng M, Hu K, Liu W, Hu X, Hu F, Huang L, Wang P, Hu Y, Huang Y, Li W, Liang C, Yin X, He Q, Yu X (2011) Proteomic analysis of excretory secretory products from *Clonorchis sinensis* adult worms: molecular characterization and serological reactivity of a excretory-secretory antigen-fructose-1,6-bisphosphatase. Parasitol Res 109:737–744

Zheng M, Hu K, Liu W, Li H, Chen J, Yu X (2013) Proteomic analysis of different period excretory secretory products from *Clonorchis sinensis* adult worms: molecular characterization, immunolocalization, and serological reactivity of two excretory secretory antigens-methionine aminopeptidase 2 and acid phosphatase. Parasitol Res 112:1287–1297

Chapter 7
Infection-Associated Hematological Malignancies

Dmitriy W. Gutkin

Abstract The association between infectious agents and hematologic human malignancies has been actively studied, and numerous viral and bacterial agents were found to play a significant role in the pathogenesis of these diseases. However, the exact molecular mechanisms of infection-induced oncogenesis are still not completely understood. Three viruses, Epstein-Barr virus (EBV), Human T-cell lymphotropic virus I (HTLV-I), and Kaposi's sarcoma herpesvirus (KSHV), and one bacteria, Helicobacter pylori (H. pylori) have been definitely associated with human hematologic malignancies, particularly, lymphoid neoplasms. Although these agents employ very different specific oncogenic mechanisms, they converge on several common intercellular pathways (such as cell-cycle regulation, proliferation and apoptosis) that eventually lead to malignant transformation. Environmental and host cofactors such as immunosuppression, genetic predisposition, and mutagens can accelerate the development of these neoplasms. The study of infectious agents and of the multiple mechanisms deployed by them has improved our current understanding of cancer biology. The emerging information may expedite the development of new targeted approaches to prevent and treat infection-associated hematologic malignancies. In this review we will focus on the molecular mechanisms that are involved in the development of the wide spectrum of hematologic malignancies, associated with EBV, HTLV-1, KSHV and H. pylori, as well as their main clinical features.

Keywords Hematological malignancy • Burkitt lymphoma • Hodgkin lymphoma • Diffuse large B-cell lymphoma • MALT lymphoma • Extranodal NK/T cell lymphoma • Angioimmunoblastic T-cell lymphoma • Immunodeficiency-associated lymphoproliferative disorder • Epstein-Barr virus • Human T-cell lymphotropic virus I • Kaposi's sarcoma herpesvirus • Helicobacter pylori

---

D.W. Gutkin (✉)
Department of Pathology, S735, University of Pittsburgh Medical Center, 3550 Terrace Street, Pittsburgh, 15261 PA, USA
e-mail: DWGutkin@hotmail.com

M.R. Shurin et al. (eds.), *Infection and Cancer: Bi-Directorial Interactions*, DOI 10.1007/978-3-319-20669-1_7

Abbreviations

AITL	Angioimmunoblastic T-cell lymphoma
BL	Burkitt lymphoma
CNS	central nervous system
DLBCL	diffuse large B cell lymphoma
EBV	Epstein-Barr virus
HHV-8	Human Herpesvirus-8
HL	Hodgkin lymphoma
HTLV-I	Human T-cell lymphotropic virus I
KSHV	Kaposi's sarcoma herpesvirus
LANA	latency-associated nuclear antigen
LYG	Lymphomatoid granulomatosis
MALT	mucosa associated lymphoid tissue
PEL	primary effusion lymphoma
PTLD	Post-transplant lymphoproliferative diseases

Introduction

Fifty years ago herpesvirus particles were identified by Epstein and co-authors in high-grade lymphoma (Epstein et al. 1964). Since that time, the associations between infectious agents and hematologic human malignancies has been actively studied, and numerous viral and bacterial agents were found to play a significant role in the pathogenesis of these diseases. However, the exact molecular mechanisms of infection-induced oncogenesis are still not completely understood.

According to the current WHO classification, hematologic malignancies are tumors of hematopoetic and lymphoid tissues, and are stratified according to their lineage as myeloid, lymphoid, and histiocytic/dendritic neoplasms (Harris et al. 2008). Interestingly, infectious agents have been implicated in the pathogenesis of only one group of these tumors, namely, lymphoid neoplasms. However, the lymphoid tumors associated with infections include a wide variety of diseases, from the extremely aggressive ones, like Burkitt lymphoma, to rather indolent ones, like extranodal MALT lymphoma.

Lymphoma is defined as a cancer of lymphocytes, typically presenting as a solid tumor, with malignant cells often originating in and involving lymph nodes. The recent classification, categorizes lymphomas into Precursor (immature cell) lymphoid neoplasms, Mature B-cell neoplasms, T-cell and NK-cell neoplasms, Hodgkin lymphoma, and Post-transplant lymphoproliferative disorders (Jaffe 2009).

Recent literature reviews have identified several common traits typical for infectious lymphomagenesis: (1) infectious agents are necessary but not sufficient for tumor development, and thus the incidence of lymphoma is much lower than infection prevalence in human populations; (2) malignancies appear in the context of persistent infections and much later than the first episode of infection; (3) the innate and adoptive immune system and features of microenvironment can play a deleterious or a protective role, depending on a specific situation. This indicates that within the context of multistep oncogenesis, infection provides only a subset of the required oncogenic hits. Additional cofactors such as immunosuppression, chronic inflammation, somatic mutations, genetic predisposition, and exposure to carcinogens are necessary for malignant transformation (Bouvard et al. 2009; Mesri et al. 2014).

Three viruses, Epstein-Barr virus (EBV), Human T-cell lymphotropic virus I (HTLV-I), Kaposi's sarcoma herpesvirus (KSHV), and one bacteria, Helicobacter pylori have been definitely associated with lymphoid neoplasms.

Numerous other infectious agents (hepatitis C virus, Parvovirus B19, Campylobacter jejuni, Chlamydia psittaci, Borrelia afzelii) have been implicated in different lymphoproliferative disorders, but their etiologic significance is less certain (Ferreri et al. 2004; Lecuit et al. 2004; Matsuo et al. 2004; Engels 2007).

Additionally, HIVinfection increases the risk of lymphoma likely due to its immunosuppressive nature, but has not been definitively linked to oncogenesis (Carbone 2003; Mbulaiteye et al. 2003).

In this review we will focus on the molecular mechanisms that are involved in the development of the wide spectrum of lymphomas, associated with EBV, HTLV-1, KSHV and H. pylori and their clinical features. Therapeutic management of these diseases is beyond the scope of the article, but can be found in several recent reviews (Carbone et al. 2008; Roschewski and Wilson 2012; Pereira and Medeiros 2014).

Epstein-Barr Virus (EBV)

EBV was originally discovered in cultured lymphoblasts from Burkitt's lymphoma patients in 1964 (Epstein et al. 1964), and has been defined as a "carcinogenic agent" by WHO since 1997 (Roschewski and Wilson 2012). EBV is a human DNA virus of the herpesviridae family. It has a doublestranded DNA genome measuring approximately 170 kb in length and encoding >80 genes. Mature virions are 120–180 nm in diameter. It is an ubiquitous virus infecting >90 % of the population worldwide (Young and Rickinson 2004). Most individuals are infected with EBV during the first 3 years of life, at which time it is typically an asymptomatic process. However, infection in older age is frequently symptomatic and can manifest as infectious mononucleosis (Parkin 2006; Javier and Butel 2008). EBV infections are most prevalent in developing countries, in populations of low socioeconomic status (Young and Rickinson 2004).

Means of transmission for EBV is through saliva. The primary site of infection is the oropharyngeal cavity, and EBV is capable of infecting both B cells and epithelial cells and switching between the two. EBV preferentially infects B-lymphocytes by binding to the cell surface CD21 receptor and HLA class II molecules as a co-receptor. Transformation of B cells is a highly efficient process requiring a large portion of the EBV genome, which becomes circular for replication and latency (Liao 2006). After primary infection, the EBV episome largely remains latent in resting memory B cells in most patients, thereby enabling persistent infection in a non-replicative (latent) form (Cohen 2000; Thorley-Lawson and Gross 2004). Latent infection is characterized by maintenance of the genome and expression of a limited number of genes, including six nuclear antigens (EBNAs-1, -2, -3a, -3b, -3c, and -LP) and three latent membrane proteins (LMP-1, -2A, and -2B). Also expressed are 2 small non-coding RNAs, EBER-1 and EBER-2, as well as BamHI-A rightward transcripts (BART) (Young and Murray 2003; Saha and Robertson 2011).

Three types of EBV latency have been described. Type I latency is characterized by the expression of EBNA-1 and two small noncoding Epstein–Barr RNAs (EBERs) (Carbone et al. 2008). EBV gene expression in latency II is limited to EBNA-1, EBERs, LMP-1, LMP-2A and LMP-2B. Latency III usually involves the unrestricted expression of all EBNAs, EBERs, and LMPs and occurs mainly in immunocompromised individuals (Young and Murray 2003). These three distinct patterns of latency programs are associated with different types of lymphomas. The pattern of latency I is seen in Burkitt's lymphoma. A pattern of latency II is found in Hodgkin's lymphoma and peripheral T-cell lymphoma. The pattern of latency III, also known as the "growth program," is commonly found in post-transplant lymphoproliferative disorders (PTLD) (Roschewski and Wilson 2012).

Etiologic significance of EBV in lymphomagenesis has been proven in vitro and in vivo.

In vitro, EBV can transform B-lymphocytes into cells that proliferate in an unregulated fashion (Thorley-Lawson and Gross 2004). In vivo EBV infection immortalizes B lymphocytes causing polyclonal proliferation, which could result in the escape of some transformed clones. (Martin and Gutkind 2008). Although all EBV proteins and RNAs may have some oncogenic potential, three of them are especially important in malignant transformation: EBNA-1, LMP-1 and LMP-2A. EBNA-1 is expressed in all EBV-associated malignances and is essential for viral replication (Levitskaya et al. 1995; Wilson et al. 1996).

Several oncogenic mechanisms of EBNA-1 are described:

- It induces chromosomal aberrations and double-strand breaks through activating reactive oxygen species (ROS) production (Gruhne et al. 2009a, b).
- It is necessary for survival of EBV-positive malignant cells possibly by inhibiting apoptosis (Kennedy et al. 2003; Saha and Robertson 2011).
- It can promote genetic instability and produce telomere dysfunction in transformed B-cells (Kamranvar et al. 2007; Kamranvar and Masucci 2011).
- It can bind to numerous sites on the host chromosome (Canaan et al. 2009; Dresang et al. 2009; Lu et al. 2010; Sompallae et al. 2010) and may alter chromatin structure or nucleosome positioning at those sites (Wang and Frappier 2009). EBNA1 binding to the host genome promotes a global change in chromatin organization that results in more open chromatin structure (Coppotelli et al. 2013), and can alter host cell gene expression (Tempera and Lieberman 2014).
- It is intricately involved in evading the host immunity. EBNA1 contains a Gly/Ala repeat sequence, through which proteasomal degradation and antigen presentation of the protein are impaired (Levitskaya et al. 1995).

It is important to reiterate that EBNA-1 is present in all types of EBV latency and thus can be involved in the development of all EBV-associated hematologic malignancies. Other two oncogenic proteins, LMP-1 and LMP-2A are expressed in latencies II and III, and also have numerous mechanisms of action.

- LMP-1 acts as a constitutively active receptor that mimics activated CD40, a member of the tumor necrosis factor receptor family (Mosialos et al. 1995) and can mediate some of the proliferation signals driving the B-cell population expansion (Arvanitakis et al. 1995; Soni et al. 2007). The cytoplasmic carboxyl terminus of LMP-1 binds to a tumor necrosis factor receptor associated factor and the tumor necrosis factor receptor associated death domain protein (Brown et al. 2001). This, in turn, induces the activation of several key signaling molecules such as PI3K, JNK and JAKs leading to the activation of transcription factors including NF-kB, AP-1 and STATs (Kilger et al. 1998).
- LMP-1 promotes genomic instability by inhibiting DNA repair pathways and inactivating the DNA damage checkpoint (Liu et al. 2004, 2005; Gruhne et al. 2009a, b; Kim et al. 2013).
- It can up-regulate the expression of the anti-apoptotic proteins Bcl-2, MCL-1, and bfl-1, and the process involves the induction of cell surface adhesion and the TNFR/CD40 pathway. This provides both growth and differentiation responses and is associated with activation of a number of signaling pathways (Gregory et al. 1991; Laherty et al. 1992).
- It can block apoptotic signals delivered through the Fas/Fas ligand and TRAIL/ death receptor pathways (Snow et al. 2006).
- It can cause hypermethylation of a set of cellular promoters (Rossi et al. 2003; Murray et al. 2004; Doerr et al. 2005; Ushmorov et al. 2006).
- It induces proinflammatory cytokines such as IL-6 and IL-8 (Eliopoulos et al. 1997, 1999).
- It participates in immune evasion of EBV-associated malignant cells (Middeldorp and Pegtel 2008; Munz and Moormann 2008).
- LMP2A is structurally and functionally related to B-cell receptor (BCR) and can activate phosphatidylinositol 3-kinase/AKT, NF-κB, NOTCH, and mitogen-activated protein kinases (Portis et al. 2004). Interestingly, LMP2A appears to have transformation effects in epithelial cells but not in B cells (Fu et al. 2013).

In addition to EBNA-1, LMP-1 and LMP-2A, several other EBV molecules can play a role in lymphomagenesis. Specifically, EBV-encoded noncoding RNAs (EBERs) and microRNAs (BART and BHRLF1 miRNAs) have been implicated in oncogenesis (Lopes et al. 2013; Vereide et al. 2013), although the targets and mechanisms of their action are not yet known (Murata et al. 2014). As small RNAs, EBERs may confer an apoptotic-resistant phenotype by up-regulating Bcl-2 expression in EBV-associated malignant tumors (Fu et al. 2013).

EBV nuclear antigen (EBNA) 3A/C also plays a role in maintenance and formation of cancer cells by silencing tumor suppressor genes (Maruo et al. 2011; Skalska et al. 2013). Undoubtedly, the list of EBV- associated mechanisms of lymphomagenesis will be growing. As mentioned above, the main cellular hosts of latent EBV infection are B lymphocytes of germinal center. Accordingly, the majority of EBV-associated lymphoid neoplasms are the tumors arising from B cells with germinal center phenotype. The most significant ones are Burkitt lymphoma and Hodgkin lymphoma.

Burkitt Lymphoma (BL)

BL is a highly aggressive lymphoma and has the fastest doubling time among human tumors (de Leval and Hasserjian 2009). BL can be subdivided into three clinical variants: endemic BL, sporadic BL, and immunodeficiency-associated BL with important differences in epidemiology, clinical presentation, and biology.

All forms of BL share a crucial common feature, a chromosomal translocation that places the c-myc oncogene (on chromosome 8) under the control of an Ig gene locus (either the heavy chain locus on chromosome 14 or one of the light chain loci on chromosome 2 or 22), thereby releasing c-myc expression from the usual tight controls (Leoncini et al. 2008). While additional cellular genetic changes are required to achieve full malignancy (Schmitz et al. 2012), de-regulated expression of the c-myc protein is the prime determinant of the BL cell phenotype. Importantly, this translocation is thought to occur as an accident of the process of somatic Ig gene mutation which is a natural accompaniment of a B cell's germinal center transit. Indeed, BL cells strongly resemble germinal centroblasts and the tumor is considered to be a classical malignancy of germinal center origin (Leoncini et al. 2008).

BL is associated with the EBV of type I latency and only EBNA1 and EBERs are expressed in the lymphomas (Young and Rickinson 2004; Rowe et al. 2009). Interestingly, there is a marked variation in geographical distribution and degree of association with EBV. About 95 % of endemic BL cases are associated with EBV and are commonly found in equatorial Africa and Papua New Guinea (van den Bosch 2004; Young and Rickinson 2004). In contrast, only 5–15 % of sporadic BL, affecting children and young adults throughout the world and 40 % of HIV-associated BL are EBV positive (Wright 1999). Subtypes of BL also differ in clinical manifestation. Typically, endemic BL presents as tumors affecting the jaw and facial bones, while sporadic BL more commonly arises in the gut and upper respiratory tract (Yustein and Dang 2007). HIV-associated BL characteristically involves the lymph nodes and bone marrow (Leoncini et al. 2008).

As it was mentioned above, EBNA-1 protein is poorly antigenic and has little to no HLA class I response. Thus, the CD8+ T-cell response to BL is largely diminished. Various pathways by which BL escapes immune detection by inhibiting both HLA class I- and II-mediated antigen presentation to T cells are essential to the disease pathogenesis (Munz and Moormann 2008; God and Haque 2010).

The disparity between the epidemiology of EBVinfection and the epidemiology of BL led to the current view that genetic or environmental factors (for example, parasitic infections such as malaria) also have a crucial role in the development of this disease (Martin and Gutkind 2008). Specifically for endemic BL it is believed that chronic malarial infection increases BL risk through expanding germinal center activity, hence increasing the risk of the c-myc/Ig gene translocation occurring (Rickinson 2014). However, epidemiologic studies suggest that malaria and EBV alone cannot account for the distribution of endemic BL in high risk regions. Other possible local cofactors (i.e. Arboviruses) could explain such characteristics (van den Bosch 2004).

Morphologically BL is composed of medium-sized cells with a diffuse growth pattern. The cells are loosely cohesive. They have round nuclei with finely clumped chromatin and multiple nucleoli. The cytoplasm is deeply basophilic and contains lipid vacuoles. The tumor has an extremely high proliferation rate as well as a high fraction of apoptosis. The tumor is highly aggressive but potentially curable (Leoncini et al. 2008).

Hodgkin Lymphoma (HL)

HL was first described by the British pathologist Thomas Hodgkin in 1832 as a primary disorder of the lymphatic glands. HL has unique clinico-pathologic features, with the characteristic Reed-Sternberg cells and its variants being the specific neoplastic cells in HL. Hodgkin's lymphoma accounts for about 1 % of all cancers and 30 % of lymphoid malignancies worldwide (Grywalska et al. 2013). EBV-positive HL has expression of EBNA-1, LMP-1, LMP-2A and the EBERs, indicative of type II latency (Deacon et al. 1993).

According to the WHO classification, HL can be histologically divided into four subtypes: lymphocyte predominant, nodular sclerosing, mixed cellularity, and lymphocyte depleted (Stein et al. 2008).

All subtypes of HL usually arise in lymph nodes, preferentially in the cervical region and the majority of them manifest clinically in young adults. Morphologically, neoplastic tissues contain scattered large mononucleated or multinucleated tumor cells (Hodgkin and Reed- Sternberg cells), residing in an abundant heterogenous admixture of non-neoplastic small lymphocytes, eosinophils, neutrophils, histiocytes, plasma cells, and fibroblasts. The neoplastic cells represent only a minority of the cell infiltrate (1–10 %). The composition of the reactive cellular infiltrate varies according to the histological subtype (Stein et al. 2008).

Intriguingly, not all subtypes harbor EBV to the same degree. EBV positivity in lymphoma tissue is found in 95 % of lymphocyte depleted HD, 70 % of mixed cellularity HD, and 10–40 % of nodular sclerosing HD. On the other hand, lymphocyte predominant Hodgkin's disease subtype is almost always EBV negative (Chapman and Rickinson 1998). The most likely role of EBV in HL pathogenesis is that the expression of LMP-1 and LMP-2A prevents apoptosis by mimicking CD40 and BCR signaling, respectively (Kapatai and Murray 2007).

EBV in HL has been reported to negatively affect prognosis. Detection of EBV DNA in peripheral blood of patients with HL correlates with prognostic factors and may serve as a useful biomarker for disease activity in EBV-associated HL (Hohaus et al. 2011).

Other B Cell Lymphomas

EBV is also found in association with number of other, less common, B cell tumors, many of which are classified histologically as subtypes of diffuse large B cell lymphoma (DLBCL). However, all arise in particular circumstances and are distinct from DLBCLs as seen in the general population.

Lymphomatoid Granulomatosis (LYG)

LYG is a rare angiodestructive EBV-driven lymphoproliferative disease comprised of atypical clonal B-cells in an inflammatory background. Evidence of immune dysregulation can be found in many patients, and individuals with known immunodeficiency are at increased risk (Wilson et al. 1996). Patients present with multiple pulmonary nodules of varying size in the mid and lower lung fields, often with evidence of central necrosis and/or cavitation. Other common sites of extranodal involvement include the central nervous system (CNS) and skin in up to 20 % of patients (Jaffe and Wilson 1997). One striking feature of LYG is that lymph nodes and spleen are almost always spared at initial diagnosis and only involved at late stages of disease (Roschewski and Wilson 2012).

Histologically, LYG consists of a small number of EBV positive B-cells admixed with a prominent inflammatory background comprised of T-cells, plasma cells, and histiocytes. The malignant B-cells usually are large in size and express LMP-1 and EBER. Vascular changes and angiodestruction are distinctive features with intimal thickening of blood vessels and accompanying necrosis in many cases (Putaluga et al. 2008).

Diffuse Large B-Cell Lymphoma of the Elderly

DLBCL of the elderly occurs in patients older than 50 years and without any known immunodeficiency. The proportion of EBV positive cases increases with increasing patient age. It's believed to be related to immunologic deterioration or senescence in immunity that is a part of the aging process. The majority of patients present with extranodal disease, involving skin, lung, stomach, with or without simultaneous lymph node involvement. (Kuze et al. 2000). LMP-1 can be detected in the majority of cases and EBNA-2 is found in about 25–35 % of cases (Oyama et al. 2003, 2007).

Histologically, the tumor consists of large transformed cells, immunoblasts and Reed-Sternberg-like cells, mixed with reactive inflammatory cells, such as small lymphocytes, plasma cells, and histiocytes (Nakamura et al. 2008).

Diffuse Large B-Cell Lymphoma Associated with Chronic Inflammation

This type of B-cell lymphomas occurs in the context of long-standing chronic inflammation and mainly involves body cavities (pleural, peritoneal). The interval between the onset of chronic inflammation and malignant lymphoma is usually over 10 years. It is strongly associated with EBV, expressing LMP-1 together with EBNA-1and EBNA-2. The EBV latency pattern is type III in more than 60 % of cases (Sasajima et al. 1993; Takakuwa et al. 2003). Chronic inflammation at the local site enables EBV transformed B cells to escape from the host immune surveillance through production of IL-10, an immunosuppressive cytokine and providing autocrine to paracrine growth through IL-6 and IL-6 receptor. Morphologically the tumor is composed of large transformed cells with round nuclei and multiple nucleoli. Massive necrosis may be present (Chan et al. 2008).

Tumors of Non-B Cell Origin

The majority of EBV-associated lymphomas are of B-cell origin, however several tumors of non-B cell origin are described. They probably arise from rare circumstances in which the virus gains entry into an unnatural host cell type, specifically CD4+ or CD8+ T cells and/or NK cells, leading to a lymphomatous transformation (Rickinson 2014). Two of these entities are well characterized: extranodal NK/T cell lymphoma and angioimmunoblastic T-cell lymphoma.

Extranodal NK/T Cell Lymphoma

This is a rare condition of NK-cell or cytotoxic T-cell origin that usually affects immunocompetent middle aged men of Asian, Native American or Central/South American descent. It mostly occurs in the upper aerodigestive tract such as the nasopharynx and paranasal cavity, but can occur in other sites such as the skin, kidney, and gastrointestinal tract. Patients with nasal involvement present due to mass effect and commonly have significant associated facial destruction and tissue necrosis. EBV is associated with virtually all cases of extranodal NK/T cell exhibiting a type II latency pattern with expression of LMP-1 and EBNA-1 (Chuang et al. 2007; Gualco et al. 2011).

It is not known how EBV enters into T or NK cells. Most likely it only occurs when mature T or NK cells are in contact with foci of high EBV replication (Rickinson 2014). EBV viral load is intimately tied to prognosis, clinical course, and disease relapse (Au et al. 2004). Histologic features include endothelial damage and angioinvasion by medium-large lymphoma cells which are usually positive for T cell marker CD3 and NK cell marker CD56 (Roschewski and Wilson 2012).

Angioimmunoblastic T-Cell Lymphoma (AITL)

AITL is a rare neoplasm but represents the most common subtype of peripheral T-cell lymphoma (Jaffe and Ralfkiaer 2001). Viral genome is detected in up to 100 % of AITL (Anagnostopoulos et al. 1992; Weiss et al. 1992). Curiously, the malignant T-cells are usually negative for EBV and it is the background B-cells that are infected (Dogan et al. 2003). The putative cell of origin in AITL is now recognized to be a CD4+ T-cell of germinal center origin, known as a follicular helper T (TFH) cell (Dogan et al. 2003; Alizadeh and Advani 2008). EBV immunoblasts that resemble R-S cells are often found in lymph nodes of patients with AITL early in the disease course, raising the hypothesis that EBV plays a role in TFH cell activation. Reports of expanded B-cell clones that give rise to EBV-driven B-cell lymphomas (such as DLBCL) in patients with AITL are not uncommon. Elevated viral load has been associated both with B-cell clonal disorders and higher risk of disease progression (Zhou et al. 2007).

The exact role of EBV in lymphomagenesis is not completely understood, but it might involve upregulation of the CD28 ligand by EBV positive B-cells which leads to up-regulation and activation of TFH cells and production of chemokines such as CXCL13 (Dunleavy et al. 2007). Clinically AITL is characterized by systemic disease with a widespread lymphadenopathy, immune-mediated hemolysis, and polyclonal hypergammaglobulinemia, and presents poor prognosis (Iannitto et al. 2008, Grywalska et al. 2013).

Immunodeficiency-Associated Lymphoproliferative Disorders

Patients with congenital, acquired, or iatrogenic defects in cellular immunity are at risk for EBV-associated lymphomas. Defects in cellular immunity allow EBV positive cells to proliferate in an unregulated fashion. These lymphomas are usually associated with a type III latency program expressing multiple latency proteins. The cellular origin of most of these lymphomas is the germinal center B cells (Murata et al. 2014).

Post-Transplant Lymphoproliferative Diseases (PTLD)

PTLD are lymphomas that occur in the setting of acquired immune deficiency after allogeneic transplantation of either solid organs or hematopoetic stem cells. The clinical presentation of PTLD varies considerably and can be disseminated or localized. Involvement is frequently extranodal and includes the transplanted organ itself and sanctuary-sites such as the CNS (Evens et al. 2010). They are usually of B-cell origin but 10–15 % of PTLDs are of T/NK-cell origin (Draoua et al. 2004). EBV has been linked to most PTLD cases, with a near 100 % association (Brink et al. 1997). The wide expression of the latent EBV-encoded proteins suggests an important role that EBV plays in the oncogenic process by rescuing these cells from apoptosis (Timms et al. 2003; Capello et al. 2005).

AIDS-Associated Lymphomas

These types of cancer occur in HIV-positive patients and may present as an initial clinical manifestation of AIDS. HIV infection impairs cellular immunity, therefore predisposing the host to the development of lymphomas (Raphael et al. 2008). Depending on cell morphology and clinical presentation AIDS-related lymphomas can include BL, DLBCL, primary CNS lymphoma, HL and other (Long et al. 2008; Raphael et al. 2008). EBV is identified in the neoplastic cells of approximately 40 % of cases, but it varies considerably with histologic subtypes. EBV is associated with almost all primary CNS lymphoma and HL, 80 % of DLBCL, and 30–50 % of BL (Raphael et al. 2008). These lymphomas are usually aggressive in nature, causing significant morbidity and mortality in immunodeficient hosts (Roschewski and Wilson 2012).

In summary, EBV is the first human virus associated with hematologic malignancies (lymphomas). The majority of these lymphomas are of B cell origin with the germinal center phenotype. They develop after a prolonged period of latent infection. EBV latency proteins and RNAs interfere with intracellular proliferative and apoptotic pathways, eventually shifting a balance towards cell proliferation and facilitating the survival of cells with somatic mutations that leads to the development of the malignant clone.

Human T-Cell Lymphotropic Virus I (HTLV-I)

HTLV-1 is a slow transforming, single stranded RNA retrovirus, a member of the delta retrovirus family. It possesses a diploid genome similar to other retroviruses. HTLV-1 displays a special tropism for CD4 cells. After infection, HTLV-I promotes clonal proliferation of infected cells (Liao 2006). It has been identified in 1980 in cultured human T-cell lymphoma cells, and is now considered as etiologic agent of Adult T-cell leukemia-lymphoma (ATLL) (Poiesz et al. 1980; Martin and Gutkind 2008). Between 10 and 20 million people worldwide are infected with HTLV-1 (Martin and Gutkind 2008), but ATLL impacts a small minority of patients. The cumulative risk of ATLL among HTLV-I carriers in Japan was estimated at about 6.6 % for men and 2.1 % for women. Most patients infected by HTLV-I are carriers and asymptomatic for the duration of life (Arisawa et al. 2000). The virus is transmitted through blood transfusions, sexual contact, and parturition.

ATLL was first described as a distinct clinical entity in 1977 based on its unique demographic distribution and clinicopathologic features (Uchiyama et al. 1977). It is geographically clustered, involving populations in the Caribbean, Japan, western Africa, and parts of South America and Central Asia. The average age of those with ATLL at the time of diagnosis is 40 years (Liao 2006).

The molecular mechanism by which HTLV-1 causes cancer is still uncertain, but it is clear that it involves the function of a virally encoded essential protein called Tax. The mechanism by which Tax induces transformation seems to involve the alteration of several cell growth regulatory pathways, as well as, epigenetic mechanisms and interference with the cellular DNA repair apparatus (Yoshida 2001). Tax strongly activates the NF-kB, AP-1 and CRE pathways that have a particularly potent proliferative effect on lymphocytes. It also interacts with several factors regulating chromatin remodeling, and causes the inhibition of DNA repair mechanisms, thus leading to genomic instability and transformation (Matsuoka and Jeang 2007; Qu and Xiao 2011).

Malignant cells are usually pleomorphic, and have a clonal rearrangement of T cell receptor genes. They have a T-cell immunophenotype, expressing CD4, CD25, and frequently FOXP3 thus showing features of regulatory T cells (Karube et al. 2004).

There are four described variants of ATLL: chronic, smoldering, acute, and lymphoma subtypes. The chronic and smoldering variants are generally more indolent, and exhibit overall survival rates at 4 years of approximately 50 %. The outcomes for the acute and lymphomatous forms are significantly worse and median survival is in the range of 12 months (Yamada et al. 2001; Tsukasaki et al. 2007). The acute variant is the most common and is characterized by leukemia with markedly elevated white blood cell count, skin rash, and generalized lymphadenopathy. Many patients have an associated T-cell immunodeficiency with opportunistic infections such as Pneumocystis jirovecii. The lymphomatous variant is characterized by a prominent lymphadenopathy, but without leukemia (Ohshima et al. 2008).

Kaposi's Sarcoma Herpesvirus (KSHV)

Kaposi's sarcoma herpesvirus (KSHV), also known as human herpesvirus 8 (HHV-8), has been recently associated with the pathogenesis of several malignant diseases, including an unusual type of lymphoma, called primary effusion lymphoma, in addition to Kaposi's sarcoma and multicentric Castleman disease (Cai et al. 2005). The KSHV genome is a linear, double-stranded DNA of approximately 165–170 kb in length (Renne et al. 1996). During latency, it may also exist in a circular, episomal form in the host nucleus (Lagunoff and Ganem 1997). Among the viruses that infect humans, KSHV is most closely related to the gamma herpesvirus family. KSHV encodes 87 proteins and at least 17 microRNAs (Russo et al. 1996). Similar to other herpes viruses, the life cycle of KSHV includes prolonged latent and lytic phases. During the latent phase, a subset of genes is expressed, such as the latency-associated nuclear antigen (LANA), viral cyclin (vCyclin), viral FLICE inhibitory protein (vFLIP), kaposins and others (Cai et al. 2005). These proteins increase proliferative signals, decrease apoptosis and induce the activation of proangiogenic and inflammatory signals (Fuentes-Gonzalez et al. 2013).

LANA is required for the replication of the latent episomal viral DNA. It is considered to be an oncogenic protein due to its ability to deregulate tumor suppressor pathways associated with p53 and pRb and to transform cells in cooperation with the cellular oncogene H-ras (Radkov et al. 2000). V-cyclin is another candidate KSHV oncogene because of its homology to the human cyclin-D/Prad oncogene. It is a constitutive activator of cyclin dependent kinase 6 (CDK6). Substrates of the vCyclin/CDK6 complex include pRb and p27 (Ojala et al. 2000). As such, vCyclin efficiently accelerates cell cycle progression (Carbone et al. 2010; Fuentes-Gonzalez et al. 2013). Several KSHV micro RNAs have also been shown to modulate host gene expression, suggesting their role in the pathogenesis of malignancies induced by KSHV (Samols et al. 2007).

Hematologic malignancy associated with KSHN, **primary effusion lymphoma(PEL)** is a very rare subgroup of B-cell lymphomas presenting as pleural, peritoneal and pericardial effusions in the absence of a solid tumor mass or recognizable nodal involvement (Carbone et al. 2010). The disease initially affects one single serous cavity, usually remains localized to body cavities throughout the clinical course of the lymphoma, and occasionally extends into tissues underlying the serous membranes, including the omentum and the outer parts of the gastrointestinal tract wall (Komanduri et al. 1996; Nador et al. 1996). The presence of KSHV has been incorporated as a diagnostic criterion for PEL (Carbone et al. 1996).

Morphologically, PEL tumor cells show immunoblastic and plasmablastic features, with large round nuclei and prominent nucleoli. Some cells resemble Reed-Sternberg cells (Said and Cesarman 2008; Carbone et al. 2010). Interestingly, up to 80 % of these tumors in addition to KSHV also carry EBV as a latency infection. The pathogenetic significance of EBV remains uncertain (Rickinson 2014), but the combined presence of these two viruses appears to be unique to PEL (Carbone et al. 2010).

Helicobacter Pylori

Helicobacter pylori (*H. pylori*) is a spiral-shaped Gram-negative bacterium that infects approximately 50 % of humans worldwide. H. pylori infection, generally acquired in childhood, is the most frequent chronic bacterial infection, and is a major cause of gastroduodenal disease, including chronic gastritis, benign peptic ulcers, gastric carcinoma and gastric mucosa associated lymphoid tissue (MALT) lymphoma, although only a very small proportion of H. pylori-infected subjects develop these complications (Eck and Fischbach 2010; Zucca et al. 1998; Ferrucci and Zucca 2007). H. pylori has been ranked as a class I carcinogen by the

International Agency for Research on Cancer since 1994 (Wang et al. 2013). MALT lymphomas represent approximately 7 % of newly- diagnosed lymphomas (Troch et al. 2009). They are a rare malignancy, with a worldwide incidence estimated at 1–1.5 cases per 105,000 per year (Pereira and Medeiros 2014).

Several arguments support the central role played by H. pylori in the pathogenesis of MALT lymphoma. Chronic infection with H. pylori is associated with the induction of gastric lymphoid follicles, representing the proposed first step in lymphomagenesis (Siddiqui et al. 2011). In addition, H. pylori infection can be demonstrated serologically in most patients, and the bacterium can be histologically identified in the gastric mucosa of the majority of gastric MALT lymphomas, with some series describing incidences as high as 92 % (Isaacson 2005). Antibiotics have been used successfully to treat these lymphomas (Stolte et al. 2002; Zullo et al. 2010; Nakamura et al. 2012; Ferreri et al. 2013). These etiologic associations indicate that chronic H. pylori infection leads to lymphoid hyperplasia, which in the presence of appropriate microenvironmental factors and a genetic predisposition, can culminate in the emergence of a malignant lymphoid clone (Wang et al. 2013).

Several bacterial factors have been identified to influence the pathogenesis of H. pylori associated diseases. One of the most important H. pylori virulence factors is CagA protein, encoded by the cytotoxin-associated gene (cag) pathogenicity island (Hatakeyama 2008; Backert et al. 2010). CagA-positive strains associate with higher grades of mucosal inflammation, severe gastritis and gastric carcinogenesis contrary to CagA-negative strains (Hatakeyama and Higashi 2005; Chen et al 2013). Seropositivity of CagA is present in 89.0–95.5 % of patients with gastric MALT lymphoma (Eck et al. 1997). In order to exert its pathogenic effect CagA needs to be delivered into the host cells. It has been shown that H. pylori has a complex secretion system that represents a needle-like structure protruding from the bacterial surface to inject CagA protein into target cells (Tegtmeyer et al. 2011). The delivered CagA functions as a typical oncoprotein in gastric MALT lymphoma pathogenesis, but the molecular mechanisms of its action are not completely understood (Wang et al. 2013; Ohnishi et al. 2008).

In host cells, CagA undergoes tyrosine phosphorylation by c-src/Lyn kinase. The phosphorylated CagA deregulates the intracellular signaling pathways and initiates the malignant transformation of B lymphocytes (Zhu et al. 2007; Lin et al. 2010; Lu et al. 2010). Lin and co-workers (Lin et al. 2010) suggested that following the phosphorylation, CagA induces the activation of Erk1/2 and mitogen activated protein kinase and the up-regulation of the anti-apoptotic proteins Bcl-2 and Bcl-XL in human B lymphocytes. Umehara and co-workers (Umehara et al. 2003) showed that CagA may inhibit the B lymphocyte apoptosis by impairing the p53 and JAK/STAT pathway. An imbalance between apoptosis and proliferation allows the lymphocytes to acquire survival ability, which contributes to the pathogenesis of lymphoma (Sibony and Jones 2012; Kuo et al. 2013).

Although MALT lymphoma is a B-cell lymphoid neoplasm, other inflammatory cells are playing an important role in its pathogenesis. It has been shown that these lymphomas are infiltrated by type 2 Th (Th2)-polarized T-cells and that tumor proliferation is enhanced by intratumoral CD4+ T-cells (Craig et al. 2010). A large proportion of these CD4+ T-cells are suppressive CD25+ forkhead box P3 (FOXP3)+ regulatory T-cells (Tregs), which are themselves recruited by malignant B-cells (Craig et al. 2010; Garcia et al. 2012). Lymphomagenesis is also facilitated by DNA-damage caused by reactive oxygen species produced by neutrophils as part of the immune response to H. pylori infection (Pereira and Medeiros 2014).

It is certain that chronic mucosal inflammation is the basic process underlying the occurrence and development of gastric MALT lymphoma. The incapability of the host immune system to clear the bacterial pathogen results in a persistent infection, which leads to lymphoid hyperplasia. Numerous lymphocytes and antigen-presenting cells, develops in the stroma under the gastric epithelium where antigens are processed and presented to lymphocytes, as part of a normal adaptive immune response. With time, the acquisition of additional genetic aberrations culminates in the activation of intracellular survival pathways, with disease progression due to proliferation and resistance to apoptosis, and the emergence of a malignant clone (Pereira and Medeiros 2014). The key events in this process are non-random chromosomal translocations involving a limited group of genes. The three most common and characteristic translocations – t(11;18) (q21;q21), t(1;14) (p22;q32), and t(14;18) (q32;q21) generate oncogenic fusion proteins that activate the NF-κB pathway, and cause overexpression of NF-κB target genes (Du 2011). As it was mentioned earlier, H. pylori virulent factors (e.g. CagA) block apoptosis, promoting the accumulation of these genetically abnormal cells that should otherwise be removed from the tissue (Sibony and Jones 2012; Kuo et al. 2013).

Clinically, gastric MALT lymphoma is an indolent, low-grade B-cell lymphoid neoplasm. Most MALT lymphomas at diagnosis are characterized by non-disseminated (early-stage) disease, with both marrow and distal nodal involvement being rare, although regional lymph node infiltration is relatively frequent (Sagaert et al. 2010; Owens and Smith 2011). Histologically the tumor is composed of small to medium-sized cells with slightly irregular nuclei and inconspicuous nucleoli, resembling cells of follicular center. These cells infiltrate around reactive follicles in a marginal zone distribution and eventually overrun some or most of the follicles. Sometimes the cells resemble monocytes or plasma cells (Isaacson 2005). Although MALT lymphoma is a low-grade disease, with large transformed cells being rare in the neoplastic infiltrate, it can undergo transformation to an aggressive diffuse large B-cell lymphoma through poorly understood mechanisms (Ferrucci and Zucca 2007). H. pylori eradication through specific antibiotherapy leads to lymphoma regression and sustained remission in the majority of cases (Zullo et al. 2010; Nakamura et al. 2012; Ferreri et al. 2013).

Conclusions

Human infection-associated hematologic malignancies are pathobiological consequences of infections that evolved powerful mechanisms to persist and replicate through deregulation of host oncogenic pathways. Environmental and host cofactors such as immunosuppression, genetic predisposition, or mutagens can accelerate the development of these neoplasms. The exact role of infectious agents varies with the clinical scenario and subtype of the disease. Although these agents employ very different specific oncogenic mechanisms, they converge on several common intercellular pathways (such as cell-cycle regulation, proliferation and apoptosis) that eventually lead to malignant transformation.

The study of infectious agents and their oncogenes and of the multiple mechanisms deployed by them to circumvent the growth-suppressive and pro apoptotic function of tumor suppressor genes has improved our current understanding of cancer biology. It helped with discovery of new classes of cellular modulators, which may induce cell cycle deregulation and disrupt host immune responses, thus facilitating immune evasion of tumors. Both tumor-promoting inflammation and immunosuppression are implicated in this process. The emerging information may expedite the development of new targeted approaches to prevent and treat infection-associated hematologic malignancies. Today, therapeutic eradication of H. pylori is the proven method of prevention and treatment of gastric MALT lymphoma. On the other hand, treatment of most of virally induced malignancies is currently unsatisfactory and novel therapies targeting viral life cycles, host cell proliferation and apoptotic pathways, as well as antitumor immune mechanisms are being developed.

References

Alizadeh AA, Advani RH (2008) Evaluation and management of angioimmunoblastic T-cell lymphoma: a review of current approaches and future strategies. Clin Adv Hematol Oncol 6(12):899–909

Anagnostopoulos I, Hummel M et al (1992) Heterogeneous Epstein-Barr virus infection patterns in peripheral T-cell lymphoma of angioimmunoblastic lymphadenopathy type. Blood 80(7):1804–1812

Arisawa K, Soda M et al (2000) Evaluation of adult T-cell leukemia/lymphoma incidence and its impact on non-Hodgkin lymphoma incidence in southwestern Japan. Int J Cancer 85(3):319–324

Arvanitakis L, Yaseen N et al (1995) Latent membrane protein-1 induces cyclin D2 expression, pRb hyperphosphorylation, and loss of TGF-beta 1-mediated growth inhibition in EBV-positive B cells. J Immunol 155(3):1047–1056

Au WY, Pang A et al (2004) Quantification of circulating Epstein-Barr virus (EBV) DNA in the diagnosis and monitoring of natural killer cell and EBV-positive lymphomas in immunocompetent patients. Blood 104(1):243–249

Backert S, Tegtmeyer N et al (2010) The versatility of Helicobacter pylori CagA effector protein functions: the master key hypothesis. Helicobacter 15(3):163–176

Bouvard V, Baan R et al (2009) A review of human carcinogens–Part B: biological agents. Lancet Oncol 10(4):321–322

Brink AA, Dukers DF et al (1997) Presence of Epstein-Barr virus latency type III at the single cell level in post-transplantation lymphoproliferative disorders and AIDS related lymphomas. J Clin Pathol 50(11):911–918

Brown KD, Hostager BS et al (2001) Differential signaling and tumor necrosis factor receptor-associated factor (TRAF) degradation mediated by CD40 and the Epstein-Barr virus oncoprotein latent membrane protein 1 (LMP1). J Exp Med 193(8):943–954

Cai X, Lu S et al (2005) Kaposi's sarcoma-associated herpesvirus expresses an array of viral microRNAs in latently infected cells. Proc Natl Acad Sci U S A 102(15):5570–5575

Canaan A, Haviv I et al (2009) EBNA1 regulates cellular gene expression by binding cellular promoters. Proc Natl Acad Sci U S A 106(52):22421–22426

Capello D, Rossi D et al (2005) Post-transplant lymphoproliferative disorders: molecular basis of disease histogenesis and pathogenesis. Hematol Oncol 23(2):61–67

Carbone A (2003) Emerging pathways in the development of AIDS-related lymphomas. Lancet Oncol 4(1):22–29

Carbone A, Gloghini A et al (1996) Kaposi's sarcoma-associated herpesvirus DNA sequences in AIDS-related and AIDS-unrelated lymphomatous effusions. Br J Haematol 94(3):533–543

Carbone A, Gloghini A et al (2008) EBV-associated lymphoproliferative disorders: classification and treatment. Oncologist 13(5):577–585

Carbone A, Cesarman E et al (2010) Understanding pathogenetic aspects and clinical presentation of primary effusion lymphoma through its derived cell lines. AIDS 24(4):479–490

Chan JKC, Aozasa K, Gaulard P (2008) DLBCL associated with chronic inflammation. In: Swerdlow ECSH, Harris NL, Jaffe ES, Pileri SA, Harald S, Thiele J, Vardiman JW (eds) WHO classification of tumours of haemotopoietic and lymphoid tissues. IARC Press, Lyon, pp 245–246

Chapman AL, Rickinson AB (1998) Epstein-Barr virus in Hodgkin's disease. Ann Oncol 9(Suppl 5):S5–S16

Chen CY, Wang FY et al (2013) Amino acid polymorphisms flanking the EPIYA-A motif of Helicobacter pylori CagA C-terminal region is associated with gastric cancer in east China: experience from a single center. J Dig Dis 14(7):358–365

Chuang HC, Lay JD et al (2007) Pathogenesis and mechanism of disease progression from hemophagocytic lymphohistiocytosis to Epstein-Barr virus-associated T-cell lymphoma: nuclear factor-kappa B pathway as a potential therapeutic target. Cancer Sci 98(9):1281–1287

Cohen JI (2000) Epstein-Barr virus infection. N Engl J Med 343(7):481–492

Coppotelli G, Mughal N et al (2013) The Epstein-Barr virus nuclear antigen-1 reprograms transcription by mimicry of high mobility group A proteins. Nucleic Acids Res 41(5):2950–2962

Craig VJ, Cogliatti SB et al (2010) B-cell receptor signaling and CD40 ligand-independent T cell help cooperate in Helicobacter-induced MALT lymphomagenesis. Leukemia 24(6):1186–1196

de Leval L, Hasserjian RP (2009) Diffuse large B-cell lymphomas and burkitt lymphoma. Hematol Oncol Clin North Am 23(4):791–827

Deacon EM, Pallesen G et al (1993) Epstein-Barr virus and Hodgkin's disease: transcriptional analysis of virus latency in the malignant cells. J Exp Med 177(2):339–349

Doerr JR, Malone CS et al (2005) Patterned CpG methylation of silenced B cell gene promoters in classical Hodgkin lymphoma-derived and primary effusion lymphoma cell lines. J Mol Biol 350(4):631–640

Dogan A, Attygalle AD et al (2003) Angioimmunoblastic T-cell lymphoma. Br J Haematol 121(5):681–691

Draoua HY, Tsao L et al (2004) T-cell post-transplantation lymphoproliferative disorders after cardiac transplantation: a single institutional experience. Br J Haematol 127(4):429–432

Dresang LR, Vereide DT et al (2009) Identifying sites bound by Epstein-Barr virus nuclear antigen 1 (EBNA1) in the human genome: defining a position-weighted matrix to predict sites bound by EBNA1 in viral genomes. J Virol 83(7):2930–2940

Du MQ (2011) MALT lymphoma: many roads lead to nuclear factor-kappab activation. Histopathology 58(1):26–38

Dunleavy K, Wilson WH et al (2007) Angioimmunoblastic T cell lymphoma: pathobiological insights and clinical implications. Curr Opin Hematol 14(4):348–353

Eck M, Fischbach W (2010) Gastric MALT-type lymphoma. Pathology, pathogenesis, diagnostics and therapy. Pathologe 31(3):188–194

Eck M, Schmausser B et al (1997) MALT-type lymphoma of the stomach is associated with Helicobacter pylori strains expressing the CagA protein. Gastroenterology 112(5):1482–1486

Eliopoulos AG, Stack M et al (1997) Epstein-Barr virus-encoded LMP1 and CD40 mediate IL-6 production in epithelial cells via an NF-kappaB pathway involving TNF receptor-associated factors. Oncogene 14(24):2899–2916

Eliopoulos AG, Gallagher NJ et al (1999) Activation of the p38 mitogen-activated protein kinase pathway by Epstein-Barr virus-encoded latent membrane protein 1 coregulates interleukin-6 and interleukin-8 production. J Biol Chem 274(23):16085–16096

Engels EA (2007) Infectious agents as causes of non-Hodgkin lymphoma. Cancer Epidemiol Biomarkers Prev 16(3):401–404

Epstein MA, Achong BG et al (1964) Virus particles in cultured lymphoblasts from Burkitt's lymphoma. Lancet 1(7335):702–703

Evens AM, Roy R et al (2010) Post-transplantation lymphoproliferative disorders: diagnosis, prognosis, and current approaches to therapy. Curr Oncol Rep 12(6):383–394

Ferreri AJ, Guidoboni M et al (2004) Evidence for an association between Chlamydia psittaci and ocular adnexal lymphomas. J Natl Cancer Inst 96(8):586–594

Ferreri AJ, Govi S et al (2013) The role of Helicobacter pylori eradication in the treatment of diffuse large B-cell and marginal zone lymphomas of the stomach. Curr Opin Oncol 25(5):470–479

Ferrucci PF, Zucca E (2007) Primary gastric lymphoma pathogenesis and treatment: what has changed over the past 10 years? Br J Haematol 136(4):521–538

Fu Q, He C et al (2013) Epstein-Barr virus interactions with the Bcl-2 protein family and apoptosis in human tumor cells. J Zhejiang Univ Sci B 14(1):8–24

Fuentes-Gonzalez AM, Contreras-Paredes A et al (2013) The modulation of apoptosis by oncogenic viruses. Virol J 10:182

Garcia M, Bellosillo B et al (2012) Study of regulatory T-cells in patients with gastric malt lymphoma: influence on treatment response and outcome. PLoS One 7(12):e51681

God JM, Haque A (2010) Burkitt lymphoma: pathogenesis and immune evasion. J Oncol 2010 pii: 516047

Gregory CD, Dive C et al (1991) Activation of Epstein-Barr virus latent genes protects human B cells from death by apoptosis. Nature 349(6310):612–614

Gruhne B, Sompallae R et al (2009a) The Epstein-Barr virus nuclear antigen-1 promotes genomic instability via induction of reactive oxygen species. Proc Natl Acad Sci U S A 106(7):2313–2318

Gruhne B, Sompallae R et al (2009b) Three Epstein-Barr virus latency proteins independently promote genomic instability by inducing DNA damage, inhibiting DNA repair and inactivating cell cycle checkpoints. Oncogene 28(45):3997–4008

Grywalska E, Markowicz J et al (2013) Epstein-Barr virus-associated lymphoproliferative disorders. Postepy Hig Med Dosw (Online) 67:481–490

Gualco G, Domeny-Duarte P et al (2011) Clinicopathologic and molecular features of 122 Brazilian cases of nodal and extranodal NK/T-cell lymphoma, nasal type, with EBV subtyping analysis. Am J Surg Pathol 35(8):1195–1203

Harris NL, Campo E, Jaffe ES, Pileri SA, Stein H, Swerdlow SH, Thiele J, Vardman JW (2008) Introduction to the WHO classification of tumours of haematopoietic and lymphoid tissues. In: Swerdlow ECSH, Harris NL, Jaffe ES, Pileri SA, Harald S, Thiele J, Vardiman JW (eds) WHO classification of tumours of haematopoietic and lymphoid tissues. IARC Press, Lyon, pp 14–15

Hatakeyama M (2008) SagA of CagA in Helicobacter pylori pathogenesis. Curr Opin Microbiol 11(1):30–37

Hatakeyama M, Higashi H (2005) Helicobacter pylori CagA: a new paradigm for bacterial carcinogenesis. Cancer Sci 96(12):835–843

Hohaus S, Santangelo R et al (2011) The viral load of Epstein-Barr virus (EBV) DNA in peripheral blood predicts for biological and clinical characteristics in Hodgkin lymphoma. Clin Cancer Res 17(9):2885–2892

Iannitto E, Ferreri AJ et al (2008) Angioimmunoblastic T-cell lymphoma. Crit Rev Oncol Hematol 68(3):264–271

Isaacson PG (2005) Update on MALT lymphomas. Best Pract Res Clin Haematol 18(1):57–68
Jaffe ES (2009) The 2008 WHO classification of lymphomas: implications for clinical practice and translational research. Hematol Am Soc Hematol Educ Program 2009:523–531
Jaffe ES, Ralfkiaer E (2001) Angioimmunolastic T-cell lymphoma. In: Jaffe NLHES, Stein H, Vardiman JW (eds) Pathology and genetics: tumours of haematopoietic and lymphoid tissues. IRAC Press, Lyon, pp 225–226
Jaffe ES, Wilson WH (1997) Lymphomatoid granulomatosis: pathogenesis, pathology and clinical implications. Cancer Surv 30:233–248
Javier RT, Butel JS (2008) The history of tumor virology. Cancer Res 68(19):7693–7706
Kamranvar SA, Masucci MG (2011) The Epstein-Barr virus nuclear antigen-1 promotes telomere dysfunction via induction of oxidative stress. Leukemia 25(6):1017–1025
Kamranvar SA, Gruhne B et al (2007) Epstein-Barr virus promotes genomic instability in Burkitt's lymphoma. Oncogene 26(35):5115–5123
Kapatai G, Murray P (2007) Contribution of the Epstein Barr virus to the molecular pathogenesis of Hodgkin lymphoma. J Clin Pathol 60(12):1342–1349
Karube K, Ohshima K et al (2004) Expression of FoxP3, a key molecule in CD4CD25 regulatory T cells, in adult T-cell leukaemia/lymphoma cells. Br J Haematol 126(1):81–84
Kennedy G, Komano J et al (2003) Epstein-Barr virus provides a survival factor to Burkitt's lymphomas. Proc Natl Acad Sci U S A 100(24):14269–14274
Kilger E, Kieser A et al (1998) Epstein-Barr virus-mediated B-cell proliferation is dependent upon latent membrane protein 1, which simulates an activated CD40 receptor. Embo J 17(6):1700–1709
Kim JH, Kim WS et al (2013) Epstein-Barr virus latent membrane protein 1 increases genomic instability through Egr-1-mediated up-regulation of activation-induced cytidine deaminase in B-cell lymphoma. Leuk Lymphoma 54(9):2035–2040
Komanduri KV, Luce JA et al (1996) The natural history and molecular heterogeneity of HIV-associated primary malignant lymphomatous effusions. J Acquir Immune Defic Syndr Hum Retrovirol 13(3):215–226
Kuo SH, Chen LT et al (2013) Detection of the Helicobacter pylori CagA protein in gastric mucosa-associated lymphoid tissue lymphoma cells: clinical and biological significance. Blood Cancer J 3:e125
Kuze T, Nakamura N et al (2000) The characteristics of Epstein-Barr virus (EBV)-positive diffuse large B-cell lymphoma: comparison between EBV(+) and EBV(−) cases in Japanese population. Jpn J Cancer Res 91(12):1233–1240
Lagunoff M, Ganem D (1997) The structure and coding organization of the genomic termini of Kaposi's sarcoma-associated herpesvirus. Virology 236(1):147–154
Laherty CD, Hu HM et al (1992) The Epstein-Barr virus LMP1 gene product induces A20 zinc finger protein expression by activating nuclear factor kappa B. J Biol Chem 267(34):24157–24160
Lecuit M, Abachin E et al (2004) Immunoproliferative small intestinal disease associated with Campylobacter jejuni. N Engl J Med 350(3):239–248
Leoncini L, Raphael M, Stein H, Harris NL, Jaffe ES, Kluin PM (2008) Burkitt lymphoma. In: Swerdlow ECSH, Harris NL, Jaffe ES, Pileri SA, Harald S, Thiele J, Vardiman JW (eds) WHO classification of tumours of haemotopoietic and lymphoid tissues. IARC Press, Lyon, pp 262–264
Levitskaya J, Coram M et al (1995) Inhibition of antigen processing by the internal repeat region of the Epstein-Barr virus nuclear antigen-1. Nature 375(6533):685–688
Liao JB (2006) Viruses and human cancer. Yale J Biol Med 79(3–4):115–122
Lin WC, Tsai HF et al (2010) Translocation of Helicobacter pylori CagA into Human B lymphocytes, the origin of mucosa-associated lymphoid tissue lymphoma. Cancer Res 70(14):5740–5748
Liu MT, Chen YR et al (2004) Epstein-Barr virus latent membrane protein 1 induces micronucleus formation, represses DNA repair and enhances sensitivity to DNA-damaging agents in human epithelial cells. Oncogene 23(14):2531–2539

Liu MT, Chang YT et al (2005) Epstein-Barr virus latent membrane protein 1 represses p53-mediated DNA repair and transcriptional activity. Oncogene 24(16):2635–2646
Long JL, Engels EA et al (2008) Incidence and outcomes of malignancy in the HAART era in an urban cohort of HIV-infected individuals. AIDS 22(4):489–496
Lopes LF, Ruiz Miyazawa KW et al (2013) Epstein-Barr virus (EBV) microRNAs: involvement in cancer pathogenesis and immunopathology. Int Rev Immunol 32(3):271–281
Lu F, Wikramasinghe P et al (2010) Genome-wide analysis of host-chromosome binding sites for Epstein-Barr Virus Nuclear Antigen 1 (EBNA1). Virol J 7:262
Martin D, Gutkind JS (2008) Human tumor-associated viruses and new insights into the molecular mechanisms of cancer. Oncogene 27(Suppl 2):S31–S42
Maruo S, Zhao B et al (2011) Epstein-Barr virus nuclear antigens 3C and 3A maintain lymphoblastoid cell growth by repressing p16INK4A and p14ARF expression. Proc Natl Acad Sci U S A 108(5):1919–1924
Matsuo K, Kusano A et al (2004) Effect of hepatitis C virus infection on the risk of non-Hodgkin's lymphoma: a meta-analysis of epidemiological studies. Cancer Sci 95(9):745–752
Matsuoka M, Jeang KT (2007) Human T-cell leukaemia virus type 1 (HTLV-1) infectivity and cellular transformation. Nat Rev Cancer 7(4):270–280
Mbulaiteye SM, Biggar RJ et al (2003) Immune deficiency and risk for malignancy among persons with AIDS. J Acquir Immune Defic Syndr 32(5):527–533
Mesri EA, Feitelson MA et al (2014) Human viral oncogenesis: a cancer hallmarks analysis. Cell Host Microbe 15(3):266–282
Middeldorp JM, Pegtel DM (2008) Multiple roles of LMP1 in Epstein-Barr virus induced immune escape. Semin Cancer Biol 18(6):388–396
Mosialos G, Birkenbach M et al (1995) The Epstein-Barr virus transforming protein LMP1 engages signaling proteins for the tumor necrosis factor receptor family. Cell 80(3):389–399
Munz C, Moormann A (2008) Immune escape by Epstein-Barr virus associated malignancies. Semin Cancer Biol 18(6):381–387
Murata T, Sato Y et al (2014) Modes of infection and oncogenesis by the Epstein-Barr virus. Rev Med Virol 24(4):242–253
Murray PG, Qiu GH et al (2004) Frequent epigenetic inactivation of the RASSF1A tumor suppressor gene in Hodgkin's lymphoma. Oncogene 23(6):1326–1331
Nador RG, Cesarman E et al (1996) Primary effusion lymphoma: a distinct clinicopathologic entity associated with the Kaposi's sarcoma-associated herpes virus. Blood 88(2):645–656
Nakamura S, Jaffe ES, Swerdlow SH (2008) EBV positive diffuse large B-cell lymphoma of the elderly. In: Swerdlow ECSH, Harris NL, Jaffe ES, Pileri SA, Harald S, Thiele J, Vardiman JW (eds) WHO classification of tumours of haemotopoietic and lymphoid tissues. IARC Press, Lyon, pp 243–244
Nakamura S, Sugiyama T et al (2012) Long-term clinical outcome of gastric MALT lymphoma after eradication of Helicobacter pylori: a multicentre cohort follow-up study of 420 patients in Japan. Gut 61(4):507–513
Ohnishi N, Yuasa H et al (2008) Transgenic expression of Helicobacter pylori CagA induces gastrointestinal and hematopoietic neoplasms in mouse. Proc Natl Acad Sci U S A 105(3):1003–1008
Ohshima K, Jaffe ES, Kikuchi M (2008) Adult T-cell leukaemia/lymphoma. In: Swerdlow ECSH, Harris NL, Jaffe ES, Pileri SA, Harald S, Thiele J, Vardiman JW (eds) WHO classification of tumors: pathology and genetics of tumors of haematopoietic and lymphoid tissues. IARC Press, Lyon, pp 281–284
Ojala PM, Yamamoto K et al (2000) The apoptotic v-cyclin-CDK6 complex phosphorylates and inactivates Bcl-2. Nat Cell Biol 2(11):819–825
Owens SR, Smith LB (2011) Molecular aspects of H. pylori-related MALT lymphoma. Pathol Res Int 2011:193149
Oyama T, Ichimura K et al (2003) Senile EBV+ B-cell lymphoproliferative disorders: a clinicopathologic study of 22 patients. Am J Surg Pathol 27(1):16–26

Oyama T, Yamamoto K et al (2007) Age-related EBV-associated B-cell lymphoproliferative disorders constitute a distinct clinicopathologic group: a study of 96 patients. Clin Cancer Res 13(17):5124–5132

Parkin DM (2006) The global health burden of infection-associated cancers in the year 2002. Int J Cancer 118(12):3030–3044

Pereira MI, Medeiros JA (2014) Role of Helicobacter pylori in gastric mucosa-associated lymphoid tissue lymphomas. World J Gastroenterol 20(3):684–698

Poiesz BJ, Ruscetti FW et al (1980) Detection and isolation of type C retrovirus particles from fresh and cultured lymphocytes of a patient with cutaneous T-cell lymphoma. Proc Natl Acad Sci U S A 77(12):7415–7419

Portis T, Ikeda M et al (2004) Epstein-Barr virus LMP2A: regulating cellular ubiquitination processes for maintenance of viral latency? Trends Immunol 25(8):422–426

Putaluga S, Wilson WH, Jaffe ES (2008) Lymphomatoid granulomatosis. In: Swerdlow ECSH, Harris NL, Jaffe ES, Pileri SA, Harald S, Thiele J, Vardiman JW (eds) WHO classification of tumors: pathology and genetics of tumors of haematopoietic and lymphoid tissues. IARC Press, Lyon, pp 247–249

Qu Z, Xiao G (2011) Human T-cell lymphotropic virus: a model of NF-kappaB-associated tumorigenesis. Viruses 3(6):714–749

Radkov SA, Kellam P et al (2000) The latent nuclear antigen of Kaposi sarcoma-associated herpesvirus targets the retinoblastoma-E2F pathway and with the oncogene Hras transforms primary rat cells. Nat Med 6(10):1121–1127

Raphael M, Said J, Borisch B, Cesarman E, Harris NL (2008) Lymphomas associated with HIV infection. In: Swerdlow ECSH, Harris NL, Jaffe ES, Pileri SA, Harald S, Thiele J, Vardiman JW (eds) WHO classification of tumours of haemotopoietic and lymphoid tissues. IARC Press, Lyon, pp 340–342

Renne R, Lagunoff M et al (1996) The size and conformation of Kaposi's sarcoma-associated herpesvirus (human herpesvirus 8) DNA in infected cells and virions. J Virol 70(11):8151–8154

Rickinson AB (2014) Co-infections, inflammation and oncogenesis: future directions for EBV research. Semin Cancer Biol 26:99–115

Roschewski M, Wilson WH (2012) EBV-associated lymphomas in adults. Best Pract Res Clin Haematol 25(1):75–89

Rossi D, Gaidano G et al (2003) Frequent aberrant promoter hypermethylation of O6-methylguanine-DNA methyltransferase and death-associated protein kinase genes in immunodeficiency-related lymphomas. Br J Haematol 123(3):475–478

Rowe M, Kelly GL et al (2009) Burkitt's lymphoma: the Rosetta Stone deciphering Epstein-Barr virus biology. Semin Cancer Biol 19(6):377–388

Russo JJ, Bohenzky RA et al (1996) Nucleotide sequence of the Kaposi sarcoma-associated herpesvirus (HHV8). Proc Natl Acad Sci U S A 93(25):14862–14867

Sagaert X, Van Cutsem E et al (2010) Gastric MALT lymphoma: a model of chronic inflammation-induced tumor development. Nat Rev Gastroenterol Hepatol 7(6):336–346

Saha A, Robertson ES (2011) Epstein-Barr virus-associated B-cell lymphomas: pathogenesis and clinical outcomes. Clin Cancer Res 17(10):3056–3063

Said J, Cesarman E (2008) Primary efusion lymphoma. In: Swerdlow ECSH, Harris NL, Jaffe ES, Pileri SA, Harald S, Thiele J, Vardiman JW (eds) WHO classification of tumors: pathology and genetics of tumors of haematopoietic and lymphoid tissues. IARC Press, Lyon, pp 260–261

Samols MA, Skalsky RL et al (2007) Identification of cellular genes targeted by KSHV-encoded microRNAs. PLoS Pathog 3(5):e65

Sasajima Y, Yamabe H et al (1993) High expression of the Epstein-Barr virus latent protein EB nuclear antigen-2 on pyothorax-associated lymphomas. Am J Pathol 143(5):1280–1285

Schmitz R, Young RM et al (2012) Burkitt lymphoma pathogenesis and therapeutic targets from structural and functional genomics. Nature 490(7418):116–120

Sibony M, Jones NL (2012) Recent advances in Helicobacter pylori pathogenesis. Curr Opin Gastroenterol 28(1):30–35

Siddiqui ST, Naz E et al (2011) Frequency of Helicobacter pylori in biopsy proven gastritis and its association with lymphoid follicle formation. J Pak Med Assoc 61(2):138–141

Skalska L, White RE et al (2013) Induction of p16(INK4a) is the major barrier to proliferation when Epstein-Barr virus (EBV) transforms primary B cells into lymphoblastoid cell lines. PLoS Pathog 9(2):e1003187

Snow AL, Lambert SL et al (2006) EBV can protect latently infected B cell lymphomas from death receptor-induced apoptosis. J Immunol 177(5):3283–3293

Sompallae R, Callegari S et al (2010) Transcription profiling of Epstein-Barr virus nuclear antigen (EBNA)-1 expressing cells suggests targeting of chromatin remodeling complexes. PLoS One 5(8):e12052

Soni V, Cahir-McFarland E et al (2007) LMP1 TRAFficking activates growth and survival pathways. Adv Exp Med Biol 597:173–187

Stein H, Delsol G, Pileri SA, Weiss LM, Poppema S, Jaffe ES (2008) Classical Hodgkin lymphoma, introduction. In: Swerdlow ECSH, Harris NL, Jaffe ES, Pileri SA, Harald S, Thiele J, Vardiman JW (eds) WHO classification of tumours of haemotopoietic and lymphoid tissues. IARC Press, Lyon, pp 326–330

Stolte M, Bayerdorffer E et al (2002) Helicobacter and gastric MALT lymphoma. Gut 50(Suppl 3):III19–III24

Takakuwa T, Luo WJ et al (2003) Establishment and characterization of unique cell lines derived from pyothorax-associated lymphoma which develops in long-standing pyothorax and is strongly associated with Epstein-Barr virus infection. Cancer Sci 94(10):858–863

Tegtmeyer N, Wessler S et al (2011) Role of the cag-pathogenicity island encoded type IV secretion system in Helicobacter pylori pathogenesis. FEBS J 278(8):1190–1202

Tempera I, Lieberman PM (2014) Epigenetic regulation of EBV persistence and oncogenesis. Semin Cancer Biol 26:22–29

Thorley-Lawson DA, Gross A (2004) Persistence of the Epstein-Barr virus and the origins of associated lymphomas. N Engl J Med 350(13):1328–1337

Timms JM, Bell A et al (2003) Target cells of Epstein-Barr-virus (EBV)-positive post-transplant lymphoproliferative disease: similarities to EBV-positive Hodgkin's lymphoma. Lancet 361(9353):217–223

Troch M, Jonak C et al (2009) A phase II study of bortezomib in patients with MALT lymphoma. Haematologica 94(5):738–742

Tsukasaki K, Utsunomiya A et al (2007) VCAP-AMP-VECP compared with biweekly CHOP for adult T-cell leukemia-lymphoma: Japan Clinical Oncology Group Study JCOG9801. J Clin Oncol 25(34):5458–5464

Uchiyama T, Yodoi J et al (1977) Adult T-cell leukemia: clinical and hematologic features of 16 cases. Blood 50(3):481–492

Umehara S, Higashi H et al (2003) Effects of Helicobacter pylori CagA protein on the growth and survival of B lymphocytes, the origin of MALT lymphoma. Oncogene 22(51):8337–8342

Ushmorov A, Leithauser F et al (2006) Epigenetic processes play a major role in B-cell-specific gene silencing in classical Hodgkin lymphoma. Blood 107(6):2493–2500

van den Bosch CA (2004) Is endemic Burkitt's lymphoma an alliance between three infections and a tumour promoter? Lancet Oncol 5(12):738–746

Vereide DT, Seto E et al (2013) Epstein-Barr virus maintains lymphomas via its miRNAs. Oncogene 33(10):1258–1264

Wang S, Frappier L (2009) Nucleosome assembly proteins bind to Epstein-Barr virus nuclear antigen 1 and affect its functions in DNA replication and transcriptional activation. J Virol 83(22):11704–11714

Wang HP, Zhu YL et al (2013) Role of Helicobacter pylori virulence factor cytotoxin-associated gene A in gastric mucosa-associated lymphoid tissue lymphoma. World J Gastroenterol 19(45):8219–8226

Weiss LM, Jaffe ES et al (1992) Detection and localization of Epstein-Barr viral genomes in angioimmunoblastic lymphadenopathy and angioimmunoblastic lymphadenopathy-like lymphoma. Blood 79(7):1789–1795

Wilson JB, Bell JL et al (1996) Expression of Epstein-Barr virus nuclear antigen-1 induces B cell neoplasia in transgenic mice. Embo J 15(12):3117–3126

Wright DH (1999) What is Burkitt's lymphoma and when is it endemic? Blood 93(2):758

Yamada Y, Tomonaga M et al (2001) A new G-CSF-supported combination chemotherapy, LSG15, for adult T-cell leukaemia-lymphoma: Japan Clinical Oncology Group Study 9303. Br J Haematol 113(2):375–382

Yoshida M (2001) Multiple viral strategies of HTLV-1 for dysregulation of cell growth control. Annu Rev Immunol 19:475–496

Young LS, Murray PG (2003) Epstein-Barr virus and oncogenesis: from latent genes to tumours. Oncogene 22(33):5108–5121

Young LS, Rickinson AB (2004) Epstein-Barr virus: 40 years on. Nat Rev Cancer 4(10):757–768

Yustein JT, Dang CV (2007) Biology and treatment of Burkitt's lymphoma. Curr Opin Hematol 14(4):375–381

Zhou Y, Attygalle AD et al (2007) Angioimmunoblastic T-cell lymphoma: histological progression associates with EBV and HHV6B viral load. Br J Haematol 138(1):44–53

Zhu Y, Wang C et al (2007) The Helicobacter pylori virulence factor CagA promotes Erk1/2-mediated Bad phosphorylation in lymphocytes: a mechanism of CagA-inhibited lymphocyte apoptosis. Cell Microbiol 9(4):952–961

Zucca E, Bertoni F et al (1998) Molecular analysis of the progression from Helicobacter pylori-associated chronic gastritis to mucosa-associated lymphoid-tissue lymphoma of the stomach. N Engl J Med 338(12):804–810

Zullo A, Hassan C et al (2010) Effects of Helicobacter pylori eradication on early stage gastric mucosa-associated lymphoid tissue lymphoma. Clin Gastroenterol Hepatol 8(2):105–110

Chapter 8
Multiple Infections and Cancer: Etiology, Mechanisms and Implications in Cancer Control

Mukesh Verma

Abstract The association between infectious agents and cancer has been known for a long time. Among all agents, the hepatitis B and C viruses (HBV and HCV, respectively), human papillomaviruses (HPV), and *Helicobacter pylori* (*H. pylori*) are associated with approximately 15 % of all human cancers. In recent years, two or more infectious agents have been identified in different neoplasms. For example, the coexistence of Epstein-Barr Virus (EBV) in human immunodeficiency virus (HIV)-related oral plasmablastic lymphomas (PBLs) was demonstrated recently. *Chlamydia trachomatis* (*C. trachomatis*) bacteria infected HPV-positive cells and induced inflammation and, ultimately, cervical cancer. Studies have been conducted to follow cancer patients infected with multiple agents, but the underlying mechanism is not completely understood. Co-infections alter the immune system and affect the host's susceptibility to cancer; the co-infecting agent sometimes is the causative cofactor. Many of these infectious agents are highly prevalent in the world; however, most infected individuals do not develop cancer and may remain lifelong carriers. Studying co-infections is important because different treatment strategies are required in such cases. This article focuses on utilizing existing information to better understand the etiology of cancer that develops as a result of multiple infections, and developing strategies to control disease and identify challenges and research opportunities in the field.

Keywords Cancer • Co-infection • Biomarkers • Cofactors • AIDS-related malignancies • Targeted therapy • Vaccines

M. Verma, Ph.D. (✉)
Methods and Technologies Branch, Epidemiology and Genomics Research Program, Division of Cancer Control and Population Sciences, National Cancer Institute, National Institutes of Health, 9601 Medical Center Drive, Suite 4E102, Rockville, MD 20850, USA
e-mail: vermam@mail.nih.gov

M.R. Shurin et al. (eds.), *Infection and Cancer: Bi-Directorial Interactions*, DOI 10.1007/978-3-319-20669-1_8

Abbreviations

BL	Burkitt lymphoma
C. trachomatis	*Chlamydia trachomatis*
EBNA1	Epstein-Barr Virus Nuclear Antigen 1
EBV	Epstein-Barr virus
H. pylori	*Helicobacter pylori*
HAART	highly active antiretroviral therapy
HBV	hepatitis B virus
HCC	hepatocellular carcinoma
HCV	hepatitis C virus
HDV	hepatitis δ (delta) virus
HIV-1	human immunodeficiency virus type 1
HPV	human papillomavirus
HTLV-1	human T-lymphotropic virus type 1
IARC	International Agency for Research on Cancer
KSHV/HHV8	Kaposi sarcoma-associated herpesvirus/human herpesvirus 8
LMP	latent membrane protein

Introduction

Most infectious agents that have been associated with cancer—mainly viruses and bacteria, but also fungi and parasites—infect the host as a single agent (Wang et al. 2011; Itatsu et al. 1999; Ferrasi et al. 2010; Minoura-Etoh et al. 2006; Cho et al. 2003; Darani and Yousefi 2012; Chen et al. 2011; Ishii et al. 1994; Prasad et al. 2013; Chemaly et al. 2013; Scudellari 2013; Frappier 2013; Read and Douglas 2014; Whitaker et al. 2013). Recently, however, more than one agent has been reported to infect individuals who later develop different kinds of cancers (Al Moustafa et al. 2009; Beilke et al. 2005; Bissinger and Berg 2013; Boy et al. 2011; Ferber et al. 2003; Gozlan et al. 2009; Khan et al. 2011; Lattario et al. 2008; Mirzamani et al. 2006; Sureau 2006). *Helicobacter pylori* (*H. pylori*) is associated with gastric cancer and hepatocellular carcinoma (HCC) or liver cancer; Epstein-Barr virus (EBV) is associated with Burkitt lymphoma (BL), Hodgkin lymphoma, nasopharyngeal carcinoma, and gastric cancer; hepatitis viruses (hepatitis B virus or HBV, hepatitis C virus or HCV, and hepatitis D virus or HDV) are associated with HCC; and parasites *Opisthorchis viverrini* and *Schistosoma haematobium* are associated with cholangiocarcinoma and HCC (Ishii et al. 1994; Sripa et al. 2012; Srivatanakul et al. 1991; Shiff et al. 2010; Pagano et al. 2004). This article summarizes single agents in different cancers and then discusses multiple infections and cancers. The emphasis is on understanding the underlying mechanisms and on developing preventive and therapeutic strategies in cases with multiple infections. The topic is clinically significant because multiple infections have been reported in several cancers, including cervical cancer, which is the third most common form of cancer among women and the fourth leading cause of death from cancer in women (Lattario et al. 2008).

Infectious Agents, Cancer and Etiology

The major problem in understanding the mechanism of virus-mediated carcinogenesis is that there is no fixed site in the genome that is preferred by cancer-associated infectious agents. The integration of HPV and HBV in the human telomerase reverse transcriptase (hTERT) gene was reported long ago by Ferber and colleagues (Ferber et al. 2003), but confirmation by other groups has yet to be reported. HPV causes cervical cancer and anogenital cancer (Wong et al. 2002; zur Hausen 1999, 2009). Early onset of sexual activity, multiparity, smoking, and many sexual partners are considered risk factors for cervical cancer. Most women infected with HPV do not develop cervical cancer, which suggests the presence of other cofactors that may trigger carcinogenesis. In some cases, other infectious agents may be the cofactors (Lattario et al. 2008). HPV has also been reported in other cancers, such as head and neck squamous cell carcinoma (HNSCC), anal and perianal skin cancer (Gillison and Shah 2001; Bjorge et al. 2002; Mork et al. 2001; van Houten et al. 2001; Wasylyk et al. 2013; Mendelsohn et al. 2010), oropharyngeal cancer (Forte et al. 2012; Sturgis and Ang 2011; Lill et al. 2011; Peres 2010), and lung cancer (Kotb and Petersen 2012; Carpagnano et al. 2011; Joh et al. 2010; Kountouri et al. 2010; Syrjanen 2002). HPV 16 and HPV18 are the most common strains of HPV found to cause cervical cancer. Although more than 80 strains of HPV have been reported, only HPV16 is associated with HNSCC. Identified risk factors for HNSCC include alcohol and tobacco consumption, sexual behavior, and diet; HPV16 is also considered a risk factor for HNSCC (van Houten et al. 2001).

HPV is the principal etiological factor leading to carcinoma of the cervix. To date more than 100 genotypes of HPV have been reported, and one-fifth of these are transmitted sexually (Szkaradkiewicz et al. 2002). Different strains of HPV can infect the same person. A population-level study was conducted in which more than 8,000 samples collected over 7 years were analyzed for different genotypes of HPV. The results indicated the presence of HPV 16, 31, 51, 52, 53, and 66, and the most common co-infection involved HPV 16 and 31 (Kovacs et al. 2008). This type of genotype distribution analysis could be useful when vaccines are prepared for nonresponders to existing HPV vaccines. Surgical treatment, vaccine, and cryotherapy are used to treat HPV-associated infections and diseases (Deligeoroglou et al. 2013; Wang et al. 2012a, b; Best et al. 2012; Bratcher and Sahasrabuddhe 2010; Major et al. 2008; Indrova et al. 2006; Janouskova et al. 2003).

Simian virus 40 (SV40) has been reported in mesothelioma, brain tumors, colon cancer, and osteosarcoma (Rizzo et al. 2001; Campello et al. 2010; Shah 2007). The SV40 large T antigen is the component of the virus involved in its replication, transcriptional control, transformation, and viral assembly and should be a target for therapy.

EBV, a herpes virus, is present in 90 % of the population, but EBV-associated cancers such as BL, nasopharyngeal carcinoma, mononucleosis, post-transplant lymphoma, oral hairy leukoplakia, and a subset of Hodgkin disease have been reported in a limited percentage of people (Melbye et al. 1996; Allen et al. 2013; Lara et al. 2013; Di Napoli et al. 2013; Ozsan et al. 2013; Tumwine et al. 2010; De Falco et al. 2009; Cerny et al. 2009). An association between EBV and gastric cancer also has been reported (Szkaradkiewicz et al. 2002; Kim do et al. 2007; Takada 2000). EBV generally is transmitted by saliva, and its primary infection occurs in the oral mucous membrane. Different sets of proteins are expressed in different cancer types, leading to different preventive and therapeutic strategies for EBV. B cells most often are infected by EBV, but the virus can infect other cells as well. In African populations, repeated malarial infection enables EBV to cause lymphomas by activating B cells and making them prone to abnormal changes. Lymphoma is a disease of the immune system, and any factors that reduce immunity may contribute to lymphoma. In addition to EBV, other viruses such as HCV and human T-lymphotropic virus type 1 (HTLV-1) also have been reported in lymphoma (Uphoff et al. 2010; Martino et al. 1990; Salahuddin et al. 1987; Purtilo 1986). Broad-spectrum antiviral agents and immune-based therapies are utilized for EBV-associated diseases (Rafailidis et al. 2010; Vouloumanou et al. 2012; Gartner and Preiksaitis 2010; Tsuchiya 2002; Vegso et al. 2011).

HTLV-1 is known to cause a lifelong chronic infection that may lead to adult T-cell leukemia/lymphoma (ATLL) (Traina-Dorge et al. 2007). Co-infection with HIV results in increased ATLL, however, co-infected subjects who were treated with highly active antiretroviral therapy (HAART) were found to live longer (Beilke et al. 2005). A model representing HTLV-1 and simian immunodeficiency virus (SIV) co-infection furthered the understanding of HTLV-1 leukemogenesis (Beilke et al. 2005).

Three viruses have been reported to be associated with HCC (Vassilopoulos and Calabrese 2012; Masgala et al. 2012; Mutimer and Lok 2012), namely HBV, HCV, and HDV (Lin et al. 2013; Chuang et al. 2009; Hu and Ludgate 2007). The International Agency for Research on Cancer (IARC) has designated HBV and HCV as human carcinogens. Infection by HBV increases the risk of HCC by 22 % and HCV by 17 % (Donato et al. 1998). The HBV vaccine and interferon and ribavirin treatment for HCV are recommended for the prevention and treatment of hepatitis virus-associated diseases (Tan et al. 2014; Tsuge and Chayama 2013; Coffin et al. 2013; Shin et al. 2012; Yu et al. 2011; Alberti and Caporaso 2011; Iavarone and Colombo 2011).

Chlamydia trachomatis (*C. trachomatis*) is associated with cervical atypia and cervical neoplasia. Genital infection with *C. trachomatis* may cause chronic cervicitis, pelvic inflammation, and endometritis (Smith et al. 2002). If left untreated, *C. trachomatis* may lead to invasive cervical cancer. In individuals infected with both HPV and *C. trachomatis,* it serves as a cofactor (Silva et al. 2013; Alberts et al. 2013; Bhatla et al. 2013). Epidemiology studies conducted in Brazil and Manila support increasing risk of squamous cell cancer in individuals with a positive titer for *C. trachomatis*. Further research suggests that *C. trachomatis* infection may cause inflammation, break the mucosal barrier, and contribute to the development of cervical cancer in the presence of HPV (Silva et al. 2013).

H. pylori is associated with gastric cancer (Alves et al. 2011; Huang et al. 2012; Neves Filho et al. 2010; Shin et al. 2011; Yan et al. 2011), and about half of the world's population carries *H. pylori* (Huang et al. 2012). Gaps remain in the knowledge regarding the natural history and determinants of chronic infection by *H. pylori*. IARC has designated *H. pylori* as a group I carcinogen for gastric cancer (Fuccio et al. 2010). Infection begins in childhood, but data supporting its spontaneous clearance with age are lacking. Longitudinal studies conducted to date have not differentiated chronic and acute infection. Long-term chronic inflammation represents one of the mechanisms of gastric cancer. An alternative mechanism of *H. pylori*-associated carcinogenesis involves promoter methylation of specific genes such as E-cadherin via interleukin1-beta activation of nitric oxide production (Huang et al. 2012). Other genes that become deactivated due to hypermethylation in *H. pylori*-associated gastric cancer are COX-2, HMLH1 (Alves et al. 2011), WWOX (Yan et al. 2011), and MTHFR (Neves Filho et al. 2010). Genome-wide DNA methylation profiles also have been completed and have identified methylated regions in gastric cancer patients (Shin et al. 2011). Chronic inflammation due to infection is linked to an increased risk of malignant transformation in the gastric epithelium. Host cells respond to infection by increased infiltration of macrophages and activation of nuclear factor kappa B (NFkB). As a result, the transcription of cytokines, chemokines, and adhesion molecules also is affected. Antibiotics are prescribed for the treatment of *H. pylori,* although exposure to antibiotics taken for other illnesses seems to explain only a small portion of detected spontaneous clearance events (Fuccio et al. 2010; Broussard et al. 2009). At times a triple-therapy combination of two antibiotics and a proton-pump function also is applied for treatment (Mégraud and Lamouliatte 2003; Mégraud 1999, 2001, 2003, 2004a, b, 2005, 2007, 2009, 2012; Mégraud et al. 2001, 2013; Mégraud and Lehours 2004, 2007; Mégraud and Marshall 2000; Tepes 2009; Correa et al. 2004).

Multiple Infection Mechanisms

The basic mechanisms in multiple infections can be categorized as immune response, genetic, and epigenetic (Vedham et al. 2014). Both the host and infectious agents undergo a variety of changes that contribute to cancer. Mechanical insights and development of preventive and therapeutic approaches are discussed next using examples of co-infections and multiple infections. Immune suppression by the infectious agent, induction of inflammatory cytokines, or a combination of the two can occur during multiple infections. Some investigators have described two modes of carcinogenesis—direct and indirect. An example of direct carcinogenesis occurs during viral infection, because viruses are capable of transformation by directly inserting their genetic material into the host genome, affecting survival and proliferation pathways (Saha et al. 2010). An example of indirect carcinogenesis is infection by *H. pylori* when the infection causes alterations in the host immune system, followed by genomic instability and, finally, carcinogenesis. Common factors that influence carcinogenesis include environment, age, behavior (alcohol consumption, diet, smoking, and exercise), race, and gender. Different factors are involved in specific cancers; therefore, it is essential to determine the biological information and risk factors that are associated with a specific cancer before planning preventive and therapeutic approaches. Examples of multiple infections and underlying carcinogenesis mechanisms are described below.

EBV and HIV Infection

In EBV- and HIV-positive plasmablastic lymphoma (PBL) patients, the most common genetic abnormality was rearrangement of the MYC gene, which is an oncogene (Boy et al. 2011). Increased copy numbers of the CCND1 gene also were observed. The CCND1 gene is a key regulatory gene involved in cell-cycle regulation, and its increased copy numbers may contribute to uncontrolled cell proliferation. These authors suggest that EBV infection causes genetic rearrangement and alters copy numbers of selected genes in PBL. This disease has been studied in an African population in which HIV infection causes immune suppression. HIV-positive individuals are prone to developing oral lymphomas that are associated with plasmacytic differentiation and have an aggressive clinical course with poor prognosis (Piras et al. 1996; Francischini et al. 2010; Rafaniello Raviele et al. 2009). MYC rearrangement resulted in poor survival in PBL patients who tested positive for HIV and EBV. Boy et al. suggested that EBV may be critical in the initiation of PBL of the oral cavity, and posterior oropharyngeal and nasopharyngeal epithelial cells are EBV reservoirs (Boy et al. 2011).

EBV shows three latency patterns in which EBV-encoded nuclear antigen-1 (EBNA1) plays a critical role (Lorenzetti et al. 2010). The first pattern, latency 1, is restricted only to the expression of the EBNA1 protein; the second pattern, latency 2, expresses the EBV-encoded RNAs (EBERs), BamHI-A rightward transcripts (BARTs), the EBNA1 protein, and the latent membrane proteins (LMP1, LMP2A, and LMP2B); and the third pattern, latency 3, expresses both EBERs and BARTs and all of the EBV latent proteins—the six nuclear antigens (EBNA1, EBNA2, EBNA3A, EBNA3B, EBNA3C, and EBNA-LP) and three membrane proteins (LMP1, LMP2A, and LMP2B). All three latency patterns are represented in B cell malignancies; in non-B cell malignancies, only the latency 2 and 3 patterns are displayed.

EBV and Plasmodium Falciparum

Malaria development involves infection by the parasite *Plasmodium falciparum* (*P. falciparum*). EBV infection has been reported in *P. falciparum*-infected individuals, and the consensus among investigators is that the immunomodulatory effects of malarial infection can cause expansion of EBV-positive B lymphocytes, ultimately leading to a high incidence of endemic BL (eBL) (Chene et al. 2009). In EBV-infected individuals, *P. falciparum* induces lytic EBV reactivation followed by B cell expansion. Preliminary data suggest that antimalarial treatment decreases eBL incidence. Because 4–9 year-olds can have a high incidence rate of eBL, antimalarial treatment should be modified for this age group. In malaria, the number of peripheral B cells is high; therefore, the mechanism of co-infection with EBV and *P. falciparum* should be explored further. When planning treatment, it should be noted that BL is a more curable form of cancer, with a high treatment success rate and rare recurrence (Lopes et al. 2003).

EBV and HPV

HPV belongs to the *Papillomaviridae* family, and EBV belongs to the *Herpesviridae* family. Both viruses have been reported in nasopharyngeal carcinoma and head and neck carcinoma (Al Moustafa et al. 2009; Mirzamani et al. 2006; Wong et al. 2002; Tyan et al. 1993). As for the mechanism, methylation of the death-associated protein kinase gene (DAPK) was reported in EBV and HPV co-infected cells (Lattario et al. 2008). Infected cells did not show any altered cytological features, and the methylation pattern confirmed the presence of these viruses and later on the development of cancer. Higher levels of DAPK methylation were found to be correlated with smoking status in female study participants. Another mechanism proposed is based on microRNA (miRNA) profiling (Kim do et al. 2007). Altered miRNA profiles were observed when samples from infected individuals were analyzed.

HBV and HDV

The hepatitis delta virus (HDV) is a subviral agent that requires HBV envelope proteins for replication. Infection of HBV-positive cells with HDV has been reported in HCC (Bissinger and Berg 2013; Sureau 2006; Agalliu et al. 2013; Wedemeyer 2010). Hepatitis B surface antigen (HBsAg) was used as the indicator of HBV positivity. Co-infection by HBV-positive cells and HDV causes severe complications compared to HBV-only infection. Treatment of dually infected patients is difficult, and high doses of interferon-alpha for a long period generally are recommended

(Gozlan et al. 2009; Niro et al. 2005). This example presents an interesting mechanism in which one infectious agent requires another infectious agent for survival.

HPV, HIV and HTLV

An increased association between HPV co-infection and cervical cancer has been observed in HIV-positive women (Meyrelles et al. 2013; Aggarwal et al. 2012; Feola et al. 2013; Beachler et al. 2012). When HIV/HPV-positive women were subjected to HAART, reduced immunosuppression was observed, and the course of such infections was similar to HIV-negative women (Roccio et al. 2012). Jalil et al. reported a significant reduction in HPV infection during the postpartum period in both HIV-positive and HIV-negative women (Jalil et al. 2013). Multiple infections in women with different HPV strains (HPV16, HPV18, HPV31, HPV51, and HPV52) were observed by Spinillo et al. (Spinillo et al. 2009). HIV-positive women have high rates of infection with multiple HPV types. HPV infection has been reported in HIV-positive men as well (Figliuolo et al. 2012; Mooij et al. 2013). In men, premalignant lesions of the penis were observed that led to penile cancer. Although HAART has been effective in controlling HIV, the incidence of HPV-associated oral lesions has increased (Anaya-Saavedra et al. 2013). One study in an African American population positive for HPV and HIV indicated that variants in the interleukin family of cytokine genes influenced HPV clearance (Sudenga et al. 2013). Introduction of a vaccine containing antigens from more than one virus has been promoted by Santana et al., who mixed HIV, HSV, and HPV antigens and observed a CD8+ T cell-dependent response (Santana et al. 2013). HPV infection was observed in Kaposi sarcoma patients who tested positive for HIV and Kaposi sarcoma-associated herpesvirus/human herpesvirus 8 (KSHV/HHV 8) (Huang et al. 1992). In a case control study, HPV infection was present in HTLV-positive women (Lopo et al. 2012). Studies also have shown that individuals who test positive for multiple strains of HPV respond more poorly to cancer treatments (Pao et al. 1996; Kjaer et al. 2001).

Co-infection by H. pylori *and Other Viruses*

Itatsu et al. reported infection by *H. pylori*, HTLV-1, and EBV in gastric cancer (Itatsu et al. 1999). Co-infection by *H. pylori* and HCV in liver cirrhosis and liver cancer and the Cag A gene of *H. pylori* may play a significant role in disease progression (Esmat et al. 2012). Wang et al. reported *H. pylori* infection in HBV-positive individuals in the Shandong Province of China, and alcohol consumption was found to be associated with an increased level of *H. pylori* infection (Wang et al. 2011). An association between EBV and gastric cancer also has been demonstrated, and almost 10 % of gastric carcinomas throughout the world are monoproliferations of EBV-positive tumors (Takada 2000; Iizasa et al. 2012).

H. pylori interferes with the liver cytochrome P450 metabolic activity (Giannini et al. 2003). Early prevention and sanitation are the two key elements in preventing infection. Improvements in sanitation and hygiene have reduced the transmission of these bacteria in developed countries. In the developing world—where the population is transitioning from low to high resources—adult life expectancy is increasing despite people having been infected with *H. pylori* as children (de Martel and Franceschi 2009).

Future Directions and Implications in Cancer Control

A number of challenges in the field of multiple infections can be addressed by conducting research on topics identified in this article. Selected areas include: specific risk factors and biomarkers for different racial and ethnic groups that make them especially high-risk populations for cancers associated with multiple infectious agents; utilizing these biomarkers and information about factors for targeted therapeutics; exploring the significance of *H. pylori* colonization and co-evolution in humans and implications for other infectious diseases; utilizing beneficiary information from infections—for example, it has been suggested that infection with agents such as *H. pylori* may decrease the risk of some cancers, including esophageal adenocarcinomas; utilizing information from the human microbiome to better understand this effect and develop mechanisms to improve prevention and treatment of malignant infections; exploring the mechanisms behind the increasing frequency of obesity and diabetes mellitus influencing incidence of HCC in developing countries with a high prevalence of HBV and/or HCV infection; developing successful prevention strategies against current and new oncoviruses to avoid infection and subsequent carcinogenesis; utilizing our understanding of the natural history of infectious agents in causing cancer and determination of these patterns to develop effective prevention and treatment mechanisms; translating epidemiologic data to improve our understanding of cancer etiology; identifying viral and host factors that predispose an individual to cancer development in multiple viral infections; identifying distinct immunologic response profiles that protect or predispose EBV-infected individuals to cancer development; and understanding HDV and HBV interaction to explain requirements of HDV to have HBV infected cells. These objectives can be attained by the combined efforts of teams of scientists working together in a consortium forum. In addition, increased public and private resources are needed to develop prevention and treatment strategies in cancers where multiple infections have been reported.

Still unanswered questions include the role of double (HBV and HDV) or triple (HBV, HCV, and HDV) viral infection in liver carcinogenesis. After such questions are addressed and more is understood about the underlying mechanisms, better therapeutic approaches can be identified. My program at the National Cancer

Institute (NCI) supports a number of cohorts in which disease has not yet developed, despite some subjects having been exposed to infectious agents. Samples from participants are collected and stored so that they can be studied retrospectively.

Attention by physicians is needed when transplanting the liver or other organs, because transplantation is followed by attacks by microorganisms, including pathogens that are associated with cancer (Fishman 2011; Terrault 2013). Viruses are the most common cause of opportunistic infections following transplant surgery. The ages of the donor and recipient play a major role in host immune suppression and development of a disease post-transplantation. Those with lymphoproliferative disorders are prone to infection by EBV. Viruses can affect disease development directly or indirectly (e.g., as immune suppression and alterations in the immune system of the host). New roles for NK cells and dendritic cells should be identified (Fishman 2013). Further research also should be conducted in the fields of viral latency, reactivation, cellular effects of viral infection, prevention of recurrence of viral infection after transplantation, and development of vaccines and other preventive and therapeutic approaches.

Current research in the field of infectious agents has expanded our understanding of the role of malignant infections, but the mechanisms of carcinogenesis still are not fully understood. Key determinants and risk factors have been identified by epidemiologic studies that modify the effects of infectious agents. Bicistronic vaccines and other novel approaches to vaccine generation should be investigated.

References

Agalliu I, Wang Z, Wang T, Dunn A, Parikh H, Myers T et al (2013) Characterization of SNPs associated with prostate cancer in men of Ashkenazic descent from the set of GWAS identified SNPs: impact of cancer family history and cumulative SNP risk prediction. PLoS One 8:e60083. doi:10.1371/journal.pone.0060083

Aggarwal R, Sachdeva RK, Naru J, Suri V, Sharma A, Nijhawan R (2012) HPV genotyping in north Indian women infected with HIV. Int J Gynecol Pathol 31:475–481. doi:10.1097/PGP.0b013e31824a1d60

Al Moustafa AE, Chen D, Ghabreau L, Akil N (2009) Association between human papillomavirus and Epstein-Barr virus infections in human oral carcinogenesis. Med Hypotheses 73:184–186. doi:10.1016/j.mehy.2009.02.025

Alberti A, Caporaso N (2011) HBV therapy: guidelines and open issues. Dig Liver Dis 43(Suppl 1):S57–S63. doi:10.1016/S1590-8658(10)60693-7

Alberts CJ, van der Loeff MF S, Papenfuss MR, da Silva RJ, Villa LL, Lazcano-Ponce E et al (2013) Association of *Chlamydia trachomatis* infection and herpes simplex virus type 2 serostatus with genital human papillomavirus infection in men: the HPV in men study. Sex Transm Dis 40:508–515. doi:10.1097/OLQ.0b013e318289c186

Allen PB, Laing G, Connolly A, O'Neill C (2013) EBV-associated colonic B-cell lymphoma following treatment with infliximab for IBD: a new problem? BMJ Case Rep 2013:1–3. doi:10.1136/bcr-2013-200423

Alves MK, Ferrasi AC, Lima VP, Ferreira MV, de Moura Campos Pardini MI, Rabenhorst SH (2011) Inactivation of COX-2, HMLH1 and CDKN2A gene by promoter methylation in gastric cancer: relationship with histological subtype, tumor location and *Helicobacter pylori* genotype. Pathobiology 78:266–276. doi:10.1159/000329475

Anaya-Saavedra G, Flores-Moreno B, Garcia-Carranca A, Irigoyen-Camacho E, Guido-Jimenez M, Ramirez-Amador V (2013) HPV oral lesions in HIV-infected patients: the impact of long-term HAART. J Oral Pathol Med 42:443–449. doi:10.1111/jop.12032

Beachler DC, Weber KM, Margolick JB, Strickler HD, Cranston RD, Burk RD et al (2012) Risk factors for oral HPV infection among a high prevalence population of HIV-positive and at-risk HIV-negative adults. Cancer Epidemiol Biomarkers Prev 21:122–133. doi:10.1158/1055-9965.EPI-11-0734

Beilke MA, Japa S, Moeller-Hadi C, Martin-Schild S (2005) Tropical spastic paraparesis/human T leukemia virus type 1-associated myelopathy in HIV type 1-coinfected patients. Clin Infect Dis 41:e57–e63. doi:10.1086/432890

Best SR, Niparko KJ, Pai SI (2012) Biology of human papillomavirus infection and immune therapy for HPV-related head and neck cancers. Otolaryngol Clin North Am 45:807–822. doi:10.1016/j.otc.2012.04.005

Bhatla N, Puri K, Joseph E, Kriplani A, Iyer VK, Sreenivas V (2013) Association of *Chlamydia trachomatis* infection with human papillomavirus (HPV) & cervical intraepithelial neoplasia—a pilot study. Indian J Med Res 137:533–539

Bissinger AL, Berg CP (2013) Is the HDV seroprevalence in patients admitted to a university hospital representative for all HBV carriers in Germany? Infection 41:1031–1032. doi:10.1007/s15010-013-0433-z

Bjorge T, Engeland A, Luostarinen T, Mork J, Gislefoss RE, Jellum E et al (2002) Human papillomavirus infection as a risk factor for anal and perianal skin cancer in a prospective study. Br J Cancer 87:61–64. doi:10.1038/sj.bjc.6600350

Boy SC, van Heerden MB, Babb C, van Heerden WF, Willem P (2011) Dominant genetic aberrations and coexistent EBV infection in HIV-related oral plasmablastic lymphomas. Oral Oncol 47:883–887. doi:10.1016/j.oraloncology.2011.06.506

Bratcher LF, Sahasrabuddhe VV (2010) The impact of antiretroviral therapy on HPV and cervical intraepithelial neoplasia: current evidence and directions for future research. Infect Agent Cancer 5:8. doi:10.1186/1750-9378-5-8

Broussard CS, Goodman KJ, Phillips CV, Smith MA, Fischbach LA, Day RS et al (2009) Antibiotics taken for other illnesses and spontaneous clearance of *Helicobacter pylori* infection in children. Pharmacoepidemiol Drug Saf 18:722–729. doi:10.1002/pds.1773

Campello C, Comar M, Zanotta N, Minicozzi A, Rodella L, Poli A (2010) Detection of SV40 in colon cancer: a molecular case-control study from northeast Italy. J Med Virol 82:1197–1200. doi:10.1002/jmv.21798

Carpagnano GE, Koutelou A, Natalicchio MI, Martinelli D, Ruggieri C, Di Taranto A et al (2011) HPV in exhaled breath condensate of lung cancer patients. Br J Cancer 105:1183–1190. doi:10.1038/bjc.2011.354

Cerny J, Wang S, Miron P, Hutchinson L, Brettler D (2009) EBV-negative Hodgkin lymphoma in ZAP-70-positive B-cell lymphocytic leukemia patient. Ann Hematol 88:495–497. doi:10.1007/s00277-008-0617-2

Chemaly RF, Ghantoji SS, Shah DP, Shah JN, El Taoum KK, Champlin RE et al (2013) Respiratory syncytial virus infections in children with cancer. J Pediatr Hematol Oncol 36(6):e376–e381. doi:10.1097/MPH.0000000000000086

Chen L, He Z, Qin L, Li Q, Shi X, Zhao S et al (2011) Antitumor effect of malaria parasite infection in a murine Lewis lung cancer model through induction of innate and adaptive immunity. PLoS One 6:e24407. doi:10.1371/journal.pone.0024407

Chene A, Donati D, Orem J, Mbidde ER, Kironde F, Wahlgren M et al (2009) Endemic Burkitt's lymphoma as a polymicrobial disease: new insights on the interaction between *Plasmodium falciparum* and Epstein-Barr virus. Semin Cancer Biol 19:411–420. doi:10.1016/j.semcancer.2009.10.002

Cho HJ, Kim JY, Yoo J, Lee SS (2003) Gastric carcinoma with lymphoid stroma: incidence of EBV and *Helicobacter pylori* infection. Appl Immunohistochem Mol Morphol 11:149–152

Chuang SC, La Vecchia C, Boffetta P (2009) Liver cancer: descriptive epidemiology and risk factors other than HBV and HCV infection. Cancer Lett 286:9–14. doi:10.1016/j.canlet.2008.10.040

Coffin CS, Osiowy C, Myers RP, Gill MJ (2013) Virology and clinical sequelae of long-term antiviral therapy in a North American cohort of hepatitis B virus (HBV)/human immunodeficiency virus type 1 (HIV-1) co-infected patients. J Clin Virol 57:103–108. doi:10.1016/j.jcv.2013.02.004

Correa P, Piazuelo MB, Camargo MC (2004) The future of gastric cancer prevention. Gastric Cancer 7:9–16. doi:10.1007/s10120-003-0265-0

Darani HY, Yousefi M (2012) Parasites and cancers: parasite antigens as possible targets for cancer immunotherapy. Future Oncol 8:1529–1535. doi:10.2217/fon.12.155

De Falco G, Antonicelli G, Onnis A, Lazzi S, Bellan C, Leoncini L (2009) Role of EBV in microRNA dysregulation in Burkitt lymphoma. Semin Cancer Biol 19:401–406. doi:10.1016/j.semcancer.2009.07.003

de Martel C, Franceschi S (2009) Infections and cancer: established associations and new hypotheses. Crit Rev Oncol Hematol 70:183–194. doi:10.1016/j.critrevonc.2008.07.021

Deligeoroglou E, Giannouli A, Athanasopoulos N, Karountzos V, Vatopoulou A, Dimopoulos K et al (2013) HPV infection: immunological aspects and their utility in future therapy. Infect Dis Obstet Gynecol 2013:540850. doi:10.1155/2013/540850

Di Napoli A, Al-Jadiri MF, Talerico C, Duranti E, Pilozzi E, Trivedi P et al (2013) Epstein-Barr virus (EBV) positive classical Hodgkin lymphoma of Iraqi children: an immunophenotypic and molecular characterization of Hodgkin/Reed-Sternberg cells. Pediatr Blood Cancer 60:2068–2072. doi:10.1002/pbc.24654

Donato F, Boffetta P, Puoti M (1998) A meta-analysis of epidemiological studies on the combined effect of hepatitis B and C virus infections in causing hepatocellular carcinoma. Int J Cancer 75:347–354

Esmat G, El-Bendary M, Zakarya S, Ela MA, Zalata K (2012) Role of *Helicobacter pylori* in patients with HCV-related chronic hepatitis and cirrhosis with or without hepatocellular carcinoma: possible association with disease progression. J Viral Hepat 19:473–479. doi:10.1111/j.1365-2893.2011.01567.x

Feola TD, Albert MB, Shahabi K, Endy T (2013) Prevalence of HPV in HIV-infected women in the Designated AIDS Center at Upstate Medical University and the potential benefit of vaccination regardless of age. J Assoc Nurses AIDS Care 24:176–179. doi:10.1016/j.jana.2012.03.003

Ferber MJ, Montoya DP, Yu C, Aderca I, McGee A, Thorland EC et al (2003) Integrations of the hepatitis B virus (HBV) and human papillomavirus (HPV) into the human telomerase reverse transcriptase (hTERT) gene in liver and cervical cancers. Oncogene 22:3813–3820. doi:10.1038/sj.onc.1206528

Ferrasi AC, Pinheiro NA, Rabenhorst SH, Caballero OL, Rodrigues MA, de Carvalho F et al (2010) *Helicobacter pylori* and EBV in gastric carcinomas: methylation status and microsatellite instability. World J Gastroenterol 16:312–319

Figliuolo G, Maia J, Jalkh AP, Miranda AE, Ferreira LC (2012) Clinical and laboratorial study of HPV infection in men infected with HIV. Int Braz J Urol 38:411–418

Fishman JA (2011) Infections in immunocompromised hosts and organ transplant recipients: essentials. Liver Transpl 17(Suppl 3):S34–S37. doi:10.1002/lt.22378

Fishman JA (2013) What's new and what's hot? Basic science at the American Transplant Congress 2012. Am J Transplant 13:275–280. doi:10.1111/ajt.12015

Forte T, Niu J, Lockwood GA, Bryant HE (2012) Incidence trends in head and neck cancers and human papillomavirus (HPV)-associated oropharyngeal cancer in Canada, 1992–2009. Cancer Causes Control 23:1343–1348. doi:10.1007/s10552-012-0013-z

Francischini E, Martins FM, Braz-Silva PH, Magalhaes MH, Ortega KL (2010) HIV-associated oral plasmablastic lymphoma and role of adherence to highly active antiretroviral therapy. Int J STD AIDS 21:68–70. doi:10.1258/ijsa.2008.008476

Frappier L (2013) A cancer-associated Epstein-Barr virus variant with epithelial tropism. Trends Microbiol 21:616–617. doi:10.1016/j.tim.2013.10.007

Fuccio L, Eusebi LH, Bazzoli F (2010) Gastric cancer, *Helicobacter pylori* infection and other risk factors. World J Gastrointest Oncol 2:342–347. doi:10.4251/wjgo.v2.i9.342

Gartner B, Preiksaitis JK (2010) EBV viral load detection in clinical virology. J Clin Virol 48:82–90. doi:10.1016/j.jcv.2010.03.016

Giannini E, Fasoli A, Botta F, Romagnoli P, Malfatti F, Chiarbonello B et al (2003) *Helicobacter pylori* infection is associated with greater impairment of cytochrome P-450 liver metabolic activity in anti-HCV positive cirrhotic patients. Dig Dis Sci 48:802–808

Gillison ML, Shah KV (2001) Human papillomavirus-associated head and neck squamous cell carcinoma: mounting evidence for an etiologic role for human papillomavirus in a subset of head and neck cancers. Curr Opin Oncol 13:183–188

Gozlan J, Lacombe K, Gault E, Raguin G, Girard PM (2009) Complete cure of HBV-HDV co-infection after 24 weeks of combination therapy with pegylated interferon and ribavirin in a patient co-infected with HBV/HCV/HDV/HIV. J Hepatol 50:432–434. doi:10.1016/j.jhep.2008.05.029

Hu J, Ludgate L (2007) HIV-HBV and HIV-HCV coinfection and liver cancer development. Cancer Treat Res 133:241–252

Huang YQ, Li JJ, Rush MG, Poiesz BJ, Nicolaides A, Jacobson M et al (1992) HPV-16-related DNA sequences in Kaposi's sarcoma. Lancet 339:515–518

Huang FY, Chan AO, Rashid A, Wong DK, Cho CH, Yuen MF (2012) *Helicobacter pylori* induces promoter methylation of E-cadherin via interleukin-1beta activation of nitric oxide production in gastric cancer cells. Cancer 118:4969–4980. doi:10.1002/cncr.27519

Iavarone M, Colombo M (2011) HBV-related HCC, clinical issues and therapy. Dig Liver Dis 43(Suppl 1):S32–S39. doi:10.1016/S1590-8658(10)60690-1

Iizasa H, Nanbo A, Nishikawa J, Jinushi M, Yoshiyama H (2012) Epstein-Barr Virus (EBV)-associated gastric carcinoma. Viruses 4:3420–3439

Indrova M, Bieblova J, Jandlova T, Vonka V, Pajtasz-Piasecka E, Reinis M (2006) Chemotherapy, IL-12 gene therapy and combined adjuvant therapy of HPV 16-associated MHC class I-proficient and -deficient tumours. Int J Oncol 28:253–259

Ishii A, Matsuoka H, Aji T, Ohta N, Arimoto S, Wataya Y et al (1994) Parasite infection and cancer: with special emphasis on *Schistosoma japonicum* infections (Trematoda). Mutat Res 305:273–281

Itatsu T, Miwa H, Ohkura R, Iwazaki R, Oide H, Murai T et al (1999) Primary gastric T-cell lymphoma accompanied by HTLV-I, HBV and *H. pylori* infection. Dig Dis Sci 44:1823–1836

Jalil EM, Bastos FI, Melli PP, Duarte G, Simoes RT, Yamamoto AY et al (2013) HPV clearance in postpartum period of HIV-positive and negative women: a prospective follow-up study. BMC Infect Dis 13:564. doi:10.1186/1471-2334-13-564

Janouskova O, Sima P, Kunke D (2003) Combined suicide gene and immunostimulatory gene therapy using AAV-mediated gene transfer to HPV-16 transformed mouse cell: decrease of oncogenicity and induction of protection. Int J Oncol 22:569–577

Joh J, Jenson AB, Moore GD, Rezazedeh A, Slone SP, Ghim SJ et al (2010) Human papillomavirus (HPV) and Merkel cell polyomavirus (MCPyV) in non small cell lung cancer. Exp Mol Pathol 89:222–226. doi:10.1016/j.yexmp.2010.08.001

Khan AU, Waqar M, Akram M, Zaib M, Wasim M, Ahmad S et al (2011) True prevalence of twin HDV-HBV infection in Pakistan: a molecular approach. Virol J 8:420. doi:10.1186/1743-422X-8-420

Kim do N, Chae HS, Oh ST, Kang JH, Park CH, Park WS et al (2007) Expression of viral microRNAs in Epstein-Barr virus-associated gastric carcinoma. J Virol 81:1033–1036. doi:10.1128/JVI.02271-06

Kjaer SK, Chackerian B, van den Brule AJ, Svare EI, Paull G, Walbomers JM et al (2001) High-risk human papillomavirus is sexually transmitted: evidence from a follow-up study of virgins starting sexual activity (intercourse). Cancer Epidemiol Biomark Prev 10:101–106

Kotb WF, Petersen I (2012) Morphology, DNA ploidy and HPV in lung cancer and head and neck cancer. Pathol Res Pract 208:1–8. doi:10.1016/j.prp.2011.10.009

Kountouri MP, Mammas IN, Spandidos DA (2010) Human papilloma virus (HPV) in lung cancer: unanswered questions. Lung Cancer 67:125. doi:10.1016/j.lungcan.2009.10.020

Kovacs K, Varnai AD, Bollmann M, Bankfalvi A, Szendy M, Speich N et al (2008) Prevalence and genotype distribution of multiple human papillomavirus infection in the uterine cervix: a 7.5-year longitudinal study in a routine cytology-based screening population in West Germany. J Med Virol 80:1814–1823. doi:10.1002/jmv.21282

Lara J, Cohen M, De Matteo E, Aversa L, Preciado MV, Chabay P (2013) Epstein-Barr virus (EBV) association and latency profile in pediatric Burkitt's lymphoma: experience of a single institution in Argentina. J Med Virol 86(5):845–850. doi:10.1002/jmv.23737

Lattario F, Furtado YL, Silveira FA, do Val IC, Almeida G, da Costa Carvalho Mda G (2008) Evaluation of DAPK gene methylation and HPV and EBV infection in cervical cells from patients with normal cytology and colposcopy. Arch Gynecol Obstet 277:505–509. doi:10.1007/s00404-007-0511-5

Lill C, Kornek G, Bachtiary B, Selzer E, Schopper C, Mittlboeck M et al (2011) Survival of patients with HPV-positive oropharyngeal cancer after radiochemotherapy is significantly enhanced. Wien Klin Wochenschr 123:215–221. doi:10.1007/s00508-011-1553-z

Lin H, Ha NB, Ahmed A, Ayoub W, Daugherty TJ, Lutchman GA et al (2013) Both HCV and HBV are major causes of liver cancer in Southeast Asians. J Immigr Minor Health 15:1023–1029. doi:10.1007/s10903-013-9871-z

Lopes V, Young LS, Murray PG (2003) Epstein-Barr virus-associated cancers: aetiology and treatment. Herpes 10:78–82

Lopo SS, Oliveira PM, Santana IU, Pena GB, Torrales MB, Mascarenhas RE et al (2012) Evidence of a higher prevalence of HPV infection in HTLV-1-infected women: a cross-sectional study. Rev Soc Bras Med Trop 45:305–308

Lorenzetti MA, De Matteo E, Gass H, Martinez Vazquez P, Lara J, Gonzalez P et al (2010) Characterization of Epstein Barr virus latency pattern in Argentine breast carcinoma. PLoS One 5:e13603. doi:10.1371/journal.pone.0013603

Major T, Sziklai I, Czegledy J, Gall T, Gergely L, Szarka K (2008) Follow-up of HPV DNA copy number in cidofovir therapy of recurrent respiratory papillomatosis. Anticancer Res 28:2169–2174

Martino GV, Grimaldi LM, Moiola L, Cichero P, Ieri R, Bonini PA (1990) Antibody against EBV in CSF from HIV-1 and HTLV-I infected patients. Acta Neurol (Napoli) 12:68–70

Masgala A, Bonovas S, Nikolopoulos GK (2012) Recent advances in the treatment of HIV/HBV and HIV/HCV co-infection. Mini Rev Med Chem 12:890–904

Mégraud F (1999) Resistance of *Helicobacter pylori* to antibiotics: the main limitation of current proton-pump inhibitor triple therapy. Eur J Gastroenterol Hepatol 11(Suppl 2):S35–S37; discussion S43–5

Mégraud F (2001) Resistance of *Helicobacter pylori* to antibiotics and its impact on treatment options. Drug Resist Updat 4:178–186. doi:10.1054/drup.2001.0203

Mégraud F (2003) *Helicobacter pylori* resistance to antibiotics: prevalence, mechanism, detection. What's new? Can J Gastroenterol 17(Suppl B):49B–52B

Mégraud F (2004a) Basis for the management of drug-resistant *Helicobacter pylori* infection. Drugs 64:1893–1904

Mégraud F (2004b) *H pylori* antibiotic resistance: prevalence, importance, and advances in testing. Gut 53:1374–1384. doi:10.1136/gut.2003.022111

Mégraud F (2005) Update on therapeutic options for *Helicobacter pylori*-related diseases. Curr Infect Dis Rep 7:115–120

Mégraud F (2007) *Helicobacter pylori* and antibiotic resistance. Gut 56:1502. doi:10.1136/gut.2007.132514

Mégraud F (2009) Frontiers in *Helicobacter* research. Clin Microbiol Infect 15:805. doi:10.1111/j.1469-0691.2009.02959.x

Mégraud F (2012) The challenge of *Helicobacter pylori* resistance to antibiotics: the comeback of bismuth-based quadruple therapy. Therap Adv Gastroenterol 5:103–109. doi:10.1177/1756283X11432492

Mégraud F, Lamouliatte H (2003) Review article: the treatment of refractory *Helicobacter pylori* infection. Aliment Pharmacol Ther 17:1333–1343
Mégraud F, Lehours P (2004) *Helicobacter pylori* and gastric cancer prevention is possible. Cancer Detect Prev 28:392–398. doi:10.1016/j.cdp.2004.07.005
Mégraud F, Lehours P (2007) *Helicobacter pylori* detection and antimicrobial susceptibility testing. Clin Microbiol Rev 20:280–322. doi:10.1128/CMR.00033-06
Mégraud F, Marshall BJ (2000) How to treat *Helicobacter pylori*. First-line, second-line, and future therapies. Gastroenterol Clin North Am 29:759–773
Mégraud F, Hazell S, Glupczynski Y (2001) Antibiotic susceptibility and resistance. In: Mobley HLT, Mendz GL, Hazell SL (eds) *Helicobacter pylori*: physiology and genetics. ASM Press, Washington, DC, pp 25–37
Mégraud F, Coenen S, Versporten A, Kist M, Lopez-Brea M, Hirschl AM et al (2013) *Helicobacter pylori* resistance to antibiotics in Europe and its relationship to antibiotic consumption. Gut 62:34–42. doi:10.1136/gutjnl-2012-302254
Melbye M, Cote TR, West D, Kessler L, Biggar RJ (1996) Nasopharyngeal carcinoma: an EBV-associated tumour not significantly influenced by HIV-induced immunosuppression. The AIDS/Cancer Working Group. Br J Cancer 73:995–997
Mendelsohn AH, Lai CK, Shintaku IP, Elashoff DA, Dubinett SM, Abemayor E et al (2010) Histopathologic findings of HPV and p16 positive HNSCC. Laryngoscope 120:1788–1794. doi:10.1002/lary.21044
Meyrelles AR, Siqueira JD, Hofer CB, Costa TP, Azevedo AP, Guimaraes BV et al (2013) HIV/HPV co-infection during pregnancy in southeastern Brazil: prevalence, HPV types, cytological abnormalities and risk factors. Gynecol Oncol 128:107–112. doi:10.1016/j.ygyno.2012.10.003
Minoura-Etoh J, Gotoh K, Sato R, Ogata M, Kaku N, Fujioka T et al (2006) *Helicobacter pylori-associated* oxidant monochloramine induces reactivation of Epstein-Barr virus (EBV) in gastric epithelial cells latently infected with EBV. J Med Microbiol 55:905–911. doi:10.1099/jmm.0.46580-0
Mirzamani N, Salehian P, Farhadi M, Tehran EA (2006) Detection of EBV and HPV in nasopharyngeal carcinoma by *in situ* hybridization. Exp Mol Pathol 81:231–234. doi:10.1016/j.yexmp.2006.04.006
Mooij SH, Boot HJ, Speksnijder AG, Stolte IG, Meijer CJ, Snijders PJ et al (2013) Oral human papillomavirus infection in HIV-negative and HIV-infected men who have sex with men: the HIV & HPV in MSM (H2M) study. AIDS 27(13):2117–2128. doi:10.1097/QAD.0b013e328362395c
Mork J, Lie AK, Glattre E, Hallmans G, Jellum E, Koskela P et al (2001) Human papillomavirus infection as a risk factor for squamous-cell carcinoma of the head and neck. N Engl J Med 344:1125–1131. doi:10.1056/NEJM200104123441503
Mutimer DJ, Lok A (2012) Management of HBV- and HCV-induced end stage liver disease. Gut 61(Suppl 1):i59–i67. doi:10.1136/gutjnl-2012-302076
Neves Filho EH, Alves MK, Lima VP, Rabenhorst SH (2010) MTHFR C677T polymorphism and differential methylation status in gastric cancer: an association with *Helicobacter pylori* infection. Virchows Arch 457:627–633. doi:10.1007/s00428-010-0996-3
Niro GA, Rosina F, Rizzetto M (2005) Treatment of hepatitis D. J Viral Hepat 12:2–9. doi:10.1111/j.1365-2893.2005.00601.x
Ozsan N, Cagirgan S, Saydam G, Gunes A, Hekimgil M (2013) Epstein-Barr virus (EBV) positive diffuse large B cell lymphoma of the elderly-experience of a single center from Turkey. Pathol Res Pract 209:471–478. doi:10.1016/j.prp.2013.04.014
Pagano JS, Blaser M, Buendia MA, Damania B, Khalili K, Raab-Traub N et al (2004) Infectious agents and cancer: criteria for a causal relation. Semin Cancer Biol 14:453–471. doi:10.1016/j.semcancer.2004.06.009
Pao CC, Yao DS, Lin CY, Lee SC, Su HT, Lin SC (1996) Genomic aberrations of human papillomavirus recovered from cervical cancers. Biochem Biophys Res Commun 222:116–120. doi:10.1006/bbrc.1996.0707

Peres J (2010) HPV-positive oropharyngeal cancer: data may justify new approach. J Natl Cancer Inst 102:1456–1459. doi:10.1093/jnci/djq403

Piras MA, Salis T, Naitana AV, Spanu L, Aceti A (1996) HIV infection and neoplasia. Lancet 348:1316–1317; author reply 8. doi:10.1016/S0140-6736(05)65801-1

Prasad R, Husain N, Karmakar S, Verma S (2013) Lung cancer and hepatitis C virus. South Asian J Cancer 2:86. doi:10.4103/2278-330X.110500

Purtilo DT (1986) Lymphotropic viruses, Epstein-Barr virus (EBV) and human T-cell lymphotropic virus-I (HTLV-I)/adult T-cell leukemia virus (ATLV), and HTLV-III/human immune deficiency virus (HIV) as etiological agents of malignant lymphoma and immune deficiency. AIDS Res 2(Suppl 1):S1–S6

Rafailidis PI, Mavros MN, Kapaskelis A, Falagas ME (2010) Antiviral treatment for severe EBV infections in apparently immunocompetent patients. J Clin Virol 49:151–157. doi:10.1016/j.jcv.2010.07.008

Rafaniello Raviele P, Pruneri G, Maiorano E (2009) Plasmablastic lymphoma: a review. Oral Dis 15:38–45. doi:10.1111/j.1601-0825.2008.01493.x

Read SA, Douglas MW (2014) Virus induced inflammation and cancer development. Cancer Lett 345:174–181. doi:10.1016/j.canlet.2013.07.030

Rizzo P, Bocchetta M, Powers A, Foddis R, Stekala E, Pass HI et al (2001) SV40 and the pathogenesis of mesothelioma. Semin Cancer Biol 11:63–71. doi:10.1006/scbi.2000.0347

Roccio M, Dal Bello B, Gardella B, Carrara M, Gulminetti R, Mariani B et al (2012) HPV infection and intraepithelial lesions: comparison between HIV positive and negative women. Curr HIV Res 10:614–619

Saha A, Kaul R, Murakami M, Robertson ES (2010) Tumor viruses and cancer biology: modulating signaling pathways for therapeutic intervention. Cancer Biol Ther 10:961–978

Salahuddin SZ, Ablashi DV, Hunter EA, Gonda MA, Sturzenegger S, Markham PD et al (1987) HTLV-III infection of EBV-genome-positive B-lymphoid cells with or without detectable T4 antigens. Int J Cancer 39:198–202

Santana VC, Diniz MO, Cariri FA, Ventura AM, Cunha-Neto E, Almeida RR et al (2013) Bicistronic DNA vaccines simultaneously encoding HIV, HSV and HPV antigens promote CD8(+) T cell responses and protective immunity. PLoS One 8:e71322. doi:10.1371/journal.pone.0071322

Scudellari M (2013) HPV: sex, cancer and a virus. Nature 503:330–332. doi:10.1038/503330a

Shah KV (2007) SV40 and human cancer: a review of recent data. Int J Cancer 120:215–223. doi:10.1002/ijc.22425

Shiff C, Naples JM, Isharwal S, Bosompem KM, Veltri RW (2010) Non-invasive methods to detect schistosome-based bladder cancer: is the association sufficient for epidemiological use? Trans R Soc Trop Med Hyg 104:3–5. doi:10.1016/j.trstmh.2009.05.013

Shin CM, Kim N, Jung Y, Park JH, Kang GH, Park WY et al (2011) Genome-wide DNA methylation profiles in noncancerous gastric mucosae with regard to *Helicobacter pylori* infection and the presence of gastric cancer. Helicobacter 16:179–188. doi:10.1111/j.1523-5378.2011.00838.x

Shin JW, Jung SW, Park BR, Kim CJ, Eum JB, Kim BG et al (2012) Prediction of response to entecavir therapy in patients with HBeAg-positive chronic hepatitis B based on on-treatment HBsAg, HBeAg and HBV DNA levels. J Viral Hepat 19:724–731. doi:10.1111/j.1365-2893.2012.01599.x

Silva J, Cerqueira F, Medeiros R (2013) *Chlamydia trachomatis* infection: implications for HPV status and cervical cancer. Arch Gynecol Onstet 289:1157–1158. doi:10.1007/s00404-013-3122-3

Smith JS, Munoz N, Herrero R, Eluf-Neto J, Ngelangel C, Franceschi S et al (2002) Evidence for *Chlamydia trachomatis* as a human papillomavirus cofactor in the etiology of invasive cervical cancer in Brazil and the Philippines. J Infect Dis 185:324–331. doi:10.1086/338569

Spinillo A, Dal Bello B, Gardella B, Roccio M, Dacco MD, Silini EM (2009) Multiple human papillomavirus infection and high grade cervical intraepithelial neoplasia among women with cytological diagnosis of atypical squamous cells of undetermined significance or low grade squamous intraepithelial lesions. Gynecol Oncol 113:115–119. doi:10.1016/j.ygyno.2008.12.037

Sripa B, Brindley PJ, Mulvenna J, Laha T, Smout MJ, Mairiang E et al (2012) The tumorigenic liver fluke *Opisthorchis viverrini*—multiple pathways to cancer. Trends Parasitol 28:395–407. doi:10.1016/j.pt.2012.07.006

Srivatanakul P, Parkin DM, Jiang YZ, Khlat M, Kao-Ian UT, Sontipong S et al (1991) The role of infection by *Opisthorchis viverrini*, hepatitis B virus, and aflatoxin exposure in the etiology of liver cancer in Thailand. A correlation study. Cancer 68:2411–2417

Sturgis EM, Ang KK (2011) The epidemic of HPV-associated oropharyngeal cancer is here: is it time to change our treatment paradigms. J Natl Compr Canc Netw 9:665–673

Sudenga SL, Wiener HW, Shendre A, Wilson CM, Tang J, Shrestha S (2013) Variants in interleukin family of cytokines genes influence clearance of high risk HPV in HIV-1 coinfected African-American adolescents. Hum Immunol 74:1696–1700. doi:10.1016/j.humimm.2013.08.010

Sureau C (2006) The role of the HBV envelope proteins in the HDV replication cycle. Curr Top Microbiol Immunol 307:113–131

Syrjanen KJ (2002) HPV infections and lung cancer. J Clin Pathol 55:885–891

Szkaradkiewicz A, Kruk-Zagajewska A, Wal M, Jopek A, Wierzbicka M, Kuch A (2002) Epstein-Barr virus and human papillomavirus infections and oropharyngeal squamous cell carcinomas. Clin Exp Med 2:137–141. doi:10.1007/s102380200019

Takada K (2000) Epstein-Barr virus and gastric carcinoma. Mol Pathol 53:255–261

Tan AT, Hoang LT, Chin D, Rasmussen E, Lopatin U, Hart S et al (2014) Reduction of HBV replication prolongs the early immunological response to IFNalpha therapy. J Hepatol 60:54–61. doi:10.1016/j.jhep.2013.08.020

Tepes B (2009) Can gastric cancer be prevented? J Physiol Pharmacol 60(Suppl 7):71–77

Terrault N (2013) Prophylaxis in HBV-infected liver transplant patients: end of the HBIG era? Am J Gastroenterol 108:949–951. doi:10.1038/ajg.2013.122

Traina-Dorge VL, Martin LN, Lorino R, Winsor EL, Beilke MA (2007) Human T cell leukemia virus type 1 up-regulation after simian immunodeficiency virus-1 coinfection in the nonhuman primate. J Infect Dis 195:562–571. doi:10.1086/510914

Tsuchiya S (2002) Diagnosis of Epstein-Barr virus-associated diseases. Crit Rev Oncol Hematol 44:227–238

Tsuge M, Chayama K (2013) Availability of monitoring serum HBV DNA plus RNA during nucleot(s)ide analogue therapy. J Gastroenterol 48:779–780. doi:10.1007/s00535-013-0801-6

Tumwine LK, Orem J, Kerchan P, Byarugaba W, Pileri SA (2010) EBV, HHV8 and HIV in B cell non Hodgkin lymphoma in Kampala, Uganda. Infect Agent Cancer 5:12. doi:10.1186/1750-9378-5-12

Tyan YS, Liu ST, Ong WR, Chen ML, Shu CH, Chang YS (1993) Detection of Epstein-Barr virus and human papillomavirus in head and neck tumors. J Clin Microbiol 31:53–56

Uphoff CC, Denkmann SA, Steube KG, Drexler HG (2010) Detection of EBV, HBV, HCV, HIV-1, HTLV-I and -II, and SMRV in human and other primate cell lines. J Biomed Biotechnol 2010:904767. doi:10.1155/2010/904767

van Houten VM, Snijders PJ, van den Brekel MW, Kummer JA, Meijer CJ, van Leeuwen B et al (2001) Biological evidence that human papillomaviruses are etiologically involved in a subgroup of head and neck squamous cell carcinomas. Int J Cancer 93:232–235. doi:10.1002/ijc.1313

Vassilopoulos D, Calabrese LH (2012) Management of rheumatic disease with comorbid HBV or HCV infection. Nat Rev Rheumatol 8:348–357. doi:10.1038/nrrheum.2012.63

Vedham V, Divi RL, Starks VL, Verma M (2014) Multiple infections and cancer: implications in epidemiology. Technol Cancer Res Treat 13:177–194. doi:10.7785/tcrt.2012.500366

Vegso G, Hajdu M, Sebestyen A (2011) Lymphoproliferative disorders after solid organ transplantation—classification, incidence, risk factors, early detection and treatment options. Pathol Oncol Res 17:443–454. doi:10.1007/s12253-010-9329-8

Vouloumanou EK, Rafailidis PI, Falagas ME (2012) Current diagnosis and management of infectious mononucleosis. Curr Opin Hematol 19:14–20. doi:10.1097/MOH.0b013e32834daa08

Wang MY, Yue JY, Zhang YX, Liu XD, Gao XZ (2011) *Helicobacter pylori* infection in asymptomatic HBV carriers, alcohol users and normal adult population in Shandong Province, China. Clin Res Hepatol Gastroenterol 35:560–562. doi:10.1016/j.clinre.2010.12.014

Wang CC, Lai CH, Huang YT, Chao A, Chou HH, Hong JH (2012a) HPV genotypes predict survival benefits from concurrent chemotherapy and radiation therapy in advanced squamous cell carcinoma of the cervix. Int J Radiat Oncol Biol Phys 84:e499–e506. doi:10.1016/j.ijrobp.2012.06.031

Wang HW, Zhang LL, Miao F, Lv T, Wang XL, Huang Z (2012b) Treatment of HPV infection-associated cervical condylomata acuminata with 5-aminolevulinic acid-mediated photodynamic therapy. Photochem Photobiol 88:565–569. doi:10.1111/j.1751-1097.2011.01060.x

Wasylyk B, Abecassis J, Jung AC (2013) Identification of clinically relevant HPV-related HNSCC: in p16 should we trust? Oral Oncol 49:e33–e37. doi:10.1016/j.oraloncology.2013.07.014

Wedemeyer H (2010) Re-emerging interest in hepatitis delta: new insights into the dynamic interplay between HBV and HDV. J Hepatol 52:627–629. doi:10.1016/j.jhep.2010.02.001

Whitaker NJ, Glenn WK, Sahrudin A, Orde MM, Delprado W, Lawson JS (2013) Human papillomavirus and Epstein Barr virus in prostate cancer: koilocytes indicate potential oncogenic influences of human papillomavirus in prostate cancer. Prostate 73:236–241. doi:10.1002/pros.22562

Wong M, Pagano JS, Schiller JT, Tevethia SS, Raab-Traub N, Gruber J (2002) New associations of human papillomavirus, Simian virus 40, and Epstein-Barr virus with human cancer. J Natl Cancer Inst 94:1832–1836

Yan J, Zhang M, Zhang J, Chen X, Zhang X (2011) *Helicobacter pylori* infection promotes methylation of WWOX gene in human gastric cancer. Biochem Biophys Res Commun 408:99–102. doi:10.1016/j.bbrc.2011.03.127

Yu LH, Li N, Cheng SQ (2011) The role of antiviral therapy for HBV-related hepatocellular carcinoma. Int J Hepatol 2011:416459. doi:10.4061/2011/416459

zur Hausen H (1999) Papillomaviruses in human cancers. Proc Assoc Am Physicians 111:581–587

zur Hausen H (2009) Papillomaviruses in the causation of human cancers—a brief historical account. Virology 384:260–265. doi:10.1016/j.virol.2008.11.046

Chapter 9
Inflammatory Mechanisms of Infection-Associated Cancer

Jotham Suez, Maayan Levy, Christoph A. Thaiss, and Eran Elinav

Abstract The concept of microbial infection as an underlying cause of cancer has been firmly established over many years of research. The notion, however, that even commensal elements of the normal microbiota are involved in tumor-promoting processes has only recently came to the focus of attention. The inflammatory response to infection and tumor growth is remarkably and unexpectedly similar. Inflammation affects all phases of cancer growth, from incipient neoplasia to tumor progression and metastasis, and it has been proven that chronic inflammation associated with viral, bacterial, and parasitic infections may increase cancer risk. Remarkably, eradication of infection and ensuing inflammation is proven in multiple etiologies to prevent cancer. One of the biggest challenges in the field involves the mechanistic elucidation of the innate and adaptive immune response to incipient neoplasia, pathogenic infections, and commensal microbial colonization. Insights gained from such studies will enable to uncover new therapeutic targets against chronic infection and its associated cancer.

Keywords Inflammation-associated tumorigenesis • Viral oncogenes • Immune suppression • Chronic inflammation • Immune evasion • Bacterial pathogens • IL-6 • IL-22 • COX2 inhibitors • Anti-microbial drugs

Introduction

As early as 1863, Rudolf Virchow described leukocytes infiltrated within tumors and proposed that cancer occurred at sites of chronic inflammation. In this chapter, we will discuss the role of microorganisms and infection in inflammation-associated-tumorigenesis. While multiple additional etiologies may result in chronic inflammation

---

Author contributed equally with all other contributors.

J. Suez • M. Levy • C.A. Thaiss • E. Elinav, M.D., Ph.D. (✉)
Department of Immunology, Weizmann Institute of Science,
100 Herzl Street, Rehovot 76100, Israel
e-mail: Eran.Elinav@weizmann.ac.il

M.R. Shurin et al. (eds.), *Infection and Cancer: Bi-Directorial Interactions*, DOI 10.1007/978-3-319-20669-1_9

and subsequently tumorigenesis (including dietary factors and obesity, inhaled pollutants [e.g. asbestos], tobacco use, autoimmunity), they are beyond the scope of this chapter and are reviewed elsewhere (Elinav et al. 2013).

In 1890, Russell (1890) suggested for the first time the possibility of bacteria-induced carcinogenesis. Moreover, in the nineteenth century, in the pioneering work of Robert Koch and Louis Pasteur it was discovered that bacteria could be found at sites of tumors. A few years later, the appearance of bacteria in cancerous tissue has been suggested (Russell 1890; L'Esperance 1931; Wuerthele-Caspe et al. 1950) while the direct causal relationship between the presence of a bacterial species in a tissue and neoplastic transformation was only postulated at a later stage (Livingston and Alexander-Jackson 1965). It is now estimated that 15–20 % of the global burden of cancer, and up to 25 % in developing countries, is associated with chronic infections (Anand et al. 2008; Parkin 2006).

Many pathogens, particularly viruses, promote cancer through well-described genetic mechanisms. Examples include infection with human papillomavirus (HPV), hepatitis B and C viruses (HBV and HCV), human herpes virus 8 (HHV-8), Epstein-Barr virus (EBV), and *Helicobacter pylori*. The common infectious agents associated with cancer are summarized in Table 9.1.

Remarkably, eradication of infection and ensuing inflammation is proven in multiple etiologies to prevent cancer, for instance in the cases of *H. pylori* eradication using antimicrobials, or vaccination against HBV or HPV, demonstrating the power of understanding the source of inflammation for intervening with the disease (Giarelli 2007).

The Inflammatory Response in Infection-Associated Tumorigenesis

The inflammatory response to infection and tumor growth is remarkably and unexpectedly similar. Both trigger innate immune responses, leading to instruction and initiation of adaptive immunity. Although the mechanisms involved in the initial recognition by the innate immune system are different in infection and tumorigenesis, both involve pattern recognition of abnormal cell surface molecules. Furthermore, inflammation affects all phases of cancer growth, from incipient neoplasia to tumor progression and metastasis, and even tumor chemotherapy. Even more strikingly, in the case of infection-associated cancer immune responses against the infectious agent and against the resultant tumor co-occur and overlap. The similarities between the inflammatory response against infections and against cancer are reflected by the commonalities in immune cell populations that are recruited and the inflammatory mediators that they secrete.

As described in Table 9.1, multiple bacteria, viruses, and eukaryotic parasites have been associated with cancer development. In addition, members of the commensal gut microbiota have also been suggested to contribute to susceptibility to tumor-promoting inflammation and its various manifestations. Consequently, better

Table 9.1 Infectious agents associated with cancer

Infectious agent	Associated pathology	Associated malignancy
Viruses		
HBV or HCV	Hepatitis	Hepatocellular carcinoma
HPV	Genital warts, cervicitis (inflammation of the cervix)	Cervical and anal carcinoma
EBV	Mononucleosis (glandular fever)	Burkitt B cell lymphoma, Hodgkin B cell lymphoma, Nasopharyngeal carcinoma
HIV-1	AIDS	Non-Hodgkin lymphoma
HHV-8	AIDS, Castleman's disease	Kaposi's sarcoma, primary effusion lymphoma
MCPyV	Immunosuppression	Merkel cell carcinoma
HTLV-1	Myelopathy (compression of the spinal cord), uveitis (inflammation of the uvea), dermatitis (inflammation of the skin)	Adult T cell leukemia
Bacteria		
Helicobacter pylori	Gastritis	Gastric adenocarcinoma, MALT lymphoma
Neisseria gonorrhoeae, Chlamydia trachomatis	Pelvic inflammatory disease	Ovarian carcinoma
Haemophilus influenzae	Chronic obstructive pulmonary disease	Lung cancer
Escherichia coli, Bacteroides	Cholecystitis (inflammation of the gall bladder), IBD	Gall bladder adenocarcinoma, CRC
Salmonella Typhimurium	Gastroenteritis	Gall bladder adenocarcinoma
Staphylococcus aureus, Enterobacter, Streptococcus	Osteomyelitis	Squamous cell carcinoma in draining sinuses
Trematodes (flukes)		
Schistosoma	Schistosomiasis (parasitic disease), chronic cystitis (inflammation of the bladder)	Bladder, liver, CRC
Opisthorchis viverrini, Clonorchis sinensis	Opisthorchiasis (parasitic disease), cholangitis (inflammation of bile ducts)	Cholangiocarcinoma

AIDS acquired immunodeficiency syndrome, *CRC* colorectal carcinoma, *EBV* Epstein-Barr virus, *HBV* hepatitis B virus, *HCV* hepatitis C virus, *HHV-8* human herpes virus 8, *HIV-1* human immunodeficiency virus 1, *HPV* human papillomavirus, *HTLV-1* human T-cell leukemia virus 1, *IBD* irritable bowel disease, *MALT* mucosa-associated lymphoid tissue, *MCPyV* Merkel cell polyomavirus

understanding of the mechanisms by which these microbial components promote tumorigenesis is emerging as an essential factor in improving cancer prophylaxis and treatment. In the following parts, we will review mechanisms by which both pathogens and commensals initiate and promote tumorigenesis, and discuss their commonalities and differences.

Carcinogenesis Triggered by Viruses

The concept that cancer could be caused by a virus dates back to 1908, when Oluf Bang and Vilhelm Ellerman demonstrated the transmissibility of chicken leukemia (Hu et al. 2004). This idea was subsequently extended to solid tumors by Peyton Rous (Van Epps 2005). To date, an association of viruses with human cancer has been demonstrated for HPV, HBV, HCV, HHV-8, EBV, Merkel cell polyomavirus, and human T-lymphotropic/leukemia virus-1 (HTLV-1) (Parkin 2006) (Table 9.1). While in many of these cases, retroviral insertion of oncogene or activation of endogenous proto-oncogenes has direct transforming potential, indirect viral oncogenicity involves the induction of a chronic inflammatory and tissue repair response, which promotes tumor initiation and tumor growth (zur Hausen 2001). Direct transformation and indirect oncogenicity are not mutually exclusive processes.

Viral oncogenes target cellular tumor suppressor pathways to support viral replication. There is strong selection to maintain viral genes that can initiate tumorigenesis, as some viruses encode oncoproteins that target RB1, p53, NF-kB, interferons and β-catenin (Munger et al. 1989; Levine 2009; Mosialos et al. 1995; Fujimuro et al. 2003; Moore and Chang 1998). Some DNA viruses reinitiate the cell cycle entry of differentiated cells to support the viral replication (Munger et al. 1989; Scheffner et al. 1990). In addition, viral oncogenes can induce genomic instability, which contribute to carcinogenesis (Duensing et al. 2000; Hein et al. 2009).

Hepatocellular carcinoma (HCC) is induced following liver cirrhosis from chronic virus-induced cell death and regeneration (Seeger and Mason 2000; Tsai and Chung 2010). HBV plays a role in HCC (Beasley et al. 1981) and is integrated into the genomes of tumor cells in almost all HBV-related cancers, nonetheless it is not clear whether this is required for cancer cell proliferation (Seeger and Mason 2000). Both the HBV protein HBx and the HCV core protein can induce NF-kB, which prevents virus-induced apoptosis and subsequently results in production of growth and survival factors that stimulate tumor progression and development. Nevertheless, in the case of HCV, a mechanism involving the inhibition of NF-kB was described, mediated by the HCV non-structural protein 5A (NS5A). NF-kB inhibition is mediated by TNFα, and as a result potentiates TNFα-induced c-Jun N-terminal kinases (JNK) activation (Park et al. 2003). It has been postulated that this inhibition may actually stimulate the development of HCC, as inhibition of NF-κB activation in hepatocytes increases their susceptibility to carcinogen-induced death, and hepatocyte death triggers the compensatory proliferation of surviving hepatocytes (Karin et al. 2006). This compensatory proliferation is probably mediated by acute Kupffer cells activation as a result of hepatocytes necrosis, leading to propagation of oncogenic mutations and expansion of transformed cells.

Another example of virus-induced carcinogenesis occurs in Merkel cell carcinoma (MCC), the most aggressive skin cancer. It has been demonstrated that human Merkel cell polyomavirus (MCV) is found in approximately 80 % of MCC tumors. MCC mainly occurs in elderly and immunocompromised individuals (Pastrana et al. 2009). Similar to SV40 and murine polyomaviruses, MCV encodes a multiply spliced tumor (T) antigen protein complex that targets several tumor suppressor proteins (Shuda et al. 2008). T antigen expression is required for the survival of virus-positive Merkel cell lines (Houben et al. 2010). The MCV T antigen can activate independent DNA replication from the integrated viral origin that will cause DNA strand breaks in the tumor cell (Shuda et al. 2008). The viral large T antigen not only targets tumor suppressor molecules, such as RB1, but it is also required for productive virus replication (Kwun et al. 2009).

Carcinogenesis Triggered by Parasites

As with viral infections, chronic inflammation may also be the result of infection with parasites, which has also been associated with increased cancer risk. Schistosomes are blood flukes that colonize the capillaries of the human bladder. *Schistosoma* infection causes chronic inflammation, which has been associated with various malignancies, including carcinomas of the bladder, uterus, liver, and colon (Mostafa et al. 1999; Madbouly et al. 2007). In particular, *Schistosoma haematobium* is a frequent cause of bladder cancer in countries with a high rate of *S. haematobium* infection (Lucas 1982). The mechanisms of *S. haematobium*-induced tumorigenesis are not fully understood. *Schistosoma* egg deposition elicits a chronic inflammatory response with infiltrating macrophages and neutrophils (Kuper et al. 2000). Phagocyte-derived N-nitrosamines and polycyclic hydrocarbons might contribute to carcinogenesis through DNA-mutagenic activity (Rosin et al. 1994). Consistently, high levels of urinary N-nitrosamines have been detected in patients with schistosomal infections (Mostafa et al. 1994). In addition, parasitic liver flukes, in particular *Opisthorchis viverrini* and *Clonorchis sinensis*, have been associated with cholangiocarcinoma development (Schwartz 1986). These parasites are endemic in East and Southeast Asia, and localize to the liver, bile duct, and gallbladder of infected individuals (Keiser and Utzinger 2005). Epidemiological studies indicated that *C. sinensis* and *O. viverrini* infection are predisposing factors for cholangiocarcinoma development in endemic areas (Flavell 1981), however, the molecular etiology remains largely unclear. It is possible that the histopathologic alteration induced by liver fluke infection, i.e. epithelial desquamation due to chronic tissue irritation, provokes aberrant epithelial hyperplasia (Sripa et al. 2007).

Carcinogenesis Triggered by Specific Bacterial Pathogens

There are multiple mechanisms by which bacteria may prompt carcinogenesis, including immune suppression, chronic inflammation, and immune evasion. Specific bacterial pathogens can lead to the development of cancer. Gallbladder cancer can be promoted by specific bacterial pathogens and is associated with chronic *Salmonellaenterica* serovars Typhi and Paratyphi infections (Caygill et al.

1994; Welton et al. 1979). Additionally, infections with *Campylobacter jejuni* (Peterson 2004), *Borrelia burgdorferi* (Ponzoni et al. 2011) and *Chlamydia psittaci* (Ferreri et al. 2012) are associated with certain lymphomas, and these are suggested to regress after antibiotic treatment.

H. pylori is a classic example of the connotation between bacterial infection and gastrointestinal malignancies through its propensity to cause lifelong inflammation, and has been linked to adenocarcinoma with both gastric cancer and mucosa-associated lymphoid tissue (MALT) lymphoma.

H. pylori is a Gram-negative bacterial pathobiont (a commensal bacterium that has pathogenic potential) that selectively colonizes the gastric epithelium. Colonization usually occurs during childhood and can lead to superficial gastritis. Remarkably, *H. pylori* has the capacity to persist for decades in the harsh gastric environment due to an inability of the host to eliminate the infection. Infection with *H. pylori*, which is classified as a carcinogen, may lead to the sequential development of gastritis, gastric ulcer, atrophy, gastric cancer and cancer mortality (Fox and Wang 2007). Approximately half of the world's population is inoculated with *H. pylori*. In most individuals, *H. pylori* colonization does not cause any symptoms (Peek and Blaser 2002) although they present increased risk to developing cancer. Uemura et al. (2001) reported that gastric cancer developed in approximately 3 % of *H. pylori*-infected patients, compared to none of the uninfected patients. Likewise, chronic inoculation with *H. pylori* predisposes to MALT lymphomas, hematopoietic malignancies characterized by clonal expansion of B cells and T helper cells that are reactive to *H. pylori*-derived antigens. Wotherspoon et al. showed that regression occurs after *H. pylori* eradication (Wotherspoon et al. 1993).

H. pylori has a number of direct effects on host epithelial tissues that could affect tumorigenesis, including induction of proliferation, the inflammatory response and apoptosis. Co-culture of *H. pylori* with epithelial cells has been shown to reduce expression of the cell-cycle regulatory protein, which leads to epithelial-cell G1 arrest (Shirin et al. 1999; Ahmed et al. 2000). The host response to *H. pylori* can also induce epithelial-cell proliferation (Levi et al. 1989). The pathogen's virulence factors likely play an important role in determining the outcome of infection. *H. pylori* strains expressing the virulence factors cytotoxin-associated gene A (CagA) or vacuolating cytotoxin A (VacA) cause an increased inflammation and cancer rates (Fox and Wang 2007). CagA encodes a 120–140-kDa protein that is translocated into host cells by the type IV *cag* secretion system after bacterial attachment. Inside the host cell, CagA is tyrosine phosphorylated. CagA acts via the activation of host-derived signaling pathways. Phosphorylated CagA targets and interacts with numerous intracellular effectors, which results in the activation of tumor- promoting pathways. Following its activation, CagA induces morphological changes in the host, as well as actin reorganization, variations in the cell cycle and autocrine effects. Alterations in cell control may ultimately lead to cellular damage and to increase risks for gastric cancer development.

The generation of transgenic mice expressing CagA has provided an evidence for a causal relationship between CagA and oncogenesis by demonstrating that transgenic expression of CagA leads to gastric epithelial cell proliferation and carcinoma (Ohnishi et al. 2008). Much remains to be learned about the mechanism by which CagA initiates carcinogenesis. In addition to CagA, components of *H. pylori* peptidoglycan can be delivered into host cells. Peptidoglycans recognized by the host intracellular pattern recognition receptor Nod1 (Chamaillard et al. 2003; Girardin et al. 2003) leading to activation of NF-kB and production of proinflammatory cytokines.

Host factors, such as the effector cytokines IL-10, TNFα and IL-1β influence the pathological outcomes of *H. pylori*infection (Crabtree et al. 1991; El-Omar et al. 2003). It was shown that TNFα and IL-1β levels are significantly higher in *H. pylori-positive* patients (El-Omar et al. 2003; Noach et al. 1994) and this might explain the recruitment and activation of neutrophils in the gastric mucosa during infection. Moreover mouse models, such as the IL-1β transgenic mice over- expressing human IL-1β in epithelial cells of the stomach, were found to develop increased dysplasia and carcinoma when infected with *Helicobacter felis* (Tu et al. 2008).

It has been shown that in chronic *H. pylori*infection, there is a lack of epithelial necrosis and induction of apoptosis (Moss et al. 1996). Two *H. pylori* proteins, urease B and membrane protein 1, have recently been shown to induce TNFα expression and transformation in cells that constitutively overexpress the oncogenic protein RAS (Suganuma et al. 2001). In addition to stimulating cytokine production, *H. pylori* activates proinflammatory cyclooxygenase (COX) enzymes (Juttner et al. 2003; Romano et al. 1998; Fu et al. 1999). COX-2 is the rate-limiting enzyme in prostaglandin synthesis. Cox-2 expression is further increased within gastric premalignant and malignant lesions (Ristimaki et al. 1997; Sung et al. 2000). An additional potential contributing factor in the inflammation-to-carcinoma progression is oxidative stress, which is generated during *H. pylori*-infection of gastric epithelial cells (Ding et al. 2007).

Since the isolation and characterization of *H. pylori*, other *Helicobacter* species have been isolated, many of which can produce serious non-gastric disease in their respective hosts. *H. hepaticus*infection has been demonstrated to exacerbate the development of cancer at both intestinal and extraintestinal sites (Fox et al. 1994; Ward et al. 1994a). When using a mouse model for liver tumors it was recognized that in addition to chronic active hepatitis, infection with *H. hepaticus* is also associated with the development of hepatocellular carcinoma (Fox et al. 1996, 2010; Ward et al. 1994b; Erdman et al. 2003).

Modulation of Cancer-Associated Inflammation by the Microbiota

While the above infectious and inflammatory risk factors for cancers are mostly perceived to be mono-infections mediated by discrete pathogens, it has been recently appreciated that complex commensal microbial ecosystems within the mammalian hosts, termed the microbiota, bear fundamental importance to physiology and susceptibility to pathophysiology. Of the sites of mammalian microbial colonization, the human gastrointestinal tract harbors a highly complex and abundant microbial community, encompassing more than 1000 bacterial species that can reach levels as high as 10^{13}–10^{14} microorganisms in the large intestine (Berg 1996). The members of this microbial ecosystem are 10-fold more numerous than the amount of host cells, accompanied by 100-fold more genes than the ones found in the host genome (Bäckhed et al. 2015). The host and its resident microbiota pose a mutually beneficial relationship.

The phylogenetic signature of the intestinal microbiota is mostly comprised of some major taxonomic units, including Bacteroidetes (mostly *Bacteroides*) and Firmicutes (including *Bacillus, Lactobacillus*, and *Clostridium*).

While some similarities exist in the taxonomic composition of the microbiota in healthy and neoplastic tissue, the tumor microbiota is generally less diverse. In addition, specific bacterial strains show a strong association with higher abundance in neoplastic tissue, including *Bacteroides spp*., *Clostridia*, *Streptococcus bovis*, *H. pylori*, and *Fusobacterium spp*. (Peek and Blaser 2002; Gold et al. 2004; Nakamura et al. 2002; Ellmerich et al. 2000; Castellarin et al. 2012). Both commensals and pathogens are associated with carcinogenesis. Colorectal cancer (CRC) patients feature a distinct taxonomic representation of the major phylogenetic units colonizing the intestine (Arthur et al. 2012; Sobhani et al. 2011). As mentioned before, the most classical example is *H. pylori* association with development of gastric cancer (Chiba et al. 2012). Scanlan et al. demonstrated that *Clostridium leptum* and *Clostridium coccoides* were specific to CRC and polyposis (Scanlan et al. 2008). Furthermore, CRC patients were found to harbor enhanced levels of *Bacteroides* and *Fusobacterium spp*. (Wu et al. 2013).

The microbiota can include potentially virulent species, termed 'pathobionts', which can cause disease when intestinal homeostasis is disrupted (Round and Mazmanian 2009; Chow and Mazmanian 2010). Chronic inflammatory disease triggers the pathogenic potential of resident commensals.

Additionally, genomic islands have been found to be directly associated with carcinogenic activities by commensals. The most prominent example is a commensal *E. coli* strain harboring a polyketide synthase (pks) pathogenicity island encoding Colibactin, an enzyme with putative genotoxic activity (Nougayrede et al. 2006; Cuevas-Ramos et al. 2010; Putze et al. 2009). These bacteria were found in abundance in the inflammatory setting of CRC and additionally shown to be involved in genomic destabilization of host cells, thus driving tumor initiation and progression, potentially representing a bona fide carcinogenic hit (Arthur et al. 2012).

The capacity of Colibactin to promote tumorigenesis *in vivo* has been recently proven by Arthur et al. in an animal model of colitis-associated CRC (Arthur et al. 2012), where inflammation alters the intestinal microbiota by favoring the proliferation of genotoxic commensals promoting CRC. The authors found an altered composition of the colonic-adherent microbiota in IL-10–deficient mice. Although IL-10–deficient mice colonized solely with either the *E. coli* strain NC101 (produces the genotoxic colibactin) or *Enterococcus faecalis* exhibited aggressive colitis and similar production of inflammatory cytokines, only the first developed invasive adenocarcinoma after azoxymethane treatment. These data support the idea that damage to host DNA by *E. coli* exerts specific carcinogenic effects independently of inflammation. Furthermore, Cuevas-Ramos and colleagues (Cuevas-Ramos et al. 2010) showed that adherent/invasive *E. coli* strains were highly abundant in the colonic mucosa of patients with colorectal carcinoma and adenoma, yet not in normal colonic mucosa.

Pathogenic gut microbial infections have also been shown to trigger development of intestinal adenomatous polyps in mice with a mutated *Apc* gene (Rao et al. 2006). Similarly, infection with other pathogenic enteric microbiota including *Citrobacter rodentium* (Newman et al. 2001) and an enterotoxigenic *Bacteroides fragilis* (ETBF) (Wu et al. 2009) have been shown to promote colon tumorigenesis in $Apc^{Min/+}$ mice.

The commensal microbiota is also associated with pathogen-induced cancer. In the case of *H. pylori*-induced gastric cancer, the presence of a complex microbiota is required for the manifestation, as lack of commensal flora in *H. pylori*-infected mice reduces gastritis and delays intraepithelial neoplasia in a hypergastrinaemic transgenic mouse model (Lofgren et al. 2011).

Mechanistic Underpinnings of Inflammation Associated with Infection-Induced Cancer

Many of the cellular and molecular processes involved in tumorigenesis may be enabled by infection-associated inflammation, including genomic instability and increased proliferation. In an attempt to generalize the mechanism by which microbial factors contribute to tumorigenesis, we can consider a model in which chronic exposure to a pathogen (or pathobiont) causes repeated tissue injury, resulting in chronic inflammation as the major driver of cancer. The tissue damage may be direct and mediated by virulence factors (as in the case of *H. pylori* or ETBF) or oncogenes, or indirect, mediated by immune recognition of the pathogen. Nevertheless, several inflammatory pathways are common to many infection-associated tumors.

Among the most important cytokines associated with the inflammatory response accompanying infection-induced cancer development is interleukin-6 (IL-6). IL-6 signals through STAT3 on its target cells, and activation of STAT3 by IL-6 produced by myeloid cells has been documented in many cases of inflammation-associated cancer. Mechanistically, STAT3 enhances premalignant cell proliferation and inhibits apoptosis (Bollrath et al. 2009; Grivennikov et al. 2009; Waldner et al. 2012). Similarly, TNFα plays an important role in the promotion of neoplasia (Park et al. 2010). In addition to STAT3 signaling, the NF-κB signaling pathway has emerged over the last two decades as a pivotal axis in the promotion of infection-associated cancer. NF-κB activation occurs in most tumors by inflammatory stimuli or oncogenic mutations (Grivennikov et al. 2010; Karin and Greten 2005; Ben-Neriah and Karin 2011). Recently, additional attention has been given to the cytokine IL-22 and its role in tissue repair and the promotion of neoplasia (Zenewicz et al. 2008; Sonnenberg et al. 2011). IL-22 is upregulated both in response to tissue damage as

Table 9.2 Cancer-promoting functions of inflammatory immune cells

Macrophages	Produce growth factors (e.g. EGF, FGF2, CSF1)
	Promote angiogenesis (e.g. by VEGFA, MMPs)
	Promote inflammation (e.g. by histamine, prostaglandins, nitric-oxide, TNF-α, IL-1)
	Recruit inflammatory cells (e.g. by CXCL chemokines)
	Promote metastasis
Dendritic cells	Promote inflammation (e.g. by histamine, prostaglandins, nitric-oxide, TNF-α, IL-1)
	Recruit inflammatory cells (e.g. by CXCL chemokines)
	Blunt tumor-specific T cell responses (e.g. by PD-L1, arginase)
	Induce T_H2 differentiation
Neutrophils	Blunt tumor-specific T cell responses
	Promote angiogenesis (e.g. by MMPs)
	Release mitogens (e.g. elastase, reactive radicals)
	Recruit inflammatory cells (e.g. by CCL chemokines)
Myeloid-derived suppressor cells (MDSC)	Blunt tumor-specific T cell responses (e.g. by arginase, nitric oxide synthase)
T_H1 cells	Blunt tumor-specific T cell responses (e.g. by IFN-γ/PD-L1)
T_H2 cells	Promote epithelial cell proliferation (e.g. IL-4, IL-13)
	Polarize myeloid cell protumorigenic activation (e.g. IL-4, IL-13)
T_H17 cells	Recruit inflammatory cells
T_{regs}	Blunt tumor-specific T cell responses
	Promote angiogenesis
B cells	Promote protumorigenic macrophage function (e.g. by TNF-α, IL-10)
	Promote metastasis (e.g. by lymphotoxin)

well as infection, and induces wound-healing responses that are often involved in local inflammatory responses to tumor development.

The inflammatory cell types involved in the inflammatory response to tumors are summarized in Table 9.2.

Anti-inflammatory and Anti-microbial Cancer Treatment

The new insights about the nature and mechanisms of the inflammatory response against infection-associated cancer has fuelled increased interest in the development of novel treatment options based on known anti-inflammatory and anti-microbial substances. In contrast to classical cancer therapy, the use of anti-inflammatory drugs to target the molecular etiology of inflammation-induced cancer has two advantages: first, it has drastically reduced side effects, and second, it is much less likely to provoke the development of drug resistance. Three different strategies have proven effective in targeting cancer-related inflammation: enhancing antitumor immune responses, blocking pro-inflammatory pathways, and blocking the recruitment of inflammatory cells.

Bolstering the anti-tumor function of immune cells provides an attractive means of harnessing the function of tumor-infiltrating cells. T cells that produce high levels of IFNγ, either naturally occurring in the tumor infiltrate or genetically redirected towards tumor antigens, are a potent anti-tumor weapon that is amenable to *ex vivo* engineering and expansion (Restifo et al. 2012). In addition, blocking T cell-inhibitory pathways presents an attractive way of boosting antitumor immune responses. An anti-cytotoxic T-lymphocyte protein 4 (CTLA4) antibody has been approved by the FDA for the treatment of patients with metastatic melanoma (Hodi et al. 2010), and anti-programmed cell death protein 1 (PD1) treatment has shown promising results in clinical trials (Brahmer et al. 2010).

As discussed above, levels of pro-inflammatory cytokines inversely correlate with survival of cancer patients. Anti-inflammatory drugs, such as COX2 inhibitors and corticosteroids, have shown potential in preventing and treating cases of colon, breast, lung, and stomach cancer (Ulrich et al. 2006). Although promising, based on their involvement in many cancer-related inflammatory processes, targeting major inflammatory transcription factors, such as NF-κB, is notoriously difficult. The role of NF-κB is dependent on cell type, disease context, and phase of the disease (Pasparakis 2009), and targeting NF-κB may lead to seemingly contradictory outcomes, such as the reported acute inflammation due to enhanced IL-1β secretion (Greten et al. 2007). An IL-6-blocking antibody has shown first success in phase II clinical trials of patients with ovarian cancer (Coward et al. 2011). In contrast, anti-TNF antibody treatment might even lead to enhanced tumorigenesis, since malignancies have been detected in anti-TNF treated patients with rheumatoid arthritis (Bongartz et al. 2006).

Prospectively, similar efficacy with fewer complications might be achieved by targeting the recruitment of immune cells into the tumor microenvironment. Pro-angiogenic TIE2 (also known as TEK)-positive tumor-associated macrophages are attracted by CXCL12 to the tumor in breast and lung carcinoma xenograft models (Kozin et al. 2010), and interfering with this recruitment by blocking CXCL12 signaling greatly enhances the effect of the vascular-disrupting therapeutic agent combretastatin 4A phosphate in mice (Welford et al. 2011). Moreover, blockade of CCL2 signaling diminishes metastasis development in a mammary cancer model (Qian et al. 2011). These examples show the potential benefit of developing anti-chemokine therapeutic modalities for targeted intervention with the assembly of the inflammatory cancer microenvironment.

Anti-microbial drugs have been exploited much less for cancer therapy, but in those cases in which they are used, they have proven highly successful. Anti-microbial therapy is far superior to conventional or anti-inflammatory tumor therapy with respect to the expected side effects, but it shares a common problem associated with conventional chemotherapy targeted at the elimination of tumor cells, namely the emergence of drug resistance. Anti-microbial agents are currently used for gastric, liver, hematopoietic, and cervical cancer. FDA regulations recommend *H. pylori* eradication treatment for patients with gastric adenocarcinoma and gastric mucosal-associated lymphatic tissue (MALT) lymphomas. Unfortunately, complete elimination is not always possible, mainly due to the emergence of antibiotics resistance (Megraud 2004). In the case of HBV and HCV-associated HCC, eradication strategies are far more complicated, due to integration of HBV DNA into the host genome, the induction of genomic instability associated with primary infection, and its synergism with metabolic and dietary factors, such as alcohol consumption. Nevertheless, dramatic progress in primary prevention via population vaccinations and HBV and HCV treatment lead to a substantial reduction in infection rates, as well as increasing rates of sustained virological response. These in turn are expected to result in a decline in the prevalence of related virally-induced malignanices (Lu et al. 2013). Likewise, primary prevention of HIV-related cancers, in particular Kaposi's sarcoma and non-Hodgkin lymphoma, is mainly based on the therapy of AIDS, i.e. antiretroviral therapy. The recent widespread application of combinational highly active anti-retroviral therapy (HAART) to HIV patients resulted in a sharp decline in the prevalence of AIDS, including its associated malignancies (Shiels et al. 2011).

Conclusions

The concept of microbial infection as an underlying cause of cancer has been firmly established over many years of research. The notion, however, that even commensal elements of the normal microbiota are involved in tumor-promoting processes, has only recently came to the focus of attention. In this Chapter, we summarized the most common infectious agents associated with cancer development, the mechanistic commonalities of the inflammatory response launched against them, and the impact of the microbiota on the inflammatory mechanisms underlying infection-associated cancers. One of the biggest challenges in the field involves the mechanistic elucidation of the innate and adaptive immune response to incipient neoplasia, pathogenic infections, and commensal microbial colonization. Insight gained from such studies will enable to uncover new therapeutic targets against chronic infection and its associated cancer.

Consequently, the speed and efficiency with which the exciting new findings in the field can be translated to clinical application will critically depend on our understanding of intervention measures applied at the most upstream parts of the discussed pathways. This will involve modulation of commensal microbial ecology by means of administering selected microbial communities that may prevent or treat disease susceptibility by altering the configuration of a given commensal ecosystem (probiotics); dietary agents influencing the community composition and live members of the microbiota (prebiotics), or 'post-biotics'- small molecules involved in host-microbiota interactions that may influence the host susceptibility to cancer. These, combined with antimicrobial agents, classical chemotherapy and/or irradiation might offer a promising regimen to be tested in the clinical setting.

References

Ahmed A et al (2000) Helicobacter pylori inhibits gastric cell cycle progression. Microbes Infect 2(10):1159–1169
Anand P et al (2008) Cancer is a preventable disease that requires major lifestyle changes. Pharm Res 25(9):2097–2116
Arthur JC et al (2012) Intestinal inflammation targets cancer-inducing activity of the microbiota. Science 338(6103):120–123
Bäckhed F et al (2005) Host-bacterial mutualism in the human intestine. Science 307(5717):1915–1920
Beasley RP et al (1981) Hepatocellular carcinoma and hepatitis B virus. A prospective study of 22 707 men in Taiwan. Lancet 2(8256):1129–1133
Ben-Neriah Y, Karin M (2011) Inflammation meets cancer, with NF-kappaB as the matchmaker. Nat Immunol 12(8):715–723
Berg RD (1996) The indigenous gastrointestinal microflora. Trends Microbiol 4(11):430–435
Bollrath J et al (2009) gp130-mediated Stat3 activation in enterocytes regulates cell survival and cell-cycle progression during colitis-associated tumorigenesis. Cancer Cell 15(2):91–102
Bongartz T et al (2006) Anti-TNF antibody therapy in rheumatoid arthritis and the risk of serious infections and malignancies: systematic review and meta-analysis of rare harmful effects in randomized controlled trials. JAMA 295(19):2275–2285
Brahmer JR et al (2010) Phase I study of single-agent anti-programmed death-1 (MDX-1106) in refractory solid tumors: safety, clinical activity, pharmacodynamics, and immunologic correlates. J Clin Oncol 28(19):3167–3175
Castellarin M et al (2012) Fusobacterium nucleatum infection is prevalent in human colorectal carcinoma. Genome Res 22(2):299–306
Caygill CP et al (1994) Cancer mortality in chronic typhoid and paratyphoid carriers. Lancet 343(8889):83–84
Chamaillard M et al (2003) An essential role for NOD1 in host recognition of bacterial peptidoglycan containing diaminopimelic acid. Nat Immunol 4(7):702–707
Chiba T, Marusawa H, Ushijima T (2012) Inflammation-associated cancer development in digestive organs: mechanisms and roles for genetic and epigenetic modulation. Gastroenterology 143(3):550–563
Chow J, Mazmanian SK (2010) A pathobiont of the microbiota balances host colonization and intestinal inflammation. Cell Host Microbe 7(4):265–276
Coward J et al (2011) Interleukin-6 as a therapeutic target in human ovarian cancer. Clin Cancer Res 17(18):6083–6096
Crabtree JE et al (1991) Mucosal tumour necrosis factor alpha and interleukin-6 in patients with Helicobacter pylori associated gastritis. Gut 32(12):1473–1477
Cuevas-Ramos G et al (2010) Escherichia coli induces DNA damage in vivo and triggers genomic instability in mammalian cells. Proc Natl Acad Sci U S A 107(25):11537–11542
Ding SZ et al (2007) Helicobacter pylori infection induces oxidative stress and programmed cell death in human gastric epithelial cells. Infect Immun 75(8):4030–4039
Duensing S et al (2000) The human papillomavirus type 16 E6 and E7 oncoproteins cooperate to induce mitotic defects and genomic instability by uncoupling centrosome duplication from the cell division cycle. Proc Natl Acad Sci U S A 97(18):10002–10007
Elinav E et al (2013) Inflammation-induced cancer: crosstalk between tumours, immune cells and microorganisms. Nat Rev Cancer 13(11):759–771
Ellmerich S et al (2000) Production of cytokines by monocytes, epithelial and endothelial cells activated by Streptococcus bovis. Cytokine 12(1):26–31
El-Omar EM et al (2003) Increased risk of noncardia gastric cancer associated with proinflammatory cytokine gene polymorphisms. Gastroenterology 124(5):1193–1201
Erdman SE et al (2003) CD4(+)CD25(+) regulatory lymphocytes require interleukin 10 to interrupt colon carcinogenesis in mice. Cancer Res 63(18):6042–6050

Ferreri AJ et al (2012) Chlamydophila psittaci eradication with doxycycline as first-line targeted therapy for ocular adnexae lymphoma: final results of an international phase II trial. J Clin Oncol 30(24):2988–2994
Flavell DJ (1981) Liver-fluke infection as an aetiological factor in bile-duct carcinoma of man. Trans R Soc Trop Med Hyg 75(6):814–824
Fox JG, Wang TC (2007) Inflammation, atrophy, and gastric cancer. J Clin Invest 117(1):60–69
Fox JG et al (1994) Helicobacter hepaticus sp. nov., a microaerophilic bacterium isolated from livers and intestinal mucosal scrapings from mice. J Clin Microbiol 32(5):1238–1245
Fox JG et al (1996) Chronic proliferative hepatitis in A/JCr mice associated with persistent Helicobacter hepaticus infection: a model of helicobacter-induced carcinogenesis. Infect Immun 64(5):1548–1558
Fox JG et al (2010) Gut microbes define liver cancer risk in mice exposed to chemical and viral transgenic hepatocarcinogens. Gut 59(1):88–97
Fu S et al (1999) Increased expression and cellular localization of inducible nitric oxide synthase and cyclooxygenase 2 in Helicobacter pylori gastritis. Gastroenterology 116(6):1319–1329
Fujimuro M et al (2003) A novel viral mechanism for dysregulation of beta-catenin in Kaposi's sarcoma-associated herpesvirus latency. Nat Med 9(3):300–306
Giarelli E (2007) Cancer vaccines: a new frontier in prevention and treatment. Oncology (Williston Park) 21(11 Suppl Nurse Ed):11–17; discussion 18
Girardin SE et al (2003) Nod1 detects a unique muropeptide from gram-negative bacterial peptidoglycan. Science 300(5625):1584–1587
Gold JS, Bayar S, Salem RR (2004) Association of Streptococcus bovis bacteremia with colonic neoplasia and extracolonic malignancy. Arch Surg 139(7):760–765
Greten FR et al (2007) NF-kappaB is a negative regulator of IL-1beta secretion as revealed by genetic and pharmacological inhibition of IKKbeta. Cell 130(5):918–931
Grivennikov S et al (2009) IL-6 and Stat3 are required for survival of intestinal epithelial cells and development of colitis-associated cancer. Cancer Cell 15(2):103–113
Grivennikov SI, Greten FR, Karin M (2010) Immunity, inflammation, and cancer. Cell 140(6):883–899
Hein J et al (2009) Simian virus 40 large T antigen disrupts genome integrity and activates a DNA damage response via Bub1 binding. J Virol 83(1):117–127
Hodi FS et al (2010) Improved survival with ipilimumab in patients with metastatic melanoma. N Engl J Med 363(8):711–723
Houben R et al (2010) Merkel cell polyomavirus-infected Merkel cell carcinoma cells require expression of viral T antigens. J Virol 84(14):7064–7072
Hu WY, Bushman FD, Siva AC (2004) RNA interference against retroviruses. Virus Res 102(1):59–64
Juttner S et al (2003) Helicobacter pylori stimulates host cyclooxygenase-2 gene transcription: critical importance of MEK/ERK-dependent activation of USF1/-2 and CREB transcription factors. Cell Microbiol 5(11):821–834
Karin M, Greten FR (2005) NF-kappaB: linking inflammation and immunity to cancer development and progression. Nat Rev Immunol 5(10):749–759
Karin M, Lawrence T, Nizet V (2006) Innate immunity gone awry: linking microbial infections to chronic inflammation and cancer. Cell 124(4):823–835
Keiser J, Utzinger J (2005) Emerging foodborne trematodiasis. Emerg Infect Dis 11(10):1507–1514
Kozin SV et al (2010) Recruitment of myeloid but not endothelial precursor cells facilitates tumor regrowth after local irradiation. Cancer Res 70(14):5679–5685
Kuper H, Adami HO, Trichopoulos D (2000) Infections as a major preventable cause of human cancer. J Intern Med 248(3):171–183
Kwun HJ et al (2009) The minimum replication origin of merkel cell polyomavirus has a unique large T-antigen loading architecture and requires small T-antigen expression for optimal replication. J Virol 83(23):12118–12128
L'Esperance ES (1931) Studies in Hodgkin's disease. Ann Surg 93(1):162–168

Levi S et al (1989) Campylobacter pylori and duodenal ulcers: the gastrin link. Lancet 1(8648):1167–1168
Levine AJ (2009) The common mechanisms of transformation by the small DNA tumor viruses: the inactivation of tumor suppressor gene products: p53. Virology 384(2):285–293
Livingston VW, Alexander-Jackson E (1965) An experimental biologic approach to the treatment of neoplastic disease; determination of actinomycin in urine and cultures as an aid to diagnosis and prognosis. J Am Med Womens Assoc 20(9):858–866
Lofgren JL et al (2011) Lack of commensal flora in Helicobacter pylori-infected INS-GAS mice reduces gastritis and delays intraepithelial neoplasia. Gastroenterology 140(1):210–220
Lu T et al (2013) Prevention of hepatocellular carcinoma in chronic viral hepatitis B and C infection. World J Gastroenterol 19(47):8887–8894
Lucas SB (1982) Squamous cell carcinoma of the bladder and schistosomiasis. East Afr Med J 59(5):345–351
Madbouly KM et al (2007) Colorectal cancer in a population with endemic Schistosoma mansoni: is this an at-risk population? Int J Colorectal Dis 22(2):175–181
Megraud F (2004) H pylori antibiotic resistance: prevalence, importance, and advances in testing. Gut 53(9):1374–1384
Moore PS, Chang Y (1998) Antiviral activity of tumor-suppressor pathways: clues from molecular piracy by KSHV. Trends Genet 14(4):144–150
Mosialos G et al (1995) The Epstein-Barr virus transforming protein LMP1 engages signaling proteins for the tumor necrosis factor receptor family. Cell 80(3):389–399
Moss SF et al (1996) Induction of gastric epithelial apoptosis by Helicobacter pylori. Gut 38(4):498–501
Mostafa MH et al (1994) Nitrate, nitrite and volatile N-nitroso compounds in the urine of Schistosoma haematobium and Schistosoma mansoni infected patients. Carcinogenesis 15(4):619–625
Mostafa MH, Sheweita SA, O'Connor PJ (1999) Relationship between schistosomiasis and bladder cancer. Clin Microbiol Rev 12(1):97–111
Munger K et al (1989) Complex formation of human papillomavirus E7 proteins with the retinoblastoma tumor suppressor gene product. EMBO J 8(13):4099–4105
Nakamura J et al (2002) Comparison of four microbial enzymes in Clostridia and Bacteroides isolated from human feces. Microbiol Immunol 46(7):487–490
Newman JV et al (2001) Bacterial infection promotes colon tumorigenesis in Apc(Min/+) mice. J Infect Dis 184(2):227–230
Noach LA et al (1994) Mucosal tumor necrosis factor-alpha, interleukin-1 beta, and interleukin-8 production in patients with Helicobacter pylori infection. Scand J Gastroenterol 29(5):425–429
Nougayrede JP et al (2006) Escherichia coli induces DNA double-strand breaks in eukaryotic cells. Science 313(5788):848–851
Ohnishi N et al (2008) Transgenic expression of Helicobacter pylori CagA induces gastrointestinal and hematopoietic neoplasms in mouse. Proc Natl Acad Sci U S A 105(3):1003–1008
Park K-J et al (2003) Hepatitis C virus NS5A protein modulates c-Jun N-terminal kinase through interaction with tumor necrosis factor receptor-associated factor 2. J Biol Chem 278(33):30711–30718
Park EJ et al (2010) Dietary and genetic obesity promote liver inflammation and tumorigenesis by enhancing IL-6 and TNF expression. Cell 140(2):197–208
Parkin DM (2006) The global health burden of infection-associated cancers in the year 2002. Int J Cancer 118(12):3030–3044
Pasparakis M (2009) Regulation of tissue homeostasis by NF-kappaB signalling: implications for inflammatory diseases. Nat Rev Immunol 9(11):778–788
Pastrana DV et al (2009) Quantitation of human seroresponsiveness to Merkel cell polyomavirus. PLoS Pathog 5(9):e1000578
Peek RM Jr, Blaser MJ (2002) Helicobacter pylori and gastrointestinal tract adenocarcinomas. Nat Rev Cancer 2(1):28–37

Peterson MC (2004) Immunoproliferative small intestinal disease associated with Campylobacter jejuni. N Engl J Med 350(16):1685–1686; author reply 1685–6
Ponzoni M et al (2011) Prevalence of Borrelia burgdorferi infection in a series of 98 primary cutaneous lymphomas. Oncologist 16(11):1582–1588
Putze J et al (2009) Genetic structure and distribution of the colibactin genomic island among members of the family Enterobacteriaceae. Infect Immun 77(11):4696–4703
Qian BZ et al (2011) CCL2 recruits inflammatory monocytes to facilitate breast-tumour metastasis. Nature 475(7355):222–225
Rao VP et al (2006) Innate immune inflammatory response against enteric bacteria Helicobacter hepaticus induces mammary adenocarcinoma in mice. Cancer Res 66(15):7395–7400
Restifo NP, Dudley ME, Rosenberg SA (2012) Adoptive immunotherapy for cancer: harnessing the T cell response. Nat Rev Immunol 12(4):269–281
Ristimaki A et al (1997) Expression of cyclooxygenase-2 in human gastric carcinoma. Cancer Res 57(7):1276–1280
Romano M et al (1998) Helicobacter pylori up-regulates cyclooxygenase-2 mRNA expression and prostaglandin E2 synthesis in MKN 28 gastric mucosal cells in vitro. J Biol Chem 273(44):28560–28563
Rosin MP, Anwar WA, Ward AJ (1994) Inflammation, chromosomal instability, and cancer: the schistosomiasis model. Cancer Res 54(7 Suppl):1929s–1933s
Round JL, Mazmanian SK (2009) The gut microbiota shapes intestinal immune responses during health and disease. Nat Rev Immunol 9(5):313–323
Russell W (1890) An address on a characteristic organism of cancer. Br Med J 2(1563):1356–1360
Scanlan PD et al (2008) Culture-independent analysis of the gut microbiota in colorectal cancer and polyposis. Environ Microbiol 10(3):789–798
Scheffner M et al (1990) The E6 oncoprotein encoded by human papillomavirus types 16 and 18 promotes the degradation of p53. Cell 63(6):1129–1136
Schwartz DA (1986) Cholangiocarcinoma associated with liver fluke infection: a preventable source of morbidity in Asian immigrants. Am J Gastroenterol 81(1):76–79
Seeger C, Mason WS (2000) Hepatitis B virus biology. Microbiol Mol Biol Rev 64(1):51–68
Shiels MS et al (2011) Proportions of Kaposi sarcoma, selected non-Hodgkin lymphomas, and cervical cancer in the United States occurring in persons with AIDS, 1980–2007. JAMA 305(14):1450–1459
Shirin H et al (1999) Helicobacter pylori inhibits the G1 to S transition in AGS gastric epithelial cells. Cancer Res 59(10):2277–2281
Shuda M et al (2008) T antigen mutations are a human tumor-specific signature for Merkel cell polyomavirus. Proc Natl Acad Sci U S A 105(42):16272–16277
Sobhani I et al (2011) Microbial dysbiosis in colorectal cancer (CRC) patients. PLoS One 6(1):e16393
Sonnenberg GF, Fouser LA, Artis D (2011) Border patrol: regulation of immunity, inflammation and tissue homeostasis at barrier surfaces by IL-22. Nat Immunol 12(5):383–390
Sripa B et al (2007) Liver fluke induces cholangiocarcinoma. PLoS Med 4(7):e201
Suganuma M et al (2001) Helicobacter pylori membrane protein 1: a new carcinogenic factor of Helicobacter pylori. Cancer Res 61(17):6356–6359
Sung JJ et al (2000) Cyclooxygenase-2 expression in Helicobacter pylori-associated premalignant and malignant gastric lesions. Am J Pathol 157(3):729–735
Tsai WL, Chung RT (2010) Viral hepatocarcinogenesis. Oncogene 29(16):2309–2324
Tu S et al (2008) Overexpression of interleukin-1beta induces gastric inflammation and cancer and mobilizes myeloid-derived suppressor cells in mice. Cancer Cell 14(5):408–419
Uemura N et al (2001) Helicobacter pylori infection and the development of gastric cancer. N Engl J Med 345(11):784–789
Ulrich CM, Bigler J, Potter JD (2006) Non-steroidal anti-inflammatory drugs for cancer prevention: promise, perils and pharmacogenetics. Nat Rev Cancer 6(2):130–140
Van Epps HL (2005) Peyton Rous: father of the tumor virus. J Exp Med 201(3):320

Waldner MJ, Foersch S, Neurath MF (2012) Interleukin-6–a key regulator of colorectal cancer development. Int J Biol Sci 8(9):1248–1253
Ward JM et al (1994a) Chronic active hepatitis and associated liver tumors in mice caused by a persistent bacterial infection with a novel Helicobacter species. J Natl Cancer Inst 86(16):1222–1227
Ward JM et al (1994b) Chronic active hepatitis in mice caused by Helicobacter hepaticus. Am J Pathol 145(4):959–968
Welford AF et al (2011) TIE2-expressing macrophages limit the therapeutic efficacy of the vascular-disrupting agent combretastatin A4 phosphate in mice. J Clin Invest 121(5):1969–1973
Welton JC, Marr JS, Friedman SM (1979) Association between hepatobiliary cancer and typhoid carrier state. Lancet 1(8120):791–794
Wotherspoon AC et al (1993) Regression of primary low-grade B-cell gastric lymphoma of mucosa-associated lymphoid tissue type after eradication of Helicobacter pylori. Lancet 342(8871):575–577
Wu S et al (2009) A human colonic commensal promotes colon tumorigenesis via activation of T helper type 17 T cell responses. Nat Med 15(9):1016–1022
Wu N et al (2013) Dysbiosis signature of fecal microbiota in colorectal cancer patients. Microb Ecol 66(2):462–470
Wuerthele-Caspe V et al (1950) Cultural properties and pathogenicity of certain microorganisms obtained from various proliferative and neoplastic diseases. Am J Med Sci 220(6):638–646
Zenewicz LA et al (2008) Innate and adaptive interleukin-22 protects mice from inflammatory bowel disease. Immunity 29(6):947–957
zur Hausen H (2001) Oncogenic DNA viruses. Oncogene 20(54):7820–7823

Part II
Infection-Associated Cancers: Specific Examples

Chapter 10
Helicobacter pylori: The Cancer Bug

Jyoti Mishra, Paolo Ruggiero, Fabio Bagnoli, Rino Rappuoli, and Markus Stein

Abstract *Helicobacter pylori* is a human pathogen that is associated with various severe gastric diseases including gastric and duodenal ulcers, mucosa-associated lymphoid tissue lymphoma, and gastric cancer. There are an estimated 800,000 new cases of gastric cancer each year that are attributed to infection with *Helicobacter pylori*. While several viruses have been linked to cancer development, *Helicobacter pylori* is the only bacterium to date that has been clearly demonstrated to be a carcinogen. Indeed, *Helicobacter pylori* is the most important risk factor for development of gastric cancer, and eradication of the bacterium during the early stages of disease significantly reduces the risk of malignant transformation. In most cases, infection occurs during childhood and, if left untreated, usually persists for life. However, most infections with *Helicobacter pylori* remain asymptomatic or cause mild gastritis. Progression to adenocarcinoma requires infection with more pathogenic strains of *Helicobacter pylori* and also depends on the genetic predisposition of the host, particularly genes involved in increased inflammatory responses. Environmental factors and diet also play an important role. While many virulence factors of *Helicobacter pylori* have been described, the CagA toxin, which is translocated into gastric epithelial cells *via* a bacterial secretion system, appears to be the most specific for the development of a pathological phenotype. In vitro, CagA interferes with many cancer-related signaling pathways, while animal models clearly demonstrate a role for *cag*A as a putative bacterial oncogene. In this chapter we describe some of the important cancer-related signaling pathways that are triggered by CagA, discuss the importance of the immune response for cancer development, and present animal models that are commonly used to study *Helicobacter pylori* pathogenesis.

Keywords Gastric cancer • Peptic ulcer disease • CagA • Vacuolating cytotoxin A • IL-6 • IL-8 • IL-1β • IL-18 • TNF-α • MDSC • Macrophages • Animal models • COX-2 inhibitors

---

J. Mishra • M. Stein (✉)
Albany College of Pharmacy and Health Sciences, Albany, NY, USA
e-mail: mstein2@dsmanitoba.ca

P. Ruggiero • F. Bagnoli • R. Rappuoli
Novartis Vaccines Research Center, Via Fiorentina 1, 5300 Siena, Italy

M.R. Shurin et al. (eds.), *Infection and Cancer: Bi-Directorial Interactions*, DOI 10.1007/978-3-319-20669-1_10

Introduction

Helicobacter pylori (Hp) is one of the most successful human bacterial pathogen, colonizing the stomach of more than half of the world's population. The highest infection rate is found in the developing world (>90 %), while infection among developed nations is declining (Rothenbacher and Brenner 2003; Frenck and Clemens 2003). This spiral-shaped bacterium has colonized humans for at least 60,000 years and ever since has co-evolved with its human host. This is evident from the genetic diversity of the bacteria that developed during the human migration from Africa to India, East Asia, and later Europe and other parts of the world. The continuous bacterial adaptation and natural selection of mutants in the Hp chromosome during human migration generated "quasi-species" that are characterized by genetic differences and are representative for certain geographic regions (Covacci and Rappuoli 1998; Linz et al. 2007).

In spite of this long co-existence between humans and Hp, the first observation of spiral bacteria in the stomach was made by the Italian anatomist Bizzozero in Turin in 1893. However, it was not before 1982 that Warren and Marshall isolated Hp from the human stomach for the first time. The two Australian scientists demonstrated a clear association between gastric colonization with Hp and peptic ulcer disease (PUD), and mainly for this seminal discovery they were awarded the Nobel Prize in 2005 (Marshall et al. 1984; Marshall and Warren 1984). Their findings rendered the long-standing medical dogma that the human stomach is sterile and that PUD is caused by stress obsolete and opened the road for further investigation into the role of *Hp* in gastric diseases including PUD, mucosa-associated lymphoid tissue (MALT-) lymphoma, and gastric cancers (el-Omar et al. 1995, 1997; Konturek et al. 2009; Malfertheiner 2011).

Gastric cancer is currently the second leading cause of cancer-related death world-wide and the fourth most common form of cancer overall (Parkin 2004). The impact of gastric cancer in the US is lower, however, with 1.3 % representing new cancer cases and totaling 1.9 % of cancer related death according to the National Cancer Institute. Gastric cancer is very aggressive and the prognosis is poor, with an average 5 year relative survival rate in the US of about 29 % (American Cancer Society). Furthermore, while Hp contributes to early pathological stages of cancer development, it is dispensable for the late phase of carcinogenesis. This statement is supported by the clinical findings that eradication of Hp at the early, but not late stages of disease has the potential to prevent carcinogenesis (Wong et al. 2004; Ong and Duggan 2004). Acknowledging the association between gastric carcinoma and Hp infection with an attributable risk of 75 %, the International Agency for Research on Cancer (IARC) identified Hp as a group 1 carcinogen in 1994 (Herrera and Parsonnet 2009).

While more than 50 % of world's population is infected with Hp, only 10–15 % of those infected develop peptic ulcers, and approximately 1–3 % progress further to gastric cancer. Even less individuals (less than 1 %) develop MALT lymphoma (Kuipers et al. 1995). There are two major types of gastric cancer, the intestinal, and the diffuse. While Hp is associated with both types, the intestinal type is much more common (Handa et al. 1996). A major difference between the two is that intestinal type cancer occurs mainly in the elderly (average age 69 years), requires inflammation, and proceeds through the distinct histological steps of chronic superficial gastritis, atrophic gastritis, intestinal metaplasia and dysplasia, prior to the development of adenocarcinoma (Correa 1992). Diffuse type adenocarcinoma, on the other hand, occurs mostly in younger individuals and follows the infiltration of individually neoplastic cells that do not form glandular structures. The severity of the pathological outcome is the result of pathogen – host interactions, which depend on strain-specific bacterial factors, the variability of host genotypic traits, and/or environmental influences. Smoking habits, diets consisting of salted and smoked foods as well as processed meats, appear to increase risk, while fresh vegetables appear to reduce the cancer risk. Several excellent review articles describing the role of Hp in cancer development have recently been published (Polk and Peek 2010; Stein et al. 2013; Hatakeyama 2014).

Hp has also been implicated in additional disease entities, although evidence for most of these is poorly established. A possible role of Hp in the pathogenesis of extra-gastric diseases including Parkinson's, cirrhosis, and a number of additional neurological, ischemic, cardiovascular, and skin diseases have been discussed previously (Pellicano et al. 2009; Hussain and Hamid 2014). A strong association of Hp with autoimmune diseases, immune thrombocytopenia and iron deficiency anemia has been reported in both children and adults, and the treatment with anti-Hp therapy led to clinical improvement in these patients (Rocha et al. 2014; Yamanouchi et al. 2014).

In contrast to its pathological implications, clinical and epidemiological evidence strongly suggest that Hp infection may also have a protective role against some autoimmune diseases. Epidemiological data suggests an increase in esophageal cancer and asthma occurrence in populations that are not infected with Hp or where infection has been treated aggressively (Hasni 2012). However, the possible association of Hp infection with the reduction of risk of esophageal adenocarcinoma still appears controversial (IARC 2012). Therefore, a strategic and effective vaccine is desirable not only to aid the prevention of Hp-related gastrointestinal diseases but also for induction of an immune response that would prevent the development of allergic reactions (Zawahir et al. 2013; Hussain and Hamid 2014; Ayala et al. 2014).

CagA, a Master Regulator of Host Cell Signaling Pathways

While the vast majority of pathogens reside in the human host only for a short period of time, Hp has evolved the ability to colonize the human stomach chronically. To achieve chronic colonization, Hp uses a specific set of virulence factors. Multiple polar inserted sheathed flagella propel the bacteria and allow the bacterium to rapidly escape the hostile niche of the stomach and reach the more tolerable mucus, which overlays the gastric epithelial cells (Suerbaum 1995). Protection from the low gastric pH is also achieved by production of urease, an enzyme that buffers the bacterial periplasm and surroundings by converting urea into ammonia and bicarbonate (Weeks et al. 2000). Most bacteria live freely in the gastric mucus layer, while some penetrate the mucus completely and attach to the gastric epithelial cells. Attachment is mediated by outer membrane proteins like SabA and BabA, which have binding specificity for Lewis antigens present on the surfaces of gastric epithelial cells (Walz et al. 2009; Yamaoka 2008a, b). Although different bacterial factors, including the vacuolating toxin VacA (Atherton et al. 1997; Cover and Blanke 2005), the neutrophil activating protein NapA (Zanotti et al. 2002), and the proinflammatory outer membrane protein OipA (Yamaoka et al. 2002a) have been described to affect Hp virulence. The most disease-specific factor that appears to be required for the development of peptic ulcer disease and gastric cancer is the CagA cytotoxin though other virulence factors contribute to chronic colonization and inflammation, which is ultimately are the driving force for cancer development.

CagA Structure and Translocation to Host Cells

CagA is a unique protein with no known homologues. It was discovered in the 1990s as a factor that was strongly associated with severe gastric diseases, including PUD and adenocarcinoma (Covacci et al. 1993, 1997). Sequencing efforts located the *cag*A gene within a ca. 40 kbp pathogenicity island, designated as *cag*-PAI. This island also contains additional open reading frames with similarity to components of type IV secretion systems (TFSSs), exemplified by *Agrobacterium tumefaciens*, as well as unique open reading frames with unknown roles (Censini et al. 1996; Akopyants et al. 1998). The biological functions of the *cag*-PAI and of CagA were discovered several years later, after various groups reported that CagA is a substrate of the *cag* T4SS (Stein et al. 2000; Segal et al. 1999; Odenbreit et al. 2000; Backert et al. 2000; Asahi et al. 2000). The export of CagA requires a functional T4SS that associates with $\alpha5\beta1$-integrins located in the basolateral host cell membrane. This association is mediated by the surface exposed T4SS component CagL, which was shown to directly interact with $\alpha5\beta1$-integrin (Kwok et al. 2007; Jimenez-Soto et al. 2009; Conradi et al. 2012). Additionally, CagA translocation requires CagA itself to bind via its aminoterminal domains to $\alpha5\beta1$-integrin (Kaplan-Turkoz et al. 2012). This suggests that CagA is not passed into the host cell through the TFSS pilus, but rather is exported to the bacterial surface, where it interacts with $\alpha5\beta1$-integrin and additionally with phosphatidylserine in the plasma membrane of the host. Phosphatidylserine is usually part of the inner leaflet of the cellular plasma membrane, but is externalized to the outer leaflet in response to Hp-cell contact. Flipping of phosphatidylserine to the outer leaflet was required for CagA internalization (Murata-Kamiya et al. 2010).

Recent X-ray crystallography studies revealed the structural features of CagA (Hayashi et al. 2012; Backert and Tegtmeyer 2012; Kaplan-Turkoz et al. 2012).

Overall, CagA consists of a structural N-terminus that is followed by a disordered C-terminal region. The larger aminoterminal region can be further subdivided into three domains, indicated by D1, D2, and D3. The D1 domain was shown to bind to tumor suppressors (see section "CagA-dependent inactivation of tumor suppressors"), while the D2 domain contains both, the basic phosphatidylserine binding domain that tethers CagA to the plasma membrane and the α5β1-integrin binding site. The D3 domain contains an N-terminal binding sequence (NBS) that interacts with the C-terminal binding sequence (CBS) within the unstructured C-terminus of CagA. This interaction induces a loop-like structure of the disordered C-terminus that exposes the CagA dimerization motifs (CM) and the tyrosine phosphorylation motifs (EPIYA), which, as described later, are both crucial for CagA signaling functions.

Effects of CagA on Cell Shape, Proliferation, and Motility

Following T4SS-dependent translocation into host epithelial cells, CagA associates via its phosphatidylserine-binding motif with the inner surface of the plasma membrane (Murata-Kamiya et al. 2010). CagA tyrosine residues are then phosphorylated on EPIYA motifs by tyrosine kinases of the c-Src family and by c-Abl (Asahi et al. 2000; Stein et al. 2002; Poppe et al. 2007; Tammer et al. 2007). It is important to note that these motifs are located within a repeat-region located in the carboxyterminal part of the protein. Different strains of Hp express CagA molecules with different number of repeats, resulting in CagA variants carrying between two to five copies of these EPIYA motifs (Yamaoka and Graham 2001). Phosphorylated EPIYA motifs then act as scaffolds to recruit SH2 domain containing proteins. Possibly the most important of these is the protein tyrosine phosphatase SHP-2 (Higashi et al. 2002). Recruitment and aberrant activation of SHP-2 by CagA has two major cancer-related effects on epithelial tissue culture cells. The first is the activation of the Ras-Erk mitogenic pathway that leads to enhanced cell-cycle progression and increased cell proliferation. The second is morphologic changes of the host cell that are characterized by dramatic cell elongation and cytoskeletal rearrangements, referred to as the "hummingbird phenotype" (Segal et al. 1999). Similar cell-morphological changes can also be observed upon engagement of the hepatocyte growth factor receptor by scatter factor (Felici et al. 2010). However, in this case SHP-2 is recruited by the GAB-1 scaffold protein. Although there is no homology between CagA and GAB-1 on the sequence or structure level, both molecules may exert similar functions and be an example of convergent evolution. CagA morphogenic activity can be explained by the ability of activated SHP-2 to dephosphorylate focal adhesion kinase (FAK) thereby decreasing its activity (Tsutsumi et al. 2006). FAK has been well described as a regulator of cell shape and cellular motility. In agreement with this function, CagA mediated FAK dephosphorylation via SHP-2 leads to reduction of focal adhesion sites and contributes to cell detachment and increased cell motility. Based on these functions, it is not surprising that SHP-2 has been proposed to be a proto-oncogenic phosphatase, and indeed, various human cancers are characterized by gain-of function mutations within the SHP-2 gene (Matozaki et al. 2009). Interestingly, sequence variations close to the EPIYA motifs (EPIYA-A, B, C, and D) as well as a difference in the number of the EPIYA motifs have been linked to variations in the signaling ability of CagA (Naito et al. 2006). Strains expressing CagA with Eastern EPIYA motifs (D-type) or simply more EPIYA motifs may be more biologically active and therefore more strongly associated with transformation, which may explain the higher incidences of gastric cancer in East Asia (Batista et al. 2011). Additionally, wild-type CagA induces neoplasms in transgenic mice, while phosphorylation-resistant CagA does not, which further demonstrates the importance of the EPIYA motifs (Ohnishi et al. 2008).

The tyrosine-phosphorylated EPIYA motif of CagA is not only specific for the SH2 domain of SHP-2 but also interacts with the SH2 domain of the C-terminal Src kinase Csk, which in response to CagA binding is redistributed from the cytoplasm to the plasma membrane (Tsutsumi et al. 2003). At the membrane, Csk phosphorylates and thus inactivates Src family kinases, which in turn causes reduced CagA phosphorylation on the EPIYA motifs. Therefore, Csk is part of a negative feedback loop that prevents excessive CagA phosphorylation and may serve as a way to balance CagA activity in order to prevent excessive damage to the epithelium and to support chronic infection.

Crk is an adapter protein that binds to CagA dependent on tyrosine phosphorylation of the EPIYA motifs (Suzuki et al. 2005). Upon binding, Crk activates various pathways including the Sos1/H-Ras/Raf1, the C3G/Rap1/B-Raf, and the Dock180/Rac1/WASP pathways (Suzuki et al. 2005). These pathways are involved in cell spreading, motility, and proliferation and therefore may further potentiate CagA-initiated mitogenic activity.

Other proteins interact with CagA independent of the CagA phosphorylation status. The adapter protein Grb2, for example, requires the EPIYA motifs to interact with CagA in a phosphorylation-independent manner (Mimuro et al. 2002). Grb2 also activates the proto-oncogenic Ras-Erk signaling pathway and therefore may synergize with CagA-dependent Shp-2 signaling.

In summary, CagA uses several mechanisms to trigger growth factor-like mitogenic cell responses, including pro-oncogenic Ras signaling, which has the potential to contribute to tumorgenesis.

Interference of CagA with Cell Polarity and Differentiation

Various studies have shown that Hp is not randomly distributed over the apical surface of the gastric epithelial cells, but rather localizes to cell-to-cell contact sites (Amieva et al. 2003; Bagnoli et al. 2005). In epithelial cells, the apical junctional complex forms the barrier between the lumen and the interstitial space and regulates many cellular functions, including cell proliferation, cell-cell adhesion, cell movement, and establishment of apical-basolateral cell polarity. Translocated CagA is targeted to these junctions *via* its N-terminus and co-localizes with tight junctional markers ZO-1 and junctional adhesion molecule (JAM-1) (Amieva et al. 2003). Ultimately, CagA perturbs the assembly and function of tight and adherence junctions, causing the loss of epithelial cell-cell adhesion and cell polarity. The molecular mechanism for this effect of CagA has been well elucidated. The crucial molecule is the serine-threonine kinase Par1b (MARK 2), a member of the cellular Par-aPKC complex, which is important for the establishment and maintenance of apical-basolateral cell polarity (Saadat et al. 2007; Zeaiter et al. 2008). The 16-amino acid CagA-dimerization motifs (CM), located within the repeat region just downstream of the EPIYA motifs, bind to the catalytic site of Par1b and inhibit its kinase activity (Nesic et al. 2010). In polarized epithelial MDCK cells, this leads to disruption of cellular polarity, characterized by the redistribution of junctional molecules from the lateral to the apical surface and vice versa. A known Par1b-substrate is the microtubule-binding protein map (Timm et al. 2008), and Par1b inhibition by CagA causes rearrangement of microtubules, which instead of their vertical alignment characteristic for the polarized epithelium, adopt a fibroblast-like arrangement (Zeaiter et al. 2008). As a result, cell compaction and the typical columnar shape are lost. Since the mitotic spindle also consists of microtubules, Par1b inhibition also interferes with mitosis and causes chromosomal instability, which is strongly associated with cellular neoplastic transformation (Umeda et al. 2009). Three-dimensional MDCK tissue culture models further demonstrated that Par1b inhibition interferes not only with apical-basolateral cell polarity, but also with lumen formation and tubulogenesis, which are both hallmarks of epithelial cell differentiation (Zeaiter et al. 2008). In addition, inhibition of Par-1b increases the Shp-2 mediated hummingbird phenotype (Yamahashi and Hatakeyama 2013)

Overall, the disruption of tight junctions, in addition to the Shp-2 dependent changes in cell shape, supports an invasive phenotype of CagA expressing cells, where cells extrude from the polarized cell monolayer and show features characteristic of an epithelial to mesenchymal transition (EMT) (Bagnoli et al. 2005; Amieva et al. 2003; Saito et al. 2010). Beyond their primary function in embryonic development, the role of EMT in tumor formation and cancer is increasingly recognized, especially in regards to invasiveness, metastatic dissemination, and acquisition of therapeutic resistance (McConkey et al. 2009; Ouyang et al. 2010; Floor et al. 2011; Nickel and Stadler 2015). Indeed, the mesenchymal markers vimentin and fibronectin and possibly the transcription factors Snail, Slug, and ZEB1, are upregulated in CagA-transfected MDCK cells and in various gastric epithelial cell lines (Bagnoli et al. 2005; Yin et al. 2010; Saito et al. 2010). Another factor possibly contributing to EMT is matrix metalloproteinase (MMP)-7, which is increased following infection of gastric cells with *cag*-PAI positive Hp strains. MMPs mediate breakdown of extracellular matrix not only during embryogenesis, but also during cancer invasion and metastasis. Increased levels of MMP-7 have been identified in gastric dysplasia and cancer cells in vitro and in vivo (Ii et al. 2006). Therefore, we could expect that deregulation of cell-matrix interactions during EMT would actively promote cancer development on the epigenetic level.

E-cadherin is a calcium-dependent cell-cell adhesion glycoprotein that is linked to the actin cytoskeleton via β-catenin (Tian et al. 2011). Together, both proteins are an integral part of the adherence junctions and participate in maintenance of cell polarity and differentiation. It has been also established that disruption of the E-cadherin/β-catenin complex strongly contributes to tumor development (Guilford et al. 1998; Carneiro et al. 2012). Previous studies have shown that the CagA multimerization motifs (CM-motifs) that are important for Par1b-inhibition also cause association of CagA with E-cadherin (Murata-Kamiya et al. 2007; Kurashima et al. 2008; Oliveira et al. 2009). This destabilizes the E-cadherin/β-catenin complex allowing β-catenin to enter the nucleus. Once in the nucleus, β-catenin then activates transcription factors of the canonical Wnt signaling pathway, which is known to mediate cell proliferation and differentiation at various developmental stages (Kikuchi et al. 2011; Niehrs 2012). Gene activation by β-catenin is a characteristic of various cancers as 30 % of gastric cancers show nuclear accumulation of β-catenin (Tsukashita et al. 2003; Cheng et al. 2004; Saito-Diaz et al. 2013). In agreement, it is well documented that CagA triggers β-catenin mediated transcription of cancer-related genes in the stomach of Mongolian gerbils and of human subjects (Franco et al. 2005). An example of such a gene is caudal type homeobox 1 (CDX1), which is a transcription factor of intestinal cells and required for the development of intestinal metaplasia (Murata-Kamiya et al. 2007).

Another host factor that associates with the CM-motifs of CagA is the hepatocyte growth factor receptor tyrosine kinase (c-Met), which causes aberrant stimulation of the PI3-kinase/Akt pathway (Suzuki et al. 2009). PI3-kinase has been shown to prevent β-catenin degradation thus further enhancing Wnt/β-catenin signaling. Thus, activation of the c-Met receptor may deregulate growth factor receptor signaling and affect the motility and invasiveness of gastric cells.

CagA-Dependent Inactivation of Tumor Suppressors

While the CagA-dependent signaling pathways described above are dependent on the C-terminal EPIYA and CM motifs, the N-terminus of CagA has been described to inactivate at least two tumor suppressors.

In eukaryotic cells, the tumor suppressor p53 promotes cell survival by inhibiting apoptotic pathways. p53 activation is achieved by association of p53 with the apoptosis-stimulating protein p53-2 (ASPP2), which usually occurs in response to DNA damage and oncogenic stimuli. Interestingly, while Hp induces apoptosis early after infection, binding of CagA to ASPP2 causes rapid proteasomal degradation of p53 at later time points, thereby inhibiting the apoptotic signaling cascade (Buti et al. 2011).

Following CagA binding, another tumor suppressor, RUNX3, is ubiquitinated and targeted to the proteasome for degradation. In the absence of CagA, RUNX3 acts as a transcription factor that suppresses tumor formation by controlling expression of genes involved in apoptosis, growth, angiogenesis, junction formation, and in the differentiation of gastric epithelial cells (Tsang et al. 2010). There is evidence suggesting that a lack of RUNX3 is causally related to the development and progression of gastric cancer, potentially correlating with metastasis and poor prognosis of gastric cancer (Li et al. 2002; Hsu et al. 2009; Wei et al. 2005).

The Role of Hp in Epigenetic Changes

The role of epigenetic mechanisms in controlling gene expression is well established and the role of DNA methylation, post-transcriptional histone modifications, nucleosome positioning along the DNA strand, and micro-RNA have all been studied extensively in this context. Not surprisingly, these studies raise the possibility that epigenetic changes may also promote EMT and cancer development (Waddington 1959; Stadler and Allis 2012; Wang et al. 2010; Jones and Baylin 2007).

DNA Methylation DNA methylation occurs most frequently on cytosine residues located within repetitive CpG dinucleotide sequences of eukaryotic promoters where an increased methylation of these sequences leads to gene silencing (Bird 2002). Promoter methylation of cancer-related genes was observed during Hp-associated chronic gastric inflammation (Maekita et al. 2006; Nardone et al. 2007; Yoshida et al. 2013). Affected genes include DNA repair factor (O6-methylguanine DNA methyltransferase) (Sepulveda et al. 2010), regulators of gastric cell proliferation (trefoil factors 1 and 2) (Tomita et al. 2011; Peterson et al. 2010), cell cycle control proteins (p16, IRX1) (Guo et al. 2011), the cell adhesion protein E-cadherin (Chan et al. 2006), the Notch ligand Delta-like 1 (Piazzi et al. 2011), and forkhead box transcriptional regulator FoxD3 (Cheng et al. 2013). FoxD3 appears to play a role in proliferation, apoptosis, and invasiveness of gastric epithelial cells and is repressed in various gastric cancer cell lines and in the majority of gastric cancers (>80 %) (Cheng et al. 2013; Yin et al. 2012). The mechanisms that trigger hypermethylation of certain promoter sequences in gastric cancer are currently unknown.

MicroRNAs MiroRNAs (miRNAs) are non-coding RNAs of 20–24 bases that silence gene expression at the posttranscriptional level by annealing to the 3'-untranslated regions of target mRNAs. miRNAs are an important mechanism how cells regulate diverse cellular processes including cell proliferation, inflammation, cell cycle progression, apoptosis, and signal transduction (Ghildiyal and Zamore 2009; Xiao and Rajewsky 2009). Due to these important functions, the aberrant expression of miRNA, similar to increased expression of cellular kinases and phosphatases, can contribute to cancer formation and has been implicated in gastric cancer, specifically (Link et al. 2012; Yin et al. 2012; Maekita et al. 2006; Nardone et al. 2007). Several studies found that infection of gastric mucosa with Hp and CagA positive strains, especially, alters the miRNA signature of the gastric cells (Noto and Peek 2011; Yoshida et al. 2013). For example, CagA induced upregulation of miRNA-584 and miRNA-1290 in transformed cells, and overexpression of both miRNAs was associated with the development of gastric epithelial metaplasia in a knock-in mouse model experiment (Zhu et al. 2012). However, numerous other miRNAs were also affected and may promote cell proliferation and mitogenic responses, and therefore, over time, the development of cancer occurs.

Histone Modification Another important epigenetic mechanism that controls gene expression is histone modification. Postranslational modifications of histones like acetylation, methylation, ubiqutination, and phosphorylation, affect DNA packing and therefore influence molecular processes like transcription and replication (Kouzarides 2007). The presence of nucleosomes, which is DNA wrapped around a histone complex, causes occlusion of promoter regions followed by silencing of gene expression (Lin et al. 2007; Schones et al. 2008). Various studies showed the Hp infection triggers histone modifications that lead to activation of cancer related pathways, like hyper-acetylation of histone H4 induced expression of the cell cycle control factor p21 (WAF), for example (Xia et al. 2008). Another study demonstrated that *cag*-PAI dependent reduction of acetylation at serine 10 and threonine 3 of histone H3 causes the activation of the c-Jun proto-oncogene (Ding et al. 2010). In addition, decreased methylation of lysine 9 on histone H3 was implicated in increased iNOS (inducible nitric oxide synthase) expression, which is frequently associated with malignant disease (Angrisano et al. 2012).

Hp and the Host Immune Response

Experimental work in recent years has provided evidence that the pathological outcome of gastric infection depends not only on bacterial virulence factors but also on the immune response of the host (Wroblewski et al. 2010; Muller et al. 2011; Salama et al. 2013; Hardbower et al. 2014). For example, it has been well established that Hp-dependent chronic inflammation of the stomach is an important requirement for the development of intestinal type adenocarcinoma (Yeh et al. 2013; Yakirevich and Resnick 2013). Furthermore, host-gene polymorphisms that lead to increased expression of proinflammatory genes like IL-1β, TNF-α (El-Omar et al. 2003), or the soluble IL-1 receptor antagonist (IL-1RN) have all been linked with increased risk for Hp-induced gastric atrophy and adenocarcinoma (Figueiredo et al. 2002; Fox and Wang 2007; Lee et al. 2014). Additionally, individuals with certain genetic variations in the IFN-γ receptor 1 gene are more susceptible to Hp infection and have a higher likelihood to develop gastric cancer (Canedo et al. 2008; Lee et al. 2014).

Although limited work on the host immune modulation by Hp has been done to date, it has been well documented that this bacterium can manipulate the host immune system sufficiently to survive and to persist in the human stomach for decades. In susceptible hosts, this results in chronic gastric inflammation and, in later stages, in the development of gastric cancer. In the presence of Hp, the inflammatory microenvironment of the stomach and the stroma of gastric tumors are frequently filled with a wide range of innate and adaptive immune cells, including neutrophils, monocytes, macrophages, dendritic cells, myeloid-derived suppressor cells, T cells, and B cells, which further demonstrates the importance of immune cells in cancer development (Muller et al. 2011; Chung and Lim 2014; Lee et al. 2014). In the following sections of this chapter, we will discuss the current knowledge concerning the interplay of these immune cells with various Hp virulence factors, and the ability of the bacterium to modulate the immune response.

The Importance of Hp Adhesins in Inflammation and Cancer Development

Adhesins are outer membrane proteins (OMP) that are primarily implicated in adherence of Hp to the gastric mucosa of the host. Their presence has also been associated with long-term bacterial colonization and the risk for developing chronic gastric inflammation and cancer (Wroblewski et al. 2010; Salama et al. 2013).

BabA The blood group antigen binding adhesion BabA (also named HopS or OMP28) was the first Hp OMP that was demonstrated to bind to human Lewis^b and other related terminal fucose residues, expressed on the surface of gastric epithelial cells. BabA increases the colonization density and enhances inflammatory responses by strongly inducing IL-8 secretion within the gastric mucosa (Rad et al. 2002). Additionally, Mongolian gerbils infected with BabA$^+$ Hp strains have shown more gastric lesions when compared to strains lacking or having low BabA expression (Ohno et al. 2011). In patients, infection with strains expressing BabA in addition to CagA and VacA has been linked to an increased risk of gastritis, gastric ulcers, gastric cancer, and MALT lymphoma (Prinz et al. 2001; Gerhard et al. 1999). Furthermore, BabA can trigger host cell signaling and the production of proinflammatory cytokines to augment T4SS-dependent responses (Ishijima et al. 2011). In a recent study, it has been reported that BabA binding to Lewis^b could also cause genomic DNA double strand breaks leading to DNA damage in host cells (Toller et al. 2011). These results suggest that BabA plays a substantial role in gastric immuno-pathology and in the development of gastric inflammation and cancer (Kalali et al. 2014; Wroblewski et al. 2010).

SabA The sialic acid binding adhesin SabA, also known as HopP or OMP17, binds to sialyl-dimeric Lewis^x (Le^x) as well and is involved in the binding of *H. pylori* to the extracellular matrix protein laminin (Mahdavi et al. 2002; Walz et al. 2005). Le^x is an established tumor antigen and its expression on gastric epithelial cells is induced by Hp during initial colonization and also during chronic gastric inflammation (Mahdavi et al. 2002). It has been reported that SabA is associated with an increased gastric cancer risk but with a reduced risk for duodenal ulcers (Yamaoka et al. 2006).

OipA The outer inflammatory protein OipA (or HopH or OMP13) is a proinflammatory OMP of Hp (Yamaoka et al. 2000; Kalali et al. 2014). Although no receptor has been identified for OipA, the protein can increase the production of IL-8, enhance neutrophils infiltration, and activate a myriad of signaling cascades. For example, the expression of OipA has been associated with induction of the cytokines IL-1, IL-17, and TNF-α, resulting in gastric mucosal inflammation in a gerbil experimental model (Sugimoto et al. 2009a). Moreover, it has been recently reported that OipA can suppress the function of dendritic cells in vitro (Teymournejad et al. 2014). As a result, OipA-related signaling events likely contribute to severe gastrointestinal pathology (Tabassam et al. 2008; Tabassam et al. 2009; Yamaoka et al. 2004, 2006; Franco et al. 2008). Indeed, the expression of OipA has been correlated with duodenal ulcer and gastric cancer in various studies (Yamaoka et al. 2002b; Franco et al. 2008; Markovska et al. 2011). For example, OipA upregulates matrix metalloproteinase 1 (MMP-1) expression and enhances β-catenin nuclear translocation, which both have been linked to the development of gastric cancer (Wu et al. 2006; Franco et al. 2008; Tabassam et al. 2009).

In summary, OMPs of Hp may contribute to gastric malignancy by increasing bacterial colonization, promoting DNA damage, stimulating the expression of cancer markers and cancer-related pathways, and by inducing proinflammatory responses.

The Role of the Hp Toxins CagA and VacA in Host Immune Modulation

CagA CagA is one of the most intensely studied Hp virulence factors and has been linked to the development of severe gastric diseases (Sepulveda 2013). In addition to modulating the oncogenic cell signaling pathways described above, CagA has also been shown to affect immune responses. In this respect it is important to note that CagA is not only translocated into gastric epithelial cells, but also into immune cells. In B lymphoid cells, CagA possibly triggers the development of mucosa-associated lymphoid tissue (MALT) lymphoma (Lin et al. 2010). Translocation of CagA into dendritic cells (DCs) impairs DC maturation and function through IL-10 mediated activation of STAT3 (Kaebisch et al. 2014), which suppresses the secretion of the cytokine IL-12p40 and enhances the expression of the suppressive cytokine IL-10 (Tanaka et al. 2010). In epithelial cells, CagA can activate IL-8 and NF-kB to stimulate proinflammatory responses (Blaser et al. 1995; Papadakos et al. 2013). This indicates that CagA can play a role in both, pro- and anti-inflammatory responses, depending on the respective cell type, and possibly on the host physio-pathological environment as well. The seemingly opposite effect may allow CagA to balance the degree of inflammation to support chronic colonization. Additionally, CagA-induced activation of STAT3 via IL-6 plays a crucial role in inflammation-induced gastric cancer development by desensitizing TGF-β signaling (Jenkins et al. 2005). Signal Transducer and Activator of transcription 3 (STAT-3) is a transcription factor for various cytokines and growth factor genes that regulate epithelial cell homeostasis. Among other effects, the constitutive activation of STAT-3 has been implicated in cellular transformation of epithelial cancers, especially (Bronte-Tinkew et al. 2009). TGF-β-deficient mice spontaneously develop gastritis, and Hp causes tissue damage by down-regulating TGF-β1 signaling in the gastric mucosa. These signaling events were also observed in Hp-infected human patient samples, which revealed defective TGF-β1-induced Smad3 phosphorylation in the gastric mucosa (Monteleone et al. 2004; Lee et al. 2014).

VacA The vacuolating cytotoxin A (VacA) is a multifunctional toxin of Hp that is frequently co-expressed with CagA and secreted by an autotransporter system into the culture supernatant (Palframan et al. 2012). After receptor-mediated cellular uptake, VacA forms anion-selective channels in the endosomal membrane and causes the formation of large vacuoles that contain makers of the late endocytic compartment. Channel formation appears to be also related with the ability of the p34 subunit of VacA to affect mitochondrial membrane permeability, which by a yet undefined mechanism triggers apoptosis via cytochrome C release. VacA also has an important role as an immune modulator and is implicated directly in Hp colonization and persistence by resisting host immunosurveillance (Jones et al. 2010). For example, VacA inhibits T cell function through interference with antigen presentation, and T cell proliferation and activation, which all may be important for long-term colonization in the host (Molinari et al. 1998; Boncristiano et al. 2003). The β2 integrin subunit (CD18) of lymphocyte function-associated antigen-1 (LFA-1) is a specific receptor for VacA in host T cells (Sewald et al. 2008). Both VacA and GGT (γ-glutamyl transpeptidase), another secreted virulence factor of Hp, effects T cells activity by promoting the preferential differentiation of naïve T cells into T regulatory (Treg) cells through naïve T cells-tolerogenic dendritic cell (DC) interaction (Oertli et al. 2013). In these cases, Treg cells suppress the acute effector immune responses against Hp, which would be required for its eradication and clearance. In agreement with these findings, Treg cells are abundantly present in gastric mucosa of Hp-infected children and in asymptomatic adult carriers (Harris et al. 2008) (Robinson et al. 2008). The exact mechanism of VacA- and GGT-specific generation of tolerogenic DCs requires further study (Salama et al. 2013). In addition, deletion mutants of Hp lacking VacA or GGT have been shown to have colonization defects relative to their parental VacA/GGT-proficient wild-type isolates (Salama et al. 2001; Chevalier et al. 1999). VacA and GGT also promote inflammation by inducing NF-kB and IL-8 expression (Raju et al. 2012; Gong et al. 2010). GGT contributes to gastric inflammation specifically through the generation of H_2O_2 and the subsequent activation of NF-kB and IL-8 in primary gastric epithelial cells (Gong et al. 2010). Furthermore, the deprivation of glutamine induced by GGT may be the cause of gastric inflammation, which increases the risk of developing gastric cancer (Kalali et al. 2014; Rimbara et al. 2013).

Hp has the capability to survive within monocytes and macrophages. Usually Hp-infected macrophages form large vesicular compartments called megasomes, which are the result of VacA mediated homotypic phagosome fusion. These fusion events are a result of interference with phagosome maturation through coronin 1 recruitment and retention (Rittig et al. 2003; Allen et al. 2000; Zheng and Jones 2003).

Innate Immunity in Hp Infection

After Hp has penetrated the protective mucous layer of the gastric mucosa, it attaches to epithelial cells where it faces resident or recruited innate immune cells of the gastric wall, constituting the first line of the host immune defense. These immune cells, in addition to epithelial cells, typically recognize the pathogen-associated molecular patterns (PAMPs) of pathogens via distinct classes of innate immune receptors (pattern recognition receptors, PRRs) including Toll-like receptors (TLRs), the RIG-like helical receptor family (RLRs), C-type lectin receptors (CLRs), and Nod-like receptors (NLRs), all of which trigger inflammatory responses (Salama et al. 2013). Hp evades the innate immune detection by escaping proinflammatory TLRs and by suppressing CLR-mediated signaling. Hp lipopolysaccharide (LPS), for example, is predominantly tetra-acylated and is 1000-fold less bioactive than the hexa-acylated LPS of *Escherichia coli*. This reduced bioactivity prevents recognition by TLR4 thereby preventing a strong TLR4 mediated proinflammatory response in addition to subsequent adaptive immunity and clearance (Moran et al. 1997; Salama et al. 2013). Hp itself expresses Lewis antigens that mimic those Lewis antigens found on gastric epithelial cells. Expression of these bacterial Lewis antigens leads to suppression of innate immune response by decreasing IL-6 production. Additionally, Lewis antigen binding to the C-type lectin DC-SIGN, which is present on dendritic cells, suppresses Th1 responses (Bergman et al. 2004; Wu et al. 2014).

Despite these immune evasive mechanisms, chronic gastric inflammation is a hallmark of infection with Hp. Inflammation is mediated by numerous cellular events that include the production of proinflammatory mediators like IL-6, IL-8, IL-1β, IL-18, and TNF-α and NF-kB (Patel et al. 2012; Kalali et al. 2014; Salama et al. 2013).

The T4SS of Hp has been reported to deliver not only CagA, but also peptidoglycan of the bacterial cell wall into the host cytoplasm. Recognition of cytoplasmic peptidoglycan by the NOD-like receptor, NOD1, leads to activation of NF-kB-dependent proinflammatory responses, which include the secretion of IL-8, β-defensin-2, TNF-α and IL-1β (Viala et al. 2004). NOD1 activation, has been recently shown to be triggered by outer membrane vesicles (OMVs) of Hp as well. OMVs are naturally released by all Gram-negative bacteria as part of their normal growth and represent a general mechanism of peptidoglycan delivery into host epithelial cells (Kaparakis et al. 2010). Intragastric delivery of OMVs purified from Hp to mice was shown to induce innate and adaptive immune responses via a NOD1-dependent pathway (Kaparakis et al. 2010). In recent studies it has been reported that intracellular peptidoglycan of Hp can also trigger the activation of the PI3-kinase-Akt pathway, which has been implicated in anti-apoptotic effects, increased cell migration, and consequently in an associated increased risk for the development of malignancies (Nagy et al. 2009). NOD-1 also promotes the assembly of inflammasomes. This process activates cysteine protease caspase-1, an enzyme controlling the production of proinflammatory cytokines IL-1β and IL-18 (Strowig et al. 2012; Kim et al. 2008; Broz and Monack 2011). Mice lacking either caspase-1, IL-18 or its receptor IL-18R clear experimental infection with *Helicobacter felis* or Hp more efficiently compared to wild type mice and exhibit more pronounced pathogen-specific T cell responses (Hitzler et al. 2012; Oertli et al. 2012). Furthermore, the stomach-specific expression of human IL-1β is sufficient to induce gastric inflammation and gastric cancer in transgenic mice infected with *H. felis* (Tu et al. 2008). Consistent with this finding, mice lacking the IL-1β receptor (IL-1R−/−) fail to develop Hp-specific Th1 and Th17 responses and cannot control an experimental infection (Hitzler et al. 2011). In conclusion, IL-1β secretion is crucial for efficient infection control by opposing the function of IL-18 through induction of excessive T cell responses.

The chemokine IL-8 recruits immune cells from the circulation to sites of inflammation. Neutrophils are the most abundant circulating innate immune cells and are recruited to the inflammatory sites early after infection and tissue damage. Following activation via IL-8, neutrophils produce antimicrobial peptides and reactive oxygen species (ROS), which protect the stomach mucosa from invading microorganisms (Fischer et al. 2009). All Hp strains carry proteins for detoxification of these ROS such as catalase, and superoxide dismutase. Additionally, the bacteria express arginase, an enzyme that limits NO production by macrophage-, neutrophil- and epithelial cell-derived nitric oxide synthase (Gobert et al. 2001; Wang et al. 2006). Induced expression of human IL-8 in transgenic mice, which either received chemical treatment or were infected with *H. felis,* was sufficient to trigger colonic and gastric carcinogenesis. Treated or infected mice showed increased levels of local and systemic CD11b+ Gr+ myeloid cells, which modulate the tumor microenvironment, suggesting that IL-8 could be a useful therapeutic target in the prevention of inflammation-associated carcinogenesis. However, this study requires further characterization of the myeloid population in these conditions (Asfaha et al. 2013).

Myeloid-derived suppressor cells (MDSCs) and dendritic cells play also a key role in persistent infection with Hp, which contributes to chronic inflammation and further drives tumor formation (Tu et al. 2008; Teymournejad et al. 2014). MDSCs have the capability of inhibiting the inflammatory T-cell responses, and can directly induce tumor progression and metastasis by producing metalloproteinases that facilitate tumor invasion (Dolcetti et al. 2008; Marigo et al. 2008). The proinflammatory cytokine IL-1β can elicit immunosuppressive effects by recruiting MDSCs (Taketo 2009). Transgenic mice expressing human IL-1β specifically in the stomach develop spontaneous inflammation and gastric tumors, which corroborates the early recruitment and activation of MDSCs (Tu et al. 2008). Dendritic cells (DCs) that are exposed *to Hp* become tolerogenic either in the gastric mucosa or in the gastric or mesenteric lymph nodes (Oertli and Muller 2012). These tolerogenic DCs fail to induce effector T-cell responses of Th1 and Th17. Instead, these cells interact with naïve T-cells, which consequently differentiate into Treg cells (Oertli et al. 2012; Kim et al. 2011; Kao et al. 2010).

Macrophages activated by Hp infection produce nitric oxide, which causes methylation of genes associated with tumor suppression like Runx3 in the epithelial cells, risking gastric cancer development. These effects can be reversed by treatment with nitric oxide specific inhibitors corroborating one of the mechanisms by which *Hp* infection causes epigenetic changes that are initiated by chronic inflammation and associated with gastric malignancies (Katayama et al. 2009; Qu et al. 2013; Nadarajan et al. 2013). K19-Wnt1 transgenic mice, which express Wnt1 in gastric epithelial cells, show microphage infiltration into the dysplastic mucosa and β-catenin activation in the epithelium. Surprisingly, depletion of macrophages from adenomatous polyposis coli Δ716 (APC Δ716) mice abolishes intestinal tumor development suggesting a role for microphages in the induction of gastric malignancies through the activation of Wnt-β-catenin signaling (Oguma et al. 2008).

Adaptive Immunity in Hp Infection

Hp induces both cellular and humoral responses by activating T cells. For efficient clearance of infection, naïve T cells differentiate into effector T cell subsets, including type 1 helper T cells (Th1) that fight intracellular viruses and tumors, type 2 (Th2) that target helminthic parasites, and type 17 (Th17), which are important in the killing of extracellular bacteria especially in the intestine (Zhu et al. 2010). The T cell population present in the gastric mucosa of patients suffering from peptic ulcer disease shows an Hp-specific Th1 cellular pattern (D'Elios et al. 1997) producing high levels of the Th1 cytokines IFN-γ and IL-12, and low levels of the Th2 cytokines IL-4 and IL-5 (Bamford et al. 1998; Karttunen et al. 1995). Furthermore, Th1 responses are also induced via IL-12 production by activated monocytes, macrophages and dendritic cells (Guiney et al. 2003; Haeberle et al. 1997; Hafsi et al. 2004; Meyer et al. 2000; Kranzer et al. 2004). This Th1 response is not helpful in resolving the infection, however, and is rather correlated with increased bacterial colonization (Eaton et al. 2001). Th17 cells are mainly activated by IL-23, and produce the cytokines IL-17, IL-21 and IL-22, which play a role in innate and adaptive immune responses against infectious agents located in the gastric mucosa (Khader et al. 2009). Th17 cells also have been linked to the etiopathology of carcinogenesis in cancers of the colon (Sugimoto et al. 2009b).

It is not fully understood yet how Hp evades host adaptive immunity for its survival and manifestation of diseases like chronic gastritis and gastric cancer. Flow cytometry analysis of human subjects revealed that Hp infection induced both Th1 and Th2 immune responses with the presence of high levels of anti-inflammatory Treg cells (Kayhan et al. 2008). In Hp infection, these Treg cells suppress memory T cell responses that contribute to the persistence of infection (Lundgren et al. 2003). However, patients with peptic ulcers showed reduced IL-10 producing Treg cell responses and increased Th1 and Th2 responses compared to those without ulcers. These results clearly demonstrate the pattern of effector T cell responses elicited by the host during infection with Hp, and the insufficient role of the Treg cells to counteract the inflammatory damage in ulcers-bearing individuals (Robinson et al. 2008; Lee et al. 2014).

Normally, IL-10 and FoxP3 expressing Treg cells are very important for peripheral tolerance. A balance between effector T cells and Treg cells is crucial for the control of chronic inflammation. Higher levels of FoxP3 mRNA and protein expression are observed in gastric lymphocytes of Hp-infected patients (Rad et al. 2006). In recent studies, the experimental depletion of Treg cells contributed to the clearance of Hp infection (Arnold et al. 2011) and promoted vaccine-induced protective immunity in mice (Hitzler et al. 2011). Furthermore, the depletion of $FoxP3^+$ Treg cells by specific anti-CD25 antibodies resulted in severe gastritis with higher levels of cytokines, serum IgG1 and IgG2c, and decreased *Hp* colonization in mice (Rad et al. 2006). These results provide sufficient clues for a role of Treg cells in *Hp* infection and pathogenesis, and establish an equilibrium between host and bacterium allowing Hp to survive, while also preventing the occurrence of destructive acute inflammation (Wroblewski et al. 2010).

Limited information is available on the role of B cells in Hp pathogenesis. The major focus is on the development of gastric MALT lymphoma arising from activated B cells. One study using splenocytes isolated from naïve mice reported that under low multiplicity of *Hp* infection, mainly B-cell populations of splenocytes were protected from undergoing spontaneous apoptosis. These activated cells also acquired a proliferative phenotype, which has implications for the development of MALT lymphoma (Bussiere et al. 2006). Although, IgG and IgM antibodies are found in the serum and in the gastric mucosa of infected human subjects, their protective functions are marginal. Some authors suggest that B cell mediated antibody responses may be counterproductive because treatment with antibodies against Hp facilitates bacterial colonization, and counteracts resistance again infection (Akhiani et al. 2004, 2005). Nevertheless, B cells can act as antigen presenting cells (APCs) to T cells. In the case of Hp infection, functional inactivation of $CD4^+$ T cells recruited to gastric mucosa may be due to expression of CTLA-4 on the cell surface, which prevents co-stimulation when APCs engage T cells receptors (Anderson et al. 2006).

In conclusion, beyond their role of stimulating the innate immune system, T cells responses to infection with Hp effect bacterial survival and are a determining factor for the severity of infection. There is a distinct functional separation between effector T cell responses to clear Hp infection and Treg cells development influencing evasion of Hp from immune surveillance. The balance of these two major host strategies determines the fate of Hp infection and disease manifestations.

Recent Hp Vaccine Development Strategies

Currently, infections with Hp are treated with a triple therapy consisting of a proton pump inhibitor and two antibiotics, typically a combination of clarithromycin plus amoxicillin or clarithromycin plus metronidazole (Georgopoulos et al. 2013;

Selgrad et al. 2012). Treatments results in the regression of Hp mediated gastric diseases, including peptic ulcers and MALT-lymphoma (Ong and Duggan 2004). Unfortunately, gastric cancer cannot be prevented after cellular transformation begins, indicating that there is a point of no return in the process of cancer development, after which the presence of Hp is not required anymore (Talley et al. 2008). Unfortunately, antibiotic resistance, specifically clarithromycin resistance is rising, resulting in increasing treatment failures (Georgopoulos et al. 2013; Selgrad et al. 2012). Additionally, recurrent infection can occur after successful treatment. Therefore, the development of an effective vaccine appears even more urgent (Niv et al. 2008).

Many vaccine candidates against Hp have been tested successfully in the mouse model. Most recently, immunization of mice with a multi-epitope vaccine formulated with the Hp protein urease could significantly decrease Hp colonization and activate antigen-specific $CD4^{+}T$ cells, as well as IgG, IgA and mucosal sIgA antibodies responses (Guo et al. 2013). It has previously been demonstrated that Hp urease is a potent inducer of innate immune responses by activating monocytes and macrophages, as well as by stimulating cytokines and the production of NO (Gobert et al. 2002). However, only a few vaccine candidates have entered clinical trials and were based on mucosal administration of urease as a vaccine, either as a recombinant protein or *Salmonella*-vectored. However the results of these trials were somewhat disappointing due to limited immunogenicity and poor efficacy in humans, indicating that our understanding of protective immune responses is still limited (Czinn and Blanchard 2011). One phase I clinical trial involving a recombinant vaccine made of CagA, VacA, and neutrophil activating protein (NAP) showed good immune responses in humans and was safe (Malfertheiner et al. 2008). However, so far no Hp vaccine has been licensed. Given the potentially protective effects of Hp against the development of autoimmune diseases, and also given the controversial reports on the prevention of esophageal cancer by Hp infection, it might be questionable as to whether a vaccine that provides sterilizing immunity would be desirable. Instead, a vaccine that targets CagA and other virulence factors to prevent disease rather than to eradicate Hp from the stomach may be a better option. Such a vaccine could be likely recommended for whole population vaccination. However, even in the case the reasons against mass vaccination were prevalent, such a vaccine would likely be highly valuable for individuals with symptomatic infection or those with a family history of severe gastric disease.

Animal Models of Hp-Induced Carcinogenesis

Atrophic gastritis was proposed to be a premalignant condition that favors the development of gastric cancer, after correlating the increased risk of developing gastric cancer with the severity of the gastritis (Sipponen 1989). The progression of atrophic gastritis toward gastric cancer includes development of intestinal metaplasia and dysplasia (Correa et al. 1975).

Atrophy of the stomach affects primarily the glands, which become sparse and small (Dixon et al. 1996). This pathological aspect is part of a natural progression of inflammatory changes that result from long term infection with Hp, as confirmed by a recent clinical trial, which also suggested that Hp-induced severe atrophy has a greater risk for gastric cancer development in concomitance with p73 overexpression (Carrasco and Corvalan 2013).

Soon after the discovery of Hp and the demonstration of its causative relationship with gastritis and peptic ulcer in humans, animal models of Hp infection were developed, reproducing disease with some aspects close to those observed in humans, and in particular gastric inflammation (Ruggiero et al. 2004). Atrophic changes were then found in a mouse model of Hp infection after long-term observation (18 months) (Lee et al. 1993). The logical consequence of these observations was to attempt to set up animal models of gastric carcinogenesis upon Hp experimental infection, in order to experimentally investigate the relationships between Hp colonization and gastric cancer development.

The present section reports the animal models that produced evidence supporting the development of gastric carcinoma upon Hp experimental infection, exposing mechanisms that have been or could be potentially be exploited to study the pathogenesis of Hp infection. These models could also be used to study the progression of the disease toward malignancy, as well as to develop prophylactic or therapeutic tools to counteract the malignant outcomes of Hp infection.

Mongolian Gerbil

The susceptibility of Mongolian gerbils to gastric colonization upon oral administration of Hp was first reported in 1991 (Yokota et al. 1991), though only a slight presence of inflammatory cells was observed. It was subsequently demonstrated that Mongolian gerbils may develop gastric ulcers and intestinal metaplasia upon Hp infection (Hirayama et al. 1996). In a long-term study in Mongolian gerbils, it was later observed that, after 62 weeks of colonization, adenocarcinoma had developed in the pyloric region of 10 out of 27 (37 %) infected animals (Watanabe et al. 1998). Similarly, in another study, 2 out of 5 (40 %) Mongolian gerbils showed three well-differentiated gastric cancers after 18 months of colonization, while uninfected control animals showed no abnormal findings (Honda et al. 1998). Another long-term study showed that mice had developed carcinoid tumor in their stomachs after 12 months of Hp gastric colonization, (18 out of 56 animals, 32 %),. While one case of poor differentiated adenocarcinoma in the pylorus was also observed, no lesions were found in control gerbils (Hirayama et al. 1999).

Thus, the studies in Mongolian gerbils provided not only a formal demonstration of causative relationship between Hp colonization and gastric cancer, but also the basis for developing valuable tools for studies on Hp-induced gastric carcinogenesis.

The reason why Mongolian gerbils spontaneously develop pre-cancerous lesions and gastric cancer after long-term colonization may be due to the higher gastric mucosa colonization observed for this species in comparison with the murine model. A higher colonization has been ascribed to the relatively greater expression of sulfatide, a putative adhesion receptor, found in the gastric mucosa of Mongolian gerbils (Osawa et al. 2001).

Subsequent studies that used the Mongolian gerbil model of Hp-induced carcinogenesis confirmed the development of gastric adenocarcinoma, however with very variable results with rates of 4 % (1/23) (Ogura et al. 2000), 18 % (3/17) (Zheng et al. 2004), and 65 % (13/20) (Futagami et al. 2006). In contrast, some studies did not observe gastric adenocarcinoma at all, although pre-cancerous lesions were present (Sun et al. 2005; Elfvin et al. 2005). Some of these discrepancies may be ascribed to the use of different Hp strains used for infection and to the intrinsic variability of the animal model.

In order to improve the Mongolian gerbil model of gastric carcinogenicity, N-methyl-N-nitrosourea (MNU) and N-methyl-N'-nitro-N-nitrosoguanidine (MNNG), which have previously shown to be carcinogenic in gerbils (Tatematsu et al. 1998), were administered at relatively low doses in addition to Hp infection. Using this protocol it was observed that Hp infection was able to significantly increase the rates of adenocarcinoma development in comparison with the administration of carcinogen alone. In particular, the administration of MNU 30 ppm for 6 weeks prior to Hp infection or 10 ppm for 20 weeks following Hp infection resulted in a gastric adenocarcinoma incidence rate of 33.3 % (6/18) or 36.8 % (7/19), respectively, when animals were examined after 40 weeks. The administration of MNU alone did not cause adenocarcinoma (Sugiyama et al. 1998). Similar results were obtained when using MNNG 300 ppm prior to Hp infection (Shimizu et al. 1999). After 50 weeks, 44.4 % (12/27) of the animals that had received Hp+MNNG developed gastric cancer, versus 5.3 % (1/19) in the control group that had received medium+MNNG. Another study employed MNNG 50 ppm after Hp infection and found that after 24 and 52 weeks, incidences of gastric cancer were 25 % (5/20) for the Hp+MNNG group versus 13.6 % (3/22) for the control group that had received medium+MNNG (Tokieda et al. 1999). Interestingly, gerbils in the Hp+MNNG group that at 24 or 52 weeks had cleared the infection (12/20) did not show gastric adenocarcinoma, while in those that had maintained Hp colonization the incidence of gastric adenocarcinoma was 62.5 % (5/8). Further studies, generally performed with MNU, confirmed these observations (Maruta et al. 2001; Kawazoe et al. 2007), suggesting that Hp infection enhances gastric carcinogenesis in combination with known carcinogens.

The advantage of the carcinogenicity model, based on administration of carcinogenic substances+Hp when compared with the administration of Hp alone, consists essentially in the shorter period of observation necessary.

There are some peculiarities of the Mongolian gerbil model that need to be pointed out. First, p53 mutations have been found in over 50 % of Hp-positive humans and in 100 % of Hp-positive monkeys in a Japanese monkey model, with no mutations in Hp-negative subjects. Paradoxically, no p53 mutations were observed in Hp-positive gerbils. This indicates a different pathway in developing dysplasia or carcinoma between gerbils and both human and non-human primates (Murakami et al. 2002) and may represent a limitation for certain studies of carcinogenicity. Second, in Mongolian gerbils, HSP70 involved in activities including apoptosis, carcinogenesis and cytoprotection from cytotoxic damage, is not expressed, with a consequent mucosal susceptibility to potentially carcinogenic factors (Otaka et al. 2006). Also, the peculiarity of Mongolian gerbil may limit – but not abolish – the usefulness of the observations done in this model. Another point that may limit the studies with Mongolian gerbils is the relatively low availability of specific reagents, for instance to investigate immune response, in particular for those markers that are species-specific.

In spite of the fact that the applicability of the Mongolian gerbil model for studying Hp-associated gastric cancer has been called into question (Chen et al. 2007), this model remains widely used for studying Hp-related carcinogenesis. For instance, it was successfully exploited to demonstrate that eradication of Hp infection might be effective in preventing Hp-related gastric carcinogenesis. This model also demonstrated that the earlier the eradication was performed, the greater was the efficacy in preventing malignant outcome (Nozaki et al. 2003). Moreover, it was shown that in Mongolian gerbils receiving MNU + Hp, the selective COX-2 inhibitor celecoxib was able to prevent the progression of intestinal metaplasia into gastric carcinoma (Futagami et al. 2006). The Mongolian gerbil model was also useful to prove that long-term proton pump inhibitor administration may promote adenocarcinoma development. In fact, 6 months after Hp infection without any additional treatment, gastric adenocarcinomas were found in 7 % (1 out 15) of the animals, while treatment with omeprazole in addition to Hp infection after the same time revealed gastric adenocarcinoma in 60 % (9/15) of the animals (Hagiwara et al. 2011).

Mouse

As previously mentioned, the first evidence that infection by *Helicobacter* species induces a chronic inflammation that progresses to gastric atrophy, the precursor lesion to gastric adenocarcinoma in humans, has come from the mouse model (Lee et al. 1993), which is currently the most widely used animal model to study Hp infection and pathogenesis. However, upon long-term Hp colonization mice do not spontaneously develop gastric adenoma, dysplasia or carcinoma (Kim et al. 2003), in spite of the fact that other *Helicobacter* species, like *H. felis*, have been shown to be able to originate gastric cancer in mice (Fox et al. 2002). Upon infection with Hp, only particular mouse strains with a predisposition to gastric cancer may develop preneoplastic or neoplastic disease, but not gastric cancer (Rogers et al. 2005). Thus, the mouse model of Hp-induced gastric carcinogenesis requires the coadministration of carcinogenic substances, such as MNU or MNNG, with experimental procedures very similar to those of the corresponding Mongolian gerbil model described earlier (Ferrero et al. 2012).

The sensitivity of mice to MNU was demonstrated in early studies (Tatematsu et al. 1993), and was a pre-requisite to experimentally evaluate the influence of Hp infection on gastric cancer development in mice. Pre-cancerous lesions were observed in mice treated with MNU and experimentally infected with Hp (Shimizu et al. 1998). A subsequent long-term study with MNU treatment and Hp infection showed gastric adenocarcinoma in 27 % (3/11) of mice treated with MNU versus 80 % (8/10) of mice receiving both, MNU and Hp, while only chronic atrophic gastritis was observed in mice receiving only Hp infection (Han et al. 2002). In contrast, a study with the same model of Hp-induced gastric carcinogenesis did not find significant differences in the incidence of adenocarcinomas between the MNU+Hp group and MNU-alone group. They also observed higher incidences of polypoid lesions and adenomatous hyperplasia in the MNU-alone group, suggesting that Hp colonization could tend to inhibit rather than exacerbate the MNU-induced gastric carcinogenesis (Nakamura et al. 2002). The two studies used comparable MNU doses, the same strain of both mice (C57BL/6) and Hp (SS1), and similar treatment regimen. Thus, it is very difficult to understand and explain the reasons for these opposite results, except to hypothesize that repeated passages of the bacterium in the laboratory could have led to undetected genotypic changes in the infecting Hp strain, or that mice could possess different microbioma that could have affected the outcome or Hp infection. These controversial results might have contributed to orient scientists toward preferentially using of the Mongolian gerbil model rather than the mouse model for studies on Hp-related carcinogenicity. Further studies have successfully exploited the mouse model, however, demonstrating that anti-inflammatory drugs were able to either provide anti-inflammatory effects and partial anti-proliferative effects or definite chemopreventive effects in Hp infection (Hahm et al. 2003). After a period of 50 weeks, tumor incidence was 64.7 % (11/17) in the MNU+Hp group and 10.5 % (2/19) in the MNU alone group, showing a significant reduction in tumor incidence to 38.9 % (7/18) in the group that received MNU+Hp and additional treatment with nimesulide.

Transgenic Mice

If on one hand the mouse model has been less exploited than the Mongolian gerbil one for studies of Hp-related carcinogenicity, the possibility of generating transgenic mice has resulted very useful for these studies still remains. Some transgenic mice carrying transgenic expression or knocked out for factors considered to be involved in gastric carcinogenesis, have been very useful to clarify some aspects of Hp-related pathogenesis and development of gastric cancer.

Polymorphisms of interleukin-1β are considered to be associated with an increased risk of solid malignancies in humans (Xu et al. 2013). It has been shown that stomach-specific expression of human IL-1β in transgenic mice is sufficient to induce neoplasia thus providing a direct link between IL-1β and gastric carcinogenesis (Tu et al. 2008). Furthermore, polymorphisms of IL-1β are also considered to be involved in Hp-induced human gastric carcinogenesis (McNamara and El-Omar

2008). Transgenic mice defective in IL-1β expression exhibited lower gastric inflammation and developed significantly less gastric tumors than wild-type mice, providing evidence that IL-1β induced by Hp infection is able to enhance gastric carcinogenesis (Shigematsu et al. 2013).

Transforming growth factor-beta (TGF-β) is involved in mediating the Treg cell-biased response to Hp, which suppresses the effective induction of Hp-specific Th17 immune responses (Kao et al. 2010). Transgenic mice, in which the dominant negative mutant of the TGF-beta type II receptor was expressed under the control of tissue-specific promoters, lacked TGF-beta signaling. These mice were exploited to further point out the possible role of this cytokine in preventing abnormal mucosal proliferation, and for suppressing or retarding carcinogenesis. In fact, upon Hp infection, these mice showed increased susceptibility to gastrointestinal carcinogenesis (Hahm et al. 2002).

Overproduction of nitric oxide via inducible nitric oxide synthase (iNOS) is suggested to be a significant pathogenic factor in Hp-induced gastritis. To test whether iNOS was involved in Hp-associated gastric carcinogenesis, iNOS-deficient mice were used (Nam et al. 2004). Upon MNU administration and Hp infection, iNOS-deficient mice showed significantly lower incidence of gastric cancer than wild-type mice after a period of 50 weeks, indicating the involvement of iNOS in Hp-induced gastric carcinogenesis.

Hypergastrinemia, and possible synergetic effects with Hp infection, were investigated in insulin-gastrin (INS-GAS) transgenic mice. First observed in a mouse model of *H. felis* infection (Wang et al. 2000), the role of hypergastrinemia in contributing to eventual parietal cell loss and progression to gastric cancer was than confirmed in the mouse model of Hp infection (Fox et al. 2003a, b). Interestingly, it was observed that cancer development was restricted to only males, as none of the Hp-infected female mice developed cancer, and, in general, male gastric tissue responded more rapidly and aggressively to Hp infection.

Although not a perfect model to study Hp-induced carcinogenicity, transgenic mice expressing CagA provided a formal demonstration of the role of CagA as a bacterial oncoprotein and of the importance of CagA tyrosine phosphorylation in the development of Hp-associated neoplasms (Ohnishi et al. 2008). In fact, as previously mentioned (section "Effects of CagA on cell shape, proliferation, and motility"), transgenic mice expressing wild-type CagA, but not those expressing a CagA phosphorylation-resistant mutant, showed gastric epithelial hyperplasia and some of the mice developed gastric polyps and adenocarcinomas of the stomach and small intestine.

Non-Human Primate Models for Hp

Hp is enzootic in at least some rhesus monkey colonies, causing atrophy, microerosions, loss of mucus reminiscent of those seen in humans, and also peptic ulcers and gastric cancer (Dubois and Berg 1997). Early studies showed that, in selected, non-infected, young rhesus monkeys, experimental administration of Hp resulted in gastric colonization, although these effects were often transient (Dubois et al. 1996). The rhesus monkey model was then exploited to study gastric carcinogenicity due to Hp infection in combination with the oral carcinogen N-ethyl-N-nitrosoguanidine (ENNG) (Liu et al. 2009). This carcinogen is similar to nitrosamines found in foods such as smoked fish and pickled vegetables. Follow-up gastroscopies and biopsies were performed at 3-month intervals for 5 years. Transcriptional analysis of biopsy specimens at 5 years revealed group-specific expression profiles, with striking changes in monkeys receiving both Hp and ENNG. A neoplasia-specific expression profile characterized by changes in multiple cancer-associated genes was also seen. Monkeys receiving Hp + ENNG developed gastritis, intestinal metaplasia, and neoplasia, while those receiving Hp alone developed gastritis only. Based on these results, synergistic effect of Hp and the carcinogen in inducing gastric neoplasia in primates was proposed.

However, the rhesus monkey model has not been largely used for studies on Hp pathogenesis for two main reasons: first, the natural Hp infection occurring in these animals makes it difficult to find monkeys to include in the studies and secondly, these studies are very expensive and require very long follow-up.

Discussion

In this chapter we have discussed that Hp infection of epithelial cells triggers a multitude of pro-oncogenic cellular responses that over time contribute to development of gastric cancer. The major pathways include *cag*-PAI, especially CagA-dependent signaling events and the modulation of the innate and adaptive immune responses, which cause chronic inflammation in the host. However, as mentioned earlier, Hp mediated mechanisms are not sufficient per se to cause malignancy; a genetic predisposition of the human host towards increased proinflammatory responses is also required along with various environmental conditions and favorable diets.

One interesting question is: which cell type carries the potential for malignant transformation? Epithelial cells in the gastric pit and mucous producing cells, the major cell types colonized by Hp, typically experience a fast turn-over that would prohibit the accumulation of genetic mutations that are required for transformation (Mueller et al. 2004). Therefore, one hypothesis has focused on the long-lived stem cells and progenitor cells, which are located in the isthmus region of the glands, as the source of malignancy (Karam 2008). Hp could potentially effect these cells through chronic inflammatory processes as well as through epigenetic changes and via oncogenic signals generated by CagA. Tumor development would then follow the accumulation of genetic and epigenetic modifications causing the loss of the homeostatic control (Ding and Zheng 2012). Indeed, Hp has been shown to interact with epithelial progenitor cells, while stem cells have been shown to be amplified as a response to the loss of parietal cells during atrophic gastritis (Necchi et al. 2007; Oh et al. 2005). Furthermore, in vivo animal studies have revealed that Hp causes pre-neoplastic lesions containing bone marrow derived cells, which are recruited to the gastric mucosa (Varon et al. 2012). These initial experiments suggest that stem and progenitor cells may indeed be the original cell types that start the process of transformation.

Another daunting question is, as to why some strains of Hp have acquired the *cag*-PAI and inject the CagA cytotoxin into the host cells. Animal experiments have demonstrated that *cag*-PAI and CagA positive strains favor bacterial colonization of the stomach (Rieder et al. 2005; Akanuma et al. 2002). Increased colonization may enhance CagA-related oncogenic stress, which may be counteracted by epithelial cells through stimulation of apoptosis. Reduction in acid-producing parietal cells may increase the pH in the stomach and thus generate a less hostile environment that favors long-term colonization. Therefore, CagA appears to have evolved as colonization factor. However, even *cag*-PAI negative strains successfully colonize the stomach of humans and Mongolian gerbils and cause chronic infection, albeit with reduced inflammatory responses, raising doubts about the colonization factor hypothesis (Saito et al. 2005). Nevertheless, by destabilizing the apical junctional complex, CagA may release nutrients from sub-epithelial tissues. Indeed previous studies have shown that Hp receives nutrients from tissue exudates rather than from the gastric lumen (Blaser 1993; Akada et al. 2003).

Certainly, Hp has not acquired CagA or the ability to increase inflammation in the gastric mucosa to cause cancer. For most part of its evolutionary period, Hp has infected individuals who died at a comparable early age and only recent medical advances have pushed human life expectancy toward a higher age, where gastric cancer will start to manifest. Therefore, intestinal type gastric cancer is probably an accidental event that develops over a long period of time in mostly elderly individuals, who show above average proinflammatory responses and are infected with more virulent Hp strains. The host-pathogen equilibrium has evolved to only tolerate a certain variability in immune responses and bacterial virulence. When this equilibrium is exceeded, it turns detrimental for the host (severe gastric diseases including adenocarcinoma) or the bacteria (clearance by the immune system). In Hp-related gastric adenocarcinoma, this appears to be the case for only a minority of infected individuals (1–3 %). However, considering the high prevalence of Hp, Hp-associated gastric diseases will remain a heavy burden for the patients as well as for health systems for the foreseeable future.

Acknowledgements We would like to thank Mr. Steven M. Lane and Dr. Eric Yager for the critical reading of this chapter. M. Stein is supported by a RO1 grant of the National Institute of Health (grant number 1R01CA160790-01A1).

References

Akada JK, Ogura K, Dailidiene D, Dailide G, Cheverud JM, Berg DE (2003) Helicobacter pylori tissue tropism: mouse-colonizing strains can target different gastric niches. Microbiology 149:1901–1909

Akanuma M, Maeda S, Ogura K, Mitsuno Y, Hirata Y, Ikenoue T, Otsuka M, Watanabe T, Yamaji Y, Yoshida H, Kawabe T, Shiratori Y, Omata M (2002) The evaluation of putative virulence factors of Helicobacter pylori for gastroduodenal disease by use of a short-term Mongolian gerbil infection model. J Infect Dis 185:341–347

Akhiani AA, Schon K, Franzen LE, Pappo J, Lycke N (2004) Helicobacter pylori-specific antibodies impair the development of gastritis, facilitate bacterial colonization, and counteract resistance against infection. J Immunol 172:5024–5033

Akhiani AA, Stensson A, Schon K, Lycke NY (2005) Helicobacter pylori-specific antibodies impair the development of gastritis, facilitate bacterial colonization, and counteract resistance against infection. J Immunol 174:8144–8153

Akopyants NS, Clifton SW, Kersulyte D, Crabtree JE, Youree BE, Reece CA, Bukanov NO, Drazek ES, Roe BA, Berg DE (1998) Analyses of the cag pathogenicity island of Helicobacter pylori. Mol Microbiol 28:37–53

Allen LA, Schlesinger LS, Kang B (2000) Virulent strains of Helicobacter pylori demonstrate delayed phagocytosis and stimulate homotypic phagosome fusion in macrophages. J Exp Med 191:115–128

Amieva MR, Vogelmann R, Covacci A, Tompkins LS, Nelson WJ, Falkow S (2003) Disruption of the epithelial apical-junctional complex by Helicobacter pylori CagA. Science 300:1430–1434

Anderson KM, Czinn SJ, Redline RW, Blanchard TG (2006) Induction of CTLA-4-mediated anergy contributes to persistent colonization in the murine model of gastric Helicobacter pylori infection. J Immunol 176:5306–5313

Angrisano T, Lembo F, Peluso S, Keller S, Chiariotti L, Pero R (2012) Helicobacter pylori regulates iNOS promoter by histone modifications in human gastric epithelial cells. Med Microbiol Immunol 201:249–257

Arnold IC, Lee JY, Amieva MR, Roers A, Flavell RA, Sparwasser T, Muller A (2011) Tolerance rather than immunity protects from Helicobacter pylori-induced gastric preneoplasia. Gastroenterology 140:199–209

Asahi M, Azuma T, Ito S, Ito Y, Suto H, Nagai Y, Tsubokawa M, Tohyama Y, Maeda S, Omata M, Suzuki T, Sasakawa C (2000) Helicobacter pylori CagA protein can be tyrosine phosphorylated in gastric epithelial cells. J Exp Med 191:593–602

Asfaha S, Dubeykovskiy AN, Tomita H, Yang X, Stokes S, Shibata W, Friedman RA, Ariyama H, Dubeykovskaya ZA, Muthupalani S, Ericksen R, Frucht H, Fox JG, Wang TC (2013) Mice that express human interleukin-8 have increased mobilization of immature myeloid cells, which exacerbates inflammation and accelerates colon carcinogenesis. Gastroenterology 144:155–166

Atherton JC, Peek RM Jr, Tham KT, Cover TL, Blaser MJ (1997) Clinical and pathological importance of heterogeneity in vacA, the vacuolating cytotoxin gene of Helicobacter pylori. Gastroenterology 112:92–99

Ayala G, Escobedo-Hinojosa WI, De La Cruz-Herrera CF, Romero I (2014) Exploring alternative treatments for Helicobacter pylori infection. World J Gastroenterol 20:1450–1469

Backert S, Tegtmeyer N (2012) Helicobacter pylori CagA tertiary structure reveals functional insights. Cell Host Microbe 12:3–5

Backert S, Ziska E, Brinkmann V, Zimny-Arndt U, Fauconnier A, Jungblut PR, Naumann M, Meyer TF (2000) Translocation of the Helicobacter pylori CagA protein in gastric epithelial cells by a type IV secretion apparatus. Cell Microbiol 2:155–164

Bagnoli F, Buti L, Tompkins L, Covacci A, Amieva MR (2005) Helicobacter pylori CagA induces a transition from polarized to invasive phenotypes in MDCK cells. Proc Natl Acad Sci U S A 102:16339–16344

Bamford KB, Fan X, Crowe SE, Leary JF, Gourley WK, Luthra GK, Brooks EG, Graham DY, Reyes VE, Ernst PB (1998) Lymphocytes in the human gastric mucosa during Helicobacter pylori have a T helper cell 1 phenotype. Gastroenterology 114:482–492

Batista SA, Rocha GA, Rocha AM, Saraiva IE, Cabral MM, Oliveira RC, Queiroz DM (2011) Higher number of Helicobacter pylori CagA EPIYA C phosphorylation sites increases the risk of gastric cancer, but not duodenal ulcer. BMC Microbiol 11:61

Bergman MP, Engering A, Smits HH, Van Vliet SJ, Van Bodegraven AA, Wirth HP, Kapsenberg ML, Vandenbroucke-Grauls CM, Van Kooyk Y, Appelmelk BJ (2004) Helicobacter pylori modulates the T helper cell 1/T helper cell 2 balance through phase-variable interaction between lipopolysaccharide and DC-SIGN. J Exp Med 200:979–990

Bird A (2002) DNA methylation patterns and epigenetic memory. Genes Dev 16:6–21

Blaser MJ (1993) Helicobacter pylori: microbiology of a 'slow' bacterial infection. Trends Microbiol 1:255–260

Blaser MJ, Perez-Perez GI, Kleanthous H, Cover TL, Peek RM, Chyou PH, Stemmermann GN, Nomura A (1995) Infection with Helicobacter pylori strains possessing cagA is associated with an increased risk of developing adenocarcinoma of the stomach. Cancer Res 55:2111–2115

Boncristiano M, Paccani SR, Barone S, Ulivieri C, Patrussi L, Ilver D, Amedei A, D'elios MM, Telford JL, Baldari CT (2003) The Helicobacter pylori vacuolating toxin inhibits T cell activation by two independent mechanisms. J Exp Med 198:1887–1897

Bronte-Tinkew DM, Terebiznik M, Franco A, Ang M, Ahn D, Mimuro H, Sasakawa C, Ropeleski MJ, Peek RM Jr, Jones NL (2009) Helicobacter pylori cytotoxin-associated gene A activates the signal transducer and activator of transcription 3 pathway in vitro and in vivo. Cancer Res 69:632–639

Broz P, Monack DM (2011) Molecular mechanisms of inflammasome activation during microbial infections. Immunol Rev 243:174–190

Bussiere FI, Chaturvedi R, Asim M, Hoek KL, Cheng Y, Gainor J, Scholz A, Khan WN, Wilson KT (2006) Low multiplicity of infection of Helicobacter pylori suppresses apoptosis of B lymphocytes. Cancer Res 66:6834–6842

Buti L, Spooner E, Van Der Veen AG, Rappuoli R, Covacci A, Ploegh HL (2011) Helicobacter pylori cytotoxin-associated gene A (CagA) subverts the apoptosis-stimulating protein of p53 (ASPP2) tumor suppressor pathway of the host. Proc Natl Acad Sci U S A 108:9238–9243

Canedo P, Corso G, Pereira F, Lunet N, Suriano G, Figueiredo C, Pedrazzani C, Moreira H, Barros H, Carneiro F, Seruca R, Roviello F, Machado JC (2008) The interferon gamma receptor 1 (IFNGR1) -56C/T gene polymorphism is associated with increased risk of early gastric carcinoma. Gut 57:1504–1508

Carneiro P, Fernandes MS, Figueiredo J, Caldeira J, Carvalho J, Pinheiro H, Leite M, Melo S, Oliveira P, Simoes-Correia J, Oliveira MJ, Carneiro F, Figueiredo C, Paredes J, Oliveira C, Seruca R (2012) E-cadherin dysfunction in gastric cancer–cellular consequences, clinical applications and open questions. Febs Lett 586:2981–2989

Carrasco G, Corvalan AH (2013) Helicobacter pylori-induced chronic gastritis and assessing risks for gastric cancer. Gastroenterol Res Pract 2013:393015

Censini S, Lange C, Xiang Z, Crabtree JE, Ghiara P, Borodovsky M, Rappuoli R, Covacci A (1996) cag, a pathogenicity island of Helicobacter pylori, encodes type I-specific and disease-associated virulence factors. Proc Natl Acad Sci U S A 93:14648–14653

Chan AO, Peng JZ, Lam SK, Lai KC, Yuen MF, Cheung HK, Kwong YL, Rashid A, Chan CK, Wong BC (2006) Eradication of Helicobacter pylori infection reverses E-cadherin promoter hypermethylation. Gut 55:463–468

Chen D, Stenstrom B, Zhao CM, Wadstrom T (2007) Does Helicobacter pylori infection per se cause gastric cancer or duodenal ulcer? Inadequate evidence in Mongolian gerbils and inbred mice. Fems Immunol Med Microbiol 50:184–189

Cheng XX, Sun Y, Chen XY, Zhang KL, Kong QY, Liu J, Li H (2004) Frequent translocalization of beta-catenin in gastric cancers and its relevance to tumor progression. Oncol Rep 11:1201–1207

Cheng AS, Li MS, Kang W, Cheng VY, Chou JL, Lau SS, Go MY, Lee CC, Ling TK, Ng EK, Yu J, Huang TH, To KF, Chan MW, Sung JJ, Chan FK (2013) Helicobacter pylori causes epigenetic dysregulation of FOXD3 to promote gastric carcinogenesis. Gastroenterology 144(122–133):E9

Chevalier C, Thiberge JM, Ferrero RL, Labigne A (1999) Essential role of Helicobacter pylori gamma-glutamyltranspeptidase for the colonization of the gastric mucosa of mice. Mol Microbiol 31:1359–1372

Chung HW, Lim JB (2014) Role of the tumor microenvironment in the pathogenesis of gastric carcinoma. World J Gastroenterol 20:1667–1680

Conradi J, Huber S, Gaus K, Mertink F, Royo Gracia S, Strijowski U, Backert S, Sewald N (2012) Cyclic RGD peptides interfere with binding of the Helicobacter pylori protein CagL to integrins alphaVbeta3 and alpha5beta1. Amino Acids 43:219–232

Correa P (1992) Human gastric carcinogenesis: a multistep and multifactorial process--First American Cancer Society Award Lecture on Cancer Epidemiology and Prevention. Cancer Res 52:6735–6740

Correa P, Haenszel W, Cuello C, Tannenbaum S, Archer M (1975) A model for gastric cancer epidemiology. Lancet 2:58–60

Covacci A, Rappuoli R (1998) Helicobacter pylori: molecular evolution of a bacterial quasi-species. Curr Opin Microbiol 1:96–102

Covacci A, Censini S, Bugnoli M, Petracca R, Burroni D, Macchia G, Massone A, Papini E, Xiang Z, Figura N et al (1993) Molecular characterization of the 128-kDa immunodominant antigen of Helicobacter pylori associated with cytotoxicity and duodenal ulcer. Proc Natl Acad Sci U S A 90:5791–5795

Covacci A, Falkow S, Berg DE, Rappuoli R (1997) Did the inheritance of a pathogenicity island modify the virulence of Helicobacter pylori? Trends Microbiol 5:205–208

Cover TL, Blanke SR (2005) Helicobacter pylori VacA, a paradigm for toxin multifunctionality. Nat Rev Microbiol 3:320–332

Czinn SJ, Blanchard T (2011) Vaccinating against Helicobacter pylori infection. Nat Rev Gastroenterol Hepatol 8:133–140

D'elios MM, Manghetti M, Almerigogna F, Amedei A, Costa F, Burroni D, Baldari CT, Romagnani S, Telford JL, Del Prete G (1997) Different cytokine profile and antigen-specificity repertoire in Helicobacter pylori-specific T cell clones from the antrum of chronic gastritis patients with or without peptic ulcer. Eur J Immunol 27:1751–1755

Ding SZ, Zheng PY (2012) Helicobacter pylori infection induced gastric cancer; advance in gastric stem cell research and the remaining challenges. Gut Pathog 4:18

Ding SZ, Fischer W, Kaparakis-Liaskos M, Liechti G, Merrell DS, Grant PA, Ferrero RL, Crowe SE, Haas R, Hatakeyama M, Goldberg JB (2010) Helicobacter pylori-induced histone modification, associated gene expression in gastric epithelial cells, and its implication in pathogenesis. PLoS One 5, E9875

Dixon MF, Genta RM, Yardley JH, Correa P (1996) Classification and grading of gastritis. The updated Sydney System. International workshop on the histopathology of gastritis, Houston 1994. Am J Surg Pathol 20:1161–1181

Dolcetti L, Marigo I, Mantelli B, Peranzoni E, Zanovello P, Bronte V (2008) Myeloid-derived suppressor cell role in tumor-related inflammation. Cancer Lett 267:216–225

Dubois A, Berg DE (1997) The nonhuman primate model for H. pylori infection. Methods Mol Med 8:253–269

Dubois A, Berg DE, Incecik ET, Fiala N, Heman-Ackah LM, Perez-Perez GI, Blaser MJ (1996) Transient and persistent experimental infection of nonhuman primates with Helicobacter pylori: implications for human disease. Infect Immun 64:2885–2891

Eaton KA, Mefford M, Thevenot T (2001) The role of T cell subsets and cytokines in the pathogenesis of Helicobacter pylori gastritis in mice. J Immunol 166:7456–7461

Elfvin A, Bolin I, Von Bothmer C, Stolte M, Watanabe H, Fandriks L, Vieth M (2005) Helicobacter pylori induces gastritis and intestinal metaplasia but no gastric adenocarcinoma in Mongolian gerbils. Scand J Gastroenterol 40:1313–1320

El-Omar EM, Penman ID, Ardill JE, Chittajallu RS, Howie C, Mccoll KE (1995) Helicobacter pylori infection and abnormalities of acid secretion in patients with duodenal ulcer disease. Gastroenterology 109:681–691

El-Omar EM, Oien K, El-Nujumi A, Gillen D, Wirz A, Dahill S, Williams C, Ardill JE, Mccoll KE (1997) Helicobacter pylori infection and chronic gastric acid hyposecretion. Gastroenterology 113:15–24

El-Omar EM, Rabkin CS, Gammon MD, Vaughan TL, Risch HA, Schoenberg JB, Stanford JL, Mayne ST, Goedert J, Blot WJ, Fraumeni JF Jr, Chow WH (2003) Increased risk of noncardia gastric cancer associated with proinflammatory cytokine gene polymorphisms. Gastroenterology 124:1193–1201

Felici A, Giubellino A, Bottaro DP (2010) Gab1 mediates hepatocyte growth factor-stimulated mitogenicity and morphogenesis in multipotent myeloid cells. J Cell Biochem 111:310–321

Ferrero RL, Wilson JE, Sutton P (2012) Mouse models of Helicobacter-induced gastric cancer: use of cocarcinogens. Methods Mol Biol 921:157–173

Figueiredo C, Machado JC, Pharoah P, Seruca R, Sousa S, Carvalho R, Capelinha AF, Quint W, Caldas C, Van Doorn LJ, Carneiro F, Sobrinho-Simoes M (2002) Helicobacter pylori and interleukin 1 genotyping: an opportunity to identify high-risk individuals for gastric carcinoma. J Natl Cancer Inst 94:1680–1687

Fischer W, Prassl S, Haas R (2009) Virulence mechanisms and persistence strategies of the human gastric pathogen Helicobacter pylori. Curr Top Microbiol Immunol 337:129–171

Floor S, Van Staveren WC, Larsimont D, Dumont JE, Maenhaut C (2011) Cancer cells in epithelial-to-mesenchymal transition and tumor-propagating-cancer stem cells: distinct, overlapping or same populations. Oncogene 30:4609–4621

Fox JG, Wang TC (2007) Inflammation, atrophy, and gastric cancer. J Clin Invest 117:60–69

Fox JG, Sheppard BJ, Dangler CA, Whary MT, Ihrig M, Wang TC (2002) Germ-line p53-targeted disruption inhibits helicobacter-induced premalignant lesions and invasive gastric carcinoma through down-regulation of Th1 proinflammatory responses. Cancer Res 62:696–702

Fox JG, Rogers AB, Ihrig M, Taylor NS, Whary MT, Dockray G, Varro A, Wang TC (2003a) Helicobacter pylori-associated gastric cancer in INS-GAS mice is gender specific. Cancer Res 63:942–950

Fox JG, Wang TC, Rogers AB, Poutahidis T, Ge Z, Taylor N, Dangler CA, Israel DA, Krishna U, Gaus K, Peek RM Jr (2003b) Host and microbial constituents influence Helicobacter pylori-induced cancer in a murine model of hypergastrinemia. Gastroenterology 124:1879–1890

Franco AT, Israel DA, Washington MK, Krishna U, Fox JG, Rogers AB, Neish AS, Collier-Hyams L, Perez-Perez GI, Hatakeyama M, Whitehead R, Gaus K, O'brien DP, Romero-Gallo J, Peek RM Jr (2005) Activation of beta-catenin by carcinogenic Helicobacter pylori. Proc Natl Acad Sci U S A 102:10646–10651

Franco AT, Johnston E, Krishna U, Yamaoka Y, Israel DA, Nagy TA, Wroblewski LE, Piazuelo MB, Correa P, Peek RM Jr (2008) Regulation of gastric carcinogenesis by Helicobacter pylori virulence factors. Cancer Res 68:379–387

Frenck RW Jr, Clemens J (2003) Helicobacter in the developing world. Microbes Infect 5:705–713

Futagami S, Suzuki K, Hiratsuka T, Shindo T, Hamamoto T, Tatsuguchi A, Ueki N, Shinji Y, Kusunoki M, Wada K, Miyake K, Gudis K, Tsukui T, Sakamoto C (2006) Celecoxib inhibits Cdx2 expression and prevents gastric cancer in Helicobacter pylori-infected Mongolian gerbils. Digestion 74:187–198

Georgopoulos SD, Papastergiou V, Karatapanis S (2013) Current options for the treatment of Helicobacter pylori. Expert Opin Pharmacother 14:211–223

Gerhard M, Lehn N, Neumayer N, Boren T, Rad R, Schepp W, Miehlke S, Classen M, Prinz C (1999) Clinical relevance of the Helicobacter pylori gene for blood-group antigen-binding adhesin. Proc Natl Acad Sci U S A 96:12778–12783

Ghildiyal M, Zamore PD (2009) Small silencing RNAs: an expanding universe. Nat Rev Genet 10:94–108

Gobert AP, Mcgee DJ, Akhtar M, Mendz GL, Newton JC, Cheng Y, Mobley HL, Wilson KT (2001) Helicobacter pylori arginase inhibits nitric oxide production by eukaryotic cells: a strategy for bacterial survival. Proc Natl Acad Sci U S A 98:13844–13849

Gobert AP, Mersey BD, Cheng Y, Blumberg DR, Newton JC, Wilson KT (2002) Cutting edge: urease release by Helicobacter pylori stimulates macrophage inducible nitric oxide synthase. J Immunol 168:6002–6006

Gong M, Ling SS, Lui SY, Yeoh KG, Ho B (2010) Helicobacter pylori gamma-glutamyl transpeptidase is a pathogenic factor in the development of peptic ulcer disease. Gastroenterology 139:564–573

Guilford P, Hopkins J, Harraway J, Mcleod M, Mcleod N, Harawira P, Taite H, Scoular R, Miller A, Reeve AE (1998) E-cadherin germline mutations in familial gastric cancer. Nature 392:402–405

Guiney DG, Hasegawa P, Cole SP (2003) Helicobacter pylori preferentially induces interleukin 12 (IL-12) rather than IL-6 or IL-10 in human dendritic cells. Infect Immun 71:4163–4166

Guo XB, Guo L, Zhi QM, Ji J, Jiang JL, Zhang RJ, Zhang JN, Zhang J, Chen XH, Cai Q, Li JF, Yan M, Gu QL, Liu BY, Zhu ZG, Yu YY (2011) Helicobacter pylori induces promoter hypermethylation and downregulates gene expression of IRX1 transcription factor on human gastric mucosa. J Gastroenterol Hepatol 26:1685–1690

Guo L, Liu K, Zhao W, Li X, Li T, Tang F, Zhang R, Wu W, Xi T (2013) Immunological features and efficacy of the reconstructed epitope vaccine CtUBE against Helicobacter pylori infection in BALB/c mice model. Appl Microbiol Biotechnol 97:2367–2378

Haeberle HA, Kubin M, Bamford KB, Garofalo R, Graham DY, El-Zaatari F, Karttunen R, Crowe SE, Reyes VE, Ernst PB (1997) Differential stimulation of interleukin-12 (IL-12) and IL-10 by live and killed Helicobacter pylori in vitro and association of IL-12 production with gamma interferon-producing T cells in the human gastric mucosa. Infect Immun 65:4229–4235

Hafsi N, Voland P, Schwendy S, Rad R, Reindl W, Gerhard M, Prinz C (2004) Human dendritic cells respond to Helicobacter pylori, promoting NK cell and Th1-effector responses in vitro. J Immunol 173:1249–1257

Hagiwara T, Mukaisho K, Nakayama T, Sugihara H, Hattori T (2011) Long-term proton pump inhibitor administration worsens atrophic corpus gastritis and promotes adenocarcinoma development in Mongolian gerbils infected with Helicobacter pylori. Gut 60:624–630

Hahm KB, Lee KM, Kim YB, Hong WS, Lee WH, Han SU, Kim MW, Ahn BO, Oh TY, Lee MH, Green J, Kim SJ (2002) Conditional loss of TGF-beta signalling leads to increased susceptibility to gastrointestinal carcinogenesis in mice. Aliment Pharmacol Ther 16(Suppl 2):115–127

Hahm KB, Song YJ, Oh TY, Lee JS, Surh YJ, Kim YB, Yoo BM, Kim JH, Han SU, Nahm KT, Kim MW, Kim DY, Cho SW (2003) Chemoprevention of Helicobacter pylori-associated gastric carcinogenesis in a mouse model: is it possible? J Biochem Mol Biol 36:82–94

Han SU, Kim YB, Joo HJ, Hahm KB, Lee WH, Cho YK, Kim DY, Kim MW (2002) Helicobacter pylori infection promotes gastric carcinogenesis in a mice model. J Gastroenterol Hepatol 17:253–261

Handa Y, Saitoh T, Kawaguchi M, Misaka R, Ohno H, Tsai CR, Tani Y, Tsurui M, Yoshida H, Morita S, Midorikawa S, Sanji T (1996) Association of Helicobacter pylori and diffuse type gastric cancer. J Gastroenterol 31(Suppl 9):29–32

Hardbower DM, Peek RM Jr, Wilson KT (2014) At the bench: Helicobacter pylori, dysregulated host responses, DNA damage, and gastric cancer. J Leukoc Biol 96(2):201–212

Harris PR, Wright SW, Serrano C, Riera F, Duarte I, Torres J, Pena A, Rollan A, Viviani P, Guiraldes E, Schmitz JM, Lorenz RG, Novak L, Smythies LE, Smith PD (2008) Helicobacter pylori gastritis in children is associated with a regulatory T-cell response. Gastroenterology 134:491–499

Hasni SA (2012) Role of Helicobacter pylori infection in autoimmune diseases. Curr Opin Rheumatol 24:429–434

Hatakeyama M (2014) Helicobacter pylori CagA and gastric cancer: a paradigm for hit-and-run carcinogenesis. Cell Host Microbe 15:306–316

Hayashi Y, Tsujii M, Wang J, Kondo J, Akasaka T, Jin Y, Li W, Nakamura T, Nishida T, Iijima H, Tsuji S, Kawano S, Hayashi N, Takehara T (2012) CagA mediates epigenetic regulation to attenuate let-7 expression in Helicobacter pylori-related carcinogenesis. Gut 62(11):1536–1546

Herrera V, Parsonnet J (2009) Helicobacter pylori and gastric adenocarcinoma. Clin Microbiol Infect 15:971–976

Higashi H, Tsutsumi R, Muto S, Sugiyama T, Azuma T, Asaka M, Hatakeyama M (2002) SHP-2 tyrosine phosphatase as an intracellular target of Helicobacter pylori CagA protein. Science 295:683–686

Hirayama F, Takagi S, Kusuhara H, Iwao E, Yokoyama Y, Ikeda Y (1996) Induction of gastric ulcer and intestinal metaplasia in Mongolian gerbils infected with Helicobacter pylori. J Gastroenterol 31:755–757

Hirayama F, Takagi S, Iwao E, Yokoyama Y, Haga K, Hanada S (1999) Development of poorly differentiated adenocarcinoma and carcinoid due to long-term Helicobacter pylori colonization in Mongolian gerbils. J Gastroenterol 34:450–454

Hitzler I, Oertli M, Becher B, Agger EM, Muller A (2011) Dendritic cells prevent rather than promote immunity conferred by a helicobacter vaccine using a mycobacterial adjuvant. Gastroenterology 141:186–196, 196 e1

Hitzler I, Kohler E, Engler DB, Yazgan AS, Muller A (2012) The role of Th cell subsets in the control of Helicobacter infections and in T cell-driven gastric immunopathology. Front Immunol 3:142

Honda S, Fujioka T, Tokieda M, Satoh R, Nishizono A, Nasu M (1998) Development of Helicobacter pylori-induced gastric carcinoma in Mongolian gerbils. Cancer Res 58:4255–4259

Hsu PI, Hsieh HL, Lee J, Lin LF, Chen HC, Lu PJ, Hsiao M (2009) Loss of RUNX3 expression correlates with differentiation, nodal metastasis, and poor prognosis of gastric cancer. Ann Surg Oncol 16:1686–1694

Hussain SA, Hamid S (2014) Helicobacter pylori in humans: where are we now? Adv Biomed Res 3:63

IARC (2012) A review of human carcinogens. Part B: biological agents. Iarc Monogr Eval Carcinog Risks Hum 100(b):1–441

Ii M, Yamamoto H, Adachi Y, Maruyama Y, Shinomura Y (2006) Role of matrix metalloproteinase-7 (matrilysin) in human cancer invasion, apoptosis, growth, and angiogenesis. Exp Biol Med (Maywood) 231:20–27

Ishijima N, Suzuki M, Ashida H, Ichikawa Y, Kanegae Y, Saito I, Boren T, Haas R, Sasakawa C, Mimuro H (2011) BabA-mediated adherence is a potentiator of the Helicobacter pylori type IV secretion system activity. J Biol Chem 286:25256–25264

Jenkins BJ, Grail D, Nheu T, Najdovska M, Wang B, Waring P, Inglese M, Mcloughlin RM, Jones SA, Topley N, Baumann H, Judd LM, Giraud AS, Boussioutas A, Zhu HJ, Ernst M (2005) Hyperactivation of Stat3 in gp130 mutant mice promotes gastric hyperproliferation and desensitizes TGF-beta signaling. Nat Med 11:845–852

Jimenez-Soto LF, Kutter S, Sewald X, Ertl C, Weiss E, Kapp U, Rohde M, Pirch T, Jung K, Retta SF, Terradot L, Fischer W, Haas R (2009) Helicobacter pylori type IV secretion apparatus exploits beta1 integrin in a novel RGD-independent manner. PloS Pathog 5, E1000684

Jones PA, Baylin SB (2007) The epigenomics of cancer. Cell 128:683–692

Jones KR, Whitmire JM, Merrell DS (2010) A tale of two toxins: Helicobacter pylori CagA and VacA modulate host pathways that impact disease. Front Microbiol 1:115

Kaebisch R, Mejias-Luque R, Prinz C, Gerhard M (2014) Helicobacter pylori cytotoxin-associated gene A impairs human dendritic cell maturation and function through IL-10-mediated activation of STAT3. J Immunol 192:316–323

Kalali B, Mejias-Luque R, Javaheri A, Gerhard M (2014) H. pylori virulence factors: influence on immune system and pathology. Mediat Inflamm 2014:426309

Kao JY, Zhang M, Miller MJ, Mills JC, Wang B, Liu M, Eaton KA, Zou W, Berndt BE, Cole TS, Takeuchi T, Owyang SY, Luther J (2010) Helicobacter pylori immune escape is mediated by dendritic cell-induced Treg skewing and Th17 suppression in mice. Gastroenterology 138:1046–1054

Kaparakis M, Turnbull L, Carneiro L, Firth S, Coleman HA, Parkington HC, Le Bourhis L, Karrar A, Viala J, Mak J, Hutton ML, Davies JK, Crack PJ, Hertzog PJ, Philpott DJ, Girardin SE, Whitchurch CB, Ferrero RL (2010) Bacterial membrane vesicles deliver peptidoglycan to NOD1 in epithelial cells. Cell Microbiol 12:372–385

Kaplan-Turkoz B, Jimenez-Soto LF, Dian C, Ertl C, Remaut H, Louche A, Tosi T, Haas R, Terradot L (2012) Structural insights into Helicobacter pylori oncoprotein CagA interaction with beta1 integrin. Proc Natl Acad Sci U S A 109:14640–14645

Karam SM (2008) Cellular origin of gastric cancer. Ann N Y Acad Sci 1138:162–168

Karttunen R, Karttunen T, Ekre HP, Macdonald TT (1995) Interferon gamma and interleukin 4 secreting cells in the gastric antrum in Helicobacter pylori positive and negative gastritis. Gut 36:341–345

Katayama Y, Takahashi M, Kuwayama H (2009) Helicobacter pylori causes runx3 gene methylation and its loss of expression in gastric epithelial cells, which is mediated by nitric oxide produced by macrophages. Biochem Biophys Res Commun 388:496–500

Kawazoe T, Sakagami T, Nakajima K, Hori K, Fukuda Y, Matsumoto T, Miwa H (2007) Role of bacterial strain diversity of Helicobacter pylori in gastric carcinogenesis induced by N-methyl-N-nitrosourea in Mongolian gerbils. Helicobacter 12:213–223

Kayhan B, Arasli M, Eren H, Aydemir S, Kayhan B, Aktas E, Tekin I (2008) Analysis of peripheral blood lymphocyte phenotypes and Th1/Th2 cytokines profile in the systemic immune responses of Helicobacter pylori infected individuals. Microbiol Immunol 52:531–538

Khader SA, Gaffen SL, Kolls JK (2009) Th17 cells at the crossroads of innate and adaptive immunity against infectious diseases at the mucosa. Mucosal Immunol 2:403–411

Kikuchi A, Yamamoto H, Sato A, Matsumoto S (2011) New insights into the mechanism of Wnt signaling pathway activation. Int Rev Cell Mol Biol 291:21–71

Kim DH, Kim SW, Song YJ, Oh TY, Han SU, Kim YB, Joo HJ, Cho YK, Kim DY, Cho SW, Kim MW, Kim JH, Hahm KB (2003) Long-term evaluation of mice model infected with Helicobacter pylori: focus on gastric pathology including gastric cancer. Aliment Pharmacol Ther 18(Suppl 1):14–23

Kim YG, Park JH, Shaw MH, Franchi L, Inohara N, Nunez G (2008) The cytosolic sensors Nod1 and Nod2 are critical for bacterial recognition and host defense after exposure to Toll-like receptor ligands. Immunity 28:246–257

Kim JM, Kim JS, Yoo DY, Ko SH, Kim N, Kim H, Kim YJ (2011) Stimulation of dendritic cells with Helicobacter pylori vacuolating cytotoxin negatively regulates their maturation via the restoration of E2F1. Clin Exp Immunol 166:34–45

Konturek PC, Konturek SJ, Brzozowski T (2009) Helicobacter pylori infection in gastric carcinogenesis. J Physiol Pharmacol 60:3–21

Kouzarides T (2007) Chromatin modifications and their function. Cell 128:693–705

Kranzer K, Eckhardt A, Aigner M, Knoll G, Deml L, Speth C, Lehn N, Rehli M, Schneider-Brachert W (2004) Induction of maturation and cytokine release of human dendritic cells by Helicobacter pylori. Infect Immun 72:4416–4423

Kuipers EJ, Uyterlinde AM, Pena AS, Roosendaal R, Pals G, Nelis GF, Festen HP, Meuwissen SG (1995) Long-term sequelae of Helicobacter pylori gastritis. Lancet 345:1525–1528

Kurashima Y, Murata-Kamiya N, Kikuchi K, Higashi H, Azuma T, Kondo S, Hatakeyama M (2008) Deregulation of beta-catenin signal by Helicobacter pylori CagA requires the CagA-multimerization sequence. Int J Cancer 122:823–831

Kwok T, Zabler D, Urman S, Rohde M, Hartig R, Wessler S, Misselwitz R, Berger J, Sewald N, Konig W, Backert S (2007) Helicobacter exploits integrin for type IV secretion and kinase activation. Nature 449:862–866

Lee A, Chen M, Coltro N, O'rourke J, Hazell S, Hu P, Li Y (1993) Long term infection of the gastric mucosa with Helicobacter species does induce atrophic gastritis in an animal model of Helicobacter pylori infection. Zentralbl Bakteriol 280:38–50

Lee K, Hwang H, Nam KT (2014) Immune response and the tumor microenvironment: how they communicate to regulate gastric cancer. Gut Liver 8:131–139

Li QL, Ito K, Sakakura C, Fukamachi H, Inoue K, Chi XZ, Lee KY, Nomura S, Lee CW, Han SB, Kim HM, Kim WJ, Yamamoto H, Yamashita N, Yano T, Ikeda T, Itohara S, Inazawa J, Abe T, Hagiwara A, Yamagishi H, Ooe A, Kaneda A, Sugimura T, Ushijima T, Bae SC, Ito Y (2002) Causal relationship between the loss of RUNX3 expression and gastric cancer. Cell 109:113–124

Lin JC, Jeong S, Liang G, Takai D, Fatemi M, Tsai YC, Egger G, Gal-Yam EN, Jones PA (2007) Role of nucleosomal occupancy in the epigenetic silencing of the MLH1 CpG island. Cancer Cell 12:432–444

Lin WC, Tsai HF, Kuo SH, Wu MS, Lin CW, Hsu PI, Cheng AL, Hsu PN (2010) Translocation of Helicobacter pylori CagA into Human B lymphocytes, the origin of mucosa-associated lymphoid tissue lymphoma. Cancer Res 70:5740–5748

Link A, Kupcinskas J, Wex T, Malfertheiner P (2012) Macro-role of microRNA in gastric cancer. Dig Dis 30:255–267

Linz B, Balloux F, Moodley Y, Manica A, Liu H, Roumagnac P, Falush D, Stamer C, Prugnolle F, Van Der Merwe SW, Yamaoka Y, Graham DY, Perez-Trallero E, Wadstrom T, Suerbaum S, Achtman M (2007) An African origin for the intimate association between humans and Helicobacter pylori. Nature 445:915–918

Liu H, Merrell DS, Semino-Mora C, Goldman M, Rahman A, Mog S, Dubois A (2009) Diet synergistically affects Helicobacter pylori-induced gastric carcinogenesis in nonhuman primates. Gastroenterology 137:1367–1379 e1–6

Lundgren A, Suri-Payer E, Enarsson K, Svennerholm AM, Lundin BS (2003) Helicobacter pylori-specific CD4+ CD25 high regulatory T cells suppress memory T-cell responses to H. pylori in infected individuals. Infect Immun 71:1755–1762

Maekita T, Nakazawa K, Mihara M, Nakajima T, Yanaoka K, Iguchi M, Arii K, Kaneda A, Tsukamoto T, Tatematsu M, Tamura G, Saito D, Sugimura T, Ichinose M, Ushijima T (2006) High levels of aberrant DNA methylation in Helicobacter pylori-infected gastric mucosae and its possible association with gastric cancer risk. Clin Cancer Res 12:989–995

Mahdavi J, Sonden B, Hurtig M, Olfat FO, Forsberg L, Roche N, Angstrom J, Larsson T, Teneberg S, Karlsson KA, Altraja S, Wadstrom T, Kersulyte D, Berg DE, Dubois A, Petersson C, Magnusson KE, Norberg T, Lindh F, Lundskog BB, Arnqvist A, Hammarstrom L, Boren T (2002) Helicobacter pylori SabA adhesin in persistent infection and chronic inflammation. Science 297:573–578

Malfertheiner P (2011) The intriguing relationship of Helicobacter pylori infection and acid secretion in peptic ulcer disease and gastric cancer. Dig Dis 29:459–464

Malfertheiner P, Schultze V, Rosenkranz B, Kaufmann SH, Ulrichs T, Novicki D, Norelli F, Contorni M, Peppoloni S, Berti D, Tornese D, Ganju J, Palla E, Rappuoli R, Scharschmidt BF, Del Giudice G (2008) Safety and immunogenicity of an intramuscular Helicobacter pylori vaccine in noninfected volunteers: a phase I study. Gastroenterology 135:787–795

Marigo I, Dolcetti L, Serafini P, Zanovello P, Bronte V (2008) Tumor-induced tolerance and immune suppression by myeloid derived suppressor cells. Immunol Rev 222:162–179

Markovska R, Boyanova L, Yordanov D, Gergova G, Mitov I (2011) Helicobacter pylori oipA genetic diversity and its associations with both disease and cagA, vacA s, m, and i alleles among Bulgarian patients. Diagn Microbiol Infect Dis 71:335–340

Marshall BJ, Warren JR (1984) Unidentified curved bacilli in the stomach of patients with gastritis and peptic ulceration. Lancet 1:1311–1315

Marshall BJ, Mcgechie DB, Francis GJ, Utley PJ (1984) Pyloric campylobacter serology. Lancet 2:281

Maruta F, Ota H, Genta RM, Sugiyama A, Tatematsu M, Katsuyama T, Kawasaki S (2001) Role of N-methyl-N-nitrosourea in the induction of intestinal metaplasia and gastric adenocarcinoma in Mongolian gerbils infected with Helicobacter pylori. Scand J Gastroenterol 36:283–290

Matozaki T, Murata Y, Saito Y, Okazawa H, Ohnishi H (2009) Protein tyrosine phosphatase SHP-2: a proto-oncogene product that promotes Ras activation. Cancer Sci 100:1786–1793

Mcconkey DJ, Choi W, Marquis L, Martin F, Williams MB, Shah J, Svatek R, Das A, Adam L, Kamat A, Siefker-Radtke A, Dinney C (2009) Role of epithelial-to-mesenchymal transition (EMT) in drug sensitivity and metastasis in bladder cancer. Cancer Metastasis Rev 28:335–344

Mcnamara D, El-Omar E (2008) Helicobacter pylori infection and the pathogenesis of gastric cancer: a paradigm for host-bacterial interactions. Dig Liver Dis 40:504–509

Meyer F, Wilson KT, James SP (2000) Modulation of innate cytokine responses by products of Helicobacter pylori. Infect Immun 68:6265–6272
Mimuro H, Suzuki T, Tanaka J, Asahi M, Haas R, Sasakawa C (2002) Grb2 is a key mediator of Helicobacter pylori CagA protein activities. Mol Cell 10:745–755
Molinari M, Salio M, Galli C, Norais N, Rappuoli R, Lanzavecchia A, Montecucco C (1998) Selective inhibition of Ii-dependent antigen presentation by Helicobacter pylori toxin VacA. J Exp Med 187:135–140
Monteleone G, Del Vecchio Blanco G, Palmieri G, Vavassori P, Monteleone I, Colantoni A, Battista S, Spagnoli LG, Romano M, Borrelli M, Macdonald TT, Pallone F (2004) Induction and regulation of Smad7 in the gastric mucosa of patients with Helicobacter pylori infection. Gastroenterology 126:674–682
Moran AP, Lindner B, Walsh EJ (1997) Structural characterization of the lipid A component of Helicobacter pylori rough- and smooth-form lipopolysaccharides. J Bacteriol 179:6453–6463
Mueller A, Merrell DS, Grimm J, Falkow S (2004) Profiling of microdissected gastric epithelial cells reveals a cell type-specific response to Helicobacter pylori infection. Gastroenterology 127:1446–1462
Muller A, Oertli M, Arnold IC (2011) H. pylori exploits and manipulates innate and adaptive immune cell signaling pathways to establish persistent infection. Cell Commun Signal 9:25
Murakami K, Fujioka T, Kodama M, Honda S, Okimoto T, Oda T, Nishizono A, Sato R, Kubota T, Kagawa J, Nasu M (2002) Analysis of p53 mutations and Helicobacter pylori infection in human and animal models. J Gastroenterol 37(Suppl 13):1–5
Murata-Kamiya N, Kurashima Y, Teishikata Y, Yamahashi Y, Saito Y, Higashi H, Aburatani H, Akiyama T, Peek RM Jr, Azuma T, Hatakeyama M (2007) Helicobacter pylori CagA interacts with E-cadherin and deregulates the beta-catenin signal that promotes intestinal transdifferentiation in gastric epithelial cells. Oncogene 26:4617–4626
Murata-Kamiya N, Kikuchi K, Hayashi T, Higashi H, Hatakeyama M (2010) Helicobacter pylori exploits host membrane phosphatidylserine for delivery, localization, and pathophysiological action of the CagA oncoprotein. Cell Host Microbe 7:399–411
Nadarajan N, Balasubramanian LK, Kuppannan S, Ramachandran C, Leelakrishnan V (2013) Runt-related transcription factor 3: single nucleotide polymorphism rs760805, gene expression, and methylation status in Helicobacter pylori -infected patients for determination of gastric cancer risk. J Gastrointest Cancer 44:444–449
Nagy TA, Frey MR, Yan F, Israel DA, Polk DB, Peek RM Jr (2009) Helicobacter pylori regulates cellular migration and apoptosis by activation of phosphatidylinositol 3-kinase signaling. J Infect Dis 199:641–651
Naito M, Yamazaki T, Tsutsumi R, Higashi H, Onoe K, Yamazaki S, Azuma T, Hatakeyama M (2006) Influence of EPIYA-repeat polymorphism on the phosphorylation-dependent biological activity of Helicobacter pylori CagA. Gastroenterology 130:1181–1190
Nakamura Y, Sakagami T, Yamamoto N, Yokota Y, Koizuka H, Hori K, Fukuda Y, Tanida N, Kobayashi T, Shimoyama T (2002) Helicobacter pylori does not promote N-methyl-N-nitrosourea-induced gastric carcinogenesis in SPF C57BL/6 mice. Jpn J Cancer Res 93:111–116
Nam KT, Oh SY, Ahn B, Kim YB, Jang DD, Yang KH, Hahm KB, Kim DY (2004) Decreased Helicobacter pylori associated gastric carcinogenesis in mice lacking inducible nitric oxide synthase. Gut 53:1250–1255
Nardone G, Compare D, De Colibus P, De Nucci G, Rocco A (2007) Helicobacter pylori and epigenetic mechanisms underlying gastric carcinogenesis. Dig Dis 25:225–229
Necchi V, Candusso ME, Tava F, Luinetti O, Ventura U, Fiocca R, Ricci V, Solcia E (2007) Intracellular, intercellular, and stromal invasion of gastric mucosa, preneoplastic lesions, and cancer by Helicobacter pylori. Gastroenterology 132:1009–1023
Nesic D, Miller MC, Quinkert ZT, Stein M, Chait BT, Stebbins CE (2010) Helicobacter pylori CagA inhibits PAR1-MARK family kinases by mimicking host substrates. Nat Struct Mol Biol 17:130–132

Nickel A, Stadler SC (2015) Role of epigenetic mechanisms in epithelial-to-mesenchymal transition of breast cancer cells. Transl Res 165(1):126–42
Niehrs C (2012) The complex world of WNT receptor signalling. Nat Rev Mol Cell Biol 13:767–779
Niv Y, Hazazi R, Waked A, Lederfein T, Achiel K (2008) Helicobacter pylori recurrence and infection rate in Israeli adults. Dig Dis Sci 53:1211–1214
Noto JM, Peek RM (2011) The role of microRNAs in Helicobacter pylori pathogenesis and gastric carcinogenesis. Front Cell Infect Microbiol 1:21
Nozaki K, Shimizu N, Ikehara Y, Inoue M, Tsukamoto T, Inada K, Tanaka H, Kumagai T, Kaminishi M, Tatematsu M (2003) Effect of early eradication on Helicobacter pylori-related gastric carcinogenesis in Mongolian gerbils. Cancer Sci 94:235–239
Odenbreit S, Puls J, Sedlmaier B, Gerland E, Fischer W, Haas R (2000) Translocation of Helicobacter pylori CagA into gastric epithelial cells by type IV secretion. Science 287:1497–1500
Oertli M, Muller A (2012) Helicobacter pylori targets dendritic cells to induce immune tolerance, promote persistence and confer protection against allergic asthma. Gut Microbes 3:566–571
Oertli M, Sundquist M, Hitzler I, Engler DB, Arnold IC, Reuter S, Maxeiner J, Hansson M, Taube C, Quiding-Jarbrink M, Muller A (2012) DC-derived IL-18 drives Treg differentiation, murine Helicobacter pylori-specific immune tolerance, and asthma protection. J Clin Invest 122:1082–1096
Oertli M, Noben M, Engler DB, Semper RP, Reuter S, Maxeiner J, Gerhard M, Taube C, Muller A (2013) Helicobacter pylori gamma-glutamyl transpeptidase and vacuolating cytotoxin promote gastric persistence and immune tolerance. Proc Natl Acad Sci U S A 110:3047–3052
Oguma K, Oshima H, Aoki M, Uchio R, Naka K, Nakamura S, Hirao A, Saya H, Taketo MM, Oshima M (2008) Activated macrophages promote Wnt signalling through tumour necrosis factor-alpha in gastric tumour cells. Embo J 27:1671–1681
Ogura K, Maeda S, Nakao M, Watanabe T, Tada M, Kyutoku T, Yoshida H, Shiratori Y, Omata M (2000) Virulence factors of Helicobacter pylori responsible for gastric diseases in Mongolian gerbil. J Exp Med 192:1601–1610
Oh JD, Karam SM, Gordon JI (2005) Intracellular Helicobacter pylori in gastric epithelial progenitors. Proc Natl Acad Sci U S A 102:5186–5191
Ohnishi N, Yuasa H, Tanaka S, Sawa H, Miura M, Matsui A, Higashi H, Musashi M, Iwabuchi K, Suzuki M, Yamada G, Azuma T, Hatakeyama M (2008) Transgenic expression of Helicobacter pylori CagA induces gastrointestinal and hematopoietic neoplasms in mouse. Proc Natl Acad Sci U S A 105:1003–1008
Ohno T, Vallstrom A, Rugge M, Ota H, Graham DY, Arnqvist A, Yamaoka Y (2011) Effects of blood group antigen-binding adhesin expression during Helicobacter pylori infection of Mongolian gerbils. J Infect Dis 203:726–735
Oliveira MJ, Costa AM, Costa AC, Ferreira RM, Sampaio P, Machado JC, Seruca R, Mareel M, Figueiredo C (2009) CagA associates with c-Met, E-cadherin, and p120-catenin in a multiproteic complex that suppresses Helicobacter pylori-induced cell-invasive phenotype. J Infect Dis 200:745–755
Ong SP, Duggan A (2004) Eradication of Helicobacter pylori in clinical situations. Clin Exp Med 4:30–38
Osawa H, Sugano K, Iwamori M, Kawakami M, Tada M, Nakao M (2001) Comparative analysis of colonization of Helicobacter pylori and glycolipids receptor density in Mongolian gerbils and mice. Dig Dis Sci 46:69–74
Otaka M, Konishi N, Odashima M, Jin M, Wada I, Matsuhashi T, Horikawa Y, Ohba R, Watanabe S (2006) Is Mongolian gerbil really adequate host animal for study of Helicobacter pylori infection-induced gastritis and cancer? Biochem Biophys Res Commun 347:297–300
Ouyang G, Wang Z, Fang X, Liu J, Yang CJ (2010) Molecular signaling of the epithelial to mesenchymal transition in generating and maintaining cancer stem cells. Cell Mol Life Sci 67:2605–2618

Palframan SL, Kwok T, Gabriel K (2012) Vacuolating cytotoxin A (VacA), a key toxin for Helicobacter pylori pathogenesis. Front Cell Infect Microbiol 2:92

Papadakos KS, Sougleri IS, Mentis AF, Hatziloukas E, Sgouras DN (2013) Presence of terminal EPIYA phosphorylation motifs in Helicobacter pylori CagA contributes to IL-8 secretion, irrespective of the number of repeats. PLoS One 8:e56291

Parkin DM (2004) International variation. Oncogene 23:6329–6340

Patel MK, Trombly MI, Kurt-Jones EA (2012) Innate immune responses to Helicobacter pylori infection: an overview. Methods Mol Biol 921:205–207

Pellicano R, Franceschi F, Saracco G, Fagoonee S, Roccarina D, Gasbarrini A (2009) Helicobacters and extragastric diseases. Helicobacter 14(Suppl 1):58–68

Peterson AJ, Menheniott TR, O'connor L, Walduck AK, Fox JG, Kawakami K, Minamoto T, Ong EK, Wang TC, Judd LM, Giraud AS (2010) Helicobacter pylori infection promotes methylation and silencing of trefoil factor 2, leading to gastric tumor development in mice and humans. Gastroenterology 139:2005–2017

Piazzi G, Fini L, Selgrad M, Garcia M, Daoud Y, Wex T, Malfertheiner P, Gasbarrini A, Romano M, Meyer RL, Genta RM, Fox JG, Boland CR, Bazzoli F, Ricciardiello L (2011) Epigenetic regulation of Delta-Like1 controls Notch1 activation in gastric cancer. Oncotarget 2:1291–1301

Polk DB, Peek RM Jr (2010) Helicobacter pylori: gastric cancer and beyond. Nat Rev Cancer 10:403–414

Poppe M, Feller SM, Romer G, Wessler S (2007) Phosphorylation of Helicobacter pylori CagA by c-Abl leads to cell motility. Oncogene 26:3462–3472

Prinz C, Schoniger M, Rad R, Becker I, Keiditsch E, Wagenpfeil S, Classen M, Rosch T, Schepp W, Gerhard M (2001) Key importance of the Helicobacter pylori adherence factor blood group antigen binding adhesin during chronic gastric inflammation. Cancer Res 61:1903–1909

Qu Y, Dang S, Hou P (2013) Gene methylation in gastric cancer. Clin Chim Acta 424:53–65

Rad R, Gerhard M, Lang R, Schoniger M, Rosch T, Schepp W, Becker I, Wagner H, Prinz C (2002) The Helicobacter pylori blood group antigen-binding adhesin facilitates bacterial colonization and augments a nonspecific immune response. J Immunol 168:3033–3041

Rad R, Brenner L, Bauer S, Schwendy S, Layland L, Da Costa CP, Reindl W, Dossumbekova A, Friedrich M, Saur D, Wagner H, Schmid RM, Prinz C (2006) CD25+/Foxp3+ T cells regulate gastric inflammation and Helicobacter pylori colonization in vivo. Gastroenterology 131:525–537

Raju D, Hussey S, Ang M, Terebiznik MR, Sibony M, Galindo-Mata E, Gupta V, Blanke SR, Delgado A, Romero-Gallo J, Ramjeet MS, Mascarenhas H, Peek RM, Correa P, Streutker C, Hold G, Kunstmann E, Yoshimori T, Silverberg MS, Girardin SE, Philpott DJ, El Omar E, Jones NL (2012) Vacuolating cytotoxin and variants in Atg16L1 that disrupt autophagy promote Helicobacter pylori infection in humans. Gastroenterology 142:1160–1171

Rieder G, Merchant JL, Haas R (2005) Helicobacter pylori cag-type IV secretion system facilitates corpus colonization to induce precancerous conditions in Mongolian gerbils. Gastroenterology 128:1229–1242

Rimbara E, Mori S, Kim H, Shibayama K (2013) Role of gamma-glutamyltranspeptidase in the pathogenesis of Helicobacter pylori infection. Microbiol Immunol 57:665–673

Rittig MG, Shaw B, Letley DP, Thomas RJ, Argent RH, Atherton JC (2003) Helicobacter pylori-induced homotypic phagosome fusion in human monocytes is independent of the bacterial vacA and cag status. Cell Microbiol 5:887–899

Robinson K, Kenefeck R, Pidgeon EL, Shakib S, Patel S, Polson RJ, Zaitoun AM, Atherton JC (2008) Helicobacter pylori-induced peptic ulcer disease is associated with inadequate regulatory T cell responses. Gut 57:1375–1385

Rocha AM, Botelho LF, Rocha DM (2014) Improvement of thrombocytopenia after treatment for Helicobacter pylori in a patient with immunologic thrombocytopenic purpura. Rev Bras Hematol Hemoter 36:162–164

Rogers AB, Taylor NS, Whary MT, Stefanich ED, Wang TC, Fox JG (2005) Helicobacter pylori but not high salt induces gastric intraepithelial neoplasia in B6129 mice. Cancer Res 65:10709–10715

Rothenbacher D, Brenner H (2003) Burden of Helicobacter pylori and H. pylori-related diseases in developed countries: recent developments and future implications. Microbes Infect 5:693–703
Ruggiero P, Alloueche A, Rappuoli R, Del Giudice G (2004) Models for bacterial infectious diseases: Helicobacter pylori. Drug Discov Today Dis Models 1:95–100
Saadat I, Higashi H, Obuse C, Umeda M, Murata-Kamiya N, Saito Y, Lu H, Ohnishi N, Azuma T, Suzuki A, Ohno S, Hatakeyama M (2007) Helicobacter pylori CagA targets PAR1/MARK kinase to disrupt epithelial cell polarity. Nature 447:330–333
Saito H, Yamaoka Y, Ishizone S, Maruta F, Sugiyama A, Graham DY, Yamauchi K, Ota H, Miyagawa S (2005) Roles of virD4 and cagG genes in the cag pathogenicity island of Helicobacter pylori using a Mongolian gerbil model. Gut 54:584–590
Saito Y, Murata-Kamiya N, Hirayama T, Ohba Y, Hatakeyama M (2010) Conversion of Helicobacter pylori CagA from senescence inducer to oncogenic driver through polarity-dependent regulation of p21. J Exp Med 207:2157–2174
Saito-Diaz K, Chen TW, Wang X, Thorne CA, Wallace HA, Page-Mccaw A, Lee E (2013) The way Wnt works: components and mechanism. Growth Factors 31:1–31
Salama NR, Otto G, Tompkins L, Falkow S (2001) Vacuolating cytotoxin of Helicobacter pylori plays a role during colonization in a mouse model of infection. Infect Immun 69:730–736
Salama NR, Hartung ML, Muller A (2013) Life in the human stomach: persistence strategies of the bacterial pathogen Helicobacter pylori. Nat Rev Microbiol 11:385–399
Schones DE, Cui K, Cuddapah S, Roh TY, Barski A, Wang Z, Wei G, Zhao K (2008) Dynamic regulation of nucleosome positioning in the human genome. Cell 132:887–898
Segal ED, Cha J, Lo J, Falkow S, Tompkins LS (1999) Altered states: involvement of phosphorylated CagA in the induction of host cellular growth changes by Helicobacter pylori. Proc Natl Acad Sci U S A 96:14559–14564
Selgrad M, Bornschein J, Rokkas T, Malfertheiner P (2012) Helicobacter pylori: gastric cancer and extragastric intestinal malignancies. Helicobacter 17(Suppl 1):30–35
Sepulveda AR (2013) Helicobacter, inflammation, and gastric cancer. Curr Pathobiol Rep 1:9–18
Sepulveda AR, Yao Y, Yan W, Park DI, Kim JJ, Gooding W, Abudayyeh S, Graham DY (2010) CpG methylation and reduced expression of O6-methylguanine DNA methyltransferase is associated with Helicobacter pylori infection. Gastroenterology 138:1836–1844
Sewald X, Gebert-Vogl B, Prassl S, Barwig I, Weiss E, Fabbri M, Osicka R, Schiemann M, Busch DH, Semmrich M, Holzmann B, Sebo P, Haas R (2008) Integrin subunit CD18 Is the T-lymphocyte receptor for the Helicobacter pylori vacuolating cytotoxin. Cell Host Microbe 3:20–29
Shigematsu Y, Niwa T, Rehnberg E, Toyoda T, Yoshida S, Mori A, Wakabayashi M, Iwakura Y, Ichinose M, Kim YJ, Ushijima T (2013) Interleukin-1beta induced by Helicobacter pylori infection enhances mouse gastric carcinogenesis. Cancer Lett 340:141–147
Shimizu N, Kaminishi M, Tatematsu M, Tsuji E, Yoshikawa A, Yamaguchi H, Aoki F, Oohara T (1998) Helicobacter pylori promotes development of pepsinogen-altered pyloric glands, a preneoplastic lesion of glandular stomach of BALB/c mice pretreated with N-methyl-N-nitrosourea. Cancer Lett 123:63–69
Shimizu N, Inada KI, Tsukamoto T, Nakanishi H, Ikehara Y, Yoshikawa A, Kaminishi M, Kuramoto S, Tatematsu M (1999) New animal model of glandular stomach carcinogenesis in Mongolian gerbils infected with Helicobacter pylori and treated with a chemical carcinogen. J Gastroenterol 34(Suppl 11):61–66
Sipponen P (1989) Atrophic gastritis as a premalignant condition. Ann Med 21:287–290
Stadler SC, Allis CD (2012) Linking epithelial-to-mesenchymal-transition and epigenetic modifications. Semin Cancer Biol 22:404–410
Stein M, Rappuoli R, Covacci A (2000) Tyrosine phosphorylation of the Helicobacter pylori CagA antigen after cag-driven host cell translocation. Proc Natl Acad Sci U S A 97:1263–1268
Stein M, Bagnoli F, Halenbeck R, Rappuoli R, Fantl WJ, Covacci A (2002) c-Src/Lyn kinases activate Helicobacter pylori CagA through tyrosine phosphorylation of the EPIYA motifs. Mol Microbiol 43:971–980

Stein M, Ruggiero P, Rappuoli R, Bagnoli F (2013) CagA: from pathogenic mechanisms to its use as an anti-cancer vaccine. Front Immunol 4:328

Strowig T, Henao-Mejia J, Elinav E, Flavell R (2012) Inflammasomes in health and disease. Nature 481:278–286

Suerbaum S (1995) The complex flagella of gastric Helicobacter species. Trends Microbiol 3:168–170; Discussion 170–171

Sugimoto M, Ohno T, Graham DY, Yamaoka Y (2009a) Gastric mucosal interleukin-17 and −18 mRNA expression in Helicobacter pylori-induced Mongolian gerbils. Cancer Sci 100:2152–2159

Sugimoto M, Wu JY, Abudayyeh S, Hoffman J, Brahem H, Al-Khatib K, Yamaoka Y, Graham DY (2009b) Unreliability of results of PCR detection of Helicobacter pylori in clinical or environmental samples. J Clin Microbiol 47:738–742

Sugiyama A, Maruta F, Ikeno T, Ishida K, Kawasaki S, Katsuyama T, Shimizu N, Tatematsu M (1998) Helicobacter pylori infection enhances N-methyl-N-nitrosourea-induced stomach carcinogenesis in the Mongolian gerbil. Cancer Res 58:2067–2069

Sun YQ, Petersson F, Monstein HJ, Soderholm JD, Rehfeld JF, Borch K (2005) Long-term morpho-functional development of Helicobacter pylori-induced gastritis in Mongolian gerbils. Scand J Gastroenterol 40:1157–1167

Suzuki M, Mimuro H, Suzuki T, Park M, Yamamoto T, Sasakawa C (2005) Interaction of CagA with Crk plays an important role in Helicobacter pylori-induced loss of gastric epithelial cell adhesion. J Exp Med 202:1235–1247

Suzuki M, Mimuro H, Kiga K, Fukumatsu M, Ishijima N, Morikawa H, Nagai S, Koyasu S, Gilman RH, Kersulyte D, Berg DE, Sasakawa C (2009) Helicobacter pylori CagA phosphorylation-independent function in epithelial proliferation and inflammation. Cell Host Microbe 5:23–34

Tabassam FH, Graham DY, Yamaoka Y (2008) OipA plays a role in Helicobacter pylori-induced focal adhesion kinase activation and cytoskeletal re-organization. Cell Microbiol 10:1008–1020

Tabassam FH, Graham DY, Yamaoka Y (2009) Helicobacter pylori activate epidermal growth factor receptor- and phosphatidylinositol 3-OH kinase-dependent Akt and glycogen synthase kinase 3beta phosphorylation. Cell Microbiol 11:70–82

Taketo MM (2009) Role of bone marrow-derived cells in colon cancer: lessons from mouse model studies. J Gastroenterol 44:93–102

Talley NJ, Fock KM, Moayyedi P (2008) Gastric cancer consensus conference recommends Helicobacter pylori screening and treatment in asymptomatic persons from high-risk populations to prevent gastric cancer. Am J Gastroenterol 103:510–514

Tammer I, Brandt S, Hartig R, Konig W, Backert S (2007) Activation of Abl by Helicobacter pylori: a novel kinase for CagA and crucial mediator of host cell scattering. Gastroenterology 132:1309–1319

Tanaka H, Yoshida M, Nishiumi S, Ohnishi N, Kobayashi K, Yamamoto K, Fujita T, Hatakeyama M, Azuma T (2010) The CagA protein of Helicobacter pylori suppresses the functions of dendritic cell in mice. Arch Biochem Biophys 498:35–42

Tatematsu M, Yamamoto M, Iwata H, Fukami H, Yuasa H, Tezuka N, Masui T, Nakanishi H (1993) Induction of glandular stomach cancers in C3H mice treated with N-methyl-N-nitrosourea in the drinking water. Jpn J Cancer Res 84:1258–1264

Tatematsu M, Yamamoto M, Shimizu N, Yoshikawa A, Fukami H, Kaminishi M, Oohara T, Sugiyama A, Ikeno T (1998) Induction of glandular stomach cancers in Helicobacter pylori-sensitive Mongolian gerbils treated with N-methyl-N-nitrosourea and N-methyl-N′-nitro-N-nitrosoguanidine in drinking water. Jpn J Cancer Res 89:97–104

Teymournejad O, Mobarez AM, Hassan ZM, Moazzeni SM, Ahmadabad HN (2014) In vitro suppression of dendritic cells by Helicobacter pylori OipA. Helicobacter 19:136–143

Tian X, Liu Z, Niu B, Zhang J, Tan TK, Lee SR, Zhao Y, Harris DC, Zheng G (2011) E-cadherin/beta-catenin complex and the epithelial barrier. J Biomed Biotechnol 2011:567305

Timm T, Balusamy K, Li X, Biernat J, Mandelkow E, Mandelkow EM (2008) Glycogen synthase kinase (GSK) 3beta directly phosphorylates Serine 212 in the regulatory loop and inhibits microtubule affinity-regulating kinase (MARK) 2. J Biol Chem 283:18873–18882

Tokieda M, Honda S, Fujioka T, Nasu M (1999) Effect of Helicobacter pylori infection on the N-methyl-N'-nitro-N-nitrosoguanidine-induced gastric carcinogenesis in Mongolian gerbils. Carcinogenesis 20:1261–1266

Toller IM, Neelsen KJ, Steger M, Hartung ML, Hottiger MO, Stucki M, Kalali B, Gerhard M, Sartori AA, Lopes M, Muller A (2011) Carcinogenic bacterial pathogen Helicobacter pylori triggers DNA double-strand breaks and a DNA damage response in its host cells. Proc Natl Acad Sci U S A 108:14944–14949

Tomita H, Takaishi S, Menheniott TR, Yang X, Shibata W, Jin G, Betz KS, Kawakami K, Minamoto T, Tomasetto C, Rio MC, Lerkowit N, Varro A, Giraud AS, Wang TC (2011) Inhibition of gastric carcinogenesis by the hormone gastrin is mediated by suppression of TFF1 epigenetic silencing. Gastroenterology 140:879–891

Tsang YH, Lamb A, Romero-Gallo J, Huang B, Ito K, Peek RM Jr, Ito Y, Chen LF (2010) Helicobacter pylori CagA targets gastric tumor suppressor RUNX3 for proteasome-mediated degradation. Oncogene 29:5643–5650

Tsukashita S, Kushima R, Bamba M, Nakamura E, Mukaisho K, Sugihara H, Hattori T (2003) Beta-catenin expression in intramucosal neoplastic lesions of the stomach. Comparative analysis of adenoma/dysplasia, adenocarcinoma and signet-ring cell carcinoma. Oncology 64:251–258

Tsutsumi R, Higashi H, Higuchi M, Okada M, Hatakeyama M (2003) Attenuation of Helicobacter pylori CagA x SHP-2 signaling by interaction between CagA and C-terminal Src kinase. J Biol Chem 278:3664–3670

Tsutsumi R, Takahashi A, Azuma T, Higashi H, Hatakeyama M (2006) Focal adhesion kinase is a substrate and downstream effector of SHP-2 complexed with Helicobacter pylori CagA. Mol Cell Biol 26:261–276

Tu S, Bhagat G, Cui G, Takaishi S, Kurt-Jones EA, Rickman B, Betz KS, Penz-Oesterreicher M, Bjorkdahl O, Fox JG, Wang TC (2008) Overexpression of interleukin-1beta induces gastric inflammation and cancer and mobilizes myeloid-derived suppressor cells in mice. Cancer Cell 14:408–419

Umeda M, Murata-Kamiya N, Saito Y, Ohba Y, Takahashi M, Hatakeyama M (2009) Helicobacter pylori CagA causes mitotic impairment and induces chromosomal instability. J Biol Chem 284:22166–22172

Varon C, Dubus P, Mazurier F, Asencio C, Chambonnier L, Ferrand J, Giese A, Senant-Dugot N, Carlotti M, Megraud F (2012) Helicobacter pylori infection recruits bone marrow-derived cells that participate in gastric preneoplasia in mice. Gastroenterology 142:281–291

Viala J, Chaput C, Boneca IG, Cardona A, Girardin SE, Moran AP, Athman R, Memet S, Huerre MR, Coyle AJ, Distefano PS, Sansonetti PJ, Labigne A, Bertin J, Philpott DJ, Ferrero RL (2004) Nod1 responds to peptidoglycan delivered by the Helicobacter pylori cag pathogenicity island. Nat Immunol 5:1166–1174

Waddington CH (1959) Canalization of development and genetic assimilation of acquired characters. Nature 183:1654–1655

Walz A, Odenbreit S, Mahdavi J, Boren T, Ruhl S (2005) Identification and characterization of binding properties of Helicobacter pylori by glycoconjugate arrays. Glycobiology 15:700–708

Walz A, Odenbreit S, Stuhler K, Wattenberg A, Meyer HE, Mahdavi J, Boren T, Ruhl S (2009) Identification of glycoprotein receptors within the human salivary proteome for the lectin-like BabA and SabA adhesins of Helicobacter pylori by fluorescence-based 2-D bacterial overlay. Proteomics 9:1582–1592

Wang TC, Dangler CA, Chen D, Goldenring JR, Koh T, Raychowdhury R, Coffey RJ, Ito S, Varro A, Dockray GJ, Fox JG (2000) Synergistic interaction between hypergastrinemia and Helicobacter infection in a mouse model of gastric cancer. Gastroenterology 118:36–47

Wang G, Alamuri P, Maier RJ (2006) The diverse antioxidant systems of Helicobacter pylori. Mol Microbiol 61:847–860

Wang FE, Zhang C, Maminishkis A, Dong L, Zhi C, Li R, Zhao J, Majerciak V, Gaur AB, Chen S, Miller SS (2010) MicroRNA-204/211 alters epithelial physiology. Faseb J 24:1552–1571

Watanabe T, Tada M, Nagai H, Sasaki S, Nakao M (1998) Helicobacter pylori infection induces gastric cancer in Mongolian gerbils. Gastroenterology 115:642–648

Weeks DL, Eskandari S, Scott DR, Sachs G (2000) A H+ -gated urea channel: the link between Helicobacter pylori urease and gastric colonization. Science 287:482–485

Wei D, Gong W, Oh SC, Li Q, Kim WD, Wang L, Le X, Yao J, Wu TT, Huang S, Xie K (2005) Loss of RUNX3 expression significantly affects the clinical outcome of gastric cancer patients and its restoration causes drastic suppression of tumor growth and metastasis. Cancer Res 65:4809–4816

Wong BC, Lam SK, Wong WM, Chen JS, Zheng TT, Feng RE, Lai KC, Hu WH, Yuen ST, Leung SY, Fong DY, Ho J, Ching CK (2004) Helicobacter pylori eradication to prevent gastric cancer in a high-risk region of China: a randomized controlled trial. Jama 291:187–194

Wroblewski LE, Peek RM Jr, Wilson KT (2010) Helicobacter pylori and gastric cancer: factors that modulate disease risk. Clin Microbiol Rev 23:713–739

Wu JY, Lu H, Sun Y, Graham DY, Cheung HS, Yamaoka Y (2006) Balance between polyoma enhancing activator 3 and activator protein 1 regulates Helicobacter pylori-stimulated matrix metalloproteinase 1 expression. Cancer Res 66:5111–5120

Wu J, Lin K, Zeng J, Liu W, Yang F, Wang X, Xiao Y, Guo X, Nie H, Zhou T, Xu C (2014) Role of DC-SIGN in Helicobacter pylori infection of gastrointestinal cells. Front Biosci (Landmark Ed) 19:825–834

Xia G, Schneider-Stock R, Diestel A, Habold C, Krueger S, Roessner A, Naumann M, Lendeckel U (2008) Helicobacter pylori regulates p21(WAF1) by histone H4 acetylation. Biochem Biophys Res Commun 369:526–531

Xiao C, Rajewsky K (2009) MicroRNA control in the immune system: basic principles. Cell 136:26–36

Xu J, Yin Z, Cao S, Gao W, Liu L, Yin Y, Liu P, Shu Y (2013) Systematic review and meta-analysis on the association between IL-1B polymorphisms and cancer risk. PLoS One 8:e63654

Yakirevich E, Resnick MB (2013) Pathology of gastric cancer and its precursor lesions. Gastroenterol Clin N Am 42:261–284

Yamahashi Y, Hatakeyama M (2013) PAR1b takes the stage in the morphogenetic and motogenetic activity of Helicobacter pylori CagA oncoprotein. Cell Adh Migr 7:11–18

Yamanouchi J, Azuma T, Yakushijin Y, Hato T, Yasukawa M (2014) Dramatic and prompt efficacy of Helicobacter pylori eradication in the treatment of severe refractory iron deficiency anemia in adults. Ann Hematol 93(10):1779–80

Yamaoka Y (2008a) Increasing evidence of the role of Helicobacter pylori SabA in the pathogenesis of gastroduodenal disease. J Infect Dev Ctries 2:174–181

Yamaoka Y (2008b) Roles of Helicobacter pylori BabA in gastroduodenal pathogenesis. World J Gastroenterol 14:4265–4272

Yamaoka Y, Graham DY (2001) Clarifications regarding the 3' repeat region of the cagA gene in Helicobacter pylori and clinical outcome. J Clin Microbiol 39:2369–2370

Yamaoka Y, Kwon DH, Graham DY (2000) A M(r) 34,000 proinflammatory outer membrane protein (oipA) of Helicobacter pylori. Proc Natl Acad Sci U S A 97:7533–7538

Yamaoka Y, Kikuchi S, El-Zimaity HM, Gutierrez O, Osato MS, Graham DY (2002a) Importance of Helicobacter pylori oipA in clinical presentation, gastric inflammation, and mucosal interleukin 8 production. Gastroenterology 123:414–424

Yamaoka Y, Kita M, Kodama T, Imamura S, Ohno T, Sawai N, Ishimaru A, Imanishi J, Graham DY (2002b) Helicobacter pylori infection in mice: role of outer membrane proteins in colonization and inflammation. Gastroenterology 123:1992–2004

Yamaoka Y, Kudo T, Lu H, Casola A, Brasier AR, Graham DY (2004) Role of interferon-stimulated responsive element-like element in interleukin-8 promoter in Helicobacter pylori infection. Gastroenterology 126:1030–1043

Yamaoka Y, Ojo O, Fujimoto S, Odenbreit S, Haas R, Gutierrez O, El-Zimaity HM, Reddy R, Arnqvist A, Graham DY (2006) Helicobacter pylori outer membrane proteins and gastroduodenal disease. Gut 55:775–781

Yeh JM, Hur C, Schrag D, Kuntz KM, Ezzati M, Stout N, Ward Z, Goldie SJ (2013) Contribution of H. pylori and smoking trends to US incidence of intestinal-type noncardia gastric adenocarcinoma: a microsimulation model. PloS Med 10:e1001451

Yin Y, Grabowska AM, Clarke PA, Whelband E, Robinson K, Argent RH, Tobias A, Kumari R, Atherton JC, Watson SA (2010) Helicobacter pylori potentiates epithelial: mesenchymal transition in gastric cancer: links to soluble HB-EGF, gastrin and matrix metalloproteinase-7. Gut 59:1037–1045

Yin Y, Li J, Chen S, Zhou T, Si J (2012) MicroRNAs as diagnostic biomarkers in gastric cancer. Int J Mol Sci 13:12544–12555

Yokota K, Kurebayashi Y, Takayama Y, Hayashi S, Isogai H, Isogai E, Imai K, Yabana T, Yachi A, Oguma K (1991) Colonization of Helicobacter pylori in the gastric mucosa of Mongolian gerbils. Microbiol Immunol 35:475–480

Yoshida T, Kato J, Maekita T, Yamashita S, Enomoto S, Ando T, Niwa T, Deguchi H, Ueda K, Inoue I, Iguchi M, Tamai H, Ushijima T, Ichinose M (2013) Altered mucosal DNA methylation in parallel with highly active Helicobacter pylori-related gastritis. Gastric Cancer 16:488–497

Zanotti G, Papinutto E, Dundon W, Battistutta R, Seveso M, Giudice G, Rappuoli R, Montecucco C (2002) Structure of the neutrophil-activating protein from Helicobacter pylori. J Mol Biol 323:125–130

Zawahir S, Czinn SJ, Nedrud JG, Blanchard TG (2013) Vaccinating against Helicobacter pylori in the developing world. Gut Microbes 4:568–576

Zeaiter Z, Cohen D, Musch A, Bagnoli F, Covacci A, Stein M (2008) Analysis of detergent-resistant membranes of Helicobacter pylori infected gastric adenocarcinoma cells reveals a role for MARK2/Par1b in CagA-mediated disruption of cellular polarity. Cell Microbiol 10:781–794

Zheng PY, Jones NL (2003) Helicobacter pylori strains expressing the vacuolating cytotoxin interrupt phagosome maturation in macrophages by recruiting and retaining TACO (coronin 1) protein. Cell Microbiol 5:25–40

Zheng Q, Chen XY, Shi Y, Xiao SD (2004) Development of gastric adenocarcinoma in Mongolian gerbils after long-term infection with Helicobacter pylori. J Gastroenterol Hepatol 19:1192–1198

Zhu J, Yamane H, Paul WE (2010) Differentiation of effector CD4 T cell populations (*). Annu Rev Immunol 28:445–489

Zhu Y, Jiang Q, Lou X, Ji X, Wen Z, Wu J, Tao H, Jiang T, He W, Wang C, Du Q, Zheng S, Mao J, Huang J (2012) MicroRNAs up-regulated by CagA of Helicobacter pylori induce intestinal metaplasia of gastric epithelial cells. PLoS One 7:e35147

Chapter 11
Oral Infection, Carcinogenesis and Cancer

Jukka H. Meurman and Antonio Bascones-Martinez

Abstract Recent research has shown statistical associations between dental infections and cancer in general but the role of oral microbiota in carcinogenesis is unclear. Oral micro-organisms up-regulate cytokines and other inflammatory mediators that affect the complex metabolic pathways and may thus indeed be involved in carcinogenesis. Microbial populations on mouth mucosa differ between healthy and malignant sites and certain oral bacterial species have been linked with malignancies. Oral microbes also have carcinogenic metabolites, such as acetaldehyde produced from ethanol. In this chapter we briefly review current knowledge about the interaction between oral microorganisms, oral infections and cancer. The focus is both on oral and ear-nose-throat cancer and malignancies in other organs.

Keywords Oral microbiota • Oral bacteria • Cancer • Oral cancer • Carcinogenesis

Abbreviations

EBV	Epstein-Barr virus
HBV	hepatitis B virus
HCV	hepatitis C virus
HPV	human papilloma virus
HTLV-1	human T-cell lymphotropic virus
KSHV	Kaposi's associated sarcoma virus

J.H. Meurman (✉)
Department of Oral and Maxillofacial Diseases, University of Helsinki and Helsinki University Hospital, Helsinki, Finland
e-mail: jukka.meurman@helsinki.fi

A. Bascones-Martinez
Department of Oral and Maxillofacial Diseases, University of Helsinki and Helsinki University Hospital, Helsinki, Finland

Complutense University of Madrid, Madrid, Spain

M.R. Shurin et al. (eds.), *Infection and Cancer: Bi-Directorial Interactions*, DOI 10.1007/978-3-319-20669-1_11

Introduction

The best known example of infection-associated malignancy is gastric cancer caused by *Helicobacter pylori* infection (Herrera and Parsonnet 2009). In general, infection-driven inflammations have been estimated to be involved in the pathogenesis of approximately 15–20 % of human tumours (Allavena et al. 2008). Hence, human microbiota may play a role in carcinogenesis (Chang and Parsonnet 2010). The worldwide incidence of cancer also attribute to infections caused by virus. These include hepatitis B virus (HBV), hepatitis C virus (HCV), human papilloma virus (HPV), and Epstein-Barr virus (EBV), human T-cell lymphotropic virus (HTLV-1) and Kaposi's associated sarcoma virus (KSHV), which have estimated to contribute to 10–15 % of the cancers worldwide (Butel 2000; Kuper et al. 2000; Martin and Gutkind 2008; Grinde and Olsen 2010; Rautava and Syrjänen 2012). No such figures exist for bacteria- or yeast-related malignancies. Nevertheless in this chapter the main focus is on oral bacterial and yeast infections.

Inflammation is indeed a key feature in many chronic diseases including cancer (Coussens and Werb 2002; Mantovani et al. 2008). Investigations of gastrointestinal malignancies have shown that the large amount of cytokines and growth factors released during inflammation may influence carcinogenesis (Fantini and Pallone 2008). Characteristic to infections and inflammations linked to cancer is their high prevalence in populations and persistence in the host (Mergaurd et al. 1989). There may be a long latency between the initial infection and tumour appearance. An infected person in fact rarely develops cancer. However, this area of research is difficult and the overall effect of human microbiome on carcinogenesis is not known. This also relates to oral microbiota and cancer. Nevertheless, infections that trigger inflammatory reactions have been suggested as major preventable causes of cancer (Parsonnet 1999; Kuper et al. 2000; Hujoel et al. 2003).

Knowledge about human microbiome is rapidly increasing. Molecular techniques have revealed its vast diversity but, at the same time, individual stability of micro-organisms residing on skin and mucosa has been observed (Turnbaugh et al. 2007; Fritz et al. 2013). Resident micro-organisms seem to play an important role in metabolism in general but it is not known how this diversity relates to function and to the rest of the genes of microbiota (Joyce and Gahan 2014). The gut microbiome is shared among family members but each person's microbial community varies in the specific bacterial lineages with a comparable degree of co-variation. A wide array of shared microbial genes has been observed among individuals comprising an extensive, identifiable core microbiome at the gene level rather than at the microbial lineage (Turnbaugh et al. 2009). Understanding this complex ecosystem will inevitably open new possibilities for diagnosis and prevention of diseases. Corresponding studies on oral microbiota are now on their way, too (Preza et al. 2009; Holgerson et al. 2013; Dimitrov and Hoeng 2013). But certain oral pathogens, such as the periodontal bacterium *Porphyromonas gingivalis*, may nevertheless be responsible for many extra-oral manifestations thus emphasizing the need for also focusing to individual strains (Han and Wang 2013).

Traditionally it has been thought that mouth mucosa (Fiscetti 2003), the tongue (Tachibana et al. 2006) and the pharynx (Brook 2005) harbour characteristic bacterial pathogens causing chronic inflammation and focal infections (Gendron et al. 2000). Many these infections derive from oral biofilms commonly linked with dental diseases, caries and periodontal disease (Desai et al. 1991; Komiyama et al. 1985; Scannapieco 1998). New research techniques have cast more light on the changes in oral microbiota in health and disease (Wade 2013). In particular our knowledge about the number of microbial species harbouring the oral cavity and their function has increased tremendously after introduction of new molecular techniques (Kejser et al. 2008; Olsen et al. 2013).

The association between oral micro-organisms and cancer is a new observation (Meurman 2010). Interestingly this association may even be causal when considering oral and oesophageal cancer, namely mediated by acetaldehyde production of oral micro-organisms (Meurman and Uittamo 2008; Moazzez et al. 2011). This chapter briefly outlines current knowledge about the role of oral microbiota in carcinogenesis with emphasis on oral cancer. Oral microbial changes caused by cancer treatment are also discussed.

Population Studies on the Association of Oral Infections with Cancer

In a Swedish cohort of 1390 subjects followed-up for 24 years missing second molar in the right mandible (odds ratio OR 2.62 (95 % confidence interval CI 1.18–5.78) and age (OR 1.91 [CI 1.06–3.43]), appeared as the principle independent predictors significantly associating with any type of cancer (Virtanen et al. 2014). In the analyses a number of explanatory factors had been taken into account. Furthermore, chronic periodontal disease associated statistically with breast cancer in women. Of the subjects with periodontal disease and any missing molars in the mandible, 5.5 % had breast cancer in comparison to 0.5 % of the subjects who had periodontal disease but no missing molars ($P<0.02$). Female gender (OR 13.08) and missing any molar in the mandible (OR 2.36) were the explanatory variables for breast cancer in this cohort (Söder et al. 2011). The missing molars were used as proxy of long-lasting dental infections because commonly these teeth are extracted due to caries or periodontal disease. In the same cohort, death in cancer was more frequent among patients with poor oral hygiene than in those whose oral hygiene had been better. Dental plaque appeared to be a significant independent predictor associated with 1.79 times the OR of death (Söder et al. 2012).

A 16-year follow-up study on 51,529 male health professionals in the US, showed that when compared with no periodontal disease, history of periodontitis was associated with increased pancreatic cancer risk (RR 1.64,CI 1.19–2.26; $P=0.002$). The crude incidence rates were 61 % versus 25 % per 100,000 person-years, and among never smokers RR was 2.09 (CI 1.18–3.71; $P=0.01$). The baseline number of natural teeth and cumulative tooth loss during follow-up were not associated with pancreatic cancer in this study (Michaud et al. 2007).

Periodontal disease in particular has been found to link statistically with oral cancer. From Taiwan from an insurance database of one million people, a study showed that in patients with periodontal disease the hazard ratio (HR) for oral cancer was 1.79 (CI 1.42–2.25) (Wen et al. 2013). A study from Germany showed a mean alveolar bone loss 4.3 mm in 178 oral cancer patients vs. 2.9 mm in their 123 controls ($P<0.001$) and the regression model resulted in OR 2.4 (CI 1.5–3.8) in this regard (Moergel et al. 2013). In this study a history of periodontal treatment was associated with significantly reduced cancer risk ($p<0.001$; OR 0.2, CI 0.1–0.5).

In the US the national health and nutrition study including13,798 subjects clinical attachment loss (CAL) was related to the presence of tumour in the mouth (OR 4.57, CI 2.25–9.30) and precancerous lesions in oral mucosa (OR 1.55, CI 1.06–2.27), respectively (Tezal et al. 2005). Similarly, OR was 5.23 (CI 2.64–10.35) for each millimeter of alveolar bone loss vs. tongue cancer (Tezal et al. 2007). Another study by the same group on 266 cases with head and neck cancer and 207 controls showed OR 4.36 (CI 3.16–6.01) for each millimeter of alveolar bone loss after adjustment for age, gender, race/ethnicity, marital status, smoking status, alcohol use, and missing teeth. The association persisted in subjects who never used tobacco and alcohol. There was a significant interaction between smoking and alveolar bone loss ($P=0.03$). Patients with periodontitis were more likely to have poorly differentiated oral cancer than those without periodontitis (32.8 % versus 11.5 %; $P=0.038$) (Tezal et al. 2009). HPVinfection also seems to play a significant role in these connections (Tezal 2012). Interestingly, an inverse association between head and neck cancer and dental caries was observed in this US patient material (Tezal et al. 2013). This unexpected result was explained by interactions between the commensal microbiota, such as oral streptococci and the host which was thought to be important for stimulating local mucosal and systemic immunity, tolerance and fine-tuning of T-cell receptor function, epithelial turnover, mucosal vascularity, and lymphoid tissue mass.

From national data from the US periodontitis was also associated with increased orodigestive cancer mortality (relative risks [RR] 2.28, CI 1.17–4.45) and the association seemed dependent on the increasing severity of periodontitis (P for trend 0.01). Mortality was in excess for colorectal (RR 3.58, CI 1.15–11.16) and possibly for pancreatic cancer (RR 4.56, CI 0.93–22.29). Greater serum *P. gingivalis* IgG values tended to be associated with the increased orodigestive cancer mortality (P for trend 0.06). It also associated with mortality in subjects with no periodontal disease (RR 2.25,CI 1.23–4.14) (Ahn et al. 2012). Interestingly, practicing no regular oral hygiene also conferred OR 2.37 (CI 1.42–3.97) for oesophageal cancer when compared with those who undertook daily tooth brushing (Abnet et al. 2008).

A German population-based study on 4233 subjects with oral leukoplakia showed OR 1.7 (0.6–5.0), 3.3 (0.8–13.1) and 5.3 (1.2–22.7), respectively, for second, third and fourth quartiles of CAL, respectively. For bleeding on probing the respective ORs were 2.0 (0.8–4.90), 2.9 (1.1–7.8) and 3.8 (1.5–9.8) (Meisel et al. 2012). Oral leukoplakia is regarded a potentially precancerous state.

Thus, the oral infections, especially periodontitis, have been clearly associated with cancer (Gondiykar et al. 2013).

Bacteria, Oral Cancer and the Effect of Treatment on Oral Microbiota

Principally, the role of bacteria in oral cancer is not known (Rajeev et al. 2012). Microbial populations on mouth mucosa differ between healthy and malignant sites, however. For example, *Streptococcus anginosus* and *Treponema denticola* seem to associate with various upper gastrointestinal tract carcinomas and also syphilis has been mentioned to be associated with cancer (Narikiyo et al. 2004). *S. anginosus*infection might be implicated in the carcinogenesis of head and neck squamous cell carcinoma in general (Shiga et al. 2001). *S. anginosus* DNA has been detected in carcinoma tissue samples but not in lymphoma, rhabdomyosarcoma or leukoplakia samples. Dental plaque could be a dominant reservoir of this bacterium (Sasaki et al. 2005). Hooper et al. (2007) studied with molecular technique oral carcinoma specimens and observed 70 distinct taxa with 52 different phylotypes isolated from tumour tissues, and 37 taxa from within non-tumorous specimens. Differences between the composition of the microbiotas within the tumorous and non-tumorous mucosae were apparent, possibly indicating selective growth of bacteria within carcinoma tissue. Most taxa isolated from within the tumour tissue represented saccharolytic and aciduric species and studies were called for to investigate if these aspects have any link to carcinogenesis (Hooper et al. 2009).

Treatment of cancer, such as radiotherapy and chemotherapy, obviously modifies oral microbial composition leading to a major imbalance of the ecosystem (Sixou et al. 1998; Pushalkar et al. 2012; Xu et al. 2014). For example, in a study from Sweden, the patients harboured enterococci in 38 % of mouth samples *vs.* none of the controls. *Lactobacillus* spp. were detected in 92 % of the subjects and the proportion of these species was high compared with the controls. Mutans streptococci were also detected in high numbers; 31 % in the patients *vs.* 23 % in controls (Almståhl et al. 2008). On the other hand in a study from China mutans streptococci were not isolated in radiotherapy patients while lactobacilli, *S. mitis* and *S. salivarius* were the predominant caries-related oral bacteria following radiotherapy (Tong et al. 2003).

Bacteria in gingival pockets in head- and neck-irradiated patients have also been investigated. A comprehensive study from Hong Kong showed that the major components of subgingival microbiota appear similar to that of gingivitis sites in the normal population although among the radiotherapy patients bacterial or fungal species uncommon in normal subjects were also detected. These species included micro-organisms such as *Gemella, Peptostreptococcus, Staphylococcus, Stomatococcus, Streptococcus, Actinomyces, Eubacterium, Lactobacillus, Propionibacterium, Neisseria, Veillonella, Bacteroides, Campylobacter, Capnocytophaga, Fusobacterium, Kingella, Porphyromonas* and *Prevotella*. Also species of microbes that are characteristic to the normal microbiota of skin (*Peptostreptococcus prevotii* and *Propionibacterium granulosum*) and gut (*Eubacterium aerofaciens, Fusobacterium mortiferum* and *Fusobacterium varium*) were detected in this material (Leung et al. 1998). The new molecular techniques have revealed patterns of microbial shifts providing data for better understanding the impact of cancer treatment in this regard (Hu et al. 2013a, b). However, it the practical importance of the bacterial diversity observed remains to be seen.

How permanent are the shifts in oral microbiota after treatment of cancer is another interesting question. Radiotherapy or cytostatic treatment caused changes in bacterial composition in oral microbiota need not be permanent. For example, in child allogenic bone marrow transplantation patients in the UK no differences were seen in the total anaerobic counts or in the proportion of the *S. oralis* group between baseline and the end of a 119-day study, or between patients and controls (Lucas et al. 1997). However, caries risk may still be increased in particular in paediatric patients surviving a malignant disease (Dens et al. 1996). Furthermore pathogens such as *Capnocytophaga* may pose a systemic risk in these patients and call for continuous attention in order to prevent bacteraemia and also to overcome problems of developing antibiotic resistance (Sixou et al. 1998). Risk for dental caries remains high after radiotherapy while periodontitis does not seem to pose corresponding problem (Al-Nawas and Grötz 2006).

In immunosuppressed patients treated with cytostatic drugs pathogenic and opportunistic micro-organisms colonizing the mouth may be dangerous. Enterobacteria and species such as *Pseudomonas, Neisseria*, and *Veillonella* have been observed in oral samples from granulocytopenic patients with leukaemia (Peterson et al.1990). It appears that the pre-treatment oral health status is important in this respect. For example, in a study on non-lymphocytic leukaemia patients in Baltimore periodontal disease status and attachment loss were positively correlated with increase in the proportional recovery of *Staphylococcus* sp. from supragingival sites and total yeasts from supra- and subgingival sites (Reynolds et al. 1989). The authors suggested that host factors such as periodontal disease may contribute to patterns of oral microbial changes during cancer chemotherapy. However, it should also be kept in mind that sampling site as such may influence the results and in oral cancer patients optimal sampling may be difficult (Rautemaa et al. 2006). Consequently, proper sampling technique for both conventional cultivation and novel molecular methods needs to be emphasized (Rusanen et al. 2009). Table 11.1 gives examples of oral microbial strains which are often isolated from cancer patients.

Table 11.1 Micro-organisms which are frequently detected in mouth samples from cancer patients.

Candida albicans
Enterococci
Lactobacilli
Viridans streptococci
Staphylococci
Enterobacteria (*Pseudomonas, Neisseria, Veillonella*)
Capnocytophaga
Fusobacteria

Oral Cancer and Yeasts

Immunosuppression enhances the selection and outgrowth of yeasts in oral microbiota (Bensadoun et al. 2011). Hence cancer patients are at risk for invasive yeast infections and life-threatening candidemias (Pompej et al. 1993). For example, *Candida albicans* was found in 54 % of subjects who received radiotherapy to the head and neck in comparison to 15 % of controls (Almstål et al. 2008). Particularly non-*albicans Candida* strains may become a problem since these yeasts often are resistant to commonly used antifungal drugs (Redding et al. 2004). Oral yeasts also convert ethanol to carcinogenic acetaldehyde, a mechanism which may play even causal role in the development of cancer (Nieminen et al. 2009). However, more studies are needed in this area.

Local antifungal first-line therapy is recommended for oral cancer patients with mucosal *Candida* infections but severe systemic infections obviously call for intravenous medication.

Pathogenic Mechanisms in Oral Infection-Linked Carcinogenesis

Infection caused inflammation may induce cellular proliferation, inhibit apoptosis, interfere with cellular signaling mechanisms, and act as tumor promoters (Lax and Thomas 2002). Up-regulation of cytokines and other inflammatory mediators affect complex metabolic pathways. For example, the receptor for advanced glycation end products (RAGE), a multi-ligand receptor expressed on various cell membranes has been suggested to play a role also in carcinogenesis. RAGE is activated by ligands in a variety of cell types and tissues and may play a role in the oral infection – systemic health associations (Katz et al. 2010). Oral infection may also directly reflect in endothelial dysfunction (Janket et al. 2008). The cytokine reactions involved have been shown to play a role in the immune-related mechanisms of cancer development (Sheu et al. 2008). Figure 11.1 shows the complex pathways thought a play a role in infection driven carcinogenesis.

Another mechanism suggested is mediated via salivary factors. Poor oral health has been shown to associate with the genotoxic salivary activity. Bloching et al. (2007) studied dental status and saliva of 100 subjects relating their oral health to in vitro salivary mutagenicity, using the *Salmonella* test, and observed a significant association ($p<0.05$) between high plaque index and high number of carious teeth with genotoxic activity in saliva. Hence the polymicrobial burden caused by oral biofilms may also possess mutagenic interactions with saliva which may act as co-factors in carcinogenesis. These examples illustrate the complexity of tumour genesis.

Several bacteria and *Candida* strains in the mouth convert ethanol to carcinogenic acetaldehyde thus explaining the epidemiological evidence between heavy

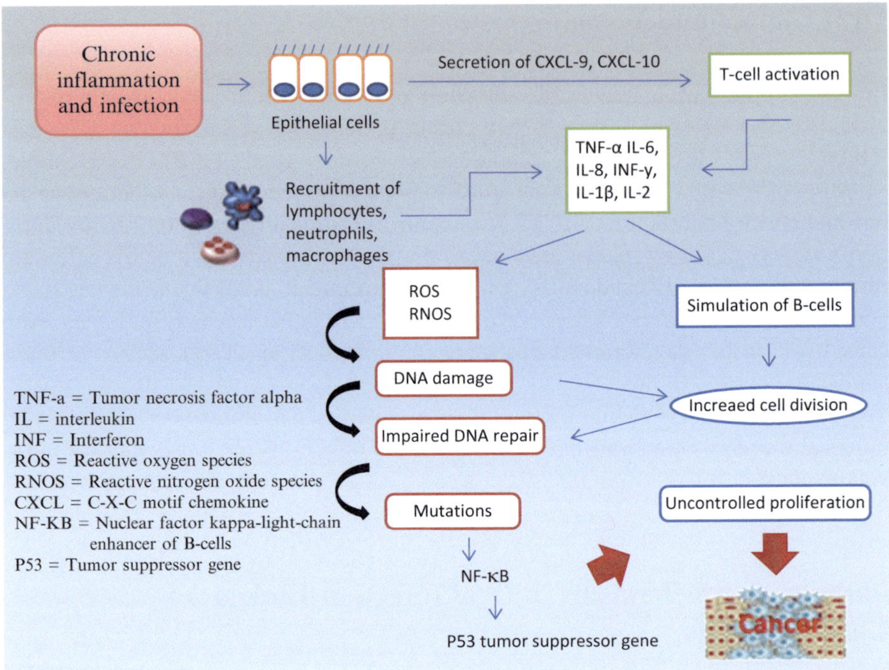

Fig. 11.1 The chronic infection and inflammation activate the T lymphocytes and release proinflammatory cytokines leading to DNA damage increased cell division and uncontrolled proliferation.

drinking, smoking and development of cancer (Homann et al. 2000a, b). Both the commonly encountered oral streptococci and yeasts possess metabolic pathways for this conversion (Kurkivuori et al. 2007; Uittamo et al. 2009; Nieminen et al. 2009). Alcohol-related carcinogenesis is well-known and the enzymes involved have been characterized. Polymorphism in these genes may partly explain why subjects differ in their liability for the development of cancer; it may be a question about higher or lower metabolic activities involved in alcohol metabolism (Marichalar-Mendia et al. 2010). Nevertheless, there is a significant dose-response relationship between intake frequency, duration and oral cancer risk (Cancela Mde et al. 2009). Similarly, smoking associates with cancer risk and smoking also causes an increase in salivary acetaldehyde concentrations thus adding to the risk related to alcohol (Morse et al. 2007). The effect of smoking and alcohol is synergistic (Salaspuro and Salaspuro 2004). Figure 11.2 depicts the ethanol acetaldehyde pathway.

Inflammation is a critical component of tumor progression (e.g., reflux esophagitis/esophageal cancer; inflammatory bowel disease/colorectal cancer) (Coussens and Werb 2002). This can also happen in the oral environment, such as in the periodontal pocket as a weak point for tumor invasion. Increasing evidence indicates that the inflammation may result from persistent mucosal or epithelial cell colonization

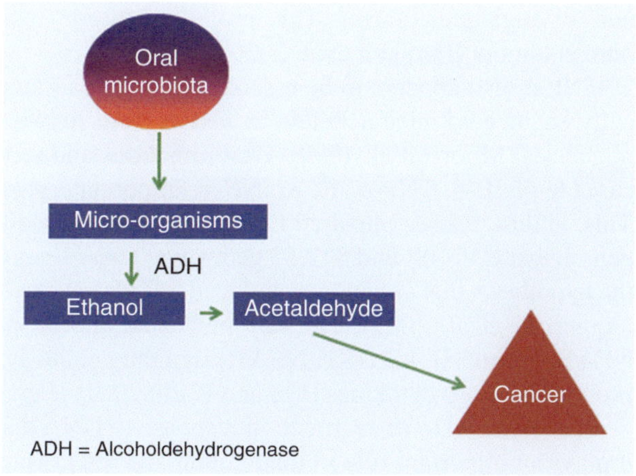

Fig. 11.2 Microorganisms metabolize ethanol to acetaldehyde leading to cancer. The alcohol dehydrogenase (ADH) play a major role in the pathway.

by microorganisms. Persistent inflammation leads to increased cellular turnover, especially in the epithelium, and provides selection pressure that results in the emergence of cells that are at high risk for malignant transformation (Moss and Blaser 2005). Meisel et al. (2012) explained how chronic periodontitis may affect the pathogenesis of precancerous lesions by showing that chronic periodontitis was a risk factor for the development of leukoplakia predisposing for oral cancer. In periodontitis, the inflammatory response caused by bacteria colonizing periodontal pockets leads to significant interleukin-8 (IL-8) and IL-6 mRNA levels induced in response to exposure to the bacteria (Yumoto et al. 1999).

Chronic inflammation causes epithelial cells to secrete CXCL 9 and CXCL 10 through the action of various inflammatory mediators, including TNF-α, IL-6, and IL-17, leading to eradication of antitumor immunity and accelerated tumor progression (Lin and Karin 2007). Tumor endothelial cells (ECs) secrete high levels of CXCL9 in all, and CXCL10 in most melanoma metastases (Amatschek et al. 2011). In the combination of the secretion of CXCL 9 and CXCL 10 with the recruitment of lymphocytes, neutrophils and macrophages which are sources of cytotoxic and genotoxic reactive nitrogen oxygen species (RNOS) the process of tumour genesis goes on. Genetic instability is a common feature of solid tumors and we and others have proposed that mutagenic RNOS generated by these tumor-infiltrating cells are, in some measure, responsible for the accumulation of mutations associated with tumour progression (Haqqani et al. 2000) (Fig. 11.1).

IL-6 is a potent pleiotropic inflammatory cytokine that is considered a key growth promoting and antiapoptotic factor (Lin and Karin 2007). Malignant transformation of oral epithelium would then be a consequence of the immune response due to macrophage and T-cell activation and cytokine release (e.g., IL-1, IL-8, and TNF-α) (Mantovani et al. 2008).

Production of interleukin-8 (IL-8) by oral epithelial cells can be expected to play a major role in the recruitment and activation of phagocytes at the infected site (Dongari-Bagtzoglou et al. 2003). IL-8 is a human CXC chemokine for neutrophils and an angiogenic factor. This proinflammatory cytokine is expressed in many human tumors (Haqqani et al. 2000).

TNF-α also appears to be essential for skin carcinogenesis, however, as genetically engineered mice deficient in TNF-α were shown to be resistant to skin carcinogenesis (Moore et al. 1999). Oral infections and sepsis promote the degradation cascade of IL-6, TNF-α, IL-8, INF-Y, important cytokines causing DNA damage This, in turn, causes impaired DNA repair and subsequent mutation; most carcinogens activate NF-κB and STAT3 pathways. These lines of evidence strongly support the hypothesis that carcinogen-induced NF-κB activation could lead carcinogenesis (Aggarwal et al. 2009). The key molecular link is provided by the inhibitor of NF-kB kinase/ NF-kB (IKK/NF-kB) signaling pathway, which is activated by many proinflammatory cytokines (Lin and Karin 2007) (Fig. 11.2).

P53 is known to be recruited in response to DNA damaging genotoxic stress and it plays an important role in maintaining the integrity of the genome. We therefore conclude that BP transcriptionally activates the human p53 gene through the induction of NF-κB activity (Pei et al. 1999). These mechanisms are anticipated to play a role in the mutations leading to oral cancer. Table 11.2 summarizes the principal mechanisms by which oral micro-organisms may enhance the development of cancer.

Prevention of Oral Microbiota-Associated Carcinogenesis

The simple answer to the question how to prevent oral microbiota-associated carcinogenesis would be avoidance of any infections and inflammations in the mouth. This, however, is not possible in practice due to the complex nature of oral biofilms. However, maintaining good level of oral hygiene can be anticipated to reduce cancer risk by the mechanisms here discussed. But there is only indirect evidence to support this (Abnet et al. 2008).

The nonessential amino acid cysteine which effectively binds acetaldehyde by forming a thiazole-carboxylic acid compound and thus eliminates the local carcinogenic effect may be one means of future prevention of oral cancer (De Vries and De Flora 1993; Van Schouten et al.2002; Salaspuro et al. 2002, 2006; Salaspuro 2007).

Table 11.2 Carcinogenic mechanisms where oral microbiota may play a role.

Induction of cell proliferation
Inhibition of apoptosis
Interference with signalling mechanisms
Tumor promoter activity
Up-regulation of cytokines and other inflammatory mediators
Effect on cellular sugar metabolism
Enhancing mutagenic activity in saliva
Metabolizing ethanol to acetaldehyde

In an animal model, cysteine was shown to reduce metastases by inhibiting the gelatinolytic activity of matrix metalloproteinases (Morini et al. 1999). Here, however, clinical studies are called for final conclusion.

Other chemicals studied in this respect are retinoids (King et al. 1982), antioxidant vitamins E and A, and carotenoids (Krisnky 1989; Tengerdy 1990), but there is no true scientific evidence. Liede et al. (1998a, b) for example could not show any effect of beta-carotene on the prevalence of oral mucosal dysplasia in their 7-year study on men.

Furthermore, many herbs and other natural remedies have been suggested to provide protection from carcinogenesis. These include garlic, cumin, cloves, cinnamon, thyme, mustard, rosemary and green tea (Lai and Roy 2004; Taylor et al. 2005). The expected effect is thought to be mediated by phytochemicals of the plants, also tested in animal models (Miller et al. 2008). Here it is interesting to cite results from a study from Japan where in women the hazard ratios of oral cancer for green tea consumption of 1–2, 3–4, and 5 or more cups per day were 0.51 (CI 0.10–2.68), 0.60 (CI 0.17–2.10), and 0.31 (CI 0.09–1.07), respectively, compared with those who daily drank less than one cup of green tea (p for trend was 0.08) (Ide et al. 2007). However, scientific evidence in general is weak of the topic and properly powered, controlled long-term studies are called for further conclusions.

Another future approach might be bacteriotherapy with probiotics, health beneficial bacteria shown to inhibit mutagenicity and provide adjuvant effect by modulation of cell-mediated immunity (Kumar et al. 2010). Probiotics may also help in preventing mucositis caused by cancer treatment, but scientific evidence still is very weak (Maria-Aggeliki et al. 2009).

Clinical Aspects of Controlling Oral Microbiota in Cancer Patients

Maintaining proper oral hygiene during treatment of cancer is of utmost importance. Life-threatening complications may arise if this is neglected (Meurman et al. 1997). Systemic infection and sepsis remains the leading causes of morbidity and mortality in immunosuppressed patients. Therefore practical guidelines have been given for the oral health care of these patients (Meurman and Scully 2012).

Chlorhexidine-containing preparations have been the standard in controlling oral microbiota in patients with cancer and other malignancies (Meurman et al. 1997; Wahlin 1989). Chlorhexidine may also prevent oral mucositis during cancer chemotherapy (Sorensen et al. 2008). However, antifungal agents must be used when yeast infections need to be controlled (Madan et al. 2008).

Proper hands-on counseling of cancer patients is often needed. Nurses, dental hygienist and auxiliary personnel at the wards must all be advised of the importance of daily oral hygiene of cancer patients. Here the oral health care personnel is in a key position in advising both the patients and hospital personnel (Chandu et al. 2002; Meurman and Grönroos 2010). Electric toothbrush may be of help (Fjeld

Table 11.3 Practical aspects in controlling oral microbiota of patients with cancer.

Diagnosing and eradication of dental infection foci is of high importance.
Importance of good oral hygiene throughout the treatment of cancer and follow-up need to be emphasized.
The patient, nurses and auxiliary personnel must be advised in how to clean the mouth and dental prostheses.
The hospital personnel must be advised how to help the patient in maintaining satisfactory oral hygiene.
Electric toothbrush with soft brush tip might be recommended.
Dental prostheses need to be checked in order to avoid mucosal damage.
Use of chlorhexidine preparations recommended in particular during immunosuppression due to cancer treatment.
Use of antifungal agents need to be prescribed in cases with oral yeast infections.
Regular dental check-ups are needed even though no subjective symptoms arise.

et al. 2014). Particular attention must also be focused on daily hygiene of dental prostheses to avoid fungal infections of the mouth (Davies et al. 2006). Worthington et al. (2007) have reviewed the prevention of oral mucositis due to treatment of cancer based on Cochrane database. Table 11.3 gives some practical guidelines for maintaining good oral health in cancer patients.

Conclusion

The mouth is a habitat of billions of micro-organisms. In cancer patients controlling oral microbiota is of utmost importance in order to avoid life-threatening systemic infections. Oral microbes may also have a role in carcinogenesis via a number of mechanisms. These include carcinogenic metabolites of oral bacteria and yeasts and chronic inflammation-mediated pathogenic cascade reactions. Hospital personnel should be advised about the importance of the mouth, both as regards the quality of life of the patient and preventing hematogenic spread of infections of the mouth. Dentists and other oral health professionals have a key role in counseling nurses, auxiliary and the medical profession, in addition to providing oral health treatment to the patients.

References

Abnet CC, Kamangar F, Islami F, Nasrollahzadeh D, Brennan P, Aghcheli K et al (2008) Tooth loss and lack of regular oral hygiene are associated with higher risk of esophageal squamous cell carcinoma. Cancer Epidemiol Biomarkers Prev 17:3062–3068

Aggarwal BB, Vijayalekshmi RV, Sung B (2009) Targeting inflammatory pathways for prevention and therapy of cancer: short-term friend, long-term foe. Clin Cancer Res 15:425–430

Ahn J, Segers S, Hayes RB (2012) Periodontal disease, Porphyromonas gingivalis serum antibody levels and orodigestive cancer mortality. Carcinogenesis 33:1055–1058

Allavena P, Garlanda C, Borrello MG, Sica A, Mantovani A (2008) Pathways connecting inflammation and cancer. Curr Opin Genet Dev 18:3–10

Almståhl A, Wikström M, Fagerberg-Mohlin B (2008) Microflora in oral ecosystems in subjects with radiation-induced hyposalivation. Oral Dis 14:541–549

Al-Nawas B, Grötz KA (2006) Prospective study of the long term change of the oral flora after radiation therapy. Support Care Cancer 14:291–296

Amatschek S, Lucas R, Eger A, Pflueger M, Hundsberger H, Knoll C, Grosse-Kracht S, Schuett W, Koszik F, Maurer D, Wiesner C (2011) CXCL9 induces chemotaxis, chemorepulsion and endothelial barrier disruption through CXCR3-mediated activation of melanoma cells. Br J Cancer 104:469–479

Bensadoun RJ, Patton LL, Lalla RV, Epstein JB (2011) Oropharyngeal candidiasis in head and neck cancer patients treated with radiation: update 2011. Support Care Cancer 19:737–744

Bloching M, Reich W, Schubert J, Grummt T, Sandner A (2007) The influence of oral hygiene on salivary quality in the Ames Test, as a marker for genotoxic effects. Oral Oncol 43:933–939

Brook I (2005) The role of bacterial interference in otitis, sinusitis and tonsillitis. Otolaryngol Head Neck Surg 33:139–146

Butel JS (2000) Viral carcinogenesis: revelation of molecular mechanisms and etiology of human disease. Carcinogenesis 21:405–426

Cancela Mde C, Ramadas K, Fayette JM, Thomas G, Muwonge R, Chapuis F, Thara S, Sankaranarayanan R, Sauvaget C (2009) Alcohol intake and oral cavity cancer risk among men in a prospective study in Kerala, India. Community Dent Oral Epidemiol 37:342–349

Chandu A, Stulner C, Bridgeman AM, Smith AC (2002) Maintenance of mouth hygiene in patients with oral cancer in the immediate post-operative period. Aust Dent J 47:170–173

Chang AH, Parsonnet J (2010) Role of bacteria in oncogenesis. Clin Microbiol Rev 23:837–857

Coussens LM, Werb Z (2002) Inflammation and cancer. Nature 420:860–867

Davies AN, Brailsford SR, Beighton D (2006) Oral candidosis in patients with advanced cancer. Oral Oncol 42:698–702

De Vries N, De Flora S (1993) N-acetyl-l-cysteine. J Cell Biochem Suppl 17F:270–277

Dens FL, Boute P, Vinckier F, Declerck D (1996) Salivary caries risk factors in long-term event-free survivors of pediatric malignant diseases. J Clin Pediatr Dent 20:241–245

Desai HG, Gill HH, Shankaran K, Mehta PR, Prabhu SR (1991) Dental plaque: a permanent reservoir of Helicobacter pylori? Scand J Gastroenterol 26:1205–1208

Dimitrov DV, Hoeng J (2013) Systems approaches to computational modeling of the oral microbiome. Front Physiol 4:172. doi:10.3389/fphys.2013.00172

Dongari-Bagtzoglou A, Kashleva H (2003) Candida albicans triggers interleukin-8 secretion by oral epithelial cells. Microb Pathog 34:169–177

Fantini MC, Pallone F (2008) Cytokines: from gut inflammation to colorectal cancer. Curr Drug Targets 9:375–380

Fischetti VA (2003) Novel method to control pathogenic bacteria on human mucous membranes. Ann N Y Acad Sci 987:207–214

Fjeld KG, Mowe M, Eide H, Willumsen T (2014) Effect of electric toothbrush on residents' oral hygiene: a randomized clinical trial in nursing homes. Eur J Oral Sci 122:142–148

Fritz JV, Desai MS, Shah P, Schneider JG, Wilmes P (2013) From meta-omics to causality: experimental models for human microbiome research. Microbiome 3:14

Gendron R, Grenier D, Maheu-Robert L (2000) The oral cavity as a reservoir of bacterial pathogens for focal infections. Microbes Infect 2:897–906

Gondivkar SM, Gondivkar RS, Gadbail AR, Chole R, Mankar M, Yuwanati M (2013) Chronic periodontitis and the risk of head and neck squamous cell carcinoma: facts and figures. Exp Oncol 35:163–167

Grinde B, Olsen I (2010) The role of viruses in oral disease. J Oral Microbiol 2:2127. doi:10.3402/jom.v2iO.2127

Han YW, Wang X (2013) Mobile microbiome: oral bacteria in extra-oral infections and inflammation. J Dent Res 92:485–491
Haqqani AS, Sandhu JK, Birnboim HC (2000) Expression of interleukin-8 promotes neutrophil infiltration and genetic instability in mutatect tumors. Neoplasia 2:561–568
Herrera V, Parsonnet J (2009) Helicobacter pylori and gastric adenocarcinoma. Clin Microbiol Infect 15:971–976
Holgerson PL, Vestman NR, Claesson R, Ohman C, Domellöf M, Tanner AC, Hernell O, Johansson I (2013) Oral microbial profile discriminates breast-fed from formula-fed infants. J Pediatr Gastroenterol Nutr 56:127–136
Homann N, Tillonen J, Salaspuro M (2000a) Microbially produced acetaldehyde from ethanol may increase the risk of colon cancer via folate deficiency. Int J Cancer 86:169–173
Homann N, Tillonen J, Meurman JH, Rintamäki H, Lindqvist C, Rautio M, Jousimies-Somer H, Salaspuro M (2000b) Increased salivary acetaldehyde levels in heavy drinkers and smokers: a microbiological approach to oral cavity cancer. Carcinogenesis 21:663–668
Hooper SJ, Crean SJ, Fardy MJ, Lewis MA, Spratt DA, Wade WG, Wilson MJ (2007) A molecular analysis of the bacteria present within oral squamous cell carcinoma. J Med Microbiol 56:1651–1659
Hooper SJ, Wilson MJ, Crean SJ (2009) Exploring the link between microorganisms and oral cancer: a systematic review of the literature. Head Neck 31:1228–1239
Hu YJ, Wang Q, Jiang YT, Ma R, Xia WW, Tang ZS, Liu Z, Liang JP, Huang ZW (2013a) Characterization of oral bacterial diversity of irradiated patients by high-throughput sequencing. Int J Oral Sci 5:21–25
Hu YJ, Shao ZY, Wang Q, Jiang YT, Ma R, Tang ZS, Liu Z, Liang JP, Huang ZW (2013b) Exploring the dynamic core microbiome of plaque microbiota during head-and-neck radiotherapy using pyrosequencing. PLoS One 8(2):e56343. doi:10.1371/journal.pone.0056343
Hujoel PP, Drangsholt M, Spiekerman C, Weiss NS (2003) An exploration of the periodontitis-cancer association. Ann Epidemiol 13:312–316
Ide R, Fujino Y, Hoshiyama Y, Mizoue T, Kubo T, Pham TM, Shirane K, Tokui N, Sakata K, Tamakoshi A, Yoshimura T, JACC Study Group (2007) A prospective study of green tea consumption and oral cancer incidence in Japan. Ann Epidemiol 17:821–826
Janket SJ, Jones JA, Meurman JH, Baird AE, Van Dyke TE (2008) Oral infection, hyperglycemia, and endothelial dysfunction. Oral Surg Oral Med Oral Pathol Oral Radiol Endod 105:173–179
Joyce SA, Gahan CG (2014) The gut microbiota and the metabolic health of the host. Curr Opin Gastroenterol 30:120–127
Katz J, Wallet S, Cha S (2010) Periodontal disease and the oral-systemic connection: is it all the RAGE? Quintessence Int 41:229–237
Keijser BJ, Zaura E, Huse SM, van der Vossen JM, Schuren FH, Montijn RC, ten Cate JM, Crielaard W (2008) Pyrosequencing analysis of the oral microflora of healthy adults. J Dent Res 87:1016–1020
King K, Jones DH, Daltrey DC, Cunliffe WJ (1982) A double-blind study of the effects of 13-cis-retinoic acid on acne, sebum excretion rate and microbial population. Br J Dermatol 107:583–590
Komiyama K, Tynan JJ, Habbick BF, Duncan DE, Liepert DJ (1985) Pseudomonas aeruginosa in the oral cavity and sputum of patients with cystic fibrosis. Oral Surg Oral Med Oral Pathol 59:590–594
Krinsky NI (1989) Antioxidant functions of carotenoids. Free Radic Biol Med 7:617–635
Kumar M, Kumar A, Nagpal R, Mohania D, Behare P, Verma V, Kumar P, Poddar D, Aggarwal PK, Henry CJ, Jain S, Yadav H (2010) Cancer-preventing attributes of probiotics: an update. Int J Food Sci Nutr 61(5):473–496
Kuper H, Adami HO, Trichopoulos D (2000) Infections as a major preventable cause of human cancer. J Intern Med 248:171–183

Kurkivuori J, Salaspuro V, Kaihovaara P, Kari K, Rautemaa R, Grönroos L, Meurman JH, Salaspuro M (2007) Acetaldehyde production from ethanol by oral streptococci. Oral Oncol 43:181–186
Lai PK, Roy J (2004) Antimicrobial and chemopreventive properties of herbs and spices. Curr Med Chem 11:1451–1460
Lax AJ, Thomas W (2002) How bacteria could cause cancer: one step at a time. Trends Microbiol 10:293–299
Leung WK, Jin LJ, Samaranayake LP, Chiu GK (1998) Subgingival microbiota of shallow periodontal pockets in individuals after head and neck irradiation. Oral Microbiol Immunol 13:1–10
Liede K, Hietanen J, Saxen L, Haukka J, Timonen T, Häyrinen-Immonen R, Heinonen OP (1998a) Long-term supplementation with alpha-tocopherol and beta-carotene and prevalence of oral mucosal lesions in smokers. Oral Dis 4:78–83
Liede KE, Alfthan G, Hietanen JH, Haukka JK, Saxen LM, Heinonen OP (1998b) Beta-carotene concentration in buccal mucosal cells with and without dysplastic oral leukoplakia after long-term beta-carotene supplementation in male smokers. Eur J Clin Nutr 52:872–876
Lin WW, Karin M (2007) A cytokine-mediated link between innate immunity, inflammation, and cancer. J Clin Invest 117:1175–1183
Lucas VS, Beighton D, Roberts GJ, Challacombe SJ (1997) Changes in the oral streptococcal flora of children undergoing allogeneic bone marrow transplantation. J Infect 35:135–141
Madan PD, Sequeira PS, Shenoy K, Shetty J (2008) The effect of three mouthwashes on radiation-induced oral mucositis in patients with head and neck malignancies: a randomized control trial. J Cancer Res Ther 4:3–8
Mantovani A, Allavena P, Sica A, Balkwill F (2008) Cancer-related inflammation. Nature 454:436–444
Maria-Aggeliki KS, Nikolaos KL, George KM, Vassilis KE (2009) The potential clinical impact of probiotic treatment for the prevention and/or anti-inflammatory therapeutic effect against radiation induced intestinal mucositis. A review. Recent Pat Inflamm Allergy Drug Discov 3:195–200
Marichalar-Mendia X, Rodriguez-Tojo MJ, Acha-Sagredo A, Rey-Barja N, Aguirre-Urizar JM (2010) Oral cancer and polymorphism of ethanol metabolising genes. Oral Oncol 46:9–13
Martin D, Gutkind JS (2008) Human tumor-associated viruses and new insights into the molecular mechanisms of cancer. Oncogene 27(Suppl 2):S31–S42
Megraud F, Brassens-Rabbe MP, Denis F, Belbouri A, Hoa DQ (1989) Seroepidemiology of Campylobacter pylori infection in various populations. J Clin Microbiol 27:1870–1873
Meisel P, Holtfreter B, Biffar R, Suemnig W, Kocher T (2012) Association of periodontitis with the risk of oral leukoplakia. Oral Oncol 48:859–863
Meurman JH (2010) Oral microbiota and cancer. J Oral Microbiol 10:2. doi:10.3402/jom.v2i0.5195
Meurman JH, Grönroos L (2010) Oral and dental health care of oral cancer patients: hyposalivation, caries and infections. Oral Oncol 46:464–467
Meurman JH, Scully C (2012) Oral cancer: comprehending the condition, causes, controversies, control and consequences: infections. Dent Update 39:298–299
Meurman JH, Uittamo J (2008) Oral micro-organisms in the etiology of cancer. Acta Odontol Scand 66:321–326
Meurman JH, Pyrhönen S, Teerenhovi L, Lindqvist C (1997) Oral sources of septicaemia in patients with malignancies. Oral Oncol 33:389–397
Michaud DS, Joshipura K, Giovannucci E, Fuchs CS (2007) A prospective study of periodontal disease and pancreatic cancer in US male health professionals. J Natl Cancer Inst 99:171–175
Miller EG, Peacock JJ, Bourland TC, Taylor SE, Wright JM, Patil BS, Miller EG (2008) Inhibition of oral carcinogenesis by citrus flavonoids. Nutr Cancer 60:69–74
Moazzez R, Thompson H, Palmer RM, Wilson RF, Proctor GB, Wade WG (2011) Effect of rinsing with ethanol-containing mouthrinses on the production of salivary acetaldehyde. Eur J Oral Sci 119:441–446

Moergel M, Kämmerer P, Kasaj A, Armouti E, Alshihri A, Weyer V, Al-Nawas B (2013) Chronic periodontitis and its possible association with oral squamous cell carcinoma – a retrospective case control study. Head Face Med 9:39. doi:10.1186/1746-160X-9-39

Moore RJ, Owens DM, Stamp G, Arnott C, Burke F, East N, Holdsworth H, Turner L, Rollins B, Pasparakis M, Kollias G, Balkwill F (1999) Mice deficient in tumor necrosis factor-alpha are resistant to skin carcinogenesis. Nat Med 5:828–831

Morini M, Cai T, Aluigi MG, Noonan DM, Masiello L, De Flora S, D'Agostini F, Albini A, Fassina G (1999) The role of the thiol N-acetylcysteine in the prevention of tumor invasion and angiogenesis. Int J Biol Markers 14:268–271

Morse DE, Psoter WJ, Cleveland D, Cohen D, Mohit-Tabatabai M, Kosis DL, Eisenberg E (2007) Smoking and drinking in relation to oral cancer and oral epithelial dysplasia. Cancer Causes Control 18:919–929

Moss SF, Blaser MJ (2005) Mechanisms of disease: inflammation and the origins of cancer. Nat Clin Pract Oncol 2:90–97

Narikiyo M, Tanabe C, Yamada Y, Igaki H, Tachimori Y, Kato H et al (2004) Frequent and preferential infection of Treponema denticola, Streptococcus mitis, and Streptococcus anginosus in esophageal cancers. Cancer Sci 95:569–574

Nieminen MT, Uittamo J, Salaspuro M, Rautemaa R (2009) Acetaldehyde production from ethanol and glucose by non-Candida albicans yeasts in vitro. Oral Oncol 45:e245–e248

Olsen I, Tribble GD, Fiehn NE, Wang BY (2013) Bacterial sex in dental plaque. J Oral Microbiol 3:5. doi:10.3402/jom.v5i0.20736

Parsonnet J (1999) Microbes and malignancy. Oxford University Press, New York, pp 3–15

Pei XH, Nakanishi Y, Takayama K, Bai F, Hara N (1999) Benzo[a]pyrene activates the human p53 gene through induction of nuclear factor kappaB activity. J Biol Chem 274:35240–35246

Peterson DE, Minah GE, Reynolds MA, Weikel DS, Overholser CD, DePaola LG, Wade JC, Suzuki JB (1990) Effect of granulocytopenia on oral microbial relationships in patients with acute leukemia. Oral Surg Oral Med Oral Pathol 70:720–723

Pompei R, Ingianni A, Cagetti MG, Rizzo A, Cotti S (1993) Evaluation of the opportunistic microbial flora and of some antimicrobial factors in the oral cavity of leukaemic patients. Microbios 75:149–157

Preza D, Olsen I, Willumsen T, Grinde B, Paster BJ (2009) Diversity and site-specificity of the oral microflora in the elderly. Eur J Clin Microbiol Infect Dis 28:1033–1040

Pushalkar S, Ji X, Li Y, Estilo C, Yegnanarayana R, Singh B, Li X, Saxena D (2012) Comparison of oral microbiota in tumor and non-tumor tissues of patients with oral squamous cell carcinoma. BMC Microbiol 12:144. doi:10.1186/1471-2180-12-144

Rajeev R, Choudhary K, Panda S, Gandhi N (2012) Role of bacteria in oral carcinogenesis. South Asian J Cancer 1:78–83

Rautava J, Syrjänen S (2012) Biology of human papillomavirus infections in head and neck carcinogenesis. Head Neck Pathol 6(Suppl 1):S3–S15

Rautemaa R, Rusanen P, Richardson M, Meurman JH (2006) Optimal sampling site for mucosal candidosis in oral cancer patients is the labial sulcus. J Med Microbiol 55:1447–1451

Redding SW, Dahiya MC, Kirkpatrick WR, Coco BJ, Patterson TF, Fothergill AW, Rinaldi MG, Thomas CR (2004) Candida glabrata is an emerging cause of oropharyngeal candidiasis in patients receiving radiation for head and neck cancer. Oral Surg Oral Med Oral Pathol Oral Radiol Endod 97:47–52

Reynolds MA, Minah GE, Peterson DE, Weikel DS, Williams LT, Overholser CD, DePaola LG, Suzuki JB (1989) Periodontal disease and oral microbial successions during myelosuppressive cancer chemotherapy. J Clin Periodontol 16:185–189

Rusanen P, Siikala E, Uittamo J, Richardson M, Rautemaa R (2009) A novel method for sampling the microbiota from the oral mucosa. Clin Oral Invest 13:243–246

Salaspuro V (2007) Pharmacological treatments and strategies for reducing oral and intestinal acetaldehyde. Novartis Found Symp 285:145–153

Salaspuro V, Salaspuro M (2004) Synergistic effect of alcohol drinking and smoking on in vivo acetaldehyde concentration in saliva. Int J Cancer 111:480–483

Salaspuro V, Hietala J, Kaihovaara P, Pihlajarinne L, Marvola M, Salaspuro M (2002) Removal of acetaldehyde from saliva by a slow-release buccal tablet of L-cysteine. Int J Cancer 97:361–364

Salaspuro VJ, Hietala JM, Marvola ML, Salaspuro MP (2006) Eliminating carcinogenic acetaldehyde by cysteine from saliva during smoking. Cancer Epidemiol Biomarkers Prev 15:146–149

Sasaki M, Yamaura C, Ohara-Nemoto Y, Tajika S, Kodama Y, Ohya T, Harada R, Kimura S (2005) Streptococcus anginosus infection in oral cancer and its infection route. Oral Dis 11:151–156

Scannapieco FA (1998) Position paper of The American Academy of Periodontology: periodontal disease as a potential risk factor for systemic diseases. J Periodontol 69:841–850

Sheu BC, Chang WC, Cheng CY, Lin HH, Chang DY, Huang SC (2008) Cytokine regulation networks in the cancer microenvironment. Front Biosci 13:6255–6268

Shiga K, Tateda M, Saijo S, Hori T, Sato I, Tateno H, Matsuura K, Takasaka T, Miyagi T (2001) Presence of Streptococcus infection in extra-oropharyngeal head and neck squamous cell carcinoma and its implication in carcinogenesis. Oncol Rep 18:245–248

Sixou JL, De Medeiros-Batista O, Gandemer V, Bonnaure-Mallet M (1998) The effect of chemotherapy on the supragingival plaque of pediatric cancer patients. Oral Oncol 34:476–483

Söder B, Yakob M, Meurman JH, Andersson LC, Klinge B, Söder PÖ (2011) Periodontal disease may associate with breast cancer. Breast Cancer Res Treat 127:497–502

Söder B, Yakob M, Meurman JH, Andersson LC, Söder PÖ (2012) The association of dental plaque with cancer mortality in Sweden. A longitudinal study. BMJ Open 2:pii: e001083. doi:10.1136/bmjopen-2012-001083

Sorensen JB, Skovsgaard T, Bork E, Damstrup L, Ingeberg S (2008) Double-blind, placebo-controlled, randomized study of chlorhexidine prophylaxis for 5-fluorouracil-based chemotherapy-induced oral mucositis with nonblinded randomized comparison to oral cooling (cryotherapy) in gastrointestinal malignancies. Cancer 112:1600–1606

Tachibana M, Yoshida A, Ansai T, Takata Y, Akifusa S, Fukuhara M et al (2006) Prevalence of periodontopathic bacteria on the tongue dorsum of elderly people. Gerodontology 23:123–126

Taylor PW, Hamilton-Miller JM, Stapleton PD (2005) Antimicrobial properties of green tea catechins. Food Sci Technol Bull 2:71–81

Tengerdy RP (1990) The role of vitamin E in immune response and disease resistance. Ann N Y Acad Sci 587:24–33

Tezal M (2012) Interaction between chronic inflammation and oral HPV infection in the etiology of head and neck cancers. Int J Otolaryngol 2012:575242. doi:10.1155/2012/575242

Tezal M, Grossi SG, Genco RJ (2005) Is periodontitis associated with oral neoplasms? J Periodontol 76:406–410

Tezal M, Sullivan MA, Reid ME, Marshall JR, Hyland A, Loree T, Lillis C, Hauck L, Wactawski-Wende J, Scannapieco FA (2007) Chronic periodontitis and the risk of tongue cancer. Arch Otolaryngol Head Neck Surg 133:450–454

Tezal M, Sullivan MA, Hyland A, Marshall JR, Stoler D, Reid ME, Loree TR, Rigual NR, Merzianu M, Hauck L, Lillis C, Wactawski-Wende J, Scannapieco FA (2009) Chronic periodontitis and the incidence of head and neck squamous cell carcinoma. Cancer Epidemiol Biomarkers Prev 18:2406–2412

Tezal M, Scannapieco FA, Wactawski-Wende J, Meurman JH, Marshall JR, Rojas IG, Stoler DL, Genco RJ (2013) Dental caries and head and neck cancers. JAMA Otolaryngol Head Neck Surg 139:1054–1060

Tong HC, Gao XJ, Dong XZ (2003) Non-mutans streptococci in patients receiving radiotherapy in the head and neck area. Caries Res 37:261–266

Turnbaugh PJ, Ley RE, Hamady M, Fraser-Liggett CM, Knight R, Gordon JI (2007) The human microbiome project. Nature 449:804–810

Turnbaugh PJ, Hamady M, Yatsunenko T, Cantarel BL, Duncan A, Ley RE, Sogin ML, Jones WJ, Roe BA, Affourtit JP, Egholm M, Henrissat B, Heath AC, Knight R, Gordon JI (2009) A core gut microbiome in obese and lean twins. Nature 457:480–484

Uittamo J, Siikala E, Kaihovaara P, Salaspuro M, Rautemaa R (2009) Chronic candidosis and oral cancer in APECED-patients: production of carcinogenic acetaldehyde from glucose and ethanol by Candida albicans. Int J Cancer 124:754–756

Van Schooten FJ, Besaratinia A, De Flora S, D'Agostini F, Izzotti A, Camoirano A, Balm AJ, Dallinga JW, Bast A, Haenen GR, Van't Veer L, Baas P, Sakai H, Van Zandwijk N (2002) Effects of oral administration of N-acetyl-L-cysteine: a multi-biomarker study in smokers. Cancer Epidemiol Biomarkers Prev 11:167–175

Virtanen E, Söder B, Andersson LC, Meurman JH, Söder PÖ (2014) History of dental infections associates with cancer in periodontally healthy subjects: a 24-year follow-up study from Sweden. J Cancer 5:79–85. doi:10.7150/jca.7402

Wade WG (2013) The oral microbiome in health and disease. Pharmacol Res 69:137–143

Wahlin YB (1989) Effects of chlorhexidine mouthrinse on oral health in patients with acute leukemia. Oral Surg Oral Med Oral Pathol 68:279–287

Wen BW, Tsai CS, Lin CL, Chang YJ, Lee CF, Hsu CH, Kao CH (2013) Cancer risk among gingivitis and periodontitis patients: a nationwide cohort study. QJM 107(4):283–290

Worthington HV, Clarkson JE, Eden OB (2007) Interventions for preventing oral mucositis for patients with cancer receiving treatment. Cochrane Database Syst Rev 4:CD000978

Xu Y, Teng F, Huang S, Lin Z, Yuan X, Zeng X, Yang F (2014) Changes of saliva microbiota in nasopharyngeal carcinoma patients under chemoradiation therapy. Arch Oral Biol 59:176–186

Yumoto H, Nakae H, Fujinaka K, Ebisu S, Matsuo T (1999) Interleukin-6 (IL-6) and IL-8 are induced in human oral epithelial cells in response to exposure to periodontopathic Eikenella corrodens. Infect Immun 67:384–394

Chapter 12
Streptococcus bovis and Colorectal Cancer

Salvatore Galdy

Abstract The role of *Streptococcus bovis* (*S. bovis*) as an aetiological agent in the development of colorectal cancer (CRC) is intriguing but uncertain. A relationship between infective endocarditis (IE) and CRC was established by McCoy and Mason in 1951 and, for the first time, an association between *S. bovis* and endocarditis was successfully recognized by Watanakunakorn in 1974. In the same year, Hoppes and Lerner hypothesized that some previous reports of endocarditis caused by penicillin-sensitive "*enterococci*", including that of McCoy, were probably unrecognized examples of *S. bovis*. In 1977, Klein and coworkers showed the prevalence of *S. bovis* in fecal cultures from patients with *S. bovis* septicemia and carcinoma of the colon was significantly increased. Thus, *S. bovis* infection should be considered a silent sign of gastrointestinal malignancy. Over the past 50 years, several case reports and studies – most retrospective – have been publishing on this topic often producing contradictory results. Currently, only *Streptococcus gallolyticus subspecies gallolyticus* (SGG) – formerly known as *S. bovis* biotype I – has been recognized to be directly related to colonic neoplasia. Hence, in order to demonstrate the presence of a colon cancer, all patients with *S. bovis/gallolyticus* infection would require an endoscopic investigation.

Keywords *Streptococcus bovis/gallolyticus* • Colorectal cancer • Infective endocarditis

Abbreviations

Cox-2	cyclooxygenase-2
CRC	colorectal cancer
IE	infective endocarditis
IL-1	interleukin 1
IL-8	interleukin 8

S. Galdy (✉)
Unit of Gastrointestinal Medical Oncology and Neuroendocrine Tumors (Director Nicola Fazio), European Institute of Oncology, Milan, Italy
e-mail: salvatore.galdy@ieo.it

M.R. Shurin et al. (eds.), *Infection and Cancer: Bi-Directorial Interactions*, DOI 10.1007/978-3-319-20669-1_12

NSAID	nonsteroidal anti-inflammatory drugs
PCR	polymerase chain reaction
S. Bovis	Streptococcus Bovis
SGG	Streptococcus Gallolyticus subspecies Gallolyticus

Colorectal Cancer: Epidemiology and Risk Factors

In the United States, colorectal cancer (CRC) is the third common cancer both in men and women, representing also the third leading cause of cancer mortality (Siegel et al. 2014). In 2012, nearly 1,4 million cases were diagnosed worldwide with more 600,000 deaths, accounting for about 10 % of human cancers and for 8.5 % of cancer mortality, respectively (www.cancerresearchuk.org). In the last 30 years, both incidence and mortality rates have declined in Western countries, probably as consequence of increased screening for and removal of premalignant polyps (Bond 2000) and widespread use of aspirin and other nonsteroidal anti-inflammatory drugs (NSAIDs) (Chan et al. 2005).

The risk for colorectal cancer increases with age and the median age of diagnosis is around 70 years (Siegel et al. 2014). Multiple factors may drive the transformation of healthy bowel mucosa to cancer, both inheritance and environmental factors such as personal and/or family history of colorectal cancer and adenomatous polyps, inherited syndromes, diet (fat intake) and obesity, physical inactivity, heavy alcohol use, smoking, inflammatory bowel disease, such as ulcerative colitis (http://www.cancer.org), and potentially infectious agents (Burnett-Hartman et al. 2008).

Streptococcus Bovis

Streptococcus bovis (*S. bovis*), a member of group D *streptococci*, is a common inhabitant of human bowel in about 5–35 % of adults (Noble 1978; Lopes et al. 2014); as regards its ability to hydrolyze esculin and viability in 40 % bile but not in 6.5 % salt broth, it may differentiated from *enterococci* (Facklam 1972).

S. bovis has long been associated with CRC, however, not all genospecies are as closely related to CRC. Historically, *S. bovis* group has been divided into 3 biotypes according to their biochemical characteristics: biotype I (mannitol fermentation positive), biotype II/1 (mannitol negative and b-glucuronidase negative), and biotype II/2 (mannitol negative and b-glucuronidase positive) (Facklam 1972). In 1989, Ruoff and colleagues showed *S. bovis* biotype I was the biotype more likely associated with both endocarditis and malignant or premalignant colonic lesions (Ruoff et al. 1989). The taxonomy of the *S. bovis* group has been evolving in the last few decades, and a new nomenclature was adopted on the basis of genetic distances and phylogenetic analyses. The new classification listed *S. bovis* biotypes I, II/1,

Table 12.1 Nomenclature of *Streptococcus bovis/ Streptococcus equinus* complex

New denomination	Former denomination	Synonym/current denomination
Streptococcus gallolyticus spp gallolyticus	*S. bovis biotype I*	*S. gallolyticus*
Streptococcus infantarius spp infantarius	*S. bovis biotype II/1*	*S. infantarius*
Streptococcus infantarius spp coli	*S. bovis biotype II/1*	*S. lutiensis*
Streptococcus gallolyticus spp pasteurianus	*S. bovis biotype II/2*	*S. pasteurianus*

and II/2 as *Streptococcus gallolyticus subspecies gallolyticus*, *Streptococcus infantarius subsp. infantarius and subsp. coli*, and *S. gallolyticus subsp. pasteurianus*, respectively (Table 12.1) (Schlegel et al. 2003).

Streptococcus Gallolyticus Subspecies Gallolyticus (Formerly *S. Bovis* Biotype I), Infective Endocarditis and Colorectal Cancer

A recent meta-analysis on the association among *S. bovis* biotypes, infective endocarditis (IE) and colonic adenomas/carcinomas revealed that patients with *S. bovis* biotype I infection had a strongly increased risk of having CRC and IE, compared to *S. bovis* biotype II-infected patients (Boleij et al. 2011b). This analysis clearly indicates that *S. bovis* should no longer be regarded as a single bacterial entity in clinical practice. IE is an infection of heart valves and mural endocardium caused by bacteria, fungi, rickettsiae and virus. Bacterial endocarditis is the most common clinical form of endocarditis and *S. bovis* is the causative agent in 10–15 % of all IE (Gupta et al. 2010); it can occur in either acute or subacute phase and usually affects immunocompromised patients or patients with valve defects. IE is usually associated with valve vegetation, continuous bacteremia, splenomegaly, remitting fever type and significant heart murmur (85 % of cases). The endocardial involvement may lead to valve leaflets rupture with heart failure in severe forms of endocarditis (Filice and Bruno 1998).

Moreover, CRC appears to occur more often among patients with *S. bovis*/IE than those with *S. bovis* infection at other sites. Only *Streptococcus gallolyticussubsp gallolyticus* (SGG) – formerly known as *S. bovis* biotype I – infection has an unambiguous association with colonic adenomas/carcinomas (prevalence range: 33–100 %) that markedly exceeds the prevalence of both colonic malignancies (0.3 %) and pre-malignancies (10–25 %) in the general population aged 60–70 (Table 12.2). More specifically, patients with SGG infection had a significantly increased risk for colorectal adenomas and carcinomas (OR = 7.3; 95 % Confidence Interval, 3.9–13.4) and for IE (OR = 16.6; 95 % CI, 8.85–31.2) (Boleij et al. 2011b). However, the new *S. bovis* taxonomy, since its publication, has been used by only a few authors; in attempt to minimize and/or avoid inconsistent results it would be

Table 12.2 Cases Series on *Streptococcus Bovis* and colorectal cancer

Trial	Design	Mean age	Total no. and biotype	No. Colon endoscopy (CE)	No. IE	No. Adenomas	No. Carcinomas	CRC/ total %	CRC/ CE %	Adenomas/ CE %	Carcinomas/CE %
Ruoff (1989)	Retrospective	67	38	12	19		15	39			
			I (17)	12	16		12	71	100		
			II/1(12)		1		1	8			
			II/2 (9)		2		2	22			
Herrero (2002)	Retrospective	65	14	9	14	1	2	21	33	11	22
			I (11)	9							
			II/1 (1)								
Tripodi (2004)	Prospective	59	30	28	30	13	1	47	50	46	4
			I (28)	27	28	13	1	47	52		
			II (2)	1	2		0				
Jean (2004)	Retrospective	61	60	19	15		11	18	47		
			I (10)		8		5	50			
			II (37)		4		6	16			
Correidora (2005)	Prospective	67	64		34		27	44			
			I (42)		31		24	57			
			II (22)		3		3	14			
Giannitsioti (2007)	Retrospective	63	142	92	142	55	5	42	65	60	5
			I (71)	46		24	3	38	59	52	7
			II/1 (5)								
			II/2 (3)								
Correidora (2008)	Prospective	NA	133				54	41			
			I (90)				51	57			
			II/1 (28)				3	11			
			II/2 (15)								

Beck (2008)	Retrospective	70	46	15	13	8	3	24	43	53	20
			I (21*)		9	5	2	33			
			II/1 (14)		4	3	1	29			
			II/2 (11)								
Vaska (2009)	Retrospective	68	20	14	8	5	5	50	71	36	36
			I (10)	9	6	2	5	70	78	22	56
			II/1 (10)	5	2	3		30	60	60	
Correidora (2012)	Prospective	66	109 (All biotype I)	98	86	57	12	57	70	58	12

Modified from Boleij et al. (2011b)

advisable using the same nomenclature in both clinical practice and scientific research. Genetic techniques such as 16S rDNA sequence analysis should be used widely for the identification of *S. bovis* group bacteria (Woo et al. 2009). At the time of writing, 10 English language studies, which have taken into account the different *S. bovis* biotypes and have reported the number of CRC and/or IE cases stratified by biotype, were available in literature (Table 12.2). This relationship has well been assessed in the prospective case-control study of Correidora and co-workers (2012), in which 109 cases of SGG bacteremia were compared with 196 symptomatic age-matched controls. The prevalence of both invasive carcinoma and advanced adenomas was significantly higher in patients with SGG bacteremia/IE than in controls, with no difference for non-advanced lesions. Advance adenoma is an adenoma with a diameter ≥1 cm concomitant with tubulovillous or villous histological findings, high-grade dysplasia, or early carcinoma.

Streptococcus Bovis/Gallolyticus and Carcinogenesis

Carcinogenesis of colorectal cancer is a multi-step process leading to cancer development as indicated by model of Faeron and Volgestein (1990). Knudson's two-hit hypothesis, which suggests an interaction between host predisposition to carcinogenesis and a second environmental hit leading to cell proliferation, is suitable for colorectal cancer too (Knudson 2001). Infection and chronic inflammation through the production of some cytokines and free radicals may activate the cyclooxygenase-2 (Cox-2) and nuclear factor-kb (NF-kb) leading to apoptosis inhibition and angiogenesis promotion (Gupta and Dubois 2001). In a rat model, in which the colonic mucosa underwent carcinogenesis by azoxymethane, Ellmerich and co-workers (2000) showed that both *S. bovis* and extracted antigens from the bacterial cell wall were able to promote the progression of pre-neoplastic lesions by increasing both cell proliferation and interleukin 8 (IL-8) levels. IL-8 is a potent angiogenic factor and acts as chemotactic factor for neutrophils promoting phagocytosis and contributing to the normal defense mechanisms of mucosal host (Köhidai and Csaba 1998); thus, its increase could be considered as an indirect sign of *S. bovis* effect on cell proliferation. Most importantly, the observation that normal rats treated with *S. bovis* did not develop aberrant crypt foci indicates that *S. bovis* might promote rather than induce the carcinogenesis of colonic tissue (Ellmerich et al. 2000).

So, many studies have shown an association between *S. bovis* bacteremia/IE and CRC, but others, which have examined the relationship between positive *S. bovis* stool carriage and CRC risk, have produced conflicting results (Klein et al. 1979; Dubrow et al. 1991; Potter et al. 1998; Chirouze et al. 2013). Contrary to conclusions of Klein and co-workers (1979), who reported a strong association between fecal carriage and colon disease in patients with *S. bovis* septicemia, most of studies, which have included patients with no sign of bacteremia/IE, showed that *S. bovis/gallolyticus* fecal carriage rates were similar between patients with CRC (11–16 %) and controls (6–13 %) (Dubrow et al. 1991; Potter et al. 1998; Chirouze et al. 2013).

Recently, however, Abudlamir and colleagues have shown that SGG DNA levels, detected by a polymerase chain reaction (PCR) assay, in tumorous tissue from 91 patients with CRC, either with or without bacteremia, were much higher (49 % and 33 %, respectively) than those in the control group (4 %; $p<0.05$). By contrast, the frequency of bacteriological isolation of *S. bovis* from feces and mucosal surfaces of CRC groups, regardless bacteremia status, was not different from the corresponding control groups ($p=0.77$). The finding of a tumoral enrichment of *S. gallolyticus* DNA in colorectal carcinoma suggests the possibility that this organism may contribute to CRC tumorigenesis, at least in terms of promoting and propagating rather than trigger factor, most conceivably by an inflammatory-mediated mechanism [mostly via interleukin 1(IL-1), cyclooxygenase-2 (COX-2) and IL-8] (Abdulamir et al. 2010). Moreover, in a Brazilian cohort of 54 rectal swab specimens - collected from patients without history of CRC and undergone DNA amplification by a real-time PCR assay – the overall prevalence of DNA *S. bovis* was 35.2 % (Lopes et al. 2014), much higher than fecal carriage rates of previous studies. Together, these data seem to confirm a discrepancy between traditional methods (fecal carriage) and more modern and sensitive techniques (DNA assays) for identifying *S. bovis* presence in bowel lumen.

Bowel Colonization and Translocation in Bloodstream

S. bovis/gallolyticus is a transient normal bacterium in the gut, characterized by a relative low adhesiveness (<15 %), occasionally colonizes colonic mucosa and/or pre-malignant lesions (Boleij et al. 2011a). Boleij and colleagues found a histone-like protein A on the cell wall of SGG able to bind heparin sulfate proteoglycans at the colon tumor cell surface during the first stages of infection (Boleij et al. 2009). Moreover, specific collagen-binding proteins (collagen types I and IV) and specific pili might explain the ability of *S. bovis/gallolyticus* to adhere to both colonic mucosa and endocardium (Sillanpää et al. 2009). Collagen type IV is note to be abundant in membranes of crypt foci of hyperplastic polyps (Galbavy et al. 2002) while damaged heart valves are rich in collagen type I (Phillippi et al. 2010). *S. bovis* is the causative agent in 10–15 % of all infective endocarditis (Gupta et al. 2010). Thus, the collagen-binding ability may be considered as the key virulence of *S. bovis/gallolyticus.*

Thereafter mucosal colonization, due to the altered intestinal wall's permeability, *S. bovis/gallolyticus* may translocate from the lumen to the bloodstream, via a paracellular mechanism without inducing epithelial IL-8 or IL-1β responses, (Boleij et al. 2012a) in contrast with previous reports (Ellmerich et al. 2000; Abdulamir et al. 2009, 2010). Actually, the in vitro model carried out by Boleij and colleagues showed that SGG was able to evade the innate immune system or at least delay the action of macrophages. These biological features may explain the stronger association between SGG/IE and CRC than other *S. bovis* strains (Boleij et al. 2012a).

Conclusions

Streptococcus bovis generally is a transient and commensal microorganism of gut microbiota of humans and other vertebrate. In some circumstances, more specifically *S. gallolyticus spp gallolyticus* (SGG) – formerly *S. bovis* biotype I – can colonize bowel mucosa adhering to both pre- and malignant lesions forming biofilms on collagen-coated surfaces. On the other hand, malignant lesions themselves could constitute a niche for the *S. bovis* overgrowth (Boleij et al. 2012a).

Some components of the *S. bovis'* bacterial cell wall are able to increase the production of specific cytokines promoting intestinal carcinogenesis (Ellmerich et al. 2000). Together, the ability of *S. bovis/gallolyticus* to escape the immune mucosal system and the altered permeability of the intestinal wall, would facilitate its passage from the lumen to the bloodstream, resulting in bacteremia and other infectious complications such as endocarditis, spondylodiscitis, abscess, etc. (Tjalsma et al. 2006); the likelihood of developing infective endocarditis is higher in elderly patients with preexisting heart valve defects. Thus, there is an unresolved debate as to whether *S. bovis* has a role in carcinogenesis or whether it is simply a colonizer in the human gut.

In the meta-analysis of Boleij and colleagues (2011b), 43 % of the *S. bovis/gallolyticus-infected* patients undergoing colonoscopy were reported to have adenomas and 18 % had carcinomas. Indeed, the prevalence of *S. bovis* infection is higher in early stages of colorectal disease. Probably, the *S. bovis/gallolyticus* may not be considered a trigger for colorectal cancer, but it might promote and accelerate the molecular pathways that lead to development of colorectal cancer from adenomas.

Based on the published data, only *S. bovis/gallolyticus* bacteremia/IE should be considered an unambiguous sign of underlying colorectal cancer. However, asymptomatic *S. bovis* bacteremia should be considered as sign of silent CRC even among younger people. In a retrospective study, 16 cases of *S. bovis* bacteremia were found among 369,477 blood donors (mean age 43.5 years). Histologically proven adenoma and carcinoma was identified in 37.5 % (6 of 16 cases) and as high as 75 % (6 of 8 cases) among those undergone colonoscopy (Lee et al. 2013). Conventional culturing methods have shown a low accuracy in detecting a *S. bovis* gut colonization in patients suffering from CRC. Therefore, other more sensitive techniques such as molecular and genetic tests (16S rDNA sequencing by PCR amplification) should be performed on feces (Lopes et al. 2014).

Another potential and useful marker for early and reliable detection of colorectal cancer would be the sero-prevalence. Only a few studies have investigated the role of humoral immune response to *S. bovis* and have correlated the *S. bovis* antibody titers with CRC. The first study that established a serological association between *S. bovis* antibodies, detected by an ELISA assay, and CRC was published in 1993 (Darjee and Gibb 1993). Unlike IgM antibodies, only IgG antibody titers were significantly higher in patients with CRC than in controls. This finding suggests that immune stimulation occurs over a long period of time, as well as the development of CRC. A subsequent study underscored a specific serological association between

SGG-IgG antibodies and colorectal lesions (both carcinoma and polyps) (Abdulamir et al. 2009). Others linked the humoral immune response to surface antigens such as histone-like protein and pilus proteins, key mediators of *S. bovis/gallolyticus* virulence, with higher specificity (78–100 %) and lower sensitivity (16–88 %) (Tjalsma et al. 2006; Boleij et al. 2012b).

As was discussed above, both CRC and *S. bovis* infection occur more often in elderly people (over 60 years). Thus, *S. bovis/gallolyticus* IgG antibodies detection, simultaneously with fecal occult blood test (FOBT), could be addressed at least to high risk individuals as a screening test. All patients with a diagnosis of *S. bovis/gallolyticus* infection should be highly recommended for an endoscopic study of the large intestine. Among *S. bovis*-infected patients, who have undergone colonoscopy, only in a few cases was reported a diagnosis of metastatic disease. Thus, the application of this advice might lead to early diagnosis of colorectal cancer in most of cases. It is estimated that about 33–100 % of patients affected by *S. bovis/gallolyticus* already have or will develop bowel cancer (Boleij et al. 2011b). After resolution of the infection, a minimum of 2-year to 4-year follow-up with colonoscopy is recommended, as there is a higher incidence of precancerous lesions and cancer of the intestine during this period (Wentling et al. 2006). Although a causal link between *S. bovis/gallolyticus* infection and CRC development has not been definitively established, this association should be no longer considered as a casualness connection.

References

Abdulamir AS, Hafidh RR, Mahdi LK et al (2009) Investigation into the controversial association of Streptococcus gallolyticus with colorectal cancer and adenoma. BMC Cancer 9:403. doi:10.1186/1471-2407-9-403

Abdulamir AS, Hafidh RR, Abu Bakar F (2010) Molecular detection, quantification, and isolation of Streptococcus gallolyticus bacteria colonizing colorectal tumors: inflammation-driven potential of carcinogenesis via IL-1, COX-2, and IL-8. Mol Cancer 9:249. doi:10.1186/1476-4598-9-249

Beck M, Frodl R, Funke G (2008) Comprehensive study of strains previously designated Streptococcus bovis consecutively isolated from human blood cultures and emended description of Streptococcus gallolyticus and Streptococcus infantarius subsp. coli. J Clin Microbiol 46:2966–2972. doi:10.1128/JCM.00078-08

Boleij A, Schaeps RM, de Kleijn S et al (2009) Surface-exposed histone-like protein a modulates adherence of Streptococcus gallolyticus to colon adenocarcinoma cells. Infect Immun 77:5519–5527. doi:10.1128/IAI.00384-09

Boleij A, Muytjens CM, Bukhari SI et al (2011a) Novel clues on the specific association of Streptococcus gallolyticus subsp gallolyticus with colorectal cancer. J Infect Dis 203:1101–1109. doi:10.1093/infdis/jiq169

Boleij A, van Gelder MM, Swinkels DW et al (2011b) Clinical Importance of Streptococcus gallolyticus infection among colorectal cancer patients: systematic review and meta-analysis. Clin Infect Dis 53:870–878. doi:10.1093/cid/cir609

Boleij A, Dutilh BE, Kortman GA et al (2012a) Bacterial responses to a simulated colon tumor microenvironment. Mol Cell Proteomics 11:851–862

Boleij A, Roelofs R, Danne C et al (2012b) Selective antibody response to Streptococcus gallolyticus pilus proteins in colorectal cancer patients. Cancer Prev Res (Phila) 5:260–265. doi:10.1158/1940-6207

Bond JH (2000) Polyp guideline: diagnosis, treatment, and surveillance for patients with colorectal polyps. Practice Parameters Committee of the American College of Gastroenterology. Am J Gastroenterol 95:3053–3063

Burnett-Hartman AN, Newcomb PA, Potter JD (2008) Infectious agents and colorectal cancer: a review of Helicobacter pylori, Streptococcus bovis, JC virus, and human papillomavirus. Cancer Epidemiol Biomarkers Prev 17:2970–2979. doi:10.1158/1055-9965

Chan AT, Giovannucci EL, Meyerhardt JA et al (2005) Long-term use of aspirin and nonsteroidal anti-inflammatory drugs and risk of colorectal cancer. JAMA 294:914–923

Chirouze C, Patry I, Duval X et al (2013) Streptococcus bovis/Streptococcus equinus complex fecal carriage, colorectal carcinoma, and infective endocarditis: a new appraisal of a complex connection. Eur J Clin Microbiol Infect Dis 32:1171–1176. doi:10.1007/s10096-013-1863-3

Corredoira JC, Alonso MP, García JF et al (2005) Clinical characteristics and significance of Streptococcus salivarius bacteremia and Streptococcus bovis bacteremia: a prospective 16-year study. Eur J Clin Microbiol Infect Dis 24:250–255

Corredoira J, Alonso MP, Coira A et al (2008) Association between Streptococcus infantarius (formerly S. bovis II/1) bacteremia and noncolonic cancer. J Clin Microbiol 46:1570. doi:10.1128/JCM.00129-08

Corredoira-Sánchez J, García-Garrote F, Rabuñal R et al (2012) Association between bacteremia due to Streptococcus gallolyticus subsp. gallolyticus (Streptococcus bovis I) and colorectal neoplasia: a case-control study. Clin Infect Dis 55:491–496. doi:10.1093/cid/cis434

Darjee R, Gibb AP (1993) Serological investigation into the association between Streptococcus bovis and colonic cancer. J Clin Pathol 46:1116–1119. Erratum in: J Clin Pathol 47:870

Dubrow R, Edberg S, Wikfors E et al (1991) Fecal carriage of Streptococcus bovis and colorectal adenomas. Gastroenterology 101:721–725

Ellmerich S, Schöller M, Duranton B et al (2000) Promotion of intestinal carcinogenesis by Streptococcus bovis. Carcinogenesis 21:753–756. doi:10.1093/carcin/21.4.753

Facklam RR (1972) Recognition of group D streptococcal species of human origin by biochemical and physiological tests. Appl Microbiol 23:1131–1139

Faeron ER, Volgestein B (1990) A genetic model for colorectal tumorigenesis. Cell 61:759–767

Filice G, Bruno R (1998) Endocardite infettiva. In: Filice G (ed) Malattie infettive, 2nd edn. McGraw-Hill Libri Italia, Milano, pp 173–181

Galbavy S, Lukac L, Porubsky J et al (2002) Collagen type IV in epithelial tumours of colon. Acta Histochem 104:331–334

Giannitsioti E, Chirouze C, Bouvet A et al (2007) Characteristics and regional variations of group D streptococcal endocarditis in France. Clin Microbiol Infect 13:770–776

Gupta RA, Dubois RN (2001) Colorectal cancer prevention and treatment by inhibition of cyclooxygenase-2. Nat Rev Cancer 1:11–21

Gupta A, Madani R, Mukhtar H (2010) Streptococcus bovis and endocarditis, a silent sign for colonic tumour. Colorectal Dis 12:164–171

Herrero IA, Rouse MS, Piper KE et al (2002) Reevaluation of Streptococcus bovis endocarditis cases from 1975 to 1985 by 16S ribosomal DNA sequence analysis. Clin Microbiol 40:3848–3850

Jean SS, Teng LJ, Hsueh PR et al (2004) Bacteremic Streptococcus bovis infections at a university hospital, 1992–2001. J Formos Med Assoc 103:118–123

Klein RS, Catalano MT, Edberg SC et al (1979) Streptococcus bovis septicemia and carcinoma of the colon. Ann Intern Med 91:560–562

Knudson A (2001) Alfred Knudson and his two-hit hypothesis. (Interview by Ezzie Hutchinson). Lancet Oncol 2:642–645

Köhidai L, Csaba G (1998) Chemotaxis and chemotactic selection induced with cytokines (IL-8, RANTES and TNF-alpha) in the unicellular Tetrahymena pyriformis. Cytokine 10:481–486. doi:10.1006/cyto.1997.0328

Lee CK, Chan HM, Ho PL et al (2013) Long-term clinical outcomes after Streptococcus bovis isolation in asymptomatic blood donors in Hong Kong. Transfusion 53:2207–2210. doi:10.1111/trf.12087

Lopes PG, Cantarelli VV, Agnes G et al (2014) Novel real-time PCR assays using TaqMan minor groove binder probes for identification of fecal carriage of Streptococcus bovis/Streptococcus equinus complex from rectal swab specimens. J Clin Microbiol 52:974–976. doi:10.1128/JCM.03253-13

Noble CJ (1978) Carriage of group D streptococci in the human bowel. J Clin Pathol 31:1182–1186

Phillippi JA, Eskay MA, Kubala AA et al (2010) Altered oxidative stress responses and increased type I collagen expression in bicuspid aortic valve patients. Ann Thorac Surg 90:1893–1898. doi:10.1016/j.athoracsur.2010.07.069

Potter MA, Cunliffe NA, Smith M et al (1998) A prospective controlled study of the association of Streptococcus bovis with colorectal carcinoma. J Clin Pathol 51:473–474

Ruoff KL, Miller SI, Garner CV et al (1989) Bacteremia with Streptococcus bovis and Streptococcus salivarius: clinical correlates of more accurate identification of isolates. J Clin Microbiol 27:305–308

Schlegel L, Grimont R, Ageron E et al (2003) Reappraisal of the taxonomy of the Streptococcus bovis/Streptococcus equinus complex and related species: description of Streptococcus gallolyticus subsp. gallolyticus subsp. nov., S. gallolyticus subsp. macedonicus subsp. nov. and S. gallolyticus subsp. pasteurianus subsp. nov. Int J Syst Evol Microbiol 53:631–645

Siegel R, Ma J, Zou Z et al (2014) Cancer statistics, 2014. CA Cancer J Clin 64:9–29. doi:10.3322/caac.21208

Sillanpää J, Nallapareddy SR, Qin X et al (2009) A collagen-binding adhesin, Acb, and ten other putative MSCRAMM and Pilus family proteins of Streptococcus gallolyticus subsp. gallolyticus (Streptococcus bovis Group, Biotype I). J Bacteriol 191:6643. doi:10.1128/JB.00909-09

Tjalsma H, Scholler-Guinard M, Lasonder E et al (2006) Profiling the humoral immune response in colon cancer patients: diagnostic antigens from Streptococcus bovis. Int J Cancer 119:2127–2135

Tripodi MF, Adinolfi LE, Ragone E et al (2004) Streptococcus bovis endocarditis and its association with chronic liver disease: an underestimated risk factor. Clin Infect Dis 38:1394–1400

Vaska VL, Faoagali JL (2009) Streptococcus bovis bacteraemia: identification within organism complex and association with endocarditis and colonic malignancy. Pathology 41:183–186. doi:10.1080/00313020802436816

Wentling GK, Metzger PP, Dozois EJ et al (2006) Unusual bacterial infections and colorectal carcinoma-Streptococcus bovis and Clostridium septicum: report of three cases. Dis Colon Rectum 49:1477–1484

Woo PC, Teng JL, Wu JK et al (2009) Guidelines for interpretation of 16S rDNA gene sequence-based results for identification of medically important aerobic Gram-positive bacteria. J Med Microbiol 58:1030–1036

Chapter 13
Human Papillomavirus-Related Cancers

Antonio Carlos de Freitas, Ana Pavla Almeida Diniz Gurgel, Bárbara Simas Chagas, Carolina Maria Medeiros do Amaral, Eliane Campos Coimbra, Élyda Gonçalves de Lima, Jacinto da Costa Silva Neto, Maria da Conceição Gomes Leitão, and Rita de Cássia Pereira de Lima

Abstract Cancer is a public health problem occupying the first and second place in number of deaths in developed and developing countries, respectively. Since the last century, the relationship between infection and cancer has been established in animals and more recently in several human cancers. Currently known that 15–20 % of cancers in the world of infectious origin, many of them related to viral infections. The human papillomavirus (HPV) stands out for its association with confirmed cervical cancer and the large volume of evidence that relate to the head and neck cancer. In addition, there is evidence of their relationship with breast cancers, lung and prostate. However, they are still required more detailed research that aim to clarify the possible mechanisms involved in these processes related to carcinogenic HPV.This chapter discusses the main molecular characteristics of HPV and its relationship with cancers using for this the infective models described by recent studies, the mechanisms of tumor progression, forms of diagnosis and therapy.

Keywords Human papillomavirus (HPV) • Cervical cancer • Head and Neck cancer • Lung cancer • Breast cancer

Abbreviations

CDK	cyclin-dependent kinase
CIN	cervical intraepithelial neoplasia

A.C. de Freitas (✉) • A.P.A.D. Gurgel • B.S. Chagas • C.M.M. do Amaral
E.C. Coimbra • É.G. de Lima • M.C.G. Leitão • R.C.P. de Lima
Laboratory of Molecular Studies and Experimental Therapy, Department of Genetics, Federal University of Pernambuco, Avenida Professor Moraes Rêgo S/N, Recife 50670-901, Brazil
e-mail: acf_ufpe@yahoo.com.br; antonio.freitas@pq.cnpq.br

J. da Costa Silva Neto
Department of Histology and Embryology, Federal University of Pernambuco
Avenida Professor Moraes Rêgo S/N, Recife 50670-901, Brazil

M.R. Shurin et al. (eds.), *Infection and Cancer: Bi-Directorial Interactions*, DOI 10.1007/978-3-319-20669-1_13

COX-2	cyclooxygenase-2
EBV	Epstein-Barr virus
EGFR	epidermal growth factor receptor
HBV	hepatitis B virus
HCV	hepatitis C virus
HNC	head and neck cancer
HPV	human papillomavirus
IRF	interferon regulatory factor
LCR	Long Control Region
MMTV	mouse mammary tumor virus
NCR	non-coding region
NSCLC	non-small cell lung cancer
ORF	open reading frame
SIL	squamous intraepithelial lesion
VEGF	vascular endothelial growth factor

Introduction

Cancer is the major cause of death in developed countries and the second leading cause of death in developing countries (WHO 2011; Jemal et al. 2011). It is estimated that about 12.7 million cases of cancer and 7.6 million deaths from cancer occurred in 2008 throughout the world. Among these, it was estimated that two million cancer cases were caused by infectious agents and 1.6 million (80 %) occurred in developing countries (de Martel et al. 2012). Infections caused by *Heliobacter pylori* (*H. pylori*), Hepatitis B virus (HBV) Hepatitis C virus (HCV) and Human papillomavirus (HPV) were together responsible for 1.9 million cancer cases worldwide (de Martel et al. 2012).

HPV infects mucosal and epithelial tissues, depending on their tropism. HPV infection causes virtually all cervical cancer cases and is involved in the carcinogenesis of some non-cervical cancers, such as vulvar, vaginal, penile, testis, anal, breast, lung, skin and head and neck cancers (Lowy and Schiller 2012). Although there are approximately 15 oncogenic genotypes, HPV16 and 18 are the most common and these are found in approximately 70 % of cervical cancers and 90 % in the other HPV-related cancers (Lowy and Schiller 2012). Moreover, HPV6 and 11 are linked to the development of 90 % of anogenital warts (Lowy and Schiller 2012).

In this chapter, there will be a discussion of HPV-related cancers. With regard to the role of HPV in cervical cancer, there will be an examination of (a) the clinical and molecular aspects of HPV-mediated carcinogenesis, (b) recent advances in diagnosis and the therapeutic field and (c) HPV genetic variants and their oncogenicity. In addition, this chapter also discusses the role of the HPV infection as a carcinogen involved in head and neck, breast and lung cancers.

Human Papillomavirus

Viral Genome

HPV are non-enveloped DNA viruses that belong to the Papillomaviridae family (Bernard et al. 2010). They have an icosaedric capsid which is formed of 72 capsomers and with a diameter of about 52–55 nm. The viral genome possesses a circular double-strand DNA with a size of 8 kb and contains eight genes and two non-coding regions (NCR1 and NCR2 or Long Control Region – LCR) (Doorbar et al. 2012; Smith et al. 2011; Zekan et al. 2011). The HPV genome is divided into four regions: early genes (E), late genes (L), and the short non-coding regions located between E5/L2 (NCR1) and Long Control Region (LCR or NCR2) located between L1 and E6 (Scheurer et al. 2005; Schiffman et al. 2007; Stanley 2010) (Fig. 13.1).

E1 and E2 genes are responsible for encoding proteins that are essential for extrachromosomal DNA replication, as well as performing the viral infection cycle (Motoyama et al. 2004). The E2 gene encodes transcriptional regulatory proteins that is complexed with E1 protein and interacts with specific binding sites in LCR (Scheurer et al. 2005), which can either inhibit or increase the transcription of the early genes (Motoyama et al. 2004). The E4 protein is expressed in the late stages of infection when the viruses are being formed and plays an important role in viral maturation and replication (zur Hausen 1996; Motoyama et al. 2004). The E5 oncogene encodes the E5 protein, which is a hydrophobic membrane-associated protein that is responsible for modulating the epidermal growth receptor involved in the cell growth (Scheurer et al. 2005). E6 and E7 genes encode oncoproteins that allow the replication of the virus, as well as the transformation and immortalization of the

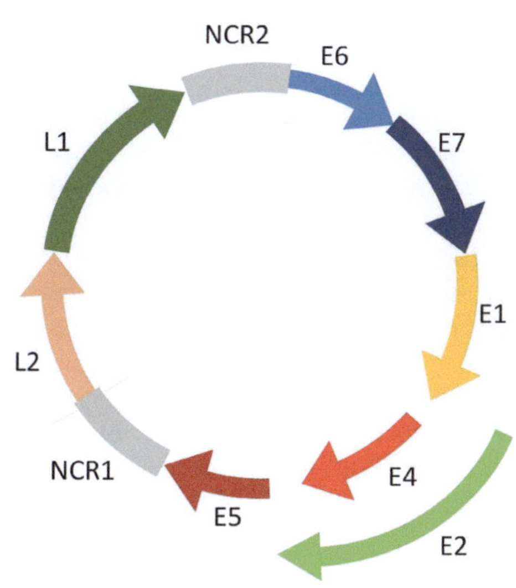

Fig. 13.1 Schematic representation of the HPV16 genome. The figure shows the early genes (E1–E7), late genes (L1 and L2) and non-coding regions (NCR1 and NCR2)

host cell (Motoyama et al. 2004). The L1 gene encodes the major viral capsid protein, while the L2 gene encodes the secondary protein of the viral capsid (Stanley 2010). The size of the LCR is 850pb and it is found between the L1 and E6 open reading frames (ORF). Early gene expression occurs through splicing, and thus generates a polycistronic RNA, which is a process regulated by the P97 promoter near the origin of the replication within the LCR (Weyn et al. 2011).

HPV Genotypes

A new HPV type is defined by the presence of less than 90 % of identity compared with the established prototypes in the L1 gene sequence, combined with the cloning and sequencing of its complete genome (Bernard et al. 2010; de Villiers et al. 2004). Previously, the term 'subtype' was used to identify HPV genomes with L1 nucleotide sequences that differ in a range of 2–10 % from the closest type, and the variants differ in nucleotide sequence of L1 by less than 2, and 5 % in LCR (Prado et al. 2005; de Villiers et al. 2004). However, recent studies have redefined variant lineages as a nucleotide sequence that differs by approximately 1 % of full genome between two or more variants of the same HPV type (Burk et al. 2013; Cornet et al. 2012; Smith et al. 2011). Moreover, the sublineage was also redefined as a nucleotide sequence that differs from 0.5 to 0.9 % within a full genome of the same HPV type (Burk et al. 2013; Cornet et al. 2012; Smith et al. 2011).

HPV are classified into 160 genotypes, on the basis of a highly conserved L1 sequences, divided into five genera: *Alphapapillomavirus* (Alpha PV), *Betapapillomavirus* (Beta PV), *Gammapapillomavirus* (Gama PV), *Mupapillomavirus* (Mu PV) and *Nupapillomavirus* (Nu PV) (Bernard et al. 2010; Burk et al. 2013; de Villiers et al. 2004). Among these, 40 genotypes infect the human genital tract, and are either classified as high risk HPV (HR HPV) or low risk HPV (LR HPV), due to their oncogenic potential (zur Hausen 2002). HR HPV are involved in vagina, vulva, cervical and penis lesions and include 16, 18, 31, 33, 35, 39, 45, 51, 52, 56, 58, 59, 68, 73, 82 genotypes (Badaracco et al. 2000; Gurgel et al. 2013; Stevens 2002). On the other hand, LR HPV were also found in genital warts and include 6, 11, 42, 44, 51, 53, 83 genotypes (Gurgel et al. 2013; zur Hausen 2002; Li et al. 2011b).

HPV16 (Alpha-9 species) and HPV18 (Alpha-7 species) are the genotypes that are most associated with cervical cancer and other HPV-related cancers. Moreover, HPV6 and HPV11, which cause cutaneous warts, is a member of the Alpha-10 species and are associated with lung cancers (Hoory et al. 2008). In contrast, the HPV that belongs to the Beta genus are generally associated with skin infections (Pfister 2003). Other HPV types belonging to the three genera, Gamma, Mu and Nu, cause cutaneous papilloma and warts, which do not usually progress to cancer (de Villiers et al. 2004).

HPV Viral Cycle

The viral cycle is well adapted to the life-cycle of the host cell, i.e., the expression patterns of genital HPV genotypes (Alpha PV) lie in keratinocyte differentiation (zur Hausen 2002). Before the HPV is able to cause infection, it is necessary to make contact with the basement membrane, which it is exposed to through micro lesions in the cervical epithelium (Doorbar 2007). It is believed that heparin sulfate proteoglicans (HSPGs) and other basement membrane components such as laminin, act as primary receptors for the anchorage of the virus (Combita et al. 2001; Doorbar et al. 2012; Giroglou et al. 2001; Johnson et al. 2009). Briefly, the HPV entry into the basal membrane cells occurs through the HSPGs and/or laminin receptor. After the binding, structural changes occur and the virus is transferred to a still unknown second receptor. These changes allow the uncoating program to begin, and this includes the cleavage of the site within the exposed N-terminus of L2. After the virus has entered the host cell, acidification of the endocytic vesicles occurs, leading to viral uncoating (Sapp and Day 2009). Subsequently, genital HPV genome migrates to the cell nucleus assuming the episome shape. E1, E2, E4, E5, E6 and E7 genes are expressed along cell cycle (Woodman et al. 2007), where each cell division propagates a viral genome to the daughter cells, during the infected keratinocyte migration from the basal to the apical layers (Fehrmann and Laimins 2003; de Freitas et al. 2012; McCance 2005). The L1 and L2 genes are expressed in the upper epithelium layers and their products create a viral capsid, which allow new viruses to be formed. The newly created viral particles are released from differentiated terminal keratinocytes, during the period of normal epithelium peeling (Stanley 2006). The basal layer primary keratinocytes act as a viral DNA reservoir for virus propagation (Fehrmann and Laimins 2003; McCance 2005; Woodman et al. 2007) (Fig. 13.2).

While a productive viral cycle leads to the production of new infective viral particles, the viral genome integration into the host genome is one of the main events in HPV-mediated carcinogenesis. During the period of integration, the E1/E2 region into viral DNA is disrupted, which impairs E1 and E2 transcription (Motoyama et al. 2004). Since the E2 product acts as a transcriptional repressor for E6 and E7 genes, its loss leads to E6 and E7 over-expression. E6 and E7-mediated degradation of p53 and pRb causes genomic instability and repression of apoptosis (zur Hausen 2002). Although the E6 and E7 oncoproteins are the most studied viral proteins involved in the cervical neoplastic process, the E5 protein is now recognized as of critical importance in ensuring the efficiency of this process. One of its best known effects is the 'half-life' increase in the epidermal growth factor receptor, which plays a role in mitogenic signalizing and cellular proliferation (Pedroza-Saavedra et al. 2010).

On the other hand, recent studies have shown that the viral cycle and transmission of cutaneous HPV genotypes (Beta and Gama PV) differ with regard to genital HPV genotypes (Nindl et al. 2013). The cutaneous HPV genome completes its viral cycle in follicle hair and is not rapidly expressed according to keratinocyte

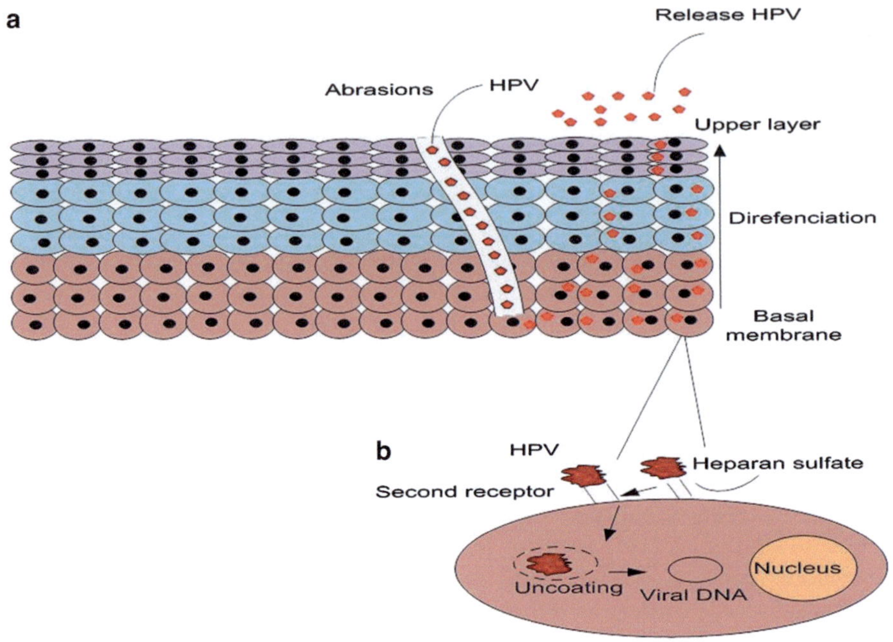

Fig. 13.2 HPV viral cycle. (**a**), HPV is introduced into the epithelium through microabrasions and infects the basal membrane cells. Following this, the early genes and late viral genes are expressed in accordance with the degree of cell differentiation from the basal membrane as far as the upper layer. Finally, the viruses are released. (**b**), HPV entry into the basal membrane cells occurs through the HSPGs and/or laminin receptors. Subsequently, several changes in conformation allow the recognition of the second receptor, the internalizing of the viral particle and viral uncoating

differentiation (Nindl et al. 2013), which may suggest the existence of an alternative viral cycle apart from the genital HPV genotypes. Furthermore, there is evidence to show that the E6 oncoprotein of these viruses does not bind to p53, but plays an indirect role in the carcinogenesis of the skin. For instance, the E6 oncoprotein binds with HIPK2, prevents HIPK2/p53 interaction and p53 phosphorylation at Ser46, and inhibits the apoptosis process (Nindl et al. 2013).

HPV and Cervical Cancer

Cervical cancer is regarded as the third major cause of death among women with cancer around the world, with about 529.800 cases recorded in 2008, of which 275.100 resulted in death (Jemal et al. 2011). The estimated figure for cervical cancer throughout the world is 11.7 %, of which Sub-Saharan Africa (24.0 %), Eastern Europe (21.4 %), and Latin America (16.1 %) have the highest percentages (Bruni et al. 2010). Developed and developing countries show different rates of morbidity and mortality caused by cervical cancer due to the nature of their control programs (WHO, 2011). For instance, in developed countries there has been a decline of approximately 80 % in cases of cervical cancer due to the effectiveness of their programs in detecting and treating precancerous lesions (WHO, 2011). In contrast, 80 % of cervical cancers worldwide are found in developing countries where most diagnoses are carried out at relatively advanced stages of the disease (Frazer 2004; WHO/ICO Information Centre on HPV and Cervical Cancer (HPV Information Centre) 2010). Although screening programs have reduced the number of cases, cervical cancer remains one of the main causes of death worldwide.

The progression of cervical lesions and cervical cancer is associated with persistent infection by specific HPV genotypes (Bosch et al. 2008; Walboomers et al. 1999; Zur Hausen 1999). HR HPV infection is commonly transmitted by sexual intercourse and can cause lesions in epithelial tissues, which can regress in 6–12 months (zur Hausen 2002). However, if these lesions are not treated, they can progress from pre-malignantconditions to cervical cancer. These cervical lesions are well characterized and detected by cytological and histopathological clinical examinations. However, if women are not treated, these pre-malignant conditions can progress into: cervical intraepithelial neoplasia – Grade 1 (CIN1) (or mild dysplasia); intraepithelial neoplasia – Grade 2 (CIN2) (or moderate dysplasia); intraepithelial neoplasia – Grade 3 (CIN3) or in situ carcinoma, which is characterized by a severe dysplasia; and squamous cell carcinoma or adenocarcinoma (zur Hausen 2002). In the Bethesda System, other terms are employed, such as: atypical squamous cells of undetermined significance (ASCUS); and squamous intraepithelial lesions (SIL), which comprises low-grade intraepithelial lesions (LSIL) and high-grade intraepithelial lesions (HSIL) (Broso and Buffetti 1993).

Pap Smear

Cervical lesions and their progression to cervical cancer can be prevented by Pap smears that detect cellular changes associated with HPV infection. The LSIL undergo morphological changes related to active replication of HPV (e.g. the koilocytosis) while the HSIL show signs of cell transformation characterized by nuclear changes. Although a conventional Pap test is the standard method used for the screening of premalignant lesions, there are still limitations in the interpretation of the morphological changes and poor interobserver reproducibility (Schiffman and Wentzensen 2010; Stoler et al. 2001).

Despite its high degree of specificity, the Pap test is relatively insensitive and shows variability in the HSIL results, which means cannot be determining factor in classifying ASCUS samples. Thus, even though the cost of the cytological test is relatively low, there is a need for frequent sampling which affects the suitability of this method for the screening of cervical lesions (Fahey et al. 1995; Stoler et al. 2001). The well-established prevention programs in developed countries have successfully lowered the incidence of cervical cancer by 75 % in the last 50 years. In the U.S., the increase in the detection of pre-invasive lesions as well as invasive cancer in the early stages can be attributed to 'the efficacy test', where the survival rate of 5 years is now approximately 92 %. Similarly, there has been a sharp decline in the incidence of cervical cancer and mortality rates in Europe (particularly in Nordic countries) and Canada owing to the implementation of cervical cytology in health care, most notably in population-based screening programs (Natunen et al. 2011). Nevertheless, false-positive results are still common since most ASCUS cases are not associated with concurrent CIN3 or cancer (Castle et al. 2010). Furthermore, the reproducibility of the diagnosis by an interobserver is difficult and about 5–17 % of women with these atypia has CIN2 or CIN3 when subjected to histological examination (Santos et al. 2003).

The establishment and maintenance of prevention programs, as well as the degree of access to effective treatments, are often unsatisfactory in developing countries, as this requires a large number of well-trained professionals and constant public funding for the required facilities (Hanumantha Rao et al. 2011; Sankaranarayanan et al. 2007).

HPV DNA Test

HPVDNA test have been suggested for screening due the equivocal Pap smear tests and follow-up of patients during the treatment of LSIL and HSIL. With regard to this, studies show that a HPV DNA test has a sensitivity of over 90 % in the detection of HSIL when compared with the Pap test (Castle et al. 2005; Walker et al. 2006); however, its specificity is insufficient because of the high rate of insignificant HPV infections (Ronco et al. 2006). Some countries such as the Netherlands and USA have adopted HPV DNA detection cytopathology for screening of cervical neoplasia. This clinical practice has improved the diagnosis of pre-malignant lesions particularly in triage of women with ASCUS and in women over 30 years old (Clifford et al. 2003; Saslow et al. 2012). The presence of persistent HPV infection for 1 or 2 years, especially type 16, is related to a 20–30 % risk of developing CIN3 in the next 5 years. In this scenario, there is a 30 % probability of progression to invasive cancer in women over 30 years old if untreated; however, if detected and treated, the chance of invasion is 1 % (McCredie et al. 2008). A positive HPV high risk associated with the result of the Pap test suggests there is a risk of progression to cancer, with a high negative predictive value, since lesions that do not have the oncogenic virus do not tend to progress (Saslow et al. 2012). In addition to their low positive predictive value and insufficient specificity, HPV DNA tests tend to be expensive and are thus not readily available in developing countries where more than 80 % of cases occur (Frazer 2004; Smith et al. 2007).

Prophylactic Vaccines

The current preventive HPV vaccines comprise the production of the major capsid L1 protein which is obtained through heterologous expression systems based on mammalian cells, plants, bacteria, insects and yeasts (Coimbra et al. 2011; Cortes-Perez et al. 2009; Kost et al. 2005).

Heterologous HPV L1 forms virus-like particles or VLPs which are macromolecular structures simulating the virions that are capable of inducing high titer neutralizing antibodies or a type-specific immune response. Currently, there are two licensed HPV prophylactic vaccines: Gardasil, manufactured by Merck & Co., Inc., which is a quadrivalent product consisting of HPV16, HPV18, HPV6 and HPV11 L1 VLPs; and Cervarix, manufactured by GlaxoSmithKline Biologicals, a bivalent product consisting of HPV16 and HPV18 L1 VLPs. These vaccines prevent infection by means of the two types (HPV16 and 18) found in almost all cases of cervical cancer and two other types (HPV6 and 11) that cause approximately 90 % of genital warts (Hoory et al. 2008; Li et al. 2011b; Pfister 2003). Hence these vaccines do not offer protection against at least 12 other oncogenic types which account for about 30 % of cervical cancer cases reported worldwide (Muñoz et al. 2006). Furthermore, the cost of licensed prophylactic vaccines is constantly increasing (up to $120 per dose, and 3 doses are required) which makes it difficult for developing countries to afford (Kling and Zeichner 2010). Thus, studies have been carried out in an attempt to obtain a platform for the production of a second generation of prophylactic vaccines based on VLPs from HPV of other types and at a feasible cost (Coimbra et al. 2011b; Karanam et al. 2009).

Overall, there are still obstacles to the implementation of the use of licensed vaccines, such as high costs and limited effectiveness in protection against HPV16, 18, 6 and 11 genotypes. Thus, the cytological test for early diagnosis remains the primary strategy for the prevention of cervical cancer, despite reports of failures and low sensitivity.

Biomarker for Cervical Cancer

The main disturbed molecules and cell pathways found in the progression of cervical cancer have been revealed as targets of HPV oncoproteins (Pinto et al. 2012; Stanley 2003). Several of these molecules have been evaluated as biomarkers that can serve the purpose of improving the screening of women with cervical dysplasia.

Some potential biomarkers of cervical carcinogenesis are targets of E6 and E7 oncoproteins (McLaughlin-Drubin and Munger 2009; Moody and Laimins 2010). For instance, the association of E7 oncoprotein with pRb abrogates the transcriptional repressor activity of pRB/E2F complexes (Sandal 2002), resulting in overexpression of E2F target genes including: Ki67, associated with cell proliferation

(Kruse et al. 2004); p16^{INK4a}, tumor suppressor protein (Klaes et al. 2001); cyclin E, required for cell-cycle progression (Baldwin et al. 2003); minichromosome maintenance proteins (MCMs) and topoisomerase II protein (TOP2) both involved in DNA replication (Malinowski 2005; Santin et al. 2005). p21 and p27, important inhibitors of cyclin-dependent kinases (CDK), as well as Bak, Bax, c-myc and others factors related to apoptotic functions, DNA replication and cell signaling are also disturbed by E6 and E7 for the maintenance of the neoplastic phenotype (Duensing et al. 2007; Duensing and Münger 2002; Eichten et al. 2004; Finzer et al. 2002; Garnett and Duerksen-Hughes 2006).

With regard to HPV E5 oncoprotein, it has been suggested that its action on cell targets are mostly important in the early stages of cervical carcinogenesis because the E5 gene is deleted when the HPV genome is integrated during the progression from a low-grade to a malignant disease (DiMaio and Mattoon 2001). However, studies show that E5 is also expressed in the later stages of cervical carcinogenesis by episomal viral genomes that coexist in an integrated form in 26–76 % of cervical cancers (Chang et al. 2001; Häfner et al. 2008). E5 seems to play an important role in carcinogenesis since it targets the epidermal growth factor receptor (EGFR), Cyclooxygenase-2 (COX-2) and vascular endothelial growth factor (VEGF), and thus disturbs the signaling pathways for cell proliferation, apoptosis and angiogenesis (Dannenberg et al. 2005; Kim et al. 2006; Pedroza-Saavedra et al. 2010; Venuti et al. 1998).

Some studies have linked an association of HR HPV persistent infection with altered expression levels of microRNAs (miRNAs) in cervical neoplasia (Lajer et al. 2012). The modulation of miRNAs has been correlated with the activity of E5, E6 and E7 oncoproteins (Greco et al. 2011; Zheng and Wang 2011). Furthermore, more than 50 % of miRNA genes are located at or near the HPV integration sites, which suggests that there is another HPV-mediated mechanism which causes an aberrant miRNA profile in cervical carcinogenesis (Calin et al. 2004). As miRNAs are a family of small non-coding RNA molecules that down-regulate the expression of their messenger RNA targets (Bartel 2009), the interplay between HPV and miRNAs suggests that altered levels of these small regulators and their mRNA targets might be promising biomarkers that could help in the diagnosis and prognosis of women with cervical dysplasia.

Several studies have demonstrated that there is an increased expression of E2F-responsive genes required for S-phase entry such as p16^{INK4a}, Ki-67, topoisomerase II-alpha and minichromosome maintenance protein-2 (ProEx) correlated with the aggravation of cervical lesions. Thus, these proteins or a combination of them (such as p16/Ki-67) are considered to be promising biomarkers for cervical cancer (Badr et al. 2008; Donà et al. 2012; Galgano et al. 2010; Panjkovic and Ivkovic-Kapicl 2006) (Fig. 13.3).

An increasing number of studies has highlighted significant correlations between miRNA expression patterns with cervical cancer. As a result, a lowering of the expression of these small molecules has been correlated with the progression of cervical cancer. Studies have found a reduced expression of miR-203 in CIN leading

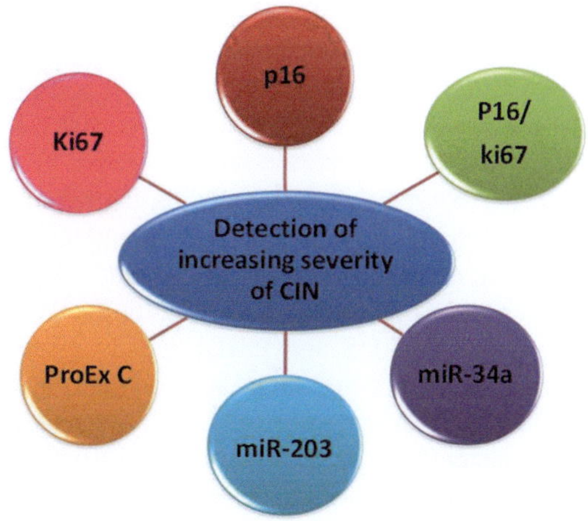

Fig. 13.3 Schematic representation of HPV targets which have the potential to detect increasing severity of CIN. HPV oncoproteins promote changes in expression level of genes (e.g. p16, Ki67, ProEx) and microRNAs (e.g. miR-203, miR34a); and in this context, many studies have evaluated the potential value of these changes in cervical cancer screening

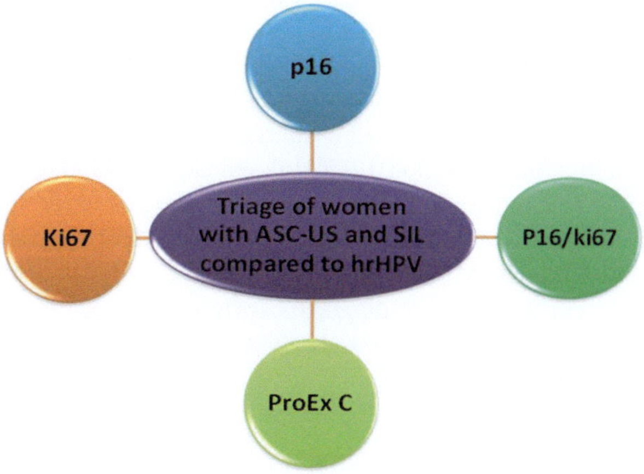

Fig. 13.4 Schematic representation of HPV targets with their potential value for ASC-US and SIL triage compared to the hrHPV test. Studies have showed that the immunohistological detection of certain molecules (e.g. p16, Ki67, ProEx C) has provided greater accuracy than HPV detection when used along with cytological test in screening of women with doubtful cytological results

to invasive cervical cancer (Hu et al. 2010) as well as an increased expression of its target, ΔNp63, which implies a suggestive correlation with cancer progression (Cheung et al. 2010). The miR-34a levels are also reduced in a way that is correlated with a reduction in the aggravation of cervical lesions. Hence, it is suggested that the modulation of miR-34a by HPV is an early-onset event in the development of cervical cancer in which potential biomarkers may be involved.

The use of the $p16^{INK4a}$, the ki-67, the combination of p16/Ki-67 and the ProEx (Fig. 13.4) for the detection of cervical dysplasia are being evaluated in a test that supplements the Pap test and also for the detection of HR HPV (including in a triage of women with equivocal cyto-histological results such as ASC-US, LSIL) and atypical squamous cells cannot exclude high-grade squamous intraepithelial lesions (ASC-H). These potential biomarkers have proved that they are capable of detecting CINs or SILs improving the diagnostic accuracy of cervical premalignant lesions (Depuydt et al. 2011; Gustinucci et al. 2012; Rokita et al. 2012; Wentzensen et al. 2012).

The p16^{INK4a} has shown a greater degree of accuracy than the Hybrid Capture II (HC2) high-risk human papillomavirus (HC2 test) for the detection of CIN2+ in the screening of women with ASC-US and LSIL due to the fact that this potential biomarker has a higher degree of specificity, without losing sensitivity. Both types of diagnosis (ASC-US and LSIL) usually result in the under treatment of affected women. Thus, it is important to highlight the potential of p16^{INK4a} for the detection of CIN2+ in triage of ASC-US cases and SIL after screening with HR HPV (Gustinucci et al. 2012; Roelens et al. 2012). The p16^{INK4a} has also shown it can be correlated with the aggravation of cervical lesions in an independent way by conducting a HC2 test (Samir et al. 2011). Moreover, studies have suggested that a combination of the HC2 test with p16^{INK4a} detection provides more detailed information about the risk of the malignant transformation of cervical lesions than with the hrHPV test by itself. As a result, the studies have raised the possibility of p16^{INK4a} being established as a marker to replace the HC2 test although they state that its clinical use is still limited by the lack of a standardized immunohistologic and cytologic scoring system (van Bogaert 2012; Kalof and Cooper 2006; Quint et al. 2013).

The use of Ki-67 as an adjunct test is currently being assessed to provide more accuracy in the diagnosis of cervical lesions, including doubtful cases, sometimes together with the HC2 test (Mimica et al. 2010). Thus, Ki-67 is being evaluated as an adjunct test in cervical biopsy diagnosis (originally classified as normal and LSIL by cytopathological diagnosis) and when compared with the HPV test, has proved to be more sensitive and specific for the LSIL identification, even in the diagnosis of uncertain cases. It was also suggested that Ki-67 should be employed together with the set of procedures for cervical cancer screening, as an adjunct to liquid based cytology to identify HSIL, as well as a surrogate marker of HPV-16 infection (Pirog et al. 2002; Sahebali et al. 2003).

In view of the proven accuracy shown by the two markers (p16 and Ki67), assessments were carried out to find out whether in combination they could screen cervical cancer with a greater degree of accuracy. Moreover, a commercial kit (the CINtecPLUS p16/Ki-67 Test) is already being evaluated (Donà et al. 2012; Wentzensen et al. 2012). The p16/Ki-67 test has shown a similar degree of sensitivity and significant improvement in specificity for CIN2+ detection in triage of ASC-US and LSIL when compared with the HC2 test. This result demonstrates that almost half the number of colposcopy patient referrals could be reduced, and is in contrast with the current practice of screening all HC2 test-positive, ASC-US and LSIL patients (Donà et al. 2012; Rokita et al. 2012; Singh et al. 2012; Wentzensen et al. 2012).

The ProEx C (immunostaining of MCM2 and top2 proteins) has led to an increase of both sensitivity and specificity when compared with HC2 test, for CIN1+ and CIN2+ detection in triage of ASC-US cases. In another study, it was found that ProEx has shown a higher specificity for the triage of ASCUS and LSIL cytology than with the HC2 test (Alaghehbandan et al. 2013). Hence, although the ProEx C can reduce the number of colposcopy referrals, they risk losing a proportion of the CIN2+ cases. This is a serious limitation which must be taken into account in the screening of ASCUS and LSIL. The effectiveness of ProEx has been shown for HSIL cases and confirmed in triage of women with ASCUS and LSIL (Badr et al. 2008; Conesa-Zamora et al. 2009; Kelly et al. 2006; Shroyer et al. 2006).

Overall the p16^{INK4a}, combination of p16/Ki-67 and ProEx detection is being evaluated as a supplementary test to the Pap test and detection of HC2 test including in triage of women with equivocal cyto-histological results, as well as ASC-US, LSIL and atypical squamous cells, although high-grade squamous intraepithelial lesions (ASC-H) cannot be excluded. However, preliminary results have shown that these potential biomarkers improve the diagnostic accuracy of cervical cancer owing to their capacity to detect CINs or SILs among premalignant lesions or equivocal cyto-histological results (Gustinucci et al. 2012; Roelens et al. 2012; Samir et al. 2011).

Early Detection Biomarker for Cervical Cancer

An accurate appraisal of the results of the CINs still represents a dilemma for gynecologists (particularly of CIN1) and this sometimes leads to the over- or under treatment of patients. The p16INK4a, KI-67 and ProEx C (Fig. 13.5) are also being assessed as predictors of the progression or regression of initial cervical lesions by determining which low-grade lesions (CIN1) need closer monitoring.

Some authors have found a correlation between the overexpression of p16INK4a and the risk of progression or regression of CINs both of CIN2 (Omori et al. 2007) and CIN1 to CIN3 (Negri et al. 2011; Ozaki et al. 2011). Although, a p16INK4a negative expression cannot definitively exclude the patient with a cervical lesion

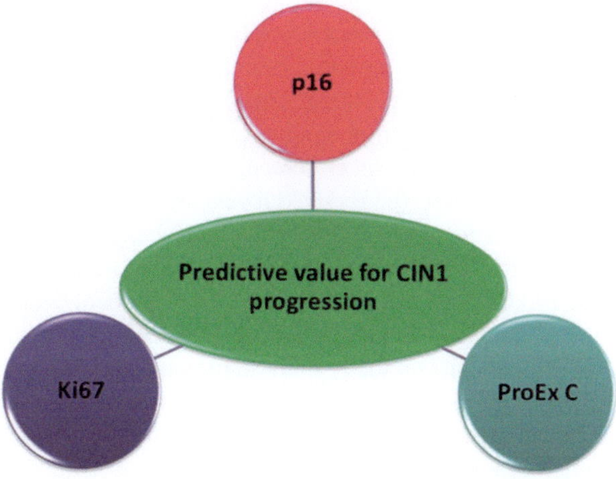

Fig. 13.5 Schematic representation of HPV targets with a predictive value to forecast CIN1 progression. The predictive value for disease progression has been suggested for some molecules such as p16, Ki67 and ProEx C. However, further validation in larger sample sizes are needed to determine the true clinical value of these potential biomarkers

from the high risk category, the immunostaining test for p16INK4a can be used as a supplementary test for the early diagnosis of cervical cancer (Izadi-Mood et al. 2012). Significantly, immunostaining of p16INK4a requires uniformity in the scoring system. Moreover, its use in the triage screening of high risk patients, especially among young patients, can assist in reducing the progression of precancerous cervical lesions in the near future.

Ki-67 has a strong, independent prognostic value for progression, owing to the fact that there is a greater predictive value for progression in the precancerous low-grade lesions than the histopathological classification criteria or the presence of the HR HPV (Kruse et al. 2004).

ProEx C, which has an equally high degree of sensitivity and a greater specificity to predict the progression of CIN2 than HC2/ProEx, is also a potential adjunct tool in the histopathological diagnosis of cervical adenocarcinoma in situ (AIS) and invasive adenocarcinoma (AC), especially in difficult cases where there are small biopsies or foci of disease.

Poor Diagnostic Biomarker for Cervical Cancer

Current studies have reported that detection of expression levels of some host molecules such as EGFR, COX-2, VEGF, p27 and miR-200a could be a useful marker to provide prognosis of cervical cancer. EGFR has been found overexpressed in 70–90 % of cervical cancer cases and correlated with poor prognosis since its higher levels were detected in HSIL when compared to LSIL (Balan et al. 2011; Kim et al. 2004; Movva et al. 2009; Nicholson et al. 2001; Soonthornthum et al. 2011). COX-2 is an enzyme expressed by regulation of EGFR thereby; COX-2 has been detected in elevated levels in CIN and cervical cancer (Greenhough et al. 2009; Venuti et al. 1998). COX-2 can be an unfavorable prognostic factor because it has been associated with resistance to chemotherapy and radiotherapy, increase of lymph node metastasis and poor survival rate in cervical cancer (Ferrandina et al. 2002; Gaffney et al. 2001). The elevated levels of VEGF (also regulated by EGFR) in cervical dysplasia is another potential biomarker for predicting the poor prognosis of cervical cancer (Dai et al. 2005). This cytokine which plays a key role in pathological angiogenesis, and lymphoangiogenesis, has been observed to be overexpressed in HSIL and cervical cancer (Cheng et al. 2000; Kang and Hong 2004; Kim et al. 2003). In contrast, the reduced levels or absence of p27 (a cyclin-dependent kinase inhibitor) in cervical cancer, has been correlated with increased aggressiveness and lymph node metastasis (Bouda et al. 2013; Huang et al. 2002; Sgambato et al. 2004). Dysregulation of miR-200a is another significant event in cervical cancer that could serve as a prognosis biomarker. This miRNA affects the metastatic potential of cervical cancer cells by the regulation of genes involved in cell motility control and has shown a significant ability to predict patient survival (Hu et al. 2010). Overall, further evaluations are needed to confirm the usefulness of potential biomarkers for predicting a poor prognosis in cervical cancer (Fig. 13.6).

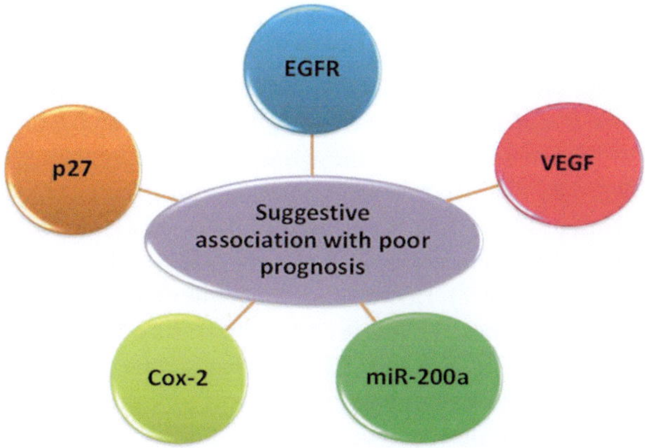

Fig. 13.6 Schematic representation of HPV targets which can serve as a prognosis biomarker in cervical cancer. EGFR, VEGF and COX-2 have been found overexpressed in cervical cancer cases and correlated with poor prognosis. Conversely, underexpression of p27 has been correlated with increased aggressiveness and lymph node metastasis. Additionally, dysregulation of miR-200a has been suggested as a prognosis biomarker for cervical cancer

Therapeutic Strategies for Cervical Dysplasia

The treatment of cervical neoplasia is based on the degree and extent of detected lesions (CIN1, CIN2, CIN3) in cervical epithelium. For instance, if the lesion is CIN1 type, it is recommended that the patient should be monitored for 1 year through repeated cytological examinations, since this type of lesion tends to regress spontaneously. Patients with high-grade lesions are usually subjected to conization for the removal of the uterine cervix to reduce the risk of progression and invasion (Wright et al. 2007). In addition to reports about the failures of the cytology test, this conventional screening is also unable to identify which premalignant lesions are likely to progress (Castle et al. 2009, 2007; Trimble et al. 2005). Thus, sometimes affected women are submitted to unnecessary treatment.

The conventional strategy for treating women with established cervical cancer consists of invasive surgery e.g. hysterectomy together with cysplatin chemotherapy and radiotherapy. However, there are many side effects arising from these therapies and their effectiveness is reduced. Further, conventional treatment can remove abnormal cells but fail to treat the HPV infection, which probably leads to a high recurrence rate among treated women (Costa et al. 2003; Nagai et al. 2000).

A number of innovative approaches to combat and treat HPV infection are currently being studied. Prophylactic HPV vaccines are unable to combat lesions and cancer and since there are a large number of HPV-infected women at risk of developing invasive cervical cancer (Day et al. 2008); for this reason, therapeutic vaccines are being evaluated that focus on key antigens of cell transformation (Garland et al. 2007; Hildesheim et al. 2007). At the same time, new strategies to improve the management of patients with cervical cancer have been tested in pre-clinical and clinical trials (Vici et al. 2014). These novel methodologies are based on the application of monoclonal antibodies and small molecule inhibitors to block major molecular pathways, which participate in cervical carcinogenesis, and are designated as "targeted therapies" (Bernard 2004; D'Abramo and Archambault 2011; Scott et al. 2012). The main targets of these therapies are molecules of which action or expression levels are disturbed by HPV oncoproteins, such as EGFR (Goncalves et al. 2008; Santin et al. 2011; Schilder et al. 2009), VEGF (Mackay et al. 2010; Schefter et al. 2012; Zighelboim et al. 2013), and COX2 (Doll et al. 2013; Ferrandina et al. 2003; Herrera et al. 2007). However, other studies have evaluated nucleic acid-based therapeutic approaches including short interfering RNA (siRNA) and antisense RNA to target HPV E6 and E7 transcripts that interfere with cellular transformation. E1/E2 HPV proteins that are involved with viral amplification and cell cycle progression have also been targeted by inhibitory peptides as a therapeutic strategy to block viral replication (White et al. 2011).

HPV Variants

There is evidence to show that mutations in the HPV genome do not occur as a result of the genetic distance (i.e. recombination), but through the fixation of single (aleatory) nucleotide polymorphisms (SNPs). Hence, these SNPs tend to become fixed in these viruses and the accumulation of mutations, such as HPV type, variant lineages and sublineage, leads to their speciation (Cornet et al. 2012).

HPV16, 18, 31, 33 and 58 have been recently reclassified into variant lineages and sublineages. For instance, HPV16 has four variant lineages (A, B, C and D) and nine sublineages (Cornet et al. 2012). In addition, HPV18 is classified into two variant lineages (A and B) and eight sublineages. HPV31 showed three viral lineages (A, B and C) and seven sublineages. With regard to the HPV33, the phylogenetic analysis showed three lineages (A, B and C) and five sublineages. HPV58 has four variant lineages (A, B, C and D) and seven sublineages (Burk et al. 2013).

Mutations screening into HPV DNA sequences is an important epidemiologic tool (Franco et al. 1999; de Freitas et al. 2012). Nucleotide substitutions can produce HPV variants that show a differential oncogenic potential (Chagas et al. 2013; Chagas et al. 2011; Gurgel et al. 2013). For instance, E6 and E7 polymorphisms may be relevant on their products activities against p53 and pRB, respectively (de Freitas et al. 2012; Sichero and Villa 2006). Similarly, polymorphisms in L1 gene may create conformational variants of its encoded peptide that present a novel set of viral neutralizing epitopes with consequences to vaccine strategies (de Freitas et al. 2012; Kast et al. 1994). Moreover, polymorphisms in LCR may alter the binding of transcriptional factors (Bernard 2002). In this scenario, several studies have been performed and demonstrated that HPV variants are associated with oncogenicity, persistence and the progression of infection (Berumen et al. 2001; Burk et al. 2003; Burroni et al. 2013; Chang et al. 2011; Chansaenroj et al. 2012; Cornet et al. 2013; Gheit et al. 2011; Ho et al. 2005; Londesborough et al. 1996; Quint et al. 2010;

Schiffman et al. 2010; Sichero et al. 2007; Villa et al. 2000; Xi et al. 2007, 2012 Yamada et al. 1997; Zehbe et al. 2011; Zuna et al. 2011).

HPV16 Variants

Several studies have shown a link between D lineage variants and the risk of developing squamous cervical lesion as well as adenocarcinoma of the cervix (Berumen et al. 2001; Burk et al. 2003; Quint et al. 2010; Zuna et al. 2011).

Variations within the E6 gene leading to amino acid changes, which may alter biological or immunogenic properties of the encoded protein (Ellis et al. 1995; Zehbe et al. 2003). With regard to the E6 functional activity, studies have showed that amino acid changes can alter their ability to abrogate serum/calcium-dependent differentiation, induce p53 degradation in vitro and regulate tumorigenesis (Chakrabarti et al. 2004). Several studies have shown a link between HPV16 E6 gene polymorphisms with viral persistence and progression of cervical lesions (Londesborough et al. 1996; Xi et al. 1997, 1998). For instance, invasive cancers were more closely associated with HPV16 E6 variants than with the prototype, in which E6 L83V SNP was most frequently found (Zehbe et al. 1998). Similarly, E6 L83V SNP showed elevated odds ratios of persistence and progression in women infected with HPV16 (Grodzki et al. 2006). In addition, E6 N127H polymorphism was identified in a Italian population and this is important due this residue is considered necessary for binding the E6 protein to p53 (Wise-Draper and Wells 2008). In Indonesia, de Boer et al. have found a variation A276G in the E6 gene with leads to N58S amino acid change in patients with CIN and cervical lesions (de Boer et al. 2004).

With regard to the E7 oncogene, Ser31Arg SNP may alter the phosphorylation of E7 oncoprotein by Casein Kinase II (CKII) (Wise-Draper and Wells 2008). In addition, another study reported a change in the nucleotide position 647 in the HPV16 E7 gene, with a change from asparagine to serine change at position 29 (N29S) (de Boer et al. 2004). This SNP is likely to be significant because of its location in an immunoreactive region (Zehbe et al. 1998). Furthermore, E7 N29S SNP was significantly more frequent in carcinomas (70 %) than in the control group (33 %) or CIN III group (50 %) (Song et al. 1997).

As mentioned above, a nucleotide sequence variation within L1 gene can play an important role in the structure of the viral capsid, immune recognition and viral neutralization and makes a significant intervention in vaccine strategies (Pande et al. 2008). For instance, the variation His202 variation within L1 protein can be assembled in virus-like particles (VLPs) more efficiently than its prototype Asp202 (Kirnbauer et al. 1993). Variations in the 83–97 residues of the L1 gene have an impact on the yield of the L1 protein (Chansaenroj et al. 2012). Several studies have found SNPs within HPV16 L1 gene (de Boer et al. 2004; Cento et al. 2009; Frati et al. 2011; Kirnbauer et al. 1993; Ntova et al. 2011; Raiol et al. 2009; Shang et al. 2011; Sichero et al. 2007; Sichero and Villa 2006; Stewart et al. 1996; Sun et al. 2011; Tornesello et al. 2004; Wheeler et al. 1997; Yamada et al. 1997; Yue et al. 2013).

Some of these polymorphisms are non-synonymous mutation and are embedded within hypervariable immunodominant regions BC, DE, EF, FG, HI loops of L1 protein. Furthermore, some of these variants are embedded within T-cell or B-cell epitopes binding region (Pillai et al. 2009).

The LCR is the binding site of cellular and viral transcription factors. The entire LCR is divided into three segments: the 5′ segment, central region and 3′ segment. The 5′ segment contains binding sites for repressing expression of the viral oncoprotein. The central region possesses an epithelial-specific enhancer and the 3′ segment contains the binding sites for replication and the p97 promoter. Hence, changes in the nucleotide sequence of LCR can alter the expressions of E6 and E7 oncogenes, as well as the viral replication of DNA. Several studies have reported a nucleotide variation within LCR (Arias-Pulido et al. 2005; Bhattacharjee and Sengupta 2006; Cento et al. 2011, 2012; Chansaenroj et al. 2012; Dong and Pfister 1999; Eriksson et al. 1999; Gagnon et al. 2007; Kämmer et al. 2002; Khouadri et al. 2006; López-Saavedra et al. 2009; Nasir et al. 2007; Pande et al. 2008; Pittayakhajonwut and Angeletti 2010; Prado et al. 2005; Raiol et al. 2009; Schmidt et al. 2001; Shang et al. 2011; Sichero et al. 2007; Tornesello et al. 2004; Veress et al. 1999). The G7521A SNP within HPV16 LCR has been found in some studies of cervical lesions and cervical cancer (Bhattacharjee and Sengupta 2006; Kämmer et al. 2000; Pande et al. 2008; Shang et al. 2011; Tornesello et al. 2004). The G7521A SNP is located in the YY1 binding site and this variation can increase the p97 promoter activity three to sixfold (Dong and Pfister 1999). Furthermore, other studies have demonstrated that SNPs within LCR are related to the increase of promoter activity and, thus, expression of E6 and E7 oncogenes. For instance, the D lineage variant showed threefold increase in the promoter activity when compared to prototype sequence (Kämmer et al. 2000). These results conformed with other studies that show a relationship between polymorphism in LCR and progression to lesion (Burroni et al. 2013; Ho et al. 2005; Londesborough et al. 1996; Schiffman et al. 2010; Sichero and Villa 2006; Villa et al. 2000; Xi et al. 1997, 1998; Yamada et al. 1997; Zehbe et al. 1998, 2011).

On the basis of the entire nucleotide sequence of LCR and/or E6, recent studies have proposed diagnostic polymorphisms to classify both HPV16 variant lineages as well as sublineages (Cornet et al. 2012; Smith et al. 2011). For instance, owing to the lineage fixation and a putative non-recombination process in HPV genome, Cornet et al. (2012) proposed that variant lineages could be detected by using 32 SNP combinations in the LCR of HPV16.

The HPV E2 protein are transcriptional transactivator proteins with very high affinities to their binding sites (Steger and Corbach 1997). The E2 nucleotide sequence is divided into three domains: transactivation domain; hinge domain and; DNA binding domain. Hence, polymorphic sites within E2 gene can alter the DNA replication as well as the expression of early genes. In this context, several studies have found SNPs within E2 gene in the tree domains (Azizi et al. 2008; Casas et al. 1999; Eriksson et al. 1999; Giannoudis et al. 2001; Graham and Herrington 2000; Tsakogiannis et al. 2012; Watts et al. 2002). Furthermore, polymorphic sites or epigenetic factor within E2 and LCR can alter the expression of E6/E7 oncoproteins.

For instance, the involvement of E2 binding site methylation in presence of intact E2, leads to the loss of E2 repressor activity in CaCx (Bhattacharjee and Sengupta 2006).

HPV18 Variants

Studies suggest that HPV18 variants lineages are linked to different levels of oncogenic potential and persistence of infection (Altekruse et al. 2003; Burk et al. 2003; Schlecht et al. 2005; Sichero et al. 2007; Xi et al. 2007). As mentioned above, HPV18 is the second most common HPV infection detected in cervical cancer and it is the type that is most closely linked to adenocarcinoma of the cervix (Beskow et al. 2005; Muñoz et al. 2003; Teshima et al. 1997). In a study undertaken in Brazil, Villa et al. found a pattern of increased risk of HSIL linked to B and C variant lineages (non-European variants) compared with the A variant lineage (European variant) (Villa et al. 2000).

Other studies have demonstrated the presence of E6 N129K SNP in Brazilian, Dutch, Indonesian and Surinamese patients. This SNP is highly conserved within oncogenic HPV variants. However, other studies have demonstrated that this change in E6 N129K SNP did not affect its capacity in promoting p53 degradation (Cerqueira et al. 2008; De la Cruz-Hernández et al. 2005).

With regard to the L1 gene, A5503G, C5701G, C6470G, C6625G, C6842G nucleotide changes have been reported in two studies (Arias-Pulido et al. 2005; Shen et al. 2013). These variants are near to the L1 C-terminal domain, which may affect immune responses of the HPV18 capsid protein (Arias-Pulido et al. 2005; Frati et al. 2011).

With regard to the LCR, A41G and T104C SNPs within HPV18 LCR, they seem to be able to achieve a high activity of the E6/E7 p97 promoter by modulating Sp1 and YY1 activities (Cerqueira et al. 2008). Moreover, the transcriptional activities in LCR was found also altered in 2.64–8.18 times in LCR variants when compared to prototype sequence (Sichero et al. 2005).

HPV31 Variants

There are only a limited number of studies of the clinical relevance of the HPV31 variants (Chagas et al. 2013). However, two recent studies have demonstrated that A and B lineage variants are associated with risk to HSIL (Schiffman et al. 2010; Xi et al. 2012).

With regard to the E6 gene, genetic variations at positions 190 (A190G), 368 (A368G), 413 (C413T) and 537 (G537G) were detected and are located in T-cell and/or B-cell epitope site (Chagas et al. 2013; Chagas et al. 2011). These variations may influence the viral peptides presentation to T-cell, which is a key mechanism of control of infection and development of cervical lesions. With regard to the E7 gene, mutations at positions 67, 136 and 184 were classified as non-synonymous, and cause amino acid changes at codons 23 (H to Y) 46 (E to Y) and 62 (K to E), respectively (Chagas et al. 2011). The observed genetics variations at positions 67 (C67T), 136 (G136A), 184 (A184G) were located in T-cell and/or B-cell epitope site (Chagas et al. 2011).

With respect to the LCR of HPV31, analysis revealed that the G7449A, G7457A, C7474T, G7525A and T7575C variations are embedded within sites for the transcriptional binding factors, which can potentially affect the expression of early genes (Cento et al. 2011).

HPV33 Variants

The differential risk for cancer development arising from viral variants is not well documented to HPV33. A recent study showed an association between alineage and cancer cases (Godínez et al. 2013).

Several studies have found a variation within LCR, E6, E7 and L1 gene of HPV33 (Cornut et al. 2010; Gagnon et al. 2007; Khouadri et al. 2006; Ntova et al. 2011; Raiol et al. 2009; Vrtačnik Bokal et al. 2010). The E6 L83V SNP of HPV33 was not associated with HPV persistence or risk of HSIL (Gagnon et al. 2007), although these same SNP in HPV16 are associated with HSIL and cervical cancer. In addition, in a Japanese population, HPV33 E6 variants were more frequently observed in CINs I/II than in CIN III/ICCs (Xin et al. 2001).

Non–prototype changes within LCR of HPV33 were significantly associated to HSIL in the Brazilian and Canadian populations (Khouadri et al. 2006). The C7732G variation, which results in the loss of a putative binding site for the cellular upstream stimulatory factor, was significantly associated to HSIL (Khouadri et al. 2006). The presence of a 78-base pair deletion in HPV33 and the presence of non-synonymous E7 variations in HPV-35 were associated to viral persistence (Gagnon et al. 2004). These studies also suggested the involvement of LCR in the carcinogenesis and showed the relationship between non-prototype variants and HSIL.

HPV58 Variants

For HPV58, the data of risk in developing cervical cancer in association with viral variants is still scarce (Cento et al. 2011; Cerqueira et al. 2003; Chan et al. 2002; Ding et al. 2010; Raiol et al. 2009; Wu et al. 2009; Xin et al. 2001). There have been a number of publications that have explored the heterogeneity of this genotype, but no accurate information on the prevalence of different variants in lesions has been provided. There was found to be a high frequency of HPV58 in East Asia and, Central and South America.

Among HPV58-positive women, a study demonstrated that the occurrence of E7 C632T (T20I) and E7 G760A (G63S) variants showed a positive association with the severity of neoplasia (Ding et al. 2010), with an odd ratio higher than 6.5-fold (Chan et al. 2002). However, no significant association was found between the E6 and E7 mutations of either HPV types or the cytological lesions.

DNA sequence alterations in HPV58 E6 gene that leads to an alteration of D86E alteration was only detected in one case of CIN III (Xin et al. 2001). With respect to E7 gene, a study identified 12 nucleotide substitutions. The G41R is located at the end of the N-terminal unstructured domain of E7 protein, while the G at position 63 (G63S) is the initial amino acid of the β-2 sheet and this variation was more frequently present in CIN II/III and has been associated with an increased oncogenic risk (Chan et al. 2002).

With regard to the HPV58 LCR, a significant association was found between T7207A, C7284G, T7345C, T7369G, T431G and T7483G and abnormal cervical cytology (Cento et al. 2011). Some mutations found in the LCR of HPV58 are embedded into transcription binding sites, which can affect the transcription of the oncogenic genes (Cento et al. 2011).

HPV and Head and Neck Cancer

Head and neck cancer (HNC) is a heterogeneous group of tumors that are classified according to their location: oral cavity, nasopharynx, oropharynx, hypopharynx, larynx, nasal cavity, paranasal sinuses and salivary glands. About 95 % of HNC have their origins in squamous cells, followed by adenocarcinomas, melanomas and other rare tumors.

The HNC is the sixth leading cause of cancer diagnosis worldwide. Among the parts of the head affected, there were more frequent cases of cancer at the base of tongue and in the tonsils. Laryngeal cancer is the second most common type and is in the respiratory tract with an estimated 160,000 new cases and approximately 83,000 deaths in 2008 (Jemal et al. 2011). Among the factors associated with HNC, are family history, poor diet, unfavorable socioeconomic conditions, chronic inflammation, exposure to chemicals and solvent vapors and infection caused by HPV (Chung et al. 2014; Dahlstrom et al. 2013; D'Souza et al. 2007; Gillison 2008). Tobacco use is the main factor responsible for the HNC cases. In some countries such as India, the Philippines, Malaysia and Myamar the use of betel, (a plant which is commonly mixed with tobacco), is the main cause of exposure to mutagens, either through smoking or chewing. In addition, alcohol aggravates the effects of smoking by diluting the mucus available in the mucosa of the upper respiratory tract, which provides a layer of protection to the epithelium lining (Chaturvedi et al. 2011; Hammarstedt et al. 2006).

According to U.S. data, in the last 30 years, the tumors associated with tobacco use have declined, possibly as a result of anti-smoking campaigns. This is in marked contrast with the incidence of tumors associated with HPV infection, which have increased significantly among white men and young women, most of this increase being attributed to changes in sexual behavior. Studies have shown that this increase is due to the infection caused by HPV16 (Chaturvedi et al. 2011; Hammarstedt et al. 2006; Jemal et al. 2013). From 1988 to 2004, there was an increase of 225 % of HNC cases in the United States with a rise from 0.8 cases per 100.000 individuals in 1988 to 2.6 in 2004. Chaturvedi et al. estimated that in 2020 the incidence of

HPV-positive HNC would be higher than cervical cancer and in 2030 half of all head and neck cancers would be related to HPV. Patients with HPV-positive HNC are less associated with exposure to tobacco, since there are 30 % of non-smokers to HPV-positive compared with less than 5 % for HPV-negative. Patients with HPV-positive HNC are also associated with lower alcoholic use compared with HPV-negative (Gillison 2008; Hong et al. 2013).

Much is known about the process of HPV infection in the cervical epithelium because there have been studies over a period of several years, but not much is known about the epithelium that makes up the tissues of the head and neck. Some studies have suggested that transmission of HPV has highlighted the practice of oral sex in the anogenital region (Beachler et al. 2012). It remains unclear if the kiss (French kiss) is also a form of contamination or transmission from mother to son. Koskimaa et al. (2012) believe that there may be non-sexual routes for the transmission of HPV.

Some studies have found that the prevalence of oral HPV infection is closely linked to older age, the male gender, smoking and sexual behavior such as oral sex and a number of sexual partners. The lowering of the age of the first experience of sexual intercourse is one of the prominent factors, as well as the number of sexual partners (D'Souza et al. 2009; Gillison et al. 2012; Smith et al. 2007).

The fact that men have a higher rate of oral HPV infection than women, may be because vaginal mucosa harbors an HPV viral load that is greater than male genital mucosa and skin, and so men who practice oral sex on women are more exposed to the virus. Another condition is reinfection, i.e., men who pick up an infection by having sex with a partner with persistent HPV infection (Chung et al. 2014; Gillison et al. 2012). Moreover, tobacco use is also associated with the prevalence of HPV, although there is no specific knowledge about which mechanisms are involved in cases of HPV-positive HNC.

HNC is molecularly divided into two groups: HPV-positive and HPV-negative tumors. The HPV-positive tumor may be considered to be an independent prognostic factor, but its analysis alone is negligible. Some studies have found differences in demographics, tissue location and types of tumor, from the group of HPV-negative patients (Gillison and Shah 2003; Marur et al. 2010).

Among the HNC, the oropharyngeal cancer are the most common and most are associated with HPV in the nasopharyngeal region. When in the presence of HPV16, the estimated risk of invasion is 14 times greater than with tumors that are HPV16-negative (Mork et al. 2001). However, the rates of HPV infection in HNC found in this study may vary depending on the method used for detection and genotyping, and thus lead to high levels of negativity (Gillison et al. 2000).

As in the case of cervical cancer, the HPV infection in HNC is necessary but not sufficient for the progression of the disease. It is necessary for proteins involved in the viral or host cycle, such as E6 and E7 oncoproteins and tumor suppressor protein $p16^{INK4a}$ to be overexpressed. For instance, HPV16 positive have overexpression of $p16^{INK4a}$ in HNC, as a result of E7 oncoprotein, which binds to and degrades the tumor suppressor protein pRb, by deregulating the E2F transcription factor and allowing a premature cell entry into the S phase (Pytynia et al. 2014; Weinberger et al. 2006). Furthermore, the E7 oncoproteins can alter the p15 and p16 proteins, which result in the blocking of the Cliclina-A/CDk2 complex, as well as the p21 and p27 proteins, that block the Ciclina-E/CDK2 complex (Pytynia et al. 2014).

The continued expression of E6 and E7 oncoproteins is a necessary condition for the maintenance of the malignant phenotype of tumor cells in HNC and this is one of the reasons why research into therapeutic targets, especially in cases of therapeutic vaccines, are targeted at the expression of the respective genes. The study conducted by Rampias et al. using oropharyngeal cancer cells, demonstrated that suppressing the expression of the viral oncogenes E6 and E7 of HPV16 resulted in a rapid restoration of the pathways of tumor suppressors (p53 and pRb) with a resulting increase in apoptotic activity (Rampias et al. 2009).

Molecular evidence that took account of the HPV status obtained by microsatellite analysis, initially separated the HNC group without producing evidence of HPV infection and found that these tumors are characterized by a high incidence of chromosomal deletion of arms 3p, 9q and/or 17p, when compared with HPV-positive. A loss of heterozygosity at 17p13 involving TP53 was also observed, whereas the loss of heterozygosity at 9p21 INK4a involves the tumor suppressor gene (encoding p16) (Braakhuis et al. 2004).

In cases of HNCHPV-negative, the inactivation of INK4a, either chromosomal or hypermethylation is considered to be the most common way to disrupt the pRB pathway (Shintani et al. 2001), unlike the HNC HPV-positive expressing E6 and E7 oncoproteins and can be attributed to a low rate of TP53 gene alteration resulting in overexpression of $p16^{INK4a}$ (Braakhuis et al. 2004). Studies have also shown that HPV-negative tumors accumulate at least twice as much as carcinogenesis and their mutations are based on the acquisition of a large number of changes in different signaling pathways, unlike the HPV-positive tumors that are modulated by the action of viral oncoproteins E6 and E7 (D'Souza et al. 2009; Stransky et al. 2011).

On the basis of a study by Weinberger et al. (Weinberger et al. 2006) that examined the positive cases for HPV16 and the status of p16 protein expression, it was possible to divide the molecular tumors into three classes: Class I: HPV-negative/p16 not expressed; Class II: HPV-positive/p16 not expressed and Class III: Express HPV-positive /p16. In the case of the patients of Class III, the overall survival rate was 79 % and Class II and Class I, 20 % and 18 %, respectively. The rate of disease-free survival was 75 %, 15 % and 13 % and local recurrence after 5 years was 14 % for Class III, 45 % and 75 %, and for Classes I and II, respectively. Thus they concluded that the patients in the Class III group had a more favorable prognosis than the others.

With regard to the Wnt and Notch signaling pathways, that are essential for the regulation of normal cell proliferation and differentiation, it is known that disturbances often result in the development of tumors. In oropharyngeal carcinoma cell lines, activation of the Wnt pathway was detected, with nuclear accumulation of beta-catetina, as a result of the action of the E6 and E7 oncoproteins of HPV16. This mechanism was combined with the down-regulation of endogenous Saiah-1 (Rampias et al. 2010). Smeets et al. (Smeets et al. 2011) also conducted experiments on the activation of Wnt HNC immortalization of normal keratinocytes in the oral cavity and E6- induced degradation of the p53 process. Evidence of the Wnt pathway activation and subsequent nuclear accumulation of beta-catenin, was also observed in an in vivo experimental study of the skin of transgenic mice (Bonilla-Delgado et al. 2012).

The Notch receptor or ligand is transmembrane proteins in the intracytoplasmic domain that are involved in normal cell proliferation and the differentiation process. Recent studies have suggested that mutations with a loss of function of Notch receptors are associated with HNC, lung cancer and squamous cell cancer (Agrawal et al. 2011; Stransky et al. 2011; Wang et al. 2011). In HNC, there is also a loss of function mutations in the NOTCH1 gene in 12–15 % of cases and 3–5 % of other members of the NOTCH family. It is believed that the E6 protein seeks to control the levels of Notch, but there has not been sufficient research in this area to elucidate these mechanisms, although transforming the Notch pathway is an important target for HPV (Rampias et al. 2010).

In addition, recent studies have shown that activation of PI3K can be triggered by the E6 and E7 oncoproteins of HR HPV through the activation of the Akt and mTOR complexes, resulting in viral survival. This is an important issue for route analysis and information to explain viral persistence; hence the progression of HPV-positive HNC can also be controlled through mutations in PI3K (Menges et al. 2006; Pim et al. 2005; Spangle and Münger 2010).

As well as changes of the kind mentioned, recent research has included the presence of polymorphisms as a risk factor in developing HPV-positive HNC and this may also reduce the survival rate. These polymorphisms have already been observed in other tumor types and result in changes in the expression of proteins involved in the tumorigenic process, such as genetic polymorphisms of matrix metalloproteinases. Studies in Brazil have shown that polymorphisms in the promoter region of tumor necrosis factor alpha (TNF-α) may be linked to a worse prognosis in patients with HNC. Another meta- analysis found an increased risk of oral cancer in Asian patients, (though not in Caucasians) in the presence of polymorphisms of the glutathione S-transferase M1 gene (GSTM1), although the effect is modulated by tobacco use (Corrêa et al. 2011; Nishizawa et al. 2007; Song et al. 2013; Vairaktaris et al. 2008).

Despite the current knowledge that exists on the interaction of the viral oncoproteins E6 and E7 of HR HPV, especially HPV16, and their association with HNC, it is necessary to conduct further research into the molecular carcinogenic process and tumor progression, in particular to determine the precise role and tumorigenic potential of oncoproteins as well as to extend the number of research subjects to stratify the tumor types and their location.

To some extent, the knowledge of HPV and its causal relationship with the HNC have improved the ability to diagnose and locate hidden primary tumors. Currently, there are several types of HPV tests for detecting the expression of E6 and E7 and p16 proteins through the use of mRNA and immunohistochemistry. However there is no consensus yet on the best method of diagnosing HPV infections. It is believed that the most suitable test currently being conducted is in the detection of the expression of E6/E7 mRNA as well as in situ hybridization (Fakhry et al. 2008).

Some studies have conducted a microarray analysis to identify large genomic alterations in gene expression associated with HNC. If a biomarker is expressed in all the tumors of the head and neck, its usefulness as a biomarker for selected treatment will be limited (Kuriakose et al. 2004; Langer 2012). According to Kim et al. before the prognosis of patients with HNC can be improved, it is necessary to identify biomarkers for homogeneous subgroups of tumors and the identification of HPV status may help decide the therapeutic methods. Kim and coworkers identified genes by employing data from a microarray analysis of the status of HPV genes. However, they did not identify significant differences between patient samples and normal HNC, although some patients with HNC genes of HPV-positive were not significant with respect to HNC HPV-negative. These groups of genes that were analyzed in HPV-positive HNC include various types of roads, fibroblasts, and collagen degradation actina (Kim et al. 2014).

Prophylactic vaccines for HPV16 and 18 are shown to be effective in preventing persistent infection and the development of cervical dysplasia. However, clinical trials have not included a study of the effects of these vaccines have had on oral HPV infection. The vaccine has the potential to have an impact on HNC because it is aimed at HPV16 and 18. One study found that vaccination against HPV16 reduces the prevalence of oral infection with an estimated 93.3 % efficiency. Whereas over 90 % of cases of HPV-positive HNC are caused by single HPV16, the use of these vaccines in campaigns can lead to the prevention of cervical cancer (Kreimer et al. 2005).

Knowledge of viral etiology in HPVHNC provides alternatives for primary prevention that are either in accordance with cervical carcinomas, or serve as a prophylactic vaccine based on primary prevention and secondary prevention and HPV detection and expression of p16. However, the development of therapeutic vaccines aimed at multiple HPV alpha-types, and with a focus on the E6 and E7 genes, can be an effective means of combating HNC.

HPV and Lung Cancer

Lung cancer is included among the ten types of the highest incidence of cancer and is a major cause of cancer-related deaths in the world (Siegel et al. 2013). The main symptoms of this disease are coughing and fatigue. These symptoms make the diagnosis difficult, because patients often confuse the causes and thus the initial stages of the disease (stage I and II) are hardly detected. In most cases, lung cancer is detected in stages III and IV (Hensing et al. 2014).

Lung cancer is divided into two main groups: non-small cell lung cancer (NSCLC) and small cell lung cancer (SCLC). NSCLC comprises 70 % of lung cancers. They are divided into three main types: squamous cell carcinoma (about 20 % of cases), adenocarcinomas (approximately 45 % of cases) and large cell carcinomas (estimated at 4 % of cases) (Sun et al. 2007).

Until the mid 1970s, the cause of lung cancer was attributed solely to smoking. In 1975, Roglic et al. hypothesized that HPV could be involved in the development of benign lesions of the bronchi. However in 1979, it was hypothesized that HPV could exert some influence not on benign lesions but on the development of squamous cell carcinoma of the bronchus (Syrjänen et al. 1989). This initial suspect drew the attention of some researchers and a good deal of work has emerged that investigates the relationship of this virus with lung cancer. Added to this, another factor has given rise to new research: the incidence of lung cancer in non-smokers is increasing every year, and it is included among the ten leading causes of death by cancer throughout the world (Rudin et al. 2009).

Squamous cell and small cell carcinoma are more closely linked with smoking than adenocarcinoma, which is the type most commonly found among non-smokers (Sun et al. 2007). In Asia, most women with lung cancer are non-smokers and most of these lung cancers are adenocarcinomas (Gabrielson 2006). Lung cancer among non-smokers is considered to be a disorder with specific and distinct clinical and genetic characteristics when compared to that caused by smoking (Thu et al. 2012). Lung tumors in non-smokers have a higher proportion of genomic alterations than tumors in smokers (Thu et al. 2012).

HPV infection has been detected in lung tumors in all continents except Africa which has an average incidence of 24.5 %. The highest incidence was found in Asia, with 35.7 %. In Asian countries, Taiwan has the highest incidence, (an average of 55.5 %), then comes Japan with 34.5 % and lastly, China accounted for 27.3 % (Klein et al. 2009).

The most prevalent HPV types of lung cancers are HPV16, 18, 31, 33, 6 and 11. LR HPV6 and 11 have been linked to squamous cell carcinomas and in all other subtypes of non-small cell lung cancers (NSCLC). Furthermore, HPV26, HPV35 with NSCLC and HPV45 SCLC were also found (Klein et al. 2009; Srinivasan et al. 2009).

Although the role of HPV in the development of lung cancer is not established, the virus has been cited as the second most important risk factor in the development of lung cancer (Klein et al. 2009). Although some studies believe that this virus is opportunistic (Coissard et al. 2005) there is some evidence that infection with HPV16 and 18 is associated with cancer in non-smoking women and in patients with lung adenocarcinoma (Cheng et al. 2001). In situ immunohistochemistry data showed that positive lung tumors to E6 were negative to p53 and the reduced expression of the WAF-1 and MDM2 genes indicated the loss of the tumor suppressor function of p53 due to interaction with the E6 protein (Cheng et al. 2007).

It has been found that the HPV16 DNA is integrated into the genome of squamous cell carcinoma (Aguayo et al. 2007). Furthermore, it was observed that overexpression of HPV16 E6 and E7, increased the expression of pro-angiogenic factors, HIF-1α and VEGF causes angiogenesis in vivo and in vitro in NSCLC (Li et al. 2011a).

HPV16 and 31 were also detected in the exhaled breath of patients with non-small cell carcinoma of the lung. These findings suggest that the condensation of exhaled air may be a possible non-invasive method of detection, of the virus in patients with lung cancer; it also suggests that other pathways may be involved in the transmission of HPV (Carpagnano et al. 2011).

A high prevalence of HPV16 and 18 DNA was detected in the circulatory system of patients with lung cancer. This suggests that infections in the lung tissues are derived from the cervix and spread to the lung tissue into the bloodstream, since it is the only connection between the lung tissue and cervix, the primary target for infection with HPV16 and 18. Moreover, it suggests that HPV16 and 18 may be risk markers for lung cancer (Chiou et al. 2003).

Although there is a good deal of evidence of the involvement of HPV in lung cancer, there are some issues that still need to be clarified with regard to the mechanisms by which this virus induces lung cancer. As mentioned earlier, the main function of the E6 oncoprotein is to bring about the inactivation of p53. However, E6 oncoprotein has the ability to interact with other cellular proteins regardless of the degradation of p53. E6 is involved in the transcriptional activation of the human telomerase reverse transcriptase gene (hTERT). The hTERT is an enzyme that controls the expression and activation of telomerase and is found in over 90 % of the cancer cells (Kyo et al. 2008). The hTERT mRNA levels were more significant in lung tumors that were positive for E6 HPV16/18 than for negative E6 tumors (Cheng et al. 2008). Interaction with the E6AP complex is required for E6 which can activate the transcription of the hTERT gene. Initially, E6 recruits E6AP and binds to it to form a complex in order to bind the hTERT promoter through a specific site the E box. The association with E6AP is essential for the activation of endogenous and exogenous hTERT. The E6-E6AP complex binds Myc to form a trimeric complex of E6-E6AP-Myc. This interaction facilitates the transcription of hTERT. Myc binds near the E box, downstream from the transcription site (Liu et al. 2005) as shown in Fig. 13.7.

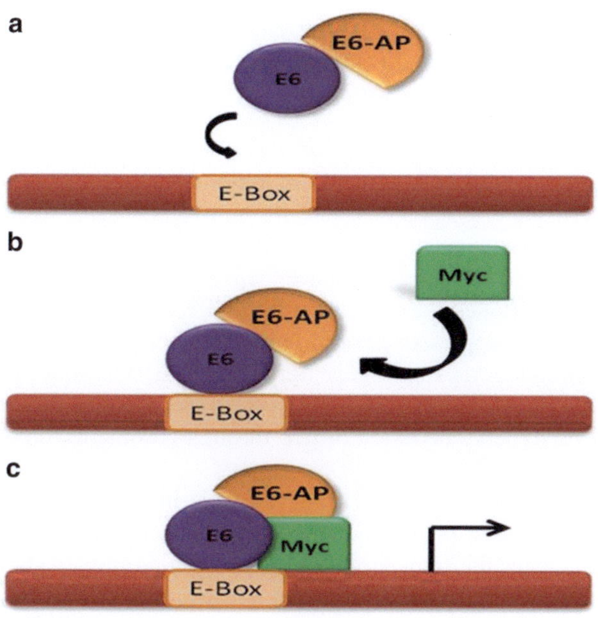

Fig. 13.7 Transcription of hTERT E6-dependent. The chart summarizes the activation of the transcription of the hTERT gene by E6 protein. (**a**), E6 protein recruits the E6AP complex. (**b**), The E6-E6AP complex binds to the hTERT promoter E-box region. (**c**), Formation of the trimeric complex (E6-E6AP-Myc) to facilitate the transcription of the hTERT gene

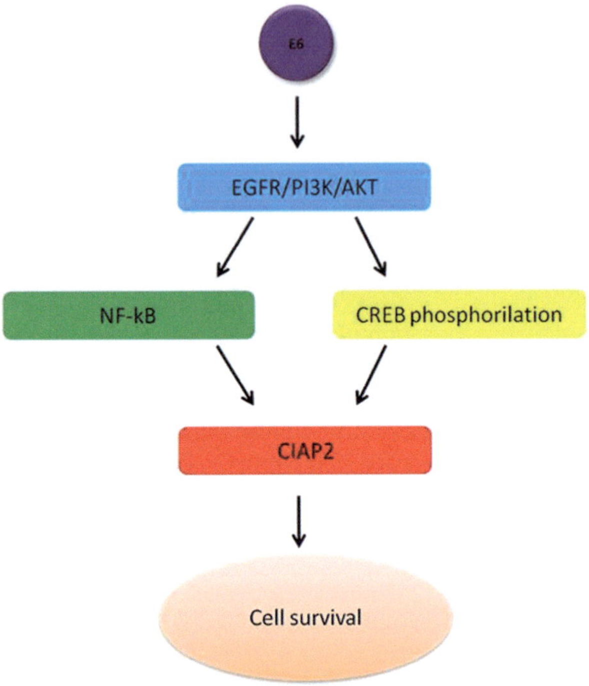

Fig. 13.8 Epidermal growth factor receptor (EGFR) upregulated by the E6 protein with CREB phosphorilation and NK-kB activation. The E6 protein actives NK-kB and CREB phosphorylation through EGFR/p13k/AkT pathway and both activate the inhibitor of apoptosis protein (cIAP2) favoring cell survival

E6 also up-regulates cIAP2, which is an inhibitor of anti-apoptotic proteins and thus confers resistance to apoptosis in lung tumors. E6 activates a EGFR/PI3K/AKT cascade which activates NF-kB and CREB phosphorylation, which activates the expression of cIAP2 and thus favors cell survival (Wu et al. 2010) (Fig. 13.8).

One study demonstrated the involvement of HPV infection in hypermethylation of the $p16^{INK4a}$ tumor suppressor gene in female lung cancer (Wu et al. 2005). This inactive hypermethylation $p16^{INK4a}$ gene transcription HPV E7 oncoprotein interacts with the E2F-Rb-HDAC complex, by releasing HDAC (histone deacetylase), which contributes to the hypermethylation of p16 through chromatin remodeling (Finzer et al. 2001; Wu et al. 2005) (Fig. 13.9).

E7 oncoprotein also enhances HIF-1 (hypoxia-inducible factor-1) dependent transcription by inducing the dissociation of HDACs from HIF-1a (Bodily et al. 2011). HIF-1 activates the transcription of genes that are involved in angiogenesis and cell survival as a vascular endothelial growth factor (VEGF). As shown in Fig. 13.10, VEGF upregulates EGFR and this up-regulation triggers the downstream mitogenic signals that support tumorigenesis. This is one of the mechanisms that are suspected to be active in HPV-infected lung tissue (Prabhu et al. 2012).

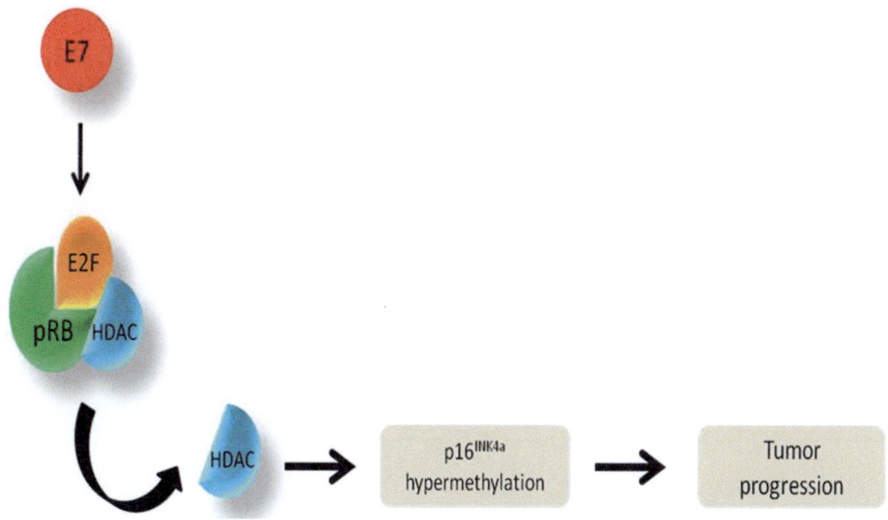

Fig. 13.9 The E7 protein contribute to the p16 ^{INK4a} hypermethylation and tumor progression. The E7 protein binds to the E2F-Rb-HDAC complex favoring HDAC dissociation. Thus, HDAC acts in p16^{INK4a} hypermethylation leading to tumor progression

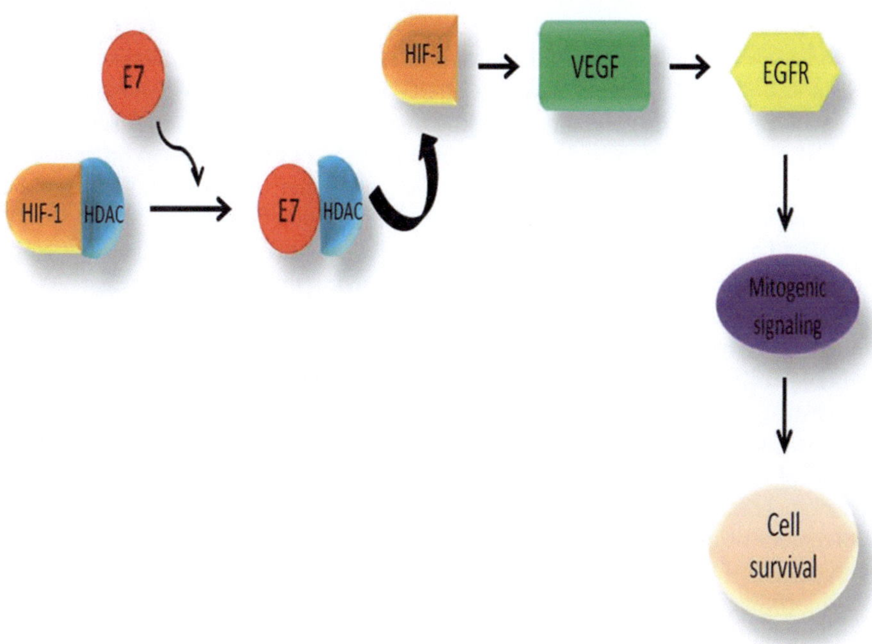

Fig. 13.10 HIF-1 E7-dependent transcription. The E7 protein binds in HDAC and causes the dissociation of the complex HIF-1 – HDAC. This interaction induces the up-regulation of VEGF, and EGFR issues mitogenic signals that lead to cell survival, tumorigenesis

It is known that estrogen facilitates HPV infection as well as playing an important role in the tumor process. In cervical carcinoma cells and the induction of aromatase, the enzyme responsible for converting androgen into estrogen, leads to an increased expression of the E6 and E7 oncogenes (Nair et al. 2005). Both normal and tumor lung cells express this enzyme and receptors for estrogens (Márquez-Garbán et al. 2009; Stabile et al. 2002). Estrogen induces cell proliferation and activates the transcription of their receptor (ERE), as well as some growth factors genes such as EGFR which stimulate mitogenic processes in lung tumors (Stabile et al. 2002).

Most of the signaling pathways that play some role in lung cancer are to some extent blocked or altered by means of HPV oncoproteins. Thus, studies of these pathways are important and necessary for a better understanding of the disease and for the development of prophylactic and therapeutic strategies (Prabhu et al. 2012).

HPV and Breast Cancer

Breast cancer is the type of cancer that affects most women around the world, and there are nearly 230.000 new cases each year (Balko et al. 2013; Herrera-Goepfert et al. 2011). Breast cancer also causes the highest number of deaths in women, with an estimated 458.000 deaths per year (Chang et al. 2012).

Breast cancer is a heterogeneous disease with respect to molecular alterations, cellular composition, and clinical outcomes. The disease results from numerous internal and external factors (Alibek et al. 2013; Parker et al. 2009). Among them, are genetic and epigenetic factors including mutations in breast cancer, susceptibility genes BRCA1 and BRCA2, ethnicity (more common in the Caucasian population), obesity, sex-steroid hormones and lifestyle are factors that are closely involved in the development of breast cancer (Alibek et al. 2013; Hedau et al. 2011). However, in 50–80 % of cases, the risk factors are not identified, and this has led to a new attempt to identify factors related to this neoplasia, such as viral infections (Klug et al. 2005). Recently, many investigations have linked breast cancer to viral infections, such as Epstein-Barr virus (EBV), mouse mammary tumor virus (MMTV) and HPV (Chang et al. 2012). For this reason, it is speculated that these viruses are co-factors with diet, oestrogens and other hormones in the initiation and progression of some types of breast cancer in genetically susceptible women (Lawson et al. 2000).

Several studies have demonstrated that HR HPV are present in more than 50 % of human breast cancers and normal mammary epithelial cells, even though the involvement of the virus in breast cancer is a controversial issue (Akil et al. 2008; Heng et al. 2009; Yasmeen et al. 2007).

Tumorigenesis can be induced by infectious agents through the induction of chronic inflammation, cellular transformation by oncogene insertion, inhibition of tumor suppressors and induction of immunosuppression. While the consequences of inflammation on tumor initiation and progression are well studied, the relationship between inflammation and viral infections in carcinogenesis is much less understood (Schäfer et al. 2013).

The notion that HPV may also play a role in human breast cancer is based on the identification of HR HPV in human breast tumors (Yasmeen et al. 2007). HR HPV infection has been found in both cervical and breast cancer of the same patients. This finding has led to the hypothesis that HPV could be transmitted to the breast through sexual activities (Akil et al. 2008). The detection of genital HPV in the nipple and areolar region suggests an alternative and perhaps more likely route of infection (de Villiers et al. 2005). Hence, it is possible that the incidence of HPV-positive breast cancer in young women is related to HR HPV genital infections, which are much more common in women who have multiple sexual partners (Akil et al. 2008).

The oncogenic mechanisms by which HPV induces cervical cancer have been intensively studied and used as a model for breast cancer (Heng et al. 2009). One hypothesis for the presence of HPV DNA in samples with breast cancer can be explained by the transport of DNA from the original site of infection to the breast tissue by the bloodstream, in patients with a history of cervical cancer; it is possible that this is involved in the carcinogenesis of breast cancer (Khammapirad et al. 2011; Widschwendter et al. 2004).

Failure of the immune system to clear persistent HPV infections can lead to the development of cancer after several decades (Moody and Laimins 2010). The E6 and E7 oncoprotein inactivates the interferon regulatory factor (IRF), so that the HPV viruses can remain as persistent, asymptomatic infections (Narisawa-Saito and Kiyono 2007). Moreover, the virus is able to become integrated into the host cell genome and use its transcription machinery to express viral proteins. Most HPV16 and 18 positive cancers contain integrated HPV genomes, which suggests that integration may, in some cases; lead to malignant progression (Alibek et al. 2013; Moody and Laimins 2010).

The HPV oncoproteins (E6 and E5) act before becoming integrated and are known to break the keratin, which leaves marks in the cytoplasm and perinuclear halo nuclear swelling, and leads to the appearance known as a koilocyte (Krawczyk et al. 2008; Lawson et al. 2009; Thomison et al. 2008). Koilocytosis is accepted as pathognomonic or characteristic of HPV infection. In this scenario, HPV infection was detected in 22 % in normal skin and 33 % of ductal carcinomas in situ (Lawson et al. 2009). Moreover, there is evidence that suggests that E6 activates telomerase, which is the enzyme responsible for maintaining telomere structure at the end of chromosomes, possibly through activating the hTERT promoter and c- Myc (Chakrabarti and Krishna 2003; McMurray and McCance 2003; Veldman et al. 2003). E6 not only increases levels of c-Myc, but is also able to combine with the Myc complex and drive the expression of its target genes. The overexpression of a c-MYC gene is the signature of most breast cancers, and there is a significant link between elevated levels of c-Myc and HPV16 infection (Alibek et al. 2013). In addition, E6 and E7 oncoprotein antagonizes the ability of BRCA1 to inhibit c-Myc E-box-mediated transactivation and human telomerase reverses transcriptase promoter activity, in a manner that is dependent on the zinc finger domains. The ability of E6 and E7 to antagonize BRCA1 does not involve proteolytic degradation of BRCA1, which suggests there are functional interactions of BRCA1 with E7 and E6

(Zhang et al. 2005). Several functions of BRCA1 are potentially relevant to the development of breast cancer, including the ability of BRCA1 to inhibit ER-α signaling (Zhang et al. 2005).

The high-risk E5 protein cooperates with E6 and E7 to cause the hyperproliferation of infected cells and is likely to facilitate malignant progression and immune modulation, and can thus involve E5 in the pivotal steps of carcinogenesis (Venuti et al. 2011). In addition to E6 and E7, E5 oncoprotein can affect receptor tyrosine kinases by associating with the cell membrane (Alibek et al. 2013). The type I receptor tyrosine kinases constitute a family of transmembrane proteins involved in various aspects of cell growth and survival and have been involved in the initiation and progression of several types of human malignancies (Rusnak et al. 2001). Several tyrosine kinase receptors have been associated with human carcinogenesis, including the ErbB tyrosine kinase family. The ErbB family of receptors comprises EGF-R (ErbB-1), ErbB-2, ErbB-3, and ErbB-4; which are the four ErbB receptor members that form homo and heterodimer complexes upon activation by a family of EGF-like ligands and thus stimulate their kinase activity and generate intracellular signals. A constitutive stimulation of these pathways through autocrine mechanisms has been linked to breast and ovarian cancer (Klapper et al. 1999; Al Moustafa et al. 2004).

Though expression of E6 and E7 is itself not sufficient for cancer development, it seems to be either directly or indirectly involved in every stage of multi-step carcinogenesis (Yugawa and Kiyono 2009). However, the data in the literature on the presence and oncogenic mechanisms of HPV in breast cancer are not yet clear; and the mechanisms of interaction between the infectious agent and host cell still need to be fully elucidated.

Conclusions

The HPV epidemic remains a challenge for researchers because the incidence continues to grow in all the world. With the improvement of molecular techniques some points were clarified resulting in new means of diagnosis and treatment. The number of trials in cervical cancer has enabled better understand the mechanisms used by HPV in the carcinogenic process, but it remains unclear if the HPV is the etiologic agent for other tumors in question or if he participates as an adjuvant and which the pathways used. Knowing well the carcinogenic process may contribute to the development of new vaccines, early diagnosis and therapeutic pathways.

References

Agrawal N, Frederick MJ, Pickering CR, Bettegowda C, Chang K, Li RJ et al (2011) Exome sequencing of head and neck squamous cell carcinoma reveals inactivating mutations in NOTCH1. Science 333(6046):1154–1157

Aguayo F, Castillo A, Koriyama C, Higashi M, Itoh T, Capetillo M et al (2007) Human papillomavirus-16 is integrated in lung carcinomas: a study in Chile. Br J Cancer 97(1):85–91

Akil N, Kassab A, Yasmeen A, Darnel AD, Bismar TA, Al Moustafa A-E (2008) Human breast cancer and sexual activities. Br J Cancer 98(2):508–509
Al Moustafa A-E, Foulkes WD, Benlimame N, Wong A, Yen L, Bergeron J et al (2004) E6/E7 proteins of HPV type 16 and ErbB-2 cooperate to induce neoplastic transformation of primary normal oral epithelial cells. Oncogene 23(2):350–358
Alaghehbandan R, Fontaine D, Bentley J, Escott N, Ghatage P, Lear A et al (2013) Performance of ProEx C and PreTect HPV-Proofer E6/E7 mRNA tests in comparison with the hybrid capture 2 HPV DNA test for triaging ASCUS and LSIL cytology. Diagn Cytopathol 41(9):767–775
Alibek K, Kakpenova A, Mussabekova A, Sypabekova M, Karatayeva N (2013) Role of viruses in the development of breast cancer. Infect Agent Cancer 8(1):32
Altekruse SF, Lacey JV Jr, Brinton LA, Gravitt PE, Silverberg SG, Barnes WA Jr et al (2003) Comparison of human papillomavirus genotypes, sexual, and reproductive risk factors of cervical adenocarcinoma and squamous cell carcinoma: Northeastern United States. Am J Obstet Gynecol 188(3):657–663
Arias-Pulido H, Peyton CL, Torrez-Martínez N, Anderson DN, Wheeler CM (2005) Human papillomavirus type 18 variant lineages in United States populations characterized by sequence analysis of LCR-E6, E2, and L1 regions. Virology 338(1):22–34
Azizi N, Brazete J, Hankins C, Money D, Fontaine J, Koushik A et al (2008) Influence of human papillomavirus type 16 (HPV-16) E2 polymorphism on quantification of HPV-16 episomal and integrated DNA in cervicovaginal lavages from women with cervical intraepithelial neoplasia. J Gen Virol 89(7):1716–1728
Badaracco G, Venuti A, Morello R, Muller A, Marcante ML (2000) Human papillomavirus in head and neck carcinomas: prevalence, physical status and relationship with clinical/pathological parameters. Anticancer Res 20(2B):1301–1305
Badr RE, Walts AE, Chung F, Bose S (2008) BD ProEx C: a sensitive and specific marker of HPV-associated squamous lesions of the cervix. Am J Surg Pathol 32(6):899–906
Balan R, Simion N, Giuşcă SE, Grigoraş A, Gheucă-Solovăstru L, Gheorghiţă V et al (2011) Immunohistochemical assessment of p16, COX-2 and EGFR in HPV-positive cervical squamous intraepithelial lesions. Rom J Morphol Embryol 52(4):1187–1194
Baldwin P, Laskey R, Coleman N (2003) Translational approaches to improving cervical screening. Nat Rev Cancer 3(3):217–226
Balko JM, Stricker TP, Arteaga CL (2013) The genomic map of breast cancer: which roads lead to better targeted therapies? Breast Cancer Res BCR 15(4):209
Bartel DP (2009) MicroRNAs: target recognition and regulatory functions. Cell 136(2):215–233
Beachler DC, D'Souza G, Sugar EA, et al (2012) Differences in oral and anal HPV natural history among HIV infected individuals. In: Proceedings of the 28th international papillomavirus conference, p. 213
Bernard H-U (2002) Gene expression of genital human papillomaviruses and considerations on potential antiviral approaches. Antivir Ther 7(4):219–237
Bernard H-U (2004) Established and potential strategies against papillomavirus infections. J Antimicrob Chemother 53(2):137–139
Bernard H-U, Burk RD, Chen Z, van Doorslaer K, Zur Hausen H, de Villiers E-M (2010) Classification of papillomaviruses (PVs) based on 189 PV types and proposal of taxonomic amendments. Virology 401(1):70–79
Berumen J, Ordoñez RM, Lazcano E, Salmeron J, Galvan SC, Estrada RA et al (2001) Asian-American variants of human papillomavirus 16 and risk for cervical cancer: a case-control study. J Natl Cancer Inst 93(17):1325–1330
Beskow AH, Engelmark MT, Magnusson JJ, Gyllensten UB (2005) Interaction of host and viral risk factors for development of cervical carcinoma in situ. Int J Cancer 117(4):690–692
Bhattacharjee B, Sengupta S (2006) HPV16 E2 gene disruption and polymorphisms of E2 and LCR: some significant associations with cervical cancer in Indian women. Gynecol Oncol 100(2):372–378
Bodily JM, Mehta KPM, Laimins LA (2011) Human papillomavirus E7 enhances hypoxia-inducible factor 1-mediated transcription by inhibiting binding of histone deacetylases. Cancer Res 71(3):1187–1195

Bonilla-Delgado J, Bulut G, Liu X, Cortés-Malagón EM, Schlegel R, Flores-Maldonado C et al (2012) The E6 oncoprotein from HPV16 enhances the canonical Wnt/β-catenin pathway in skin epidermis in vivo. Mol Cancer Res MCR 10(2):250–258

Bosch FX, Burchell AN, Schiffman M, Giuliano AR, de Sanjose S, Bruni L et al (2008) Epidemiology and natural history of human papillomavirus infections and type-specific implications in cervical neoplasia. Vaccine 26(Suppl 10):K1–K16

Bouda J, Hes O, Koprivova M, Pesek M, Svoboda T, Boudova L (2013) P27 as a prognostic factor of early cervical carcinoma. Int J Gynecol Cancer 23(1):164–169

Braakhuis BJM, Snijders PJF, Keune W-JH, Meijer CJLM, Ruijter-Schippers HJ, Leemans CR et al (2004) Genetic patterns in head and neck cancers that contain or lack transcriptionally active human papillomavirus. J Natl Cancer Inst 96(13):998–1006

Broso PR, Buffetti G (1993) The Papanicolaou classification in the Bethesda system (National Cancer Institute, Bethesda, Maryland). Minerva Ginecol 45(11):557–563

Bruni L, Diaz M, Castellsagué M, Ferrer E, Bosch FX, de Sanjosé S (2010) Cervical human papillomavirus prevalence in 5 continents: meta-analysis of 1 million women with normal cytological findings. J Infect Dis 202(12):1789–1799

Burk RD, Terai M, Gravitt PE, Brinton LA, Kurman RJ, Barnes WA et al (2003) Distribution of human papillomavirus types 16 and 18 variants in squamous cell carcinomas and adenocarcinomas of the cervix. Cancer Res 63(21):7215–7220

Burk RD, Harari A, Chen Z (2013) Human papillomavirus genome variants. Virology 445(1-2):232–243

Burroni E, Bisanzi S, Sani C, Puliti D, Carozzi F (2013) Codon 72 polymorphism of p53 and HPV type 16 E6 variants as risk factors for patients with squamous epithelial lesion of the uterine cervix. J Med Virol 85(1):83–90

Calin GA, Sevignani C, Dumitru CD, Hyslop T, Noch E, Yendamuri S et al (2004) Human microRNA genes are frequently located at fragile sites and genomic regions involved in cancers. Proc Natl Acad Sci U S A 101(9):2999–3004

Carpagnano GE, Koutelou A, Natalicchio MI, Martinelli D, Ruggieri C, Di Taranto A et al (2011) HPV in exhaled breath condensate of lung cancer patients. Br J Cancer 105(8):1183–1190

Casas L, Galvan SC, Ordoñez RM, Lopez N, Guido M, Berumen J (1999) Asian-American variants of human papillomavirus type 16 have extensive mutations in the E2 gene and are highly amplified in cervical carcinomas. Int J Cancer 83(4):449–455

Castle PE, Solomon D, Schiffman M, Wheeler CM (2005) Human papillomavirus type 16 infections and 2-year absolute risk of cervical precancer in women with equivocal or mild cytologic abnormalities. J Natl Cancer Inst 97(14):1066–1071

Castle PE, Stoler MH, Solomon D, Schiffman M (2007) The relationship of community biopsy-diagnosed cervical intraepithelial neoplasia grade 2 to the quality control pathology-reviewed diagnoses: an ALTS report. Am J Clin Pathol 127(5):805–815

Castle PE, Schiffman M, Wheeler CM, Solomon D (2009) Evidence for frequent regression of cervical intraepithelial neoplasia-grade 2. Obstet Gynecol 113(1):18–25

Castle PE, Fetterman B, Thomas Cox J, Shaber R, Poitras N, Lorey T et al (2010) The age-specific relationships of abnormal cytology and human papillomavirus DNA results to the risk of cervical precancer and cancer. Obstet Gynecol 116(1):76–84

Cento V, Ciccozzi M, Ronga L, Perno CF, Ciotti M (2009) Genetic diversity of human papillomavirus type 16 E6, E7, and L1 genes in Italian women with different grades of cervical lesions. J Med Virol 81(9):1627–1634

Cento V, Rahmatalla N, Ciccozzi M, Perno CF, Ciotti M (2011) Intratype variations of HPV 31 and 58 in Italian women with abnormal cervical cytology. J Med Virol 83(10):1752–1761

Cento V, Rahmatalla N, Ciccozzi M, Lo Presti A, Perno CF, Ciotti M (2012) Human papillomaviruses 53 and 66: clinical aspects and genetic analysis. Virus Res 163(1):212–222

Cerqueira DM, Camara GNDL, Da Cruz MR, Silva EO, Brígido MDM, Martins CRF (2003) Variants of human papillomavirus types 53, 58 and 66 identified in Central Brazil. Virus Genes 26(1):83–87

Cerqueira DM, Raiol T, Véras NMC, Von Gal Milanezi N, Amaral FA, De Macedo Brígido M et al (2008) New variants of human papillomavirus type 18 identified in central Brazil. Virus Genes 37(2):282–287

Chagas BS, Batista MVA, Guimarães V, Balbino VQ, Crovella S, Freitas AC (2011) New variants of E6 and E7 oncogenes of human papillomavirus type 31 identified in Northeastern Brazil. Gynecol Oncol 123(2):284–288

Chagas BS, Batista MV, Crovella S, Gurgel AP, Silva Neto JC, Serra IG et al (2013) Novel E6 and E7 oncogenes variants of human papillomavirus type 31 in Brazilian women with abnormal cervical cytology. Infect Genet Evol 16C:13–18

Chakrabarti O, Krishna S (2003) Molecular interactions of "high risk" human papillomaviruses E6 and E7 oncoproteins: implications for tumour progression. J Biosci 28(3):337–348

Chakrabarti O, Veeraraghavalu K, Tergaonkar V, Liu Y, Androphy EJ, Stanley MA et al (2004) Human papillomavirus type 16 E6 amino acid 83 variants enhance E6-mediated MAPK signaling and differentially regulate tumorigenesis by notch signaling and oncogenic Ras. J Virol 78(11):5934–5945

Chan PKS, Lam C-W, Cheung T-H, Li WWH, Lo KWK, Chan MYM et al (2002) Association of human papillomavirus type 58 variant with the risk of cervical cancer. J Natl Cancer Inst 94(16):1249–1253

Chang JL, Tsao YP, Liu DW, Huang SJ, Lee WH, Chen SL (2001) The expression of HPV-16 E5 protein in squamous neoplastic changes in the uterine cervix. J Biomed Sci 8(2):206–213

Chang Y-J, Chen H-C, Lee B-H, You S-L, Lin C-Y, Pan M-H et al (2011) Unique variants of human papillomavirus genotypes 52 and 58 and risk of cervical neoplasia. Int J Cancer 129(4):965–973

Chang P, Wang T, Yao Q, Lv Y, Zhang J, Guo W et al (2012) Absence of human papillomavirus in patients with breast cancer in north-west China. Med Oncol Northwood Lond Engl 29(2):521–525

Chansaenroj J, Theamboonlers A, Junyangdikul P, Swangvaree S, Karalak A, Poovorawan Y (2012) Whole genome analysis of human papillomavirus type 16 multiple infection in cervical cancer patients. Asian Pac J Cancer Prev APJCP 13(2):599–606

Chaturvedi AK, Katki HA, Hildesheim A, Rodríguez AC, Quint W, Schiffman M et al (2011) Human papillomavirus infection with multiple types: pattern of coinfection and risk of cervical disease. J Infect Dis 203(7):910–920

Cheng WF, Chen CA, Lee CN, Wei LH, Hsieh FJ, Hsieh CY (2000) Vascular endothelial growth factor and prognosis of cervical carcinoma. Obstet Gynecol 96(5 Pt 1):721–726

Cheng YW, Chiou HL, Sheu GT, Hsieh LL, Chen JT, Chen CY et al (2001) The association of human papillomavirus 16/18 infection with lung cancer among nonsmoking Taiwanese women. Cancer Res 61(7):2799–2803

Cheng Y-W, Wu M-F, Wang J, Yeh K-T, Goan Y-G, Chiou H-L et al (2007) Human papillomavirus 16/18 E6 oncoprotein is expressed in lung cancer and related with p53 inactivation. Cancer Res 67(22):10686–10693

Cheng Y-W, Wu T-C, Chen C-Y, Chou M-C, Ko J-L, Lee H (2008) Human telomerase reverse transcriptase activated by E6 oncoprotein is required for human papillomavirus-16/18-infected lung tumorigenesis. Clin Cancer Res 14(22):7173–7179

Cheung ANY, Tsun K-L, Ng K-M, Szeto E, Siu MK, Wong ES et al (2010) P634A4 and TAp73 immunocytochemistry in liquid-based cervical cytology – potential biomarkers for diagnosis and progress prediction of cervical neoplasia. Mod Pathol 23(4):559–566

Chiou H-L, Wu M-F, Liaw Y-C, Cheng Y-W, Wong R-H, Chen C-Y et al (2003) The presence of human papillomavirus type 16/18 DNA in blood circulation may act as a risk marker of lung cancer in Taiwan. Cancer 97(6):1558–1563

Chung CH, Bagheri A, D'Souza G (2014) Epidemiology of oral human papillomavirus infection. Oral Oncol 50(5):364–369

Clifford GM, Smith JS, Plummer M, Muñoz N, Franceschi S (2003) Human papillomavirus types in invasive cervical cancer worldwide: a meta-analysis. Br J Cancer 88(1):63–73

Coimbra EC, Gomes FB, Campos JF, D'arc M, Carvalho JC, Mariz FC et al (2011) Production of L1 protein from different types of HPV in Pichia pastoris using an integrative vector. Braz J Med Biol Res 44(12):1209–1214

Coissard CJ, Besson G, Polette MC, Monteau M, Birembaut PL, Clavel CE (2005) Prevalence of human papillomaviruses in lung carcinomas: a study of 218 cases. Mod Pathol 18(12):1606–1609

Combita AL, Touzé A, Bousarghin L, Sizaret PY, Muñoz N, Coursaget P (2001) Gene transfer using human papillomavirus pseudovirions varies according to virus genotype and requires cell surface heparan sulfate. FEMS Microbiol Lett 204(1):183–188

Conesa-Zamora P, Doménech-Peris A, Orantes-Casado FJ, Ortiz-Reina S, Sahuquillo-Frías L, Acosta-Ortega J et al (2009) Effect of human papillomavirus on cell cycle-related proteins p16, Ki-67, Cyclin D1, p53, and ProEx C in precursor lesions of cervical carcinoma: a tissue microarray study. Am J Clin Pathol 132(3):378–390

Cornet I, Gheit T, Franceschi S, Vignat J, Burk RD, Sylla BS et al (2012) Human papillomavirus type 16 genetic variants: phylogeny and classification based on E6 and LCR. J Virol 86(12):6855–6861

Cornet I, Gheit T, Clifford GM, Combes J-D, Dalstein V, Franceschi S et al (2013) Human papillomavirus type 16 E6 variants in France and risk of viral persistence. Infect Agent Cancer 8(1):4

Cornut G, Gagnon S, Hankins C, Money D, Pourreaux K, Franco EL et al (2010) Polymorphism of the capsid L1 gene of human papillomavirus types 31, 33, and 35. J Med Virol 82:1168–1178

Corrêa GTB, Bandeira GA, Cavalcanti BG, de Carvalho Fraga CA, dos Santos EP, Silva TF et al (2011) Association of -308 TNF-α promoter polymorphism with clinical aggressiveness in patients with head and neck squamous cell carcinoma. Oral Oncol 47(9):888–894

Cortes-Perez NG, Kharrat P, Langella P, Bermúdez-Humarán LG (2009) Heterologous production of human papillomavirus type-16 L1 protein by a lactic acid bacterium. BMC Res Notes 2(1):167

Costa S, De Simone P, Venturoli S, Cricca M, Zerbini ML, Musiani M et al (2003) Factors predicting human papillomavirus clearance in cervical intraepithelial neoplasia lesions treated by conization. Gynecol Oncol 90(2):358–365

D'Abramo C, Archambault J (2011) Small molecule inhibitors of human papillomavirus protein – protein interactions. Open Virol J 5:80–95

D'Souza G, Kreimer AR, Viscidi R, Pawlita M, Fakhry C, Koch WM et al (2007) Case–control study of human papillomavirus and oropharyngeal cancer. N Engl J Med 356(19):1944–1956

D'Souza G, Agrawal Y, Halpern J, Bodison S, Gillison ML (2009) Oral sexual behaviors associated with prevalent oral human papillomavirus infection. J Infect Dis 199(9):1263–1269

Dahlstrom KR, Calzada G, Hanby JD, Garden AS, Glisson BS, Li G et al (2013) An evolution in demographics, treatment, and outcomes of oropharyngeal cancer at a major cancer center: a staging system in need of repair. Cancer 119(1):81–89

Dai Y, Zhang X, Peng Y, Wang Z (2005) The expression of cyclooxygenase-2, VEGF and PGs in CIN and cervical carcinoma. Gynecol Oncol 97(1):96–103

Dannenberg AJ, Lippman SM, Mann JR, Subbaramaiah K, DuBois RN (2005) Cyclooxygenase-2 and epidermal growth factor receptor: pharmacologic targets for chemoprevention. J Clin Oncol 23(2):254–266

Day PM, Gambhira R, Roden RBS, Lowy DR, Schiller JT (2008) Mechanisms of human papillomavirus type 16 neutralization by 12 cross-neutralizing and 11 type-specific antibodies. J Virol 82(9):4638–4646

De Boer MA, Peters LAW, Aziz MF, Siregar B, Cornain S, Vrede MA et al (2004) Human papillomavirus type 16 E6, E7, and L1 variants in cervical cancer in Indonesia, Suriname, and The Netherlands. Gynecol Oncol 94(2):488–494

De Freitas AC, Gurgel APAD, Chagas BS, Coimbra EC, do Amaral CMM (2012) Susceptibility to cervical cancer: an overview. Gynecol Oncol [Internet]. 2012 Apr 4 [cited 2012 May 6]. Available from: http://www.ncbi.nlm.nih.gov/pubmed/22484226

De la Cruz-Hernández E, García-Carrancá A, Mohar-Betancourt A, Dueñas-González A, Contreras-Paredes A, Pérez-Cardenas E et al (2005) Differential splicing of E6 within human papillomavirus type 18 variants and functional consequences. J Gen Virol 86(9):2459–2468

De Martel C, Ferlay J, Franceschi S, Vignat J, Bray F, Forman D et al (2012) Global burden of cancers attributable to infections in 2008: a review and synthetic analysis. Lancet Oncol 13(6):607–615

De Villiers E-M, Fauquet C, Broker TR, Bernard H-U, zur Hausen H (2004) Classification of papillomaviruses. Virology 324(1):17–27

De Villiers E-M, De Sandstrom RE, zur Haus H, Buck CE (2005) Presence of papillomavirus sequences in condylomatous lesions of the mamillae and in invasive carcinoma of the breast. Breast Cancer Res 7(1):R1–R11

Depuydt CE, Makar AP, Ruymbeke MJ, Benoy IH, Vereecken AJ, Bogers JJ (2011) BD-ProExC as adjunct molecular marker for improved detection of CIN2+ after HPV primary screening. Cancer Epidemiol Biomark Prev 20(4):628–637

DiMaio D, Mattoon D (2001) Mechanisms of cell transformation by papillomavirus E5 proteins. Oncogene 20(54):7866–7873

Ding T, Wang X, Ye F, Cheng X, Ma D, Lu W et al (2010) Distribution of human papillomavirus 58 and 52 E6/E7 variants in cervical neoplasia in Chinese women. Gynecol Oncol 119(3):436–443

Doll CM, Winter K, Gaffney DK, Ryu JK, Jhingran A, Dicker AP et al (2013) COX-2 expression and survival in patients with locally advanced cervical cancer treated with chemoradiotherapy and celecoxib: a quantitative immunohistochemical analysis of RTOG C0128. Int J Gynecol Cancer 23(1):176–183

Donà MG, Vocaturo A, Giuliani M, Ronchetti L, Rollo F, Pescarmona E et al (2012) p16/Ki-67 dual staining in cervico-vaginal cytology: correlation with histology, Human Papillomavirus detection and genotyping in women undergoing colposcopy. Gynecol Oncol 126(2):198–202

Dong XP, Pfister H (1999) Overlapping YY1- and aberrant SP1-binding sites proximal to the early promoter of human papillomavirus type 16. J Gen Virol 80(Pt 8):2097–2101

Doorbar J (2007) Papillomavirus life cycle organization and biomarker selection. Dis Markers 23(4):297–313

Doorbar J, Quint W, Banks L, Bravo IG, Stoler M, Broker TR et al (2012) The biology and life-cycle of human papillomaviruses. Vaccine 30(Suppl 5):F55–F70

Duensing S, Münger K (2002) The human papillomavirus type 16 E6 and E7 oncoproteins independently induce numerical and structural chromosome instability. Cancer Res 62(23):7075–7082

Duensing A, Liu Y, Perdreau SA, Kleylein-Sohn J, Nigg EA, Duensing S (2007) Centriole overduplication through the concurrent formation of multiple daughter centrioles at single maternal templates. Oncogene 26(43):6280–6288

Eichten A, Rud DS, Grace M, Piboonniyom S-O, Zacny V, Münger K (2004) Molecular pathways executing the "trophic sentinel" response in HPV-16 E7-expressing normal human diploid fibroblasts upon growth factor deprivation. Virology 319(1):81–93

Ellis JR, Keating PJ, Baird J, Hounsell EF, Renouf DV, Rowe M et al (1995) The association of an HPV16 oncogene variant with HLA-B7 has implications for vaccine design in cervical cancer. Nat Med 1(5):464–470

Eriksson A, Herron JR, Yamada T, Wheeler CM (1999) Human papillomavirus type 16 variant lineages characterized by nucleotide sequence analysis of the E5 coding segment and the E2 hinge region. J Gen Virol 80(3):595–600

Fahey MT, Irwig L, Macaskill P (1995) Meta-analysis of Pap test accuracy. Am J Epidemiol 141(7):680–689

Fakhry C, Westra WH, Li S, Cmelak A, Ridge JA, Pinto H et al (2008) Improved survival of patients with human papillomavirus-positive head and neck squamous cell carcinoma in a prospective clinical trial. J Natl Cancer Inst 100(4):261–269

Fehrmann F, Laimins LA (2003) Human papillomaviruses: targeting differentiating epithelial cells for malignant transformation. Oncogene 22(33):5201–5207

Ferrandina G, Lauriola L, Distefano MG, Zannoni GF, Gessi M, Legge F et al (2002) Increased cyclooxygenase-2 expression is associated with chemotherapy resistance and poor survival in cervical cancer patients. J Clin Oncol 20(4):973–981

Ferrandina G, Ranelletti FO, Legge F, Lauriola L, Salutari V, Gessi M et al (2003) Celecoxib modulates the expression of cyclooxygenase-2, ki67, apoptosis-related marker, and microvessel density in human cervical cancer: a pilot study. Clin Cancer Res 9(12):4324–4331

Finzer P, Kuntzen C, Soto U, zur Hausen H, Rösl F (2001) Inhibitors of histone deacetylase arrest cell cycle and induce apoptosis in cervical carcinoma cells circumventing human papillomavirus oncogene expression. Oncogene 20(35):4768–4776

Finzer P, Aguilar-Lemarroy A, Rösl F (2002) The role of human papillomavirus oncoproteins E6 and E7 in apoptosis. Cancer Lett 188(1-2):15–24

Franco EL, Rohan TE, Villa LL (1999) Epidemiologic evidence and human papillomavirus infection as a necessary cause of cervical cancer. J Natl Cancer Inst 91(6):506–511

Frati E, Bianchi S, Colzani D, Zappa A, Orlando G, Tanzi E (2011) Genetic variability in the major capsid L1 protein of human papillomavirus type 16 (HPV-16) and 18 (HPV-18). Infect Genet Evol 11(8):2119–2124

Frazer IH (2004) Prevention of cervical cancer through papillomavirus vaccination. Nat Rev Immunol 4(1):46–54

Gabrielson E (2006) Worldwide trends in lung cancer pathology. Respirol Carlton Vic 11(5):533–538

Gaffney DK, Holden J, Davis M, Zempolich K, Murphy KJ, Dodson M (2001) Elevated cyclooxygenase-2 expression correlates with diminished survival in carcinoma of the cervix treated with radiotherapy. Int J Radiat Oncol Biol Phys 49(5):1213–1217

Gagnon S, Hankins C, Tremblay C, Forest P, Pourreaux K, Coutlée F et al (2004) Viral polymorphism in human papillomavirus types 33 and 35 and persistent and transient infection in the genital tract of women. J Infect Dis 190(9):1575–1585

Gagnon S, Hankins C, Money D, Pourreaux K, Franco E, Coutlée F (2007) Polymorphism of the L1 capsid gene and persistence of human papillomavirus type 52 infection in women at high risk or infected by HIV. J Acquir Immune Defic Syndr 44(1):61–65

Galgano MT, Castle PE, Atkins KA, Brix WK, Nassau SR, Stoler MH (2010) Using biomarkers as objective standards in the diagnosis of cervical biopsies. Am J Surg Pathol 34(8):1077–1087

Garland SM, Hernandez-Avila M, Wheeler CM, Perez G, Harper DM, Leodolter S et al (2007) Quadrivalent vaccine against human papillomavirus to prevent anogenital diseases. N Engl J Med 356(19):1928–1943

Garnett TO, Duerksen-Hughes PJ (2006) Modulation of apoptosis by human papillomavirus (HPV) oncoproteins. Arch Virol 151(12):2321–2335

Gheit T, Cornet I, Clifford GM, Iftner T, Munk C, Tommasino M et al (2011) Risks for persistence and progression by human papillomavirus type 16 variant lineages among a population-based sample of Danish women. Cancer Epidemiol Biomark Prev 20(7):1315–1321

Giannoudis A, van Duin M, Snijders PJF, Herrington CS (2001) Variation in the E2-binding domain of HPV 16 is associated with high-grade squamous intraepithelial lesions of the cervix. Br J Cancer 84(8):1058–1063

Gillison ML (2008) Human papillomavirus-related diseases: oropharynx cancers and potential implications for adolescent HPV vaccination. J Adolesc Health 43(4 Suppl):S52–S60

Gillison ML, Shah KV (2003) Chapter 9: role of mucosal human papillomavirus in nongenital cancers. JNCI Monogr 2003(31):57–65

Gillison ML, Koch WM, Capone RB, Spafford M, Westra WH, Wu L et al (2000) Evidence for a causal association between human papillomavirus and a subset of head and neck cancers. J Natl Cancer Inst 92(9):709–720

Gillison ML, Broutian T, Pickard RKL, Tong Z, Xiao W, Kahle L et al (2012) Prevalence of oral HPV infection in the United States, 2009–2010. JAMA 307(7):693–703

Giroglou T, Florin L, Schäfer F, Streeck RE, Sapp M (2001) Human papillomavirus infection requires cell surface heparan sulfate. J Virol 75(3):1565–1570

Godínez JM, Heideman DAM, Gheit T, Alemany L, Snijders PJF, Tommasino M et al (2013) Differential presence of papillomavirus variants in cervical cancer: an analysis for HPV33, HPV45 and HPV58. Infect Genet Evol 13:96–104

Goncalves A, Fabbro M, Lhommé C, Gladieff L, Extra J-M, Floquet A et al (2008) A phase II trial to evaluate gefitinib as second- or third-line treatment in patients with recurring locoregionally advanced or metastatic cervical cancer. Gynecol Oncol 108(1):42–46

Graham DA, Herrington CS (2000) HPV-16 E2 gene disruption and sequence variation in CIN 3 lesions and invasive squamous cell carcinomas of the cervix: relation to numerical chromosome abnormalities. Mol Pathol 53(4):201–206

Greco D, Kivi N, Qian K, Leivonen S-K, Auvinen P, Auvinen E (2011) Human papillomavirus 16 E5 modulates the expression of host microRNAs. PLoS One 6(7):e21646

Greenhough A, Smartt HJM, Moore AE, Roberts HR, Williams AC, Paraskeva C et al (2009) The COX-2/PGE2 pathway: key roles in the hallmarks of cancer and adaptation to the tumour microenvironment. Carcinogenesis 30(3):377–386

Grodzki M, Besson G, Clavel C, Arslan A, Franceschi S, Birembaut P et al (2006) Increased risk for cervical disease progression of French women infected with the human papillomavirus type 16 E6-350G variant. Cancer Epidemiol Biomark Prev 15(4):820–822

Gurgel AP, Chagas BS, do Amaral CM, Albuquerque EM, Serra IG, Costa Silva Neto J et al (2013) Prevalence and genetic variability in capsid L1 gene of rare human papillomaviruses (HPV) found in cervical lesions of women from North-East Brazil. BioMed Res Int 2013:546354

Gustinucci D, Passamonti B, Cesarini E, Butera D, Palmieri EA, Bulletti S et al (2012) Role of p16(INK4a) cytology testing as an adjunct to enhance the diagnostic specificity and accuracy in human papillomavirus-positive women within an organized cervical cancer screening program. Acta Cytol 56(5):506–514

Häfner N, Driesch C, Gajda M, Jansen L, Kirchmayr R, Runnebaum IB et al (2008) Integration of the HPV16 genome does not invariably result in high levels of viral oncogene transcripts. Oncogene 27(11):1610–1617

Hammarstedt L, Lindquist D, Dahlstrand H, Romanitan M, Dahlgren LO, Joneberg J et al (2006) Human papillomavirus as a risk factor for the increase in incidence of tonsillar cancer. Int J Cancer 119(11):2620–2623

Hanumantha Rao N, Baji Babu P, Rajendra L, Sriraman R, Pang Y-YS, Schiller JT et al (2011) Expression of codon optimized major capsid protein (L1) of human papillomavirus type 16 and 18 in Pichia pastoris; purification and characterization of the virus-like particles. Vaccine 29(43):7326–7334

Hedau S, Kumar U, Hussain S, Shukla S, Pande S, Jain N et al (2011) Breast cancer and human papillomavirus infection: no evidence of HPV etiology of breast cancer in Indian women. BMC Cancer 11(1):27

Heng B, Glenn WK, Ye Y, Tran B, Delprado W, Lutze-Mann L et al (2009) Human papilloma virus is associated with breast cancer. Br J Cancer 101(8):1345–1350

Hensing T, Chawla A, Batra R, Salgia R (2014) A personalized treatment for lung cancer: molecular pathways, targeted therapies, and genomic characterization. Adv Exp Med Biol 799:85–117

Herrera FG, Chan P, Doll C, Milosevic M, Oza A, Syed A et al (2007) A prospective phase I-II trial of the cyclooxygenase-2 inhibitor celecoxib in patients with carcinoma of the cervix with biomarker assessment of the tumor microenvironment. Int J Radiat Oncol Biol Phys 67(1):97–103

Herrera-Goepfert R, Khan NA, Koriyama C, Akiba S, Pérez-Sánchez VM (2011) High-risk human papillomavirus in mammary gland carcinomas and non-neoplastic tissues of Mexican women: no evidence supporting a cause and effect relationship. Breast Edinb Scotl 20(2):184–189

Hildesheim A, Herrero R, Wacholder S, Rodriguez AC, Solomon D, Bratti MC et al (2007) Effect of human papillomavirus 16/18 L1 viruslike particle vaccine among young women with preexisting infection: a randomized trial. JAMA 298(7):743–753

Ho C-M, Yang S-S, Chien T-Y, Huang S-H, Jeng C-J, Chang S-F (2005) Detection and quantitation of human papillomavirus type 16, 18 and 52 DNA in the peripheral blood of cervical cancer patients. Gynecol Oncol 99(3):615–621

Hong AM, Martin A, Chatfield M, Jones D, Zhang M, Armstrong B et al (2013) Human papillomavirus, smoking status and outcomes in tonsillar squamous cell carcinoma. Int J Cancer 132(12):2748–2754

Hoory T, Monie A, Gravitt P, Wu T-C (2008) Molecular epidemiology of human papillomavirus. J Formos Med Assoc 107(3):198–217

WHO/ICO Information Centre on HPV and Cervical Cancer (HPV Information Centre) (2010) Human papillomavirus and related cancers in Brazil [Internet]. 2010 summary report. Available at www.who.int/hpvcentre

Hu X, Schwarz JK, Lewis JS Jr, Huettner PC, Rader JS, Deasy JO et al (2010) A microRNA expression signature for cervical cancer prognosis. Cancer Res 70(4):1441–1448

Huang L-W, Chao S-L, Hwang J-L, Chou Y-Y (2002) Down-regulation of p27 is associated with malignant transformation and aggressive phenotype of cervical neoplasms. Gynecol Oncol 85(3):524–528

Izadi-Mood N, Asadi K, Shojaei H, Sarmadi S, Ahmadi SA, Sani S et al (2012) Potential diagnostic value of P16 expression in premalignant and malignant cervical lesions. J Res Med Sci 17(5):428–433

Jemal A, Bray F, Center MM, Ferlay J, Ward E, Forman D (2011) Global cancer statistics. CA Cancer J Clin 61(2):69–90

Jemal A, Simard EP, Dorell C, Noone A-M, Markowitz LE, Kohler B et al (2013) Annual report to the nation on the status of cancer, 1975–2009, featuring the burden and trends in Human Papillomavirus (HPV)–associated cancers and HPV vaccination coverage levels. J Natl Cancer Inst 105(3):175–201

Johnson KM, Kines RC, Roberts JN, Lowy DR, Schiller JT, Day PM (2009) Role of heparan sulfate in attachment to and infection of the murine female genital tract by human papillomavirus. J Virol 83(5):2067–2074

Kalof AN, Cooper K (2006) p16INK4a immunoexpression: surrogate marker of high-risk HPV and high-grade cervical intraepithelial neoplasia. Adv Anat Pathol 13(4):190–194

Kämmer C, Warthorst U, Torrez-Martinez N, Wheeler CM, Pfister H (2000) Sequence analysis of the long control region of human papillomavirus type 16 variants and functional consequences for P97 promoter activity. J Gen Virol 81(8):1975–1981

Kämmer C, Tommasino M, Syrjänen S, Delius H, Hebling U, Warthorst U et al (2002) Variants of the long control region and the E6 oncogene in European human papillomavirus type 16 isolates: implications for cervical disease. Br J Cancer 86(2):269–273

Kang J-O, Hong S-E (2004) The prognostic effect of VEGF expression in squamous cell carcinoma of the cervix treated with radiation therapy alone. J Korean Med Sci 19(5):693–697

Karanam B, Jagu S, Huh WK, Roden RBS (2009) Developing vaccines against minor capsid antigen L2 to prevent papillomavirus infection. Immunol Cell Biol 87(4):287–299

Kast W, Brandt R, Sidney J, Drijfhout J, Kubo R, Grey H et al (1994) Role of HLA-A motifs in identification of potential CTL epitopes in human papillomavirus type 16 E6 and E7 proteins. J Immunol 152(8):3904–3912

Kelly D, Kincaid E, Fansler Z, Rosenthal DL, Clark DP (2006) Detection of cervical high-grade squamous intraepithelial lesions from cytologic samples using a novel immunocytochemical assay (ProEx C). Cancer 108(6):494–500

Khammapirad T, Prueksadee J, Diaz-Arrastia C, Botting SK, Leonard M, Bonoan-Deomampo L et al (2011) Intraductal papilloma of the breast in association with preoncogenic gene of breast cancer. Asian Pac J Trop Biomed 1(2):161–163

Khouadri S, Villa LL, Simon G, Koushik A, Richardson H, Ferreira S et al (2006) Human papillomavirus type 33 polymorphisms and high-grade squamous intraepithelial lesions of the uterine cervix. J Infect Dis 194(7):886–894

Kim M-H, Seo S-S, Song Y-S, Kang D-H, Park I-A, Kang S-B et al (2003) Expression of cyclooxygenase-1 and -2 associated with expression of VEGF in primary cervical cancer and at metastatic lymph nodes. Gynecol Oncol 90(1):83–90

Kim GE, Kim YB, Cho NH, Chung H-C, Pyo HR, Lee JD et al (2004) Synchronous coexpression of epidermal growth factor receptor and cyclooxygenase-2 in carcinomas of the uterine cervix: a potential predictor of poor survival. Clin Cancer Res 10(4):1366–1374

Kim S-H, Juhnn Y-S, Kang S, Park S-W, Sung M-W, Bang Y-J et al (2006) Human papillomavirus 16 E5 up-regulates the expression of vascular endothelial growth factor through the activation of epidermal growth factor receptor, MEK/ ERK1,2 and PI3K/Akt. Cell Mol Life Sci CMLS 63(7–8):930–938

Kim K-Y, Zhang X, Cha I-H (2014) The identification of human papillomavirus status specific biomarker in head and neck cancer. Head Neck

Kirnbauer R, Taub J, Greenstone H, Roden R, Dürst M, Gissmann L et al (1993) Efficient self-assembly of human papillomavirus type 16 L1 and L1-L2 into virus-like particles. J Virol 67(12):6929–6936

Klaes R, Friedrich T, Spitkovsky D, Ridder R, Rudy W, Petry U et al (2001) Overexpression of p16(INK4A) as a specific marker for dysplastic and neoplastic epithelial cells of the cervix uteri. Int J Cancer 92(2):276–284

Klapper LN, Glathe S, Vaisman N, Hynes NE, Andrews GC, Sela M et al (1999) The ErbB-2/HER2 oncoprotein of human carcinomas may function solely as a shared coreceptor for multiple stroma-derived growth factors. Proc Natl Acad Sci U S A 96(9):4995–5000

Klein F, Amin Kotb WFM, Petersen I (2009) Incidence of human papilloma virus in lung cancer. Lung Cancer 65(1):13–18

Kling M, Zeichner JA (2010) The role of the human papillomavirus (HPV) vaccine in developing countries. Int J Dermatol 49(4):377–379

Klug SJ, Hetzer M, Blettner M (2005) Screening for breast and cervical cancer in a large German city: participation, motivation and knowledge of risk factors. Eur J Public Health 15(1):70–77

Koskimaa HM, Waterboer T, Pawlita M, Grénman S, Syrjänen K, Syrjänen S (2012) Human papillomavirus genotypes present in the oral mucosa of newborns and their concordance with maternal cervical human papillomavirus genotypes. J Pediatr 160:837–843

Kost TA, Condreay JP, Jarvis DL (2005) Baculovirus as versatile vectors for protein expression in insect and mammalian cells. Nat Biotechnol 23(5):567–575

Krawczyk E, Suprynowicz FA, Liu X, Dai Y, Hartmann DP, Hanover J et al (2008) Koilocytosis: a cooperative interaction between the human papillomavirus E5 and E6 oncoproteins. Am J Pathol 173(3):682–688

Kreimer AR, Clifford GM, Boyle P, Franceschi S (2005) Human papillomavirus types in head and neck squamous cell carcinomas worldwide: a systematic review. Cancer Epidemiol Biomark Prev 14(2):467–475

Kruse A-J, Baak JPA, Janssen EA, Kjellevold K-H, Fiane B, Lovslett K et al (2004) Ki67 predicts progression in early CIN: validation of a multivariate progression-risk model. Cell Oncol 26(1–2):13–20

Kuriakose MA, Chen WT, He ZM, Sikora AG, Zhang P, Zhang ZY et al (2004) Selection and validation of differentially expressed genes in head and neck cancer. Cell Mol Life Sci CMLS 61(11):1372–1383

Kyo S, Takakura M, Fujiwara T, Inoue M (2008) Understanding and exploiting hTERT promoter regulation for diagnosis and treatment of human cancers. Cancer Sci 99(8):1528–1538

Lajer CB, Garnæs E, Friis-Hansen L, Norrild B, Therkildsen MH, Glud M et al (2012) The role of miRNAs in human papilloma virus (HPV)-associated cancers: bridging between HPV-related head and neck cancer and cervical cancer. Br J Cancer 106(9):1526–1534

Langer CJ (2012) Exploring biomarkers in head and neck cancer. Cancer 118(16):3882–3892

Lawson JS, Tran D, Rawlinson WD (2000) From Bittner to Barr: a viral, diet and hormone breast cancer aetiology hypothesis. Breast Cancer Res 3(2):81

Lawson JS, Glenn WK, Heng B, Ye Y, Tran B, Lutze-Mann L et al (2009) Koilocytes indicate a role for human papilloma virus in breast cancer. Br J Cancer 101(8):1351–1356

Li G, He L, Zhang E, Shi J, Zhang Q, Le AD, et al (2011a) Overexpression of human papillomavirus (HPV) type 16 oncoproteins promotes angiogenesis via enhancing HIF-1α and VEGF expression in non-small cell lung cancer cells. Cancer Lett [Internet]. 2011a July 22 [cited 2011 Sept 17]. Available from: http://www.ncbi.nlm.nih.gov/pubmed/21868151

Li N, Franceschi S, Howell-Jones R, Snijders PJF, Clifford GM (2011b) Human papillomavirus type distribution in 30,848 invasive cervical cancers worldwide: Variation by geographical region, histological type and year of publication. Int J Cancer 128(4):927–935

Liu X, Yuan H, Fu B, Disbrow GL, Apolinario T, Tomaic V et al (2005) The E6AP ubiquitin ligase is required for transactivation of the hTERT promoter by the human papillomavirus E6 oncoprotein. J Biol Chem 280(11):10807–10816

Londesborough P, Ho L, Terry G, Cuzick J, Wheeler C, Singer A (1996) Human papillomavirus genotype as a predictor of persistence and development of high-grade lesions in women with minor cervical abnormalities. Int J Cancer 69(5):364–368

López-Saavedra A, González-Maya L, Ponce-de-León S, García-Carrancá A, Mohar A, Lizano M (2009) Functional implication of sequence variation in the long control region and E2 gene among human papillomavirus type 18 variants. Arch Virol 154(5):747–754

Lowy DR, Schiller JT (2012) Reducing HPV-associated cancer globally. Cancer Prev Res 5(1):18–23

Mackay HJ, Tinker A, Winquist E, Thomas G, Swenerton K, Oza A et al (2010) A phase II study of sunitinib in patients with locally advanced or metastatic cervical carcinoma: NCIC CTG Trial IND.184. Gynecol Oncol 116(2):163–167

Malinowski DP (2005) Molecular diagnostic assays for cervical neoplasia: emerging markers for the detection of high-grade cervical disease. BioTechnique 38(Suppl):17–23

Márquez-Garbán DC, Chen H-W, Goodglick L, Fishbein MC, Pietras RJ (2009) Targeting aromatase and estrogen signaling in human non-small cell lung cancer. Ann N Y Acad Sci 1155:194–205

Marur S, D'Souza G, Westra WH, Forastiere AA (2010) HPV-associated head and neck cancer: a virus-related cancer epidemic. Lancet Oncol 11(8):781–789

McCance DJ (2005) Transcriptional regulation by human papillomaviruses. Curr Opin Genet Dev 15(5):515–519

McCredie MRE, Sharples KJ, Paul C, Baranyai J, Medley G, Jones RW et al (2008) Natural history of cervical neoplasia and risk of invasive cancer in women with cervical intraepithelial neoplasia 3: a retrospective cohort study. Lancet Oncol 9(5):425–434

McLaughlin-Drubin ME, Munger K (2009) Oncogenic activities of human papillomaviruses. Virus Res 143(2):195–208

McMurray HR, McCance DJ (2003) Human papillomavirus type 16 E6 activates TERT gene transcription through induction of c-Myc and release of USF-mediated repression. J Virol 77(18):9852–9861

Menges CW, Baglia LA, Lapoint R, McCance DJ (2006) Human papillomavirus type 16 E7 up-regulates AKT activity through the retinoblastoma protein. Cancer Res 66(11):5555–5559

Mimica M, Tomić S, Kardum G, Hofman ID, Kaliterna V, Pejković L (2010) Ki-67 quantitative evaluation as a marker of cervical intraepithelial neoplasia and human papillomavirus infection. Int J Gynecol Cancer 20(1):116–119

Moody CA, Laimins LA (2010) Human papillomavirus oncoproteins: pathways to transformation. Nat Rev Cancer 10(8):550–560

Mork J, Lie AK, Glattre E, Hallmans G, Jellum E, Koskela P et al (2001) Human papillomavirus infection as a risk factor for squamous-cell carcinoma of the head and neck. N Engl J Med 344(15):1125–1131

Motoyama S, Ladines-Llave CA, Luis Villanueva S, Maruo T (2004) The role of human papilloma virus in the molecular biology of cervical carcinogenesis. Kobe J Med Sci 50(1–2):9–19

Movva S, Rodriguez L, Arias-Pulido H, Verschraegen C (2009) Novel chemotherapy approaches for cervical cancer. Cancer 115(14):3166–3180

Muñoz N, Bosch FX, de Sanjosé S, Herrero R, Castellsagué X, Shah KV et al (2003) Epidemiologic classification of human papillomavirus types associated with cervical cancer. N Engl J Med 348(6):518–527

Muñoz N, Castellsagué X, de González AB, Gissmann L (2006) Chapter 1: HPV in the etiology of human cancer. Vaccine 24(Suppl 3):S3/1–S3/10

Nagai Y, Maehama T, Asato T, Kanazawa K (2000) Persistence of human papillomavirus infection after therapeutic conization for CIN 3: is it an alarm for disease recurrence? Gynecol Oncol 79(2):294–299

Nair HB, Luthra R, Kirma N, Liu Y-G, Flowers L, Evans D et al (2005) Induction of aromatase expression in cervical carcinomas: effects of endogenous estrogen on cervical cancer cell proliferation. Cancer Res 65(23):11164–11173

Narisawa-Saito M, Kiyono T (2007) Basic mechanisms of high-risk human papillomavirus-induced carcinogenesis: roles of E6 and E7 proteins. Cancer Sci 98(10):1505–1511

Nasir L, Gault E, Morgan IM, Chambers G, Ellsmore V, Campo MS (2007) Identification and functional analysis of sequence variants in the long control region and the E2 open reading frame of bovine papillomavirus type 1 isolated from equine sarcoids. Virology 364(2):355–361

Natunen K, Lehtinen J, Namujju P, Sellors J, Lehtinen M (2011) Aspects of prophylactic vaccination against cervical cancer and other human papillomavirus-related cancers in developing countries. Infect Dis Obstet Gynecol 2011:675858

Negri G, Bellisano G, Carico E, Faa G, Kasal A, Antoniazzi S et al (2011) Usefulness of p16ink4a, ProEX C, and Ki-67 for the diagnosis of glandular dysplasia and adenocarcinoma of the cervix uteri. Int J Gynecol Pathol 30(4):407–413

Nicholson RI, Gee JM, Harper ME (2001) EGFR and cancer prognosis. Eur J Cancer Oxf Engl 1990 37(Suppl 4):S9–S15

Nindl I, Stockfleth E, Hofmann TG (2013) Report on the 6th international conference of HPV, polyomavirus, and UV radiation in skin cancer. J Invest Dermatol 133(7):1702–1705

Nishizawa R, Nagata M, Noman AA, Kitamura N, Fujita H, Hoshina H et al (2007) The 2G allele of promoter region of Matrix metalloproteinase-1 as an essential pre-condition for the early onset of oral squamous cell carcinoma. BMC Cancer 7(1):187

Ntova CK, Kottaridi C, Chranioti A, Spathis A, Kassanos D, Paraskevaidis E et al (2011) Genetic variability and phylogeny of high risk HPV type 16, 18, 31, 33 and 45 L1 gene in Greek women. Int J Mol Sci 13(12):1–17

Omori M, Hashi A, Nakazawa K, Yuminamochi T, Yamane T, Hirata S et al (2007) Estimation of prognoses for cervical intraepithelial neoplasia 2 by p16INK4a immunoexpression and high-risk HPV in situ hybridization signal types. Am J Clin Pathol 128(2):208–217

Ozaki S, Zen Y, Inoue M (2011) Biomarker expression in cervical intraepithelial neoplasia: potential progression predictive factors for low-grade lesions. Hum Pathol 42(7):1007–1012

Pande S, Jain N, Prusty BK, Bhambhani S, Gupta S, Sharma R et al (2008) Human papillomavirus type 16 variant analysis of E6, E7, and L1 genes and long control region in biopsy samples from cervical cancer patients in North India. J Clin Microbiol 46(3):1060–1066

Panjkovic M, Ivkovic-Kapicl T (2006) Ki-67 expression in squamous intraepithelial lesions of the uterine cervix. Arch Oncol 14(1–2):23–25

Parker JS, Mullins M, Cheang MCU, Leung S, Voduc D, Vickery T et al (2009) Supervised risk predictor of breast cancer based on intrinsic subtypes. J Clin Oncol 27(8):1160–1167

Pedroza-Saavedra A, Lam EW-F, Esquivel-Guadarrama F, Gutierrez-Xicotencatl L (2010) The human papillomavirus type 16 E5 oncoprotein synergizes with EGF-receptor signaling to enhance cell cycle progression and the down-regulation of p27(Kip1). Virology 400(1):44–52

Pfister H (2003) Chapter 8: human papillomavirus and skin cancer. J Natl Cancer Inst Monogr 31:52–56

Pillai MR, Hariharan R, Babu JM, Lakshmi S, Chiplunkar SV, Patkar M et al (2009) Molecular variants of HPV-16 associated with cervical cancer in Indian population. Int J Cancer 125(1):91–103

Pim D, Massimi P, Dilworth SM, Banks L (2005) Activation of the protein kinase B pathway by the HPV-16 E7 oncoprotein occurs through a mechanism involving interaction with PP2A. Oncogene 24(53):7830–7838

Pinto AP, Degen M, Villa LL, Cibas ES (2012) Immunomarkers in gynecologic cytology: the search for the ideal "biomolecular Papanicolaou test". Acta Cytol 56(2):109–121

Pirog EC, Baergen RN, Soslow RA, Tam D, DeMattia AE, Chen Y-T et al (2002) Diagnostic accuracy of cervical low-grade squamous intraepithelial lesions is improved with MIB-1 immunostaining. Am J Surg Pathol 26(1):70–75

Pittayakhajonwut D, Angeletti PC (2010) Viral trans-factor independent replication of human papillomavirus genomes. Virol J 7:123
Prabhu PR, Jayalekshmi D, Pillai MR (2012) Lung cancer and Human Papilloma Viruses (HPVs): examining the molecular evidence. J Oncol 2012:e750270
Prado JC, Calleja-Macias IE, Bernard H-U, Kalantari M, Macay SA, Allan B et al (2005) Worldwide genomic diversity of the human papillomaviruses-53, 56, and 66, a group of high-risk HPVs unrelated to HPV-16 and HPV-18. Virology 340(1):95–104
Pytynia KB, Dahlstrom KR, Sturgis EM (2014) Epidemiology of HPV-associated oropharyngeal cancer. Oral Oncol 50(5):380–386
Quint KD, de Koning MNC, van Doorn L-J, Quint WGV, Pirog EC (2010) HPV genotyping and HPV16 variant analysis in glandular and squamous neoplastic lesions of the uterine cervix. Gynecol Oncol 117(2):297–301
Quint KD, de Koning MNC, Quint WGV, Pirog EC (2013) Progression of cervical low grade squamous intraepithelial lesions: in search of prognostic biomarkers. Eur J Obstet Gynecol Reprod Biol 170(2):501–506
Raiol T, Wyant PS, de Amorim RM, Cerqueira DM, von Milanezi N, de Brígido M et al (2009) Genetic variability and phylogeny of the high-risk HPV-31, -33, -35, -52, and -58 in central Brazil. J Med Virol 81(4):685–692
Rampias T, Sasaki C, Weinberger P, Psyrri A (2009) E6 and e7 gene silencing and transformed phenotype of human papillomavirus 16-positive oropharyngeal cancer cells. J Natl Cancer Inst 101(6):412–423
Rampias T, Boutati E, Pectasides E, Sasaki C, Kountourakis P, Weinberger P et al (2010) Activation of Wnt signaling pathway by human papillomavirus E6 and E7 oncogenes in HPV16-positive oropharyngeal squamous carcinoma cells. Mol Cancer Res MCR 8(3):433–443
Roelens J, Reuschenbach M, von Knebel DM, Wentzensen N, Bergeron C, Arbyn M (2012) p16INK4a immunocytochemistry versus human papillomavirus testing for triage of women with minor cytologic abnormalities: a systematic review and meta-analysis. Cancer Cytopathol 120(5):294–307
Rokita W, Skawiński D, Zmelonek-Znamirowska A, Kedzia W, Karowicz-Bilińska A, Spaczyński R et al (2012) Results of pap smears and immunocytochemical detection of the p16 and Ki67 proteins in women with cervical intraepithelial neoplasia and cervical cancer. Ginekol Pol 83(11):822–826
Ronco G, Segnan N, Giorgi-Rossi P, Zappa M, Casadei GP, Carozzi F et al (2006) Human papillomavirus testing and liquid-based cytology: results at recruitment from the new technologies for cervical cancer randomized controlled trial. J Natl Cancer Inst 98(11):765–774
Rudin CM, Avila-Tang E, Samet JM (2009) Lung cancer in never smokers: a call to action. Clin Cancer Res 15(18):5622–5625
Rusnak DW, Affleck K, Cockerill SG, Stubberfield C, Harris R, Page M et al (2001) The characterization of novel, dual ErbB-2/EGFR, tyrosine kinase inhibitors potential therapy for cancer. Cancer Res 61(19):7196–7203
Sahebali S, Depuydt CE, Segers K, Vereecken AJ, Van Marck E, Bogers JJ (2003) Ki-67 immunocytochemistry in liquid based cervical cytology: useful as an adjunctive tool? J Clin Pathol 56(9):681–686
Samir R, Asplund A, Tot T, Pekar G, Hellberg D (2011) High-risk HPV infection and CIN grade correlates to the expression of c-myc, CD4+, FHIT, E-cadherin, Ki-67, and p16INK4a. J Low Genit Tract Dis 15(4):280–286
Sandal T (2002) Molecular aspects of the mammalian cell cycle and cancer. Oncologist 7(1):73–81
Sankaranarayanan R, Esmy PO, Rajkumar R, Muwonge R, Swaminathan R, Shanthakumari S et al (2007) Effect of visual screening on cervical cancer incidence and mortality in Tamil Nadu, India: a cluster-randomised trial. Lancet 370(9585):398–406
Santin AD, Zhan F, Bignotti E, Siegel ER, Cané S, Bellone S et al (2005) Gene expression profiles of primary HPV16- and HPV18-infected early stage cervical cancers and normal cervical epithelium: identification of novel candidate molecular markers for cervical cancer diagnosis and therapy. Virology 331(2):269–291

Santin AD, Sill MW, McMeekin DS, Leitao MM Jr, Brown J, Sutton GP et al (2011) Phase II trial of cetuximab in the treatment of persistent or recurrent squamous or non-squamous cell carcinoma of the cervix: a Gynecologic Oncology Group study. Gynecol Oncol 122(3):495–500
Santos AL, Derchain SF, Martins MR, Sarian LO, Martinez EZ, Syrjänen KJ (2003) Human papillomavirus viral load in predicting high-grade CIN in women with cervical smears showing only atypical squamous cells or low-grade squamous intraepithelial lesion. São Paulo Med J 121(6):238–243
Sapp M, Day PM (2009) Structure, attachment and entry of polyoma- and papillomaviruses. Virology 384(2):400–409
Saslow D, Solomon D, Lawson HW, Killackey M, Kulasingam SL, Cain J et al (2012) American cancer society, American society for colposcopy and cervical pathology, and American society for clinical pathology screening guidelines for the prevention and early detection of cervical cancer. CA Cancer J Clin 62(3):147–172
Schäfer G, Kabanda S, van Rooyen B, Marušič MB, Banks L, Parker MI (2013) The role of inflammation in HPV infection of the oesophagus. BMC Cancer 13(1):185
Schefter TE, Winter K, Kwon JS, Stuhr K, Balaraj K, Yaremko BP et al (2012) A phase II study of bevacizumab in combination with definitive radiotherapy and cisplatin chemotherapy in untreated patients with locally advanced cervical carcinoma: preliminary results of RTOG 0417. Int J Radiat Oncol Biol Phys 83(4):1179–1184
Scheurer ME, Tortolero-Luna G, Adler-Storthz K (2005) Human papillomavirus infection: biology, epidemiology, and prevention. Int J Gynecol Cancer 15(5):727–746
Schiffman M, Wentzensen N (2010) From human papillomavirus to cervical cancer. Obstet Gynecol 116(1):177–185
Schiffman M, Castle PE, Jeronimo J, Rodriguez AC, Wacholder S (2007) Human papillomavirus and cervical cancer. Lancet 370(9590):890–907
Schiffman M, Rodriguez AC, Chen Z, Wacholder S, Herrero R, Hildesheim A et al (2010) A population-based prospective study of carcinogenic human papillomavirus variant lineages, viral persistence, and cervical neoplasia. Cancer Res 70(8):3159–3169
Schilder RJ, Sill MW, Lee Y-C, Mannel R (2009) A phase II trial of erlotinib in recurrent squamous cell carcinoma of the cervix: a Gynecologic Oncology Group Study. Int J Gynecol Cancer 19(5):929–933
Schlecht NF, Burk RD, Palefsky JM, Minkoff H, Xue X, Massad LS et al (2005) Variants of human papillomaviruses 16 and 18 and their natural history in human immunodeficiency virus-positive women. J Gen Virol 86(Pt 10):2709–2720
Schmidt M, Kedzia W, Goździcka-Józefiak A (2001) Intratype HPV16 sequence variation within LCR of isolates from asymptomatic carriers and cervical cancers. J Clin Virol 23(1–2):65–77
Scott AM, Wolchok JD, Old LJ (2012) Antibody therapy of cancer. Nat Rev Cancer 12(4):278–287
Sgambato A, Zannoni GF, Faraglia B, Camerini A, Tarquini E, Spada D et al (2004) Decreased expression of the CDK inhibitor p27Kip1 and increased oxidative DNA damage in the multistep process of cervical carcinogenesis. Gynecol Oncol 92(3):776–783
Shang Q, Wang Y, Fang Y, Wei L, Chen S, Sun Y et al (2011) Human papillomavirus type 16 variant analysis of E6, E7, and L1 genes and long control region in identification of cervical carcinomas in patients in Northeast China. J Clin Microbiol 49(7):2656–2663
Shen M, Ding X, Li T, Chen G, Zhou X (2013) Sequence variation analysis of HPV-18 isolates in southwest China. PLoS One 8(2):e56614
Shintani S, Nakahara Y, Mihara M, Ueyama Y, Matsumura T (2001) Inactivation of the p14(ARF), p15(INK4B) and p16(INK4A) genes is a frequent event in human oral squamous cell carcinomas. Oral Oncol 37(6):498–504
Shroyer KR, Homer P, Heinz D, Singh M (2006) Validation of a novel immunocytochemical assay for topoisomerase II-alpha and minichromosome maintenance protein 2 expression in cervical cytology. Cancer 108(5):324–330
Sichero L, Villa LL (2006) Epidemiological and functional implications of molecular variants of human papillomavirus. Braz J Med Biol Res 39(6):707–717

Sichero L, Franco EL, Villa LL (2005) Different P105 promoter activities among natural variants of human papillomavirus type 18. J Infect Dis 191(5):739–742

Sichero L, Ferreira S, Trottier H, Duarte-Franco E, Ferenczy A, Franco EL et al (2007) High grade cervical lesions are caused preferentially by non-European variants of HPVs 16 and 18. Int J Cancer 120(8):1763–1768

Siegel R, Naishadham D, Jemal A (2013) Cancer statistics, 2013. CA Cancer J Clin 63(1):11–30

Singh M, Mockler D, Akalin A, Burke S, Shroyer AL, Shroyer KR (2012) Immunocytochemical colocalization of P16(INK4a) and Ki-67 predicts CIN2/3 and AIS/adenocarcinoma. Cancer Cytopathol 120(1):26–34

Smeets SJ, van der Plas M, Schaaij-Visser TBM, van Veen EAM, van Meerloo J, Braakhuis BJM et al (2011) Immortalization of oral keratinocytes by functional inactivation of the p53 and pRb pathways. Int J Cancer 128(7):1596–1605

Smith JS, Lindsay L, Hoots B, Keys J, Franceschi S, Winer R et al (2007) Human papillomavirus type distribution in invasive cervical cancer and high-grade cervical lesions: a meta-analysis update. Int J Cancer 121(3):621–632

Smith B, Chen Z, Reimers L, van Doorslaer K, Schiffman M, DeSalle R et al (2011) Sequence imputation of HPV16 genomes for genetic association studies. PLoS One 6(6):e21375

Song YS, Kee SH, Kim JW, Park NH, Kang SB, Chang WH et al (1997) Major sequence variants in E7 gene of human papillomavirus type 16 from cervical cancerous and noncancerous lesions of Korean women. Gynecol Oncol 66(2):275–281

Song X, Sturgis EM, Liu J, Jin L, Wang Z, Zhang C et al (2013) MicroRNA variants increase the risk of HPV-associated squamous cell carcinoma of the oropharynx in never smokers. PLoS One 8(2):e56622

Soonthornthum T, Arias-Pulido H, Joste N, Lomo L, Muller C, Rutledge T et al (2011) Epidermal growth factor receptor as a biomarker for cervical cancer. Ann Oncol 22(10):2166–2178

Spangle JM, Münger K (2010) The human papillomavirus type 16 E6 oncoprotein activates mTORC1 signaling and increases protein synthesis. J Virol 84(18):9398–9407

Srinivasan M, Taioli E, Ragin CC (2009) Human papillomavirus type 16 and 18 in primary lung cancers–a meta-analysis. Carcinogenesis 30(10):1722–1728

Stabile LP, Davis ALG, Gubish CT, Hopkins TM, Luketich JD, Christie N et al (2002) Human non-small cell lung tumors and cells derived from normal lung express both estrogen receptor alpha and beta and show biological responses to estrogen. Cancer Res 62(7):2141–2150

Stanley M (2003) Chapter 17: genital human papillomavirus infections–current and prospective therapies. J Natl Cancer Inst Monogr 31:117–124

Stanley M (2006) Immune responses to human papillomavirus. Vaccine 24(Suppl 1):S16–S22

Stanley M (2010) Prophylactic human papillomavirus vaccines: will they do their job? J Intern Med 267(3):251–259

Steger G, Corbach S (1997) Dose-dependent regulation of the early promoter of human papillomavirus type 18 by the viral E2 protein. J Virol 71(1):50–58

Stevens LM (2002) JAMA patient page papillomavirus. JAMA 287(18):2452

Stewart AC, Eriksson AM, Manos MM, Muñoz N, Bosch FX, Peto J et al (1996) Intratype variation in 12 human papillomavirus types: a worldwide perspective. J Virol 70(5):3127–3136

Stoler MH, Schiffman M, Atypical Squamous Cells of Undetermined Significance-Low-grade Squamous Intraepithelial Lesion Triage Study (ALTS) Group (2001) Interobserver reproducibility of cervical cytologic and histologic interpretations: realistic estimates from the ASCUS-LSIL Triage Study. JAMA 285(11):1500–1505

Stransky N, Egloff AM, Tward AD, Kostic AD, Cibulskis K, Sivachenko A et al (2011) The mutational landscape of head and neck squamous cell carcinoma. Science 333(6046):1157–1160

Sun S, Schiller JH, Gazdar AF (2007) Lung cancer in never smokers–a different disease. Nat Rev Cancer 7(10):778–790

Sun Z, Ren G, Cui X, Zhou W, Liu C, Ruan Q (2011) Genetic diversity of HPV-16 E6, E7, and L1 genes in women with cervical lesions in Liaoning province. China Int J Gynecol Cancer 21(3):551–558

Syrjänen K, Syrjänen S, Kellokoski J, Kärjä J, Mäntyjärvi R (1989) Human papillomavirus (HPV) type 6 and 16 DNA sequences in bronchial squamous cell carcinomas demonstrated by in situ DNA hybridization. Lung 167(1):33–42
Teshima H, Beaudenon S, Koi S, Katase K, Hasumi K, Masubuchi K et al (1997) Human papillomavirus type 18 DNA sequences in adenocarcinoma and adenosquamous carcinoma of the uterine cervix. Arch Gynecol Obstet 259(4):169–177
Thomison J 3rd, Thomas LK, Shroyer KR (2008) Human papillomavirus: molecular and cytologic/histologic aspects related to cervical intraepithelial neoplasia and carcinoma. Hum Pathol 39(2):154–166
Thu KL, Vucic EA, Chari R, Zhang W, Lockwood WW, English JC et al (2012) Lung adenocarcinoma of never smokers and smokers harbor differential regions of genetic alteration and exhibit different levels of genomic instability. PLoS One 7(3):e33003
Tornesello ML, Duraturo ML, Salatiello I, Buonaguro L, Losito S, Botti G et al (2004) Analysis of human papillomavirus type-16 variants in Italian women with cervical intraepithelial neoplasia and cervical cancer. J Med Virol 74(1):117–126
Trimble CL, Piantadosi S, Gravitt P, Ronnett B, Pizer E, Elko A et al (2005) Spontaneous regression of high-grade cervical dysplasia: effects of human papillomavirus type and HLA phenotype. Clin Cancer Res 11(13):4717–4723
Tsakogiannis D, Ruether IGA, Kyriakopoulou Z, Pliaka V, Theoharopoulou A, Skordas V et al (2012) Sequence variation analysis of the E2 gene of human papilloma virus type 16 in cervical lesions from women in Greece. Arch Virol 157(5):825–832
Vairaktaris E, Vassiliou S, Nkenke E, Serefoglou Z, Derka S, Tsigris C et al (2008) A metalloproteinase-9 polymorphism which affects its expression is associated with increased risk for oral squamous cell carcinoma. Eur J Surg Oncol 34(4):450–455
Van Bogaert L-J (2012) P16INK4a immunocytochemistry/immunohistochemistry: need for scoring uniformization to be clinically useful in gynecological pathology. Ann Diagn Pathol 16(5):422–426
Veldman T, Liu X, Yuan H, Schlegel R (2003) Human papillomavirus E6 and Myc proteins associate in vivo and bind to and cooperatively activate the telomerase reverse transcriptase promoter. Proc Natl Acad Sci 100(14):8211–8216
Venuti A, Salani D, Poggiali F, Manni V, Bagnato A (1998) The E5 oncoprotein of human papillomavirus type 16 enhances endothelin-1-induced keratinocyte growth. Virology 248(1):1–5
Venuti A, Paolini F, Nasir L, Corteggio A, Roperto S, Campo MS et al (2011) Papillomavirus E5: the smallest oncoprotein with many functions. Mol Cancer 10(1):140
Veress G, Szarka K, Dong XP, Gergely L, Pfister H (1999) Functional significance of sequence variation in the E2 gene and the long control region of human papillomavirus type 16. J Gen Virol 80(4):1035–1043
Vici P, Mariani L, Pizzuti L, Sergi D, Di Lauro L, Vizza E et al (2014) Emerging biological treatments for uterine cervical carcinoma. J Cancer 5(2):86–97
Villa LL, Sichero L, Rahal P, Caballero O, Ferenczy A, Rohan T et al (2000) Molecular variants of human papillomavirus types 16 and 18 preferentially associated with cervical neoplasia. J Gen Virol 81(Pt 12):2959–2968
Vrtačnik Bokal E, Kocjan BJ, Poljak M, Bogovac Ž, Jančar N (2010) Genomic variants of human papillomavirus genotypes 16, 18, and 33 in women with cervical cancer in Slovenia. J Obstet Gynaecol Res 36(6):1204–1213
Walboomers JM, Jacobs MV, Manos MM, Bosch FX, Kummer JA, Shah KV et al (1999) Human papillomavirus is a necessary cause of invasive cervical cancer worldwide. J Pathol 189(1):12–19
Walker JL, Wang SS, Schiffman M, Solomon D, ASCUS LSIL, Triage Study Group (2006) Predicting absolute risk of CIN3 during post-colposcopic follow-up: results from the ASCUS-LSIL Triage Study (ALTS). Am J Obstet Gynecol 195(2):341–348
Wang W, Xia X, Wang S, Sima N, Li Y, Han Z, et al (2011) Oncolytic adenovirus armed with human papillomavirus E2 gene in combination with radiation demonstrates synergistic enhancements of antitumor efficacy. Cancer Gene Ther [Internet]. 2011 Aug 19 [cited 2011 Sep 17]. Available from: http://www.ncbi.nlm.nih.gov/pubmed/21852812

Watts KJ, Thompson CH, Cossart YE, Rose BR (2002) Sequence variation and physical state of human papillomavirus type 16 cervical cancer isolates from Australia and New Caledonia. Int J Cancer 97(6):868–874

Weinberger PM, Yu Z, Haffty BG, Kowalski D, Harigopal M, Brandsma J et al (2006) Molecular classification identifies a subset of human papillomavirus–associated oropharyngeal cancers with favorable prognosis. J Clin Oncol 24(5):736–747

Wentzensen N, Schwartz L, Zuna RE, Smith K, Mathews C, Gold MA et al (2012) Performance of p16/Ki-67 immunostaining to detect cervical cancer precursors in a colposcopy referral population. Clin Cancer Res 18(15):4154–4162

Weyn C, Thomas D, Jani J, Guizani M, Donner C, Van Rysselberge M et al (2011) Evidence of human papillomavirus in the placenta. J Infect Dis 203(3):341–343

Wheeler CM, Yamada T, Hildesheim A, Jenison SA (1997) Human papillomavirus type 16 sequence variants: identification by E6 and L1 lineage-specific hybridization. J Clin Microbiol 35(1):11–19

White PW, Faucher A-M, Goudreau N (2011) Small molecule inhibitors of the human papillomavirus E1-E2 interaction. Curr Top Microbiol Immunol 348:61–88

WHO (2011) Strengthening cervical cancer prevention and control [Internet]. [cited 2011 Sep 20]. Available from: http://www.who.int/reproductivehealth/publications/cancers/rhr_10_13/en/index.html

Widschwendter A, Brunhuber T, Wiedemair A, Mueller-Holzner E, Marth C (2004) Detection of human papillomavirus DNA in breast cancer of patients with cervical cancer history. J Clin Virol 31(4):292–297

Wise-Draper TM, Wells SI (2008) Papillomavirus E6 and E7 proteins and their cellular targets. Front Biosci 13:1003–1017

Woodman CBJ, Collins SI, Young LS (2007) The natural history of cervical HPV infection: unresolved issues. Nat Rev Cancer 7(1):11–22

Wright TC Jr, Massad LS, Dunton CJ, Spitzer M, Wilkinson EJ, Solomon D (2007) 2006 consensus guidelines for the management of women with abnormal cervical cancer screening tests. Am J Obstet Gynecol 197(4):346–355

Wu M-F, Cheng Y-W, Lai J-C, Hsu M-C, Chen J-T, Liu W-S et al (2005) Frequent p16INK4a promoter hypermethylation in human papillomavirus-infected female lung cancer in Taiwan. Int J Cancer 113(3):440–445

Wu E, Zha X, Yu X, Zhang G, Wu Y, Fan Y et al (2009) Profile of physical status and gene variation of human papillomavirus 58 genome in cervical cancer. J Gen Virol 90(5):1229–1237

Wu H-H, Wu J-Y, Cheng Y-W, Chen C-Y, Lee M-C, Goan Y-G et al (2010) cIAP2 upregulated by E6 oncoprotein via epidermal growth factor receptor/phosphatidylinositol 3-kinase/AKT pathway confers resistance to cisplatin in human papillomavirus 16/18-infected lung cancer. Clin Cancer Res 16(21):5200–5210

Xi LF, Koutsky LA, Galloway DA, Kiviat NB, Kuypers J, Hughes JP et al (1997) Genomic variation of human papillomavirus type 16 and risk for high grade cervical intraepithelial neoplasia. J Natl Cancer Inst 89(11):796–802

Xi LF, Critchlow CW, Wheeler CM, Koutsky LA, Galloway DA, Kuypers J et al (1998) Risk of anal carcinoma in situ in relation to human papillomavirus type 16 variants. Cancer Res 58(17):3839–3844

Xi LF, Koutsky LA, Hildesheim A, Galloway DA, Wheeler CM, Winer RL et al (2007) Risk for high-grade cervical intraepithelial neoplasia associated with variants of human papillomavirus types 16 and 18. Cancer Epidemiol Biomark Prev 16(1):4–10

Xi LF, Schiffman M, Koutsky LA, Hulbert A, Lee S-K, Defilippis V et al (2012) Association of human papillomavirus type 31 variants with risk of cervical intraepithelial neoplasia grades 2–3. Int J Cancer 131(10):2300–2307

Xin CY, Matsumoto K, Yoshikawa H, Yasugi T, Onda T, Nakagawa S et al (2001) Analysis of E6 variants of human papillomavirus type 33, 52 and 58 in Japanese women with cervical intraepithelial neoplasia/cervical cancer in relation to their oncogenic potential. Cancer Lett 170(1):19–24

Yamada T, Manos MM, Peto J, Greer CE, Munoz N, Bosch FX et al (1997) Human papillomavirus type 16 sequence variation in cervical cancers: a worldwide perspective. J Virol 71(3):2463–2472

Yasmeen A, Bismar TA, Kandouz M, Foulkes WD, Desprez P-Y, Al Moustafa A-E (2007) E6/E7 of HPV type 16 promotes cell invasion and metastasis of human breast cancer cells. Cell Cycle 6(16):2038–2042

Yue Y, Yang H, Wu K, Yang L, Chen J, Huang X et al (2013) Genetic variability in L1 and L2 genes of HPV-16 and HPV-58 in Southwest China. PLoS One 8(1):e55204

Yugawa T, Kiyono T (2009) Molecular mechanisms of cervical carcinogenesis by high-risk human papillomaviruses: novel functions of E6 and E7 oncoproteins. Rev Med Virol 19(2):97–113

Zehbe I, Wilander E, Delius H, Tommasino M (1998) Human papillomavirus 16 E6 variants are more prevalent in invasive cervical carcinoma than the prototype. Cancer Res 58(4):829–833

Zehbe I, Mytilineos J, Wikström I, Henriksen R, Edler L, Tommasino M (2003) Association between human papillomavirus 16 E6 variants and human leukocyte antigen class I polymorphism in cervical cancer of Swedish women. Hum Immunol 64(5):538–542

Zehbe I, Lichtig H, Westerback A, Lambert PF, Tommasino M, Sherman L (2011) Rare human papillomavirus 16 E6 variants reveal significant oncogenic potential. Mol Cancer 10:77

Zekan J, Sirotkovic-Skerlev M, Skerlev M (2011) Oncogenic aspects of HPV infections of the female genital tract. In: Seligmann H (ed) DNA replication-current advances [Internet]. InTech; 2011 [cited 2014 May 21]. Available from: http://www.intechopen.com/books/dna-replication-current-advances/oncogenic-aspects-of-hpv-infections-of-the-female-genital-tract

Zhang Y, Fan S, Meng Q, Ma Y, Katiyar P, Schlegel R et al (2005) BRCA1 interaction with human papillomavirus oncoproteins. J Biol Chem 280(39):33165–33177

Zheng Z-M, Wang X (2011) Regulation of cellular miRNA expression by human papillomaviruses. Biochim Biophys Acta 1809(11–12):668–677

Zighelboim I, Wright JD, Gao F, Case AS, Massad LS, Mutch DG et al (2013) Multicenter phase II trial of topotecan, cisplatin and bevacizumab for recurrent or persistent cervical cancer. Gynecol Oncol 130(1):64–68

Zuna RE, Tuller E, Wentzensen N, Mathews C, Allen RA, Shanesmith R et al (2011) HPV16 variant lineage, clinical stage, and survival in women with invasive cervical cancer. Infect Agent Cancer 6(1):19

Zur Hausen H (1996) Papillomavirus infections – a major cause of human cancers. Biochim Biophys Acta 1288(2):F55–F78

Zur Hausen H (1999) Papillomaviruses in human cancers. Proc Assoc Am Physicians 111(6):581–587

Zur Hausen H (2002) Papillomaviruses and cancer: from basic studies to clinical application. Nat Rev Cancer 2(5):342–350

Part III
Cancer-Associated Infections

Chapter 14
Infectious Diseases in Cancer Patients: An Overview

Tatiana Zorina and Alexis Styche

Abstract The predisposition of cancer patients to infectious diseases which contribute to the gravity of their prognosis is well documented. The current success in therapy of both malignancies and infections is unprecedented. However, the overall co-morbidity of these conditions is still a major problem in management of these patients. Paradoxically, to some degree the problem of containing infectious complications is directly associated with the vigor of the anti-cancer therapeutic regimens. The objective of this chapter is to provide an up to date overview of our understanding of the infectious complications in cancer patients based on the type of infection and immune responses.

Keywords Cancer • Infection • Tumor immunology

Abbreviations

HBV	Hepatitis B virus
HCV	Hepatitis C virus
HIV	Human immunodeficiency virus
HAART	Highly active antiretroviral therapy
AIDS	Acquired immune deficiency syndrome
Treg	T regulatory cell
DC	Dendritic cell
IL-10	Interleukin 10
TGF-β	Transforming growth factor-β
MSC	Mesenchymal stem cell

T. Zorina, M.D., Ph.D. (✉)
Department of Bioscience Technologies, Thomas Jefferson University,
130 S. 9th St., Edison Bld, Suite 2121, Philadelphia, PA 19107, USA
e-mail: tatiana.zorina@jefferson.edu

A. Styche
Children's Hospital of Pittsburgh, Philadelphia, PA, USA

M.R. Shurin et al. (eds.), *Infection and Cancer: Bi-Directorial Interactions*, DOI 10.1007/978-3-319-20669-1_14

Introduction

The recently published book *Infectious Complications in Cancer Patients* by Valentina Stosor and Teresa Zembower offers an extensive coverage on this topic (Stosor and Zembower 2014). The objective of this chapter is to provide an overview of the most current status of this subject and to outline the major aspects of infectious diseases as one of the major causes of morbidity and mortality in cancer patients with emphasis on the latest publications in the field.

The spectrum of the infectious agents and clinical manifestations of the diseases, as well as the range of the diagnostic markers and technologies for their detection, are evolving features in the never ending battle against genetically non-self malignant and infectious invaders. This conflict is due to development of new infection- and cancer-defense mechanisms on one hand and the rapid progress of therapeutic interventions on the other.

The oncogenesis and development of the infectious diseases are both ultimately rooted in the compromised ability of the immune system to defend against non-self, either malignant or infectious, components. The underlying mechanisms of these two groups of disorders are complex and often interconnected. In some instances this pattern turns into a vicious circle. This is demonstrated in the cancers with viral etiology. Patients with HPV related squamous cell carcinoma, or HIV associated Kaposi sarcoma or HCV-induced hepatocellular carcinoma eventually become prone to viral, fungal, bacterial and other infections.

Genesis of the infectious complications in the cancer patients is based on local tumor-induced effects and on the mechanisms of generalized immunosuppression induced by both cancer-related processes and by the iatrogenic outcomes of the therapeutic regimens (Sutton 2014). The specific characteristics of the course and outcomes of the infectious diseases in cancer patients are defined by numerous factors, and are discussed in this chapter based on the type of infection and the immune system components involved.

Bacterial Infections

Bacteria represent the most common pathogen in cancer-associated infections, and bacterial sepsis continues to be a leading cause of morbidity and toxic death in children receiving intensive therapy for cancer (Alexander et al. 2012). It is well acknowledged that oncologic patients are more susceptible to bacterial infections and their systemic spread due to tumor-related and iatrogenic immunosuppression, which are comprised of the classic clinical combination of severe neutropenia, fever, hypotension and headache (Rasool Hassan et al. 2010). Acute bacterial infections negatively impact survival and increase mortality in adult and pediatric patients with solid cancers and hematologic malignancy (Lanoix et al. 2011; Attie et al. 2014; de Oliveira et al. 2014).

Further elucidation of the specific mechanisms and markers of the oncogenesis and cancer-associated infections allows development of target specific therapies. For example, it has been recognized that helicobacter pylori is one of the major causes of gastric cancer. The recently identified gastric stem/progenitor cell markers Lgr5, Villin-promoter, TFF2-mRNA and Mist and the gastric cancer stem cell markers such as CD44, CD90, CD133, Musashi-1 reveal novel information on tumor cell behavior and disease progression implicated for therapeutics (Ding and Zheng 2012).

New insight is currently emerging concerning the complexity of the infection-cancer-infection mechanisms interplay. The two major concepts being (i) the oncogenic role of the inflammatory cytokines and (ii) chemokine axis-based induction of the metastasis niches. It has been recently reported that systemic inflammation triggered by gastrointestinal tract bacteria plays a pivotal role in oncogenesis in the prostate gland (Poutahidis et al. 2013). Other recent research confirming the concept that inflammation is a critical component of tumor progression came from the study addressing correlation between oral bacterial infections and cancer. Inflammation caused by periodontal infections has been linked to cancer of the lung, kidney, pancreas, and hematologic and oral cancers (Pendyala et al. 2013).

The concept that metastatic spread of cancer can be promoted by bacterial infections was proposed based on studies on the CXCR4/ubiquitin axis. It was demonstrated that acute bacterial infections commonly seen in patients with cancer are linked to increased metastasis to the lung (Smith and Kang 2013). A model of the bacteria-induced acute lung inflammation was used to study its effect on lung metastasis in mice. Acute lung infection dramatically increased cancer cell homing to the lung and lung metastasis. It was also confirmed that the ubiquitin-CXCR4 axis plays an important role in these changes that were described as "preparation of a favorable metastatic niche" (Yan et al. 2013).

Allogeneic stem cell transplantation is a part of therapeutic modules in numerous cancers. Previous studies demonstrated that approximately 15–30 % of allogeneic hematopoietic stem cell transplantation recipients develop Clostridium difficile infection during transplantation, greatly exceeding rates in most other patient populations (Alonso and Kamboj 2014; Kinnebrew et al. 2014).

Fungal Infections

Fungi are the second most common pathogens after bacteria for causing cancer-associated infections in adult and pediatric patients with both solid organ- and hematologic malignancies. All organs from most commonly lung (Kim et al. 2014) to less frequently but associated with high mortality rates in pediatric cancer patients the central nervous system (Carter et al. 2015), are affected. Almost 50 % of the adult patients with acute myeloid leukemia that are subjected to chemotherapy develop invasive fungal infections (Neofytos et al. 2013). In pediatric oncology invasive fungal infections are also a frequent and potentially fatal complication

(Mor et al. 2011). Although solid tumors comprise the vast majority of cancers the morbidity and mortality due to fungal infections is higher in patients with hematologic malignancies. Invasive Aspergillus infection is one of the most common fungal infections that have been observed in patients with hematologic malignancy or those subjected to allogeneic hematopoietic stem cell transplantation (de Naurois et al. 2010). About 70 % of the pediatric oncology patients developing fungal infections have myeloid leukemia and acute lymphoblastic leukemia as their primary diagnosis (Mor et al. 2011).

The prolonged neutropenia of either primary or secondary genesis due to intensive chemotherapy regimens is the major predisposing factor for invasive fungal infections (de Naurois et al. 2010; Mousset et al. 2014). In addition, the fungal infection can develop as a complication of protracted use of venous catheters (Lai et al. 2004).

Aspergillosis is reported to have the highest incidence with Candidiasis second (Neofytos et al. 2013; Mor et al. 2011; de Naurois et al. 2010) and other fungal types of infections occurring less frequently as complications in adult and pediatric cancer patients (Kim et al. 2014; Caselli et al. 2014).

The increased intensity of chemotherapeutic regimens and rising resistance of fungal infections to antifungal drugs has prompted the search for new optimal prophylactic and therapeutic approaches. Antifungal drugs are often given prophylactically to patients with persistent fever, and are shown to decrease mortality rates (Johansen and Gotzsche 2014). Trimethoprim/sulfamethoxazole prophylactically administered before chemotherapy regimens has been shown to be effective in prevention of Pneumocystis pneumonia in children with solid tumors, leukemias and lymphomas (Caselli et al. 2014). However, caution in the choice of primary and secondary antifungal prophylaxis is recommended. The use of secondary prophylaxis may reduce systemic fungal infection frequency but at the same time increase the risk of colonization and infection with azole-resistant fungal strains (Gedik et al. 2014). Since it has been demonstrated that prolonged duration of neutropenia is one of the risk factors for onset of invasive Aspergillosis, addition of granulocyte-colony-stimulating factor into therapeutic and prophylactic protocols is recommended (van de Peppel et al. 2014).

Viral Infections

Viral infections are increasingly recognized as serious causes of morbidity and mortality in cancer patients. New technologies are emerging for their detection with respect to additional challenges in accurate and timely diagnosis and administration of appropriate antiviral therapy in this group of patients (Babady et al. 2012). Of special importance are the still open questions about interplay of the antiviral and anti-tumor immunity mechanisms, effectiveness and timing of vaccination and its relevance to chemotherapy regimens in the pediatric and adult oncology practice. Despite the significant advances in management of pediatric cancer and influenza, the epidemiology and outcomes of influenza in pediatric oncology patients have not altered over the past several decades (Carr et al. 2012). Although influenza usually presents as a mild illness, children with hematological conditions and solid tumors are at increased risk for complications, which may lead to delay in anticancer therapy and increase in hospitalization and antibiotic usage (Ozdemir et al. 2011). Though influenza is the most common viral infection, is only one on the long list of viruses, such as H1N1, rhinovirus, parainfluenza virus, adenovirus, respiratory syncytial virus, human parechovirus, bocavirus, metapneumovirus, and human coronavirus occurring in about 30 % of neutropenic pediatric patients. Co-detection of these viruses is not uncommon, occurring in over 20 % of infections, in various combinations (Benites et al. 2014).

Cancer patients have unique problems associated with hepatitis B (HBV) and hepatitis C (HCV) virus infections. Of special concern is the risk of reactivation of infection due to viral replication as a result of immunosuppressive effects of the chemotherapy in general and of the new emerging modalities for targeted therapies. To date different outcomes in this respect of therapies using Alemtuzumab, Brentuximab, Imatinib, Cetuximab, Panitumumab, Ppilimumab were reported, and further randomized trials are needed to establish algorithms for this issue (Yazici et al. 2014). Also of consideration is the problem of balancing the chemotherapy and antiviral medication regimens and the timing of their administration. For some patients chemotherapy has to be postponed until completion of the antiviral course of therapy, while others cannot be subjected to the viral infection therapy while under treatment for their cancer (Borchardt and Torres 2014). One large study from Japan with evaluation of over 1,000 patients with breast cancer reported that chemotherapy for breast cancer patients with HCV infection is feasible, and according to their experience, viral load doesn't change during chemotherapy (Miura et al. 2013). Further elucidation is needed to clarify whether this outcome is cancer type or other factor-specific.

Cancer patients are at substantially increased risk of Herpes Zoster and related complications. The risk of Herpes Zoster infection in cancer patients compared to the general population is from 2 to 8-fold higher; with more than double the incidence in patients with hematologic malignancies as in those with solid tumors, including brain, lung, breast, esophageal, gastric and colorectal cancers (Hata et al. 2011; Habel et al. 2013). It has also been suggested that Herpes Zoster can be used as a marker for risk of malignancy, since there is a higher incidence of malignancy following an episode of Herpes Zoster in both men and women in all age groups 18 years and over (Iglar et al. 2013).

The mechanisms of HIV and oncogenesis are interconnected in adult and pediatric patients, and their co-morbidity is complex and multifactorial. Both, the HIV-associated immune activation and inflammation on one hand, and accelerated immune senescence on the other, favor cancer development. HIV-infected patients are at enhanced risk of several cancers, including lung and anal cancers, hepatocellular carcinoma, Hodgkin's lymphoma, Kaposi's sarcoma and several other cancers, compared to the general population (Sigel et al. 2011). Management of cancers in HIV patients is currently under intense research and is specific for each malignancy.

Highly active antiretroviral therapy (HAART) changed the course of HIVinfection. HAART reduced AIDS-defined-malignancies, but increased incidence of several non-AIDS-defined-malignancies (Chiappini et al. 2014). Adenocarcinoma of the lung is the most prevalent non-AIDS-defining cancer in the HAART era, and has up to four times greater incidence in HIV-infected individuals than in the general population. Two major problems are currently associated with management of lung cancer in HIV patents. Its diagnosis is often late because of onset in a younger population and its being clinically masked by or as a pulmonary infection, which are common among HIV-infected individuals. In addition, although there is increasing experience in using radiation and chemotherapy for HIV-infected patients who do not have surgical options, there is a need for more prospective studies because this population is frequently excluded from participating in cancer trials (Mani et al. 2012). This presents the necessity of including cancer screening in HIV-infected patients (Sigel et al. 2011).

Vaccination

All infections increase morbidity and mortality in cancer patients. Regimens for anti-infection therapy interfere with chemotherapy and other treatment modalities in this group of patients. In many cases this leads to the necessity of postponing the anti-tumor therapies to manage an infection, or vice versa, to delay the anti-infection medication until the chemotherapy is completed. In light of these negative impacts together with rising antibiotic resistance, development of preventive approaches to preclude the almost inevitable infectious complications, especially in neutropenic patients, is highly desirable. Need for infection prophylaxis is also growing in hematopoietic cell transplant recipients. Due to overall improvement of this treatment during the last two decades the survival duration has increased but the incidence of post-transplantation infections, particularly those caused by respiratory viruses, has concomitantly increased with lifespan. The lack of directed antiviral therapy for most viruses, has promoted the use of inactivated influenza vaccine for hematopoietic cell transplant recipients (Shah et al. 2012).

Reports on benefits and limitations of vaccination in cancer patients are numerous and contradictory. The overall trend is that influenza vaccination is beneficial in immunocompromised patients and significantly lowers the odds of influenza-like illness in patients with HIV infection, patients with cancer, and transplant recipients (Beck et al. 2012). A multicenter observational study has found that in cancer and hematopoietic stem cell transplant recipients with Influenza A (H1N1) the incidence of pneumonia was 66 % with an 18 % 30-day mortality, however no deaths were observed among vaccinated patients (Dignani et al. 2014). The study by Kim et al. addressed the immunogenicity of influenza vaccine in colorectal cancer patients based on antibody titers in blood samples. The data showed an acceptable immune response to an influenza vaccine without significant adverse effects, supporting the recommendation for annual influenza vaccination in colorectal cancer patients (Kim et al. 2013). A simple prognostic measure was proposed for evaluation of the vaccination response to H1N1 virus. The study demonstrated that an absolute lymphocyte count above the lower normal limits for age prior to vaccination predicted the positive response to influenza vaccination in pediatric cancer patients treated with chemotherapy (Mavinkurve-Groothuis et al. 2013).

However, some caution was suggested in respect to the effectiveness of vaccination in cancer patients undergoing chemotherapy. The conclusion from the study by Shehata and colleagues was that although the active immunization in cancer patients has been shown to confer protective immunity against several infections at similar rates to healthy individuals, the immune responses to influenza vaccination in patients receiving chemotherapy were consistently weaker (Shehata and Karim 2014). Further clarification on optimal timing for vaccination in cancer patients is needed.

In addition, in one recent study the negative findings on the effectiveness of influenza vaccination in ovarian cancer patients were reported. The data in this study have shown that patients with ovarian cancer are almost uniformly unable to mount a meaningful antibody response to influenza vaccination and that despite CDC recommendations that patients undergoing chemotherapy receive influenza vaccine, there is little evidence to support its serologic effectiveness in these patients (Chu et al. 2013).

The Immune System and Cancer-Associated Infections

All components of innate and adaptive immunity are involved in shaping the course of malignancy, associated infectious diseases, and the ultimate prognosis for the patient. In addition to cellular immunity dysfunctions, these changes include alterations in cytokine profiles and humoral immunity reactions. Of special focus in this chapter are the latest updates from the rapidly evolving fields of study on the Dendritic and T regulatory cells, and neutropenia-related complications. Elucidation of the mechanisms of the immune system dysfunction on the type of cancer, nature of the infection and recently even individual patient levels allowed development of target specific therapies.

Dendritic cells (DCs) and T cells are among the major players sustaining physiologic immune conditions. Both are comprised of two functionally distinct populations: cells promoting immunity and those sustaining immune tolerance. Within T cells the functional subtypes can be easily identified as T effector and T regulatory (Treg) cells, respectively. Classification of different subgroups among dendritic cells is much more complicated. Originally DCs were discovered as cells of myeloid origin with mainly antigen-presenting function and thus as cells which promote mechanisms in adaptive immunity (Steinman 1991). Later it was recognized that depending on the DCs' maturation status (Steinman et al. 2003), hematopoietic lineage of origin (Ma et al. 2012) and expression of co-stimulatory and MHC molecules they can play either immunostimulatory or tolerogenic, immunosuppressive roles (Hurwitz and Watkins 2012; Zhong et al. 2014). In addition, the change from immunostimulatory to tolerogenic status can be induced in DCs via different mechanisms by a number of malignancies. Also, it should be taken into consideration that over 60 % of malignancies arise in older populations, and hence the age-related decline in their hematopoietic cells' ability for self-renewal also contributes to an immunocompromised condition in cancer patients over 65 years old (Lipschitz 1995; Balducci et al. 2001; Shurin et al. 2007).

The ability of malignant cells to switch the DCs function from immunostimulatory into immunosuppressive cells is one of their essential immunomodulatory features. With the onset of malignancy complicated networks of expression in the ligand/receptor axis and changes in signal transduction lead to alteration of the functional profiles of the DC subpopulations. This has a twofold outcome. On one hand it is part of the tumors defense mechanism, protecting the malignant mass from immune reactions and allowing its unhindered growth. On the other hand, the tolerogenic function of the DCs leads to generalized immunosuppression, which contributes to the onset of infectious diseases. During the last decade new insights into particular mechanisms contributing to the tumor-induced changes in DC function and the resultant increased susceptibility to infectious diseases were reported. They include both local and systemic mechanisms employed by the tumor to affect the DCs functions. Collectively they result in the inhibition of the DCs differentiation from hematopoietic progenitors (Shurin et al. 2001; Tourkova et al. 2004; Hargadon 2013) and changes in their phenotype and function (Zhong et al. 2014; Aalamian-Matheis et al. 2007; Karthaus et al. 2012; Chao et al. 2015).

Dendritic cells often accumulate in and around tumors. However, their mere presence does not reflect their input in overall immunity and hence doesn't have prognostic value. They could either be part of robust anti-tumor immunity (Esche et al. 1998; Iida et al. 2008) or they may play an essential role in tumor immune escape (Zhong et al. 2014; Wu et al. 2014; Dudek et al. 2013). DC phenotyping is a valuable tool in defining their maturation and functional status. Multiple groups have reported data contributing to an optimal phenotyping panel of markers for tolerogenic/regulatory DC in cancer. It was shown that in prostate cancer expression of CD83, CD86 and CD40 co-stimulatory molecules are decreased (Aalamian-Matheis et al. 2007). It was also reported that evaluation of the tumor-infiltrating DCs should be performed utilizing a larger panel including not only S-100 and CD1a, but also DC-SIGN, and DC-LAMP markers (Karthaus et al. 2012). Detection of DC-SIGN is of importance because it is an intercellular adhesion molecule which can interact with carbohydrate structures on some cancer cells. This interaction leads to immunosuppressive responses in DCs via inhibition of their maturation (Chao et al. 2015). Also it was suggested that different levels of expression of the MHC and co-stimulatory molecules have representative patterns for different subsets of regulatory DCs (Zhong et al. 2014).

The major cancer-related factors triggering immunosuppressive DC function are Interleukin IL-10 (Shurin et al. 2002; Lindenberg et al. 2013; Kaebisch et al. 2014); TGF-β (Liu et al. 2012; Caux et al. 1999) and IL-17 (Wu et al. 2014). Recent insight into the mechanisms of a DC-related immunocompromised condition in cancer and cancer-associated infections has led to development of numerous therapeutic approaches utilizing DC-based vaccines (Mayordomo et al. 1995a, b; Tuting et al.

1997; Mayordomo et al. 1997; Burgdorf 2010; Cui et al. 2013; Duncan et al. 2013; Fadul et al. 2011; Niu et al. 2014; Ridolfi et al. 2010; Rosenblatt et al. 2013; Van Tendeloo et al. 2010; Suso et al. 2011) and other immunotherapies (Fang et al. 2014; Gao et al. 2014; Ghansah et al. 2013; Kobayashi et al. 2014; Lu et al. 2014; Wimmers et al. 2014; Yuan et al. 2014; Zhao et al. 2014).

T regulatory (Treg) cells play an essential role in negative regulation of the immune responses in both physiologic and diseased conditions. Evidence is mounting in support of their contribution to the processes of oncogenesis. A growing number of cancers have been shown to possess the ability to utilize different mechanisms attracting Treg cells and utilizing them in defense against immune reactions in an affected individual. The role of the Treg cells in enhancing the morbidity in cancer-associated infections is via local and generalized immunosuppression mechanisms. The trafficking patterns of the Treg cells can be altered by the tumor-secreted cytokines and chemokine molecules inducing their recruitment into the vicinity of the tumor and providing a malignant mass with 'a shield' that suppresses the anti-tumor immune responses. This allows for uninhibited tumor growth and also results in compromised anti-infection immunity. Among the list of this cohort of tumors are ovarian carcinoma (Curiel et al. 2004), breast tumors (Gobert et al. 2009), mesothelioma (Hegmans et al. 2006), Hodgkin lymphoma (Ishida et al. 2006), myelodysplastic syndromes (Kotsianidis et al. 2009), gastric cancers (Mizukami et al. 2008; Ohtani et al. 2009)., the malignant plural mesothelioma and effusions (Qin et al. 2009; Shimizu et al. 2009), and lung adenocarcinoma (Wald et al. 2006). CCL22, CCR4, CXCL9, CXCR4, CCR5, CXCL12 and CCL17 are among the chemokine molecules with altered expression which were reported to contribute to distorted trafficking patterns of Treg cells in cancer.

These findings triggered a new direction in the search for immunomodulatory therapies which target chemokine molecule expression in an attempt to rectify the Treg cells trafficking and function and to result in impeding tumor growth and ameliorating general immunosuppression and infection complications. The effectiveness of Treg cell depletion in the regression of oral fibrosarcoma with an increase in survival was demonstrated utilizing a mouse model. Anti-CD25 Ab administration resulted in depletion of the Tregs and a complete regression of tumor (Whelan et al. 2014). The following studies proposed further modifications in the use of anti-CD25 Abs. It was shown that targeting the $CD25^{Low}$ subset assures the discerning depletion of the Treg cells among other CD25+ T cell populations (Weiss et al. 2012). CD73 is a cell surface enzyme that suppresses immunity. The adoptive experiments in the Treg-deficient mice have demonstrated that protumorogenic effects of Treg cells depend on the expression of CD73. In vivo blockade of CD73 with a selective inhibitor or anti-CD73 mAb reduced tumor growth and metastasis (Stagg et al. 2011). Prostaglandin E2 (PGE2) and TGF-β play a role in induction of Treg cells. Interrupting the TGF-β and PGE2 signaling pathways was suggested as an alternative approach for reduction of the tumor protective function of Treg cells (Baratelli et al. 2010). Interest in adaptation of mesenchymal stem cell (MSC) transplantation for cancer therapy lead to the finding that migration of MSCs to the sites of tumor growth, which produced inhibition of Stat3 signaling, was associated with normalization of Treg numbers and dramatically reduced pulmonary and hepatic metastases

(Ling et al. 2010). Foxp3 is an X-linked nuclear transcription factor that is considered a defining marker for the Treg population (Sakaguchi 2004; Hori et al. 2003). Recently it was shown that Foxp3 not only plays a dominant role in the development and function of Treg cells, but also is a tumor suppressor factor (Heinze et al. 2011). This finding is of special interest since it was demonstrated that Foxp3-based tumor cytotoxicity is mediated via different signaling pathways compared to that enhancing the Treg cell function. This discrimination allows development of target-specific therapies aimed at augmentation of the Treg cells anti-tumor effects and not in altering their immunosuppressive function (Heinze et al. 2011).

Neutropenia. Neutrophils are among the cells acting as a first line of defense against infection mediated by innate immunity. Neutropenia, defined as an absolute neutrophil count below 1.5 × 109/L (Newburger and Dale 2013), is a common complication in cancer. Neutropenia can develop as a result of malignancy affecting hematopoiesis directly, or via skewed production of growth factors and cytokines. In addition, neutropenia is a growing concern as a result of widespread use of aggressive chemotherapy (Nesher and Rolston 2013).

Several aspects of neutropenia are currently under thorough investigation in patients with cancer-associated infections. It was shown that in this group of patients neutropenia is associated with increased infection-related morbidity and mortality, and is directly correlated with the occurrence of sepsis (Cho et al. 2013). Cancer patients with febrile neutropenia episodes have a higher incidence of secondary infections. The presence of a central intravenous catheter, diarrhea and invasive Aspergillosis are among the predisposing risk factors (Azap et al. 2012). Cancer patients with febrile neutropenia which developed in association with hematologic neoplasms, who are undergoing high-dose chemotherapy regimens and develop bloodstream infection with Gram-negative multidrug-resistant bacteria, are reported to have prolonged hospital stays (Rosa and Goldani 2014).

In addition, determining the mechanisms underlying the onset of neutropenia are of interest in respect to discerning infectious versus sterile inflammation, which is frequently a challenge in cancer patients. A model to identify the infected from non-infected patients based on levels of a list of variables, which include acute phase proteins, cytokines, measures of coagulation, metabolism, organ stress and iron turnover was proposed for diagnostic evaluation in febrile neutropenic hematology patients (Wenneras et al. 2014). Evaluation of the risk of onset of neutropenia also plays a central role in development of the optimal prophylactic chemotherapeutic regimens and their duration (Weycker et al. 2014).

Conclusions

The susceptibility of cancer patients to various infections, with bacterial, fungal and viral being the most common, is due to both cancer-induced factors and iatrogenic outcomes of aggressive therapeutic regimens. The mechanisms evoking the immunocompromised condition in cancer of both origins and those contributing to higher incidence and more aggressive course of infectious complications in this group of patients are complex and interconnected. The highest risk for onset of infectious complications and associated morbidity and mortality is seen in patients receiving chemotherapy for treatment of hematologic malignancies. A common occurrence for these patients is development of neutropenia, which is due to chemically induced suppression of hemopoiesis already compromised by malignancy. Ironically, in some cases the incidence of infections has concomitantly increased with a lifespan prolonged due to successful anti-cancer therapy. Thus, the ultimate goal of discerning the dichotomy of chemotherapy-mediated anti-cancer and immunosuppressive effects is of direct impact on management of infections in these patients. It is especially important in light of increased antibiotic resistance. New insight in the DC and Treg cell-mediated reactions leading to morbidity in cancer patients with infectious complications permits development of novel target-specific immunomodulatory therapeutic approaches.

References

Aalamian-Matheis M, Chatta GS, Shurin MR, Huland E, Huland H, Shurin GV (2007) Inhibition of dendritic cell generation and function by serum from prostate cancer patients: correlation with serum-free PSA. Adv Exp Med Biol 601:173–182

Alexander S, Nieder M, Zerr DM, Fisher BT, Dvorak CC, Sung L (2012) Prevention of bacterial infection in pediatric oncology: what do we know, what can we learn? Pediatr Blood Cancer 59(1):16–20. doi:10.1002/pbc.23416

Alonso CD, Kamboj M (2014) Clostridium difficile infection (CDI) in solid organ and hematopoietic stem cell transplant recipients. Curr Infect Dis Rep 16(8):414-014-0414-0. doi:10.1007/s11908-014-0414-0

Attie R, Chinen LT, Yoshioka EM, Silva MC, de Lima VC (2014) Acute bacterial infection negatively impacts cancer specific survival of colorectal cancer patients. World J Gastroenterol 20(38):13930–13935. doi:10.3748/wjg.v20.i38.13930

Azap A, Bozkurt GY, Yuksel MK et al (2012) Secondary infections in cancer patients with febrile neutropenia. Turk J Haematol 29(3):254–258. doi:10.5152/tjh.2011.75

Babady NE, Mead P, Stiles J et al (2012) Comparison of the luminex xTAG RVP fast assay and the idaho technology FilmArray RP assay for detection of respiratory viruses in pediatric patients at a cancer hospital. J Clin Microbiol 50(7):2282–2288. doi:10.1128/JCM.06186-11

Balducci L, Hardy CL, Lyman GH (2001) Hematopoietic growth factors in the older cancer patient. Curr Opin Hematol 8(3):170–187

Baratelli F, Lee JM, Hazra S et al (2010) PGE(2) contributes to TGF-beta induced T regulatory cell function in human non-small cell lung cancer. Am J Transl Res 2(4):356–367

Beck CR, McKenzie BC, Hashim AB, Harris RC, University of Nottingham Influenza and the ImmunoCompromised (UNIIC) Study Group, Nguyen-Van-Tam JS (2012) Influenza vaccination for immunocompromised patients: systematic review and meta-analysis by etiology. J Infect Dis 206(8):1250–1259. doi:jis487 [pii]

Benites EC, Cabrini DP, Silva AC et al (2014) Acute respiratory viral infections in pediatric cancer patients undergoing chemotherapy. J Pediatr (Rio J) 90(4):370–376. doi:10.1016/j.jped.2014.01.006

Borchardt RA, Torres HA (2014) Challenges in managing hepatitis C virus infection in cancer patients. World J Gastroenterol 20(11):2771–2776. doi:10.3748/wjg.v20.i11.2771

Burgdorf SK (2010) Dendritic cell vaccination of patients with metastatic colorectal cancer. Dan Med Bull 57(9):B4171. doi:B4171 [pii]

Carr SB, Adderson EE, Hakim H, Xiong X, Yan X, Caniza M (2012) Clinical and demographic characteristics of seasonal influenza in pediatric patients with cancer. Pediatr Infect Dis J 31(11):e202–e207. doi:10.1097/INF.0b013e318267f7d9

Carter JH, Lenahan JL, Ishak GE, Geyer JR, Pollard J, Englund JA (2015) Medical management of invasive fungal infections of the central nervous system in pediatric cancer patients. Pediatr Blood Cancer 62(6):1095–1098

Caselli D, Petris MG, Rondelli R et al (2014) Single-day trimethoprim/sulfamethoxazole prophylaxis for pneumocystis pneumonia in children with cancer. J Pediatr 164(2):389–392.e1. doi:10.1016/j.jpeds.2013.10.021

Caux C, Massacrier C, Dubois B et al (1999) Respective involvement of TGF-beta and IL-4 in the development of langerhans cells and non-langerhans dendritic cells from CD34+ progenitors. J Leukoc Biol 66(5):781–791

Chao PZ, Hsieh MS, Cheng CW, Hsu TJ, Lin YT, Lai CH, Liao CC, Chen WY, Leung TK, Lee FP, Lin YF, Chen CH (2015) Dendritic cells respond to nasopharygeal carcinoma cells through annexin A2-recognizing DC-SIGN. Oncotarget 6(1):159–170

Chiappini E, Berti E, Gianesin K et al (2014) Pediatric human immunodeficiency virus infection and cancer in the highly active antiretroviral treatment (HAART) era. Cancer Lett 347(1):38–45. doi:10.1016/j.canlet.2014.02.002

Cho SY, Lee DG, Choi SM et al (2013) Impact of vancomycin resistance on mortality in neutropenic patients with enterococcal bloodstream infection: a retrospective study. BMC Infect Dis 13:504-2334-13-504. doi:10.1186/1471-2334-13-504

Chu CS, Boyer JD, Jawad A et al (2013) Immunologic consequences of chemotherapy for ovarian cancer: impaired responses to the influenza vaccine. Vaccine 31(46):5435–5442. doi:10.1016/j.vaccine.2013.09.001

Cui Y, Yang X, Zhu W, Li J, Wu X, Pang Y (2013) Immune response, clinical outcome and safety of dendritic cell vaccine in combination with cytokine-induced killer cell therapy in cancer patients. Oncol Lett 6(2):537–541. doi:10.3892/ol.2013.1376

Curiel TJ, Coukos G, Zou L et al (2004) Specific recruitment of regulatory T cells in ovarian carcinoma fosters immune privilege and predicts reduced survival. Nat Med 10(9):942–949. Accessed 28 Dec 2009

de Naurois J, Novitzky-Basso I, Gill MJ et al (2010) Management of febrile neutropenia: ESMO clinical practice guidelines. Ann Oncol 21(Suppl 5):v252–v256. doi:10.1093/annonc/mdq196

de Oliveira CP, Atta EH, da Silva AR (2014) Predictors of 7- and 30-day mortality in pediatric intensive care unit patients with cancer and hematologic malignancy infected with gram-negative bacteria. Braz J Infect Dis 18(6):591–599. doi:10.1016/j.bjid.2014.05.012

Dignani MC, Costantini P, Salgueira C et al (2014) Pandemic 2009 influenza A (H1N1) virus infection in cancer and hematopoietic stem cell transplant recipients; a multicenter observational study. F1000Res 3:221. doi:10.12688/f1000research.5251.1

Ding SZ, Zheng PY (2012) Helicobacter pylori infection induced gastric cancer; advance in gastric stem cell research and the remaining challenges. Gut Pathog 4(1):18-4749-4-18. doi:10.1186/1757-4749-4-18

Dudek AM, Martin S, Garg AD, Agostinis P (2013) Immature, semi-mature, and fully mature dendritic cells: toward a DC-cancer cells interface that augments anticancer immunity. Front Immunol 4:438. doi:10.3389/fimmu.2013.00438

Duncan BB, Highfill SL, Qin H et al (2013) A pan-inhibitor of DASH family enzymes induces immune-mediated regression of murine sarcoma and is a potent adjuvant to dendritic cell vaccination and adoptive T-cell therapy. J Immunother 36(8):400–411. doi:10.1097/CJI.0b013e3182a80213

Esche C, Subbotin VM, Maliszewski C, Lotze MT, Shurin MR (1998) FLT3 ligand administration inhibits tumor growth in murine melanoma and lymphoma. Cancer Res 58(3):380–383

Fadul CE, Fisher JL, Hampton TH et al (2011) Immune response in patients with newly diagnosed glioblastoma multiforme treated with intranodal autologous tumor lysate-dendritic cell vaccination after radiation chemotherapy. J Immunother 34(4):382–389. doi:10.1097/CJI.0b013e318215e300

Fang H, Ang B, Xu X et al (2014) TLR4 is essential for dendritic cell activation and anti-tumor T-cell response enhancement by DAMPs released from chemically stressed cancer cells. Cell Mol Immunol 11(2):150–159. doi:10.1038/cmi.2013.59

Gao D, Li C, Xie X et al (2014) Autologous tumor lysate-pulsed dendritic cell immunotherapy with cytokine-induced killer cells improves survival in gastric and colorectal cancer patients. PLoS One 9(4):e93886. doi:10.1371/journal.pone.0093886

Gedik H, Simsek F, Yildirmak T et al (2014) Primary or secondary antifungal prophylaxis in patients with hematological maligancies: efficacy and damage. Ther Clin Risk Manag 10:305–312. doi:10.2147/TCRM.S59683

Ghansah T, Vohra N, Kinney K et al (2013) Dendritic cell immunotherapy combined with gemcitabine chemotherapy enhances survival in a murine model of pancreatic carcinoma. Cancer Immunol Immunother 62(6):1083–1091. doi:10.1007/s00262-013-1407-9

Gobert M, Treilleux I, Bendriss-Vermare N et al (2009) Regulatory T cells recruited through CCL22/CCR4 are selectively activated in lymphoid infiltrates surrounding primary breast tumors and lead to an adverse clinical utcome. Cancer Res 69(5):2000–2009. Accessed 20 Nov 2009

Habel LA, Ray GT, Silverberg MJ et al (2013) The epidemiology of herpes zoster in patients with newly diagnosed cancer. Cancer Epidemiol Biomarkers Prev 22(1):82–90. doi:10.1158/1055-9965.EPI-12-0815

Hargadon KM (2013) Tumor-altered dendritic cell function: implications for anti-tumor immunity. Front Immunol 4:192. doi:10.3389/fimmu.2013.00192

Hata A, Kuniyoshi M, Ohkusa Y (2011) Risk of herpes zoster in patients with underlying diseases: a retrospective hospital-based cohort study. Infection 39(6):537–544. doi:10.1007/s15010-011-0162-0

Hegmans JPJJ, Hemmes A, Hammad H, Boon L, Hoogsteden HC, Lambrecht BN (2006) Mesothelioma environment comprises cytokines and T-regulatory cells that suppress immune responses. Eur Respir J 27(6):1086–1095. Accessed 20 Nov 2009

Heinze E, Chan G, Mory R et al (2011) Tumor suppressor and T-regulatory functions of Foxp3 are mediated through separate signaling pathways. Oncol Lett 2(4):665–668. doi:10.3892/ol.2011.307

Hori S, Nomura T, Sakaguchi S (2003) Control of regulatory T cell development by the transcription factor Foxp3. Science 299(5609):1057–1061

Hurwitz AA, Watkins SK (2012) Immune suppression in the tumor microenvironment: a role for dendritic cell-mediated tolerization of T cells. Cancer Immunol Immunother 61(2):289–293. doi:10.1007/s00262-011-1181-5

Iglar K, Kopp A, Glazier RH (2013) Herpes zoster as a marker of underlying malignancy. Open Med 7(2):e68–e73

Iida N, Nakamoto Y, Baba T et al (2008) Tumor cell apoptosis induces tumor-specific immunity in a CC chemokine receptor 1- and 5-dependent manner in mice. J Leukoc Biol 84(4):1001–1010. doi:10.1189/jlb.1107791

Ishida T, Ishii T, Inagaki A et al (2006) Specific recruitment of CC chemokine receptor 4-positive regulatory T cells in hodgkin lymphoma fosters immune privilege. Cancer Res 66(11):5716–5722

Johansen HK, Gotzsche PC (2014) Amphotericin B versus fluconazole for controlling fungal infections in neutropenic cancer patients. Cochrane Database Syst Rev 9:CD000239. doi:10.1002/14651858.CD000239.pub2

Kaebisch R, Mejias-Luque R, Prinz C, Gerhard M (2014) Helicobacter pylori cytotoxin-associated gene A impairs human dendritic cell maturation and function through IL-10-mediated activation of STAT3. J Immunol 192(1):316–323. doi:10.4049/jimmunol.1302476

Karthaus N, Torensma R, Tel J (2012) Deciphering the message broadcast by tumor-infiltrating dendritic cells. Am J Pathol 181(3):733–742. doi:10.1016/j.ajpath.2012.05.012
Kim DH, Lee YY, Shin US, Moon SM (2013) Immunogenicity of influenza vaccine in colorectal cancer patients. Cancer Res Treat 45(4):303–312. doi:10.4143/crt.2013.45.4.303
Kim YI, Kang HC, Lee HS et al (2014) Invasive pulmonary mucormycosis with concomitant lung cancer presented with massive hemoptysis by huge pseudoaneurysm of pulmonary artery. Ann Thorac Surg 98(5):1832–1835. doi:10.1016/j.athoracsur.2013.12.051
Kinnebrew MA, Lee YJ, Jenq RR et al (2014) Early clostridium difficile infection during allogeneic hematopoietic stem cell transplantation. PLoS One 9(3):e90158. doi:10.1371/journal.pone.0090158
Kobayashi M, Chiba A, Izawa H et al (2014) The feasibility and clinical effects of dendritic cell-based immunotherapy targeting synthesized peptides for recurrent ovarian cancer. J Ovarian Res 7:48–2215-7-48. doi:10.1186/1757-2215-7-48
Kotsianidis I, Bouchliou I, Nakou E et al (2009) Kinetics, function and bone marrow trafficking of CD4+ CD25+ FOXP3+ regulatory T cells in myelodysplastic syndromes (MDS). Leukemia 23(3):510–518. Accessed 20 Nov 2009
Lai CH, Chi CY, Chen HP, Lai CJ, Fung CP, Liu CY (2004) Port-A catheter-associated nocardia bacteremia detected by gallium inflammation scan: a case report and literature review. Scand J Infect Dis 36(10):775–777. doi:CEJYGEBR728167MB
Lanoix JP, Pluquet E, Lescure FX et al (2011) Bacterial infection profiles in lung cancer patients with febrile neutropenia. BMC Infect Dis 11:183-2334-11-183. doi:10.1186/1471-2334-11-183
Lindenberg JJ, van de Ven R, Lougheed SM et al (2013) Functional characterization of a STAT3-dependent dendritic cell-derived CD14 cell population arising upon IL-10-driven maturation. Oncoimmunology 2(4):e23837. doi:10.4161/onci.23837
Ling X, Marini F, Konopleva M et al (2010) Mesenchymal stem cells overexpressing IFN-beta inhibit breast cancer growth and metastases through Stat3 signaling in a syngeneic tumor model. Cancer Microenviron 3(1):83–95. doi:10.1007/s12307-010-0041-8
Lipschitz DA (1995) Age-related declines in hematopoietic reserve capacity. Semin Oncol 22(1 Suppl 1):3–5
Liu CZ, Zhang L, Chang XH et al (2012) Overexpression and immunosuppressive functions of transforming growth factor 1, vascular endothelial growth factor and interleukin-10 in epithelial ovarian cancer. Chin J Cancer Res 24(2):130–137. doi:10.1007/s11670-012-0130-y
Lu Y, Zhang M, Wang S et al (2014) p38 MAPK-inhibited dendritic cells induce superior antitumour immune responses and overcome regulatory T-cell-mediated immunosuppression. Nat Commun 5:4229. doi:10.1038/ncomms5229
Ma Y, Shurin GV, Gutkin DW, Shurin MR (2012) Tumor associated regulatory dendritic cells. Semin Cancer Biol 22(4):298–306. doi:10.1016/j.semcancer.2012.02.010
Mani D, Haigentz M Jr, Aboulafia DM (2012) Lung cancer in HIV infection. Clin Lung Cancer 13(1):6–13. doi:10.1016/j.cllc.2011.05.005
Mavinkurve-Groothuis AM, van der Flier M, Stelma F, van Leer-Buter C, Preijers FW, Hoogerbrugge PM (2013) Absolute lymphocyte count predicts the response to new influenza virus H1N1 vaccination in pediatric cancer patients. Clin Vaccine Immunol 20(1):118–121. doi:10.1128/CVI.00585-12
Mayordomo JI, Zorina T, Storkus WJ et al (1995a) Bone marrow-derived dendritic cells pulsed with synthetic tumour peptides elicit protective and therapeutic antitumour immunity. Nat Med 1(12):1297–1302
Mayordomo JI, Storkus WJ, Kast WM, Zorina T, DeLeo AB, Lotze MT (1995b) Bone-marrow-derived dendritic cells serve as potent adjuvant for peptide-based antitumor vaccines. J Cell Biochem (Suppl 21A):21
Mayordomo JI, Zorina T, Storkus WJ et al (1997) Bone marrow-derived dendritic cells serve as potent adjuvants for peptide-based antitumor vaccines. Stem Cells 15(2):94–103
Miura Y, Theriault RL, Naito Y et al (2013) The safety of chemotherapy for breast cancer patients with hepatitis C virus infection. J Cancer 4(6):519–523. doi:10.7150/jca.6231

Mizukami Y, Kono K, Kawaguchi Y et al (2008) CCL17 and CCL22 chemokines within tumor microenvironment are related to accumulation of Foxp3+ regulatory T cells in gastric cancer. Int J Cancer 122(10):2286–2293. Accessed 20 Nov 2009

Mor M, Gilad G, Kornreich L, Fisher S, Yaniv I, Levy I (2011) Invasive fungal infections in pediatric oncology. Pediatr Blood Cancer 56(7):1092–1097. doi:10.1002/pbc.23005

Mousset S, Buchheidt D, Heinz W et al (2014) Treatment of invasive fungal infections in cancer patients-updated recommendations of the infectious diseases working party (AGIHO) of the german society of hematology and oncology (DGHO). Ann Hematol 93(1):13–32. doi:10.1007/s00277-013-1867-1

Neofytos D, Lu K, Hatfield-Seung A et al (2013) Epidemiology, outcomes, and risk factors of invasive fungal infections in adult patients with acute myelogenous leukemia after induction chemotherapy. Diagn Microbiol Infect Dis 75(2):144–149. doi:10.1016/j.diagmicrobio.2012.10.001

Nesher L, Rolston KV (2013) Neutropenic enterocolitis, a growing concern in the era of widespread use of aggressive chemotherapy. Clin Infect Dis 56(5):711–717. doi:10.1093/cid/cis998

Newburger PE, Dale DC (2013) Evaluation and management of patients with isolated neutropenia. Semin Hematol 50(3):198–206. doi:10.1053/j.seminhematol.2013.06.010

Niu J, Ren Y, Zhang T et al (2014) Retrospective comparative study of the effects of dendritic cell vaccine and cytokine-induced killer cell immunotherapy with that of chemotherapy alone and in combination for colorectal cancer. Biomed Res Int 2014:214727. doi:10.1155/2014/214727

Ohtani H, Jin Z, Takegawa S, Nakayama T, Yoshie O (2009) Abundant expression of CXCL9 (MIG) by stromal cells that include dendritic cells and accumulation of CXCR3+ T cells in lymphocyte-rich gastric carcinoma. J Pathol 217(1):21–31. Accessed 20 Nov 2009

Ozdemir N, Celkan T, Midilli K et al (2011) Novel influenza a (H1N1) infection in a pediatric hematology oncology clinic during the 2009–2010 pandemia. Pediatr Hematol Oncol 28(4):288–293. doi:10.3109/08880018.2010.550986

Pendyala G, Joshi S, Chaudhari S, Gandhage D (2013) Links demystified: periodontitis and cancer. Dent Res J (Isfahan) 10(6):704–712

Poutahidis T, Cappelle K, Levkovich T et al (2013) Pathogenic intestinal bacteria enhance prostate cancer development via systemic activation of immune cells in mice. PLoS One 8(8):e73933. doi:10.1371/journal.pone.0073933

Qin XJ, Shi HZ, Deng JM, Liang QL, Jiang J, Ye ZJ (2009) CCL22 recruits CD4-positive CD25-positive regulatory T cells into malignant pleural effusion. Clin Cancer Res. 15(7):2231–2237. Accessed 20 Nov 2009

Rasool Hassan BA, Yusoff ZB, Othman SB (2010) Fever/clinical signs and association with neutropenia in solid cancer patients–bacterial infection as the main cause. Asian Pac J Cancer Prev 11(5):1273–1277

Ridolfi L, Petrini M, Fiammenghi L et al (2010) Unexpected high response rate to traditional therapy after dendritic cell-based vaccine in advanced melanoma: update of clinical outcome and subgroup analysis. Clin Dev Immunol 2010:504979. doi:10.1155/2010/504979

Rosa RG, Goldani LZ (2014) Factors associated with hospital length of stay among cancer patients with febrile neutropenia. PLoS One 9(10):e108969. doi:10.1371/journal.pone.0108969

Rosenblatt J, Avivi I, Vasir B et al (2013) Vaccination with dendritic cell/tumor fusions following autologous stem cell transplant induces immunologic and clinical responses in multiple myeloma patients. Clin Cancer Res 19(13):3640–3648. doi:10.1158/1078-0432.CCR-13-0282

Sakaguchi S (2004) Naturally arising CD4+ regulatory t cells for immunologic self-tolerance and negative control of immune responses. Annu Rev Immunol 22:531–562

Shah DP, Ghantoji SS, Mulanovich VE, Ariza-Heredia EJ, Chemaly RF (2012) Management of respiratory viral infections in hematopoietic cell transplant recipients. Am J Blood Res 2(4):203–218

Shehata MA, Karim NA (2014) Influenza vaccination in cancer patients undergoing systemic therapy. Clin Med Insights Oncol 8:57–64. doi:10.4137/CMO.S13774

Shimizu Y, Dobashi K, Imai H et al (2009) CXCR4+ FOXP3+ CD25+ lymphocytes accumulate in CXCL12-expressing malignant pleural mesothelioma. Int J Immunopathol Pharmacol 22(1):43–51. Accessed 20 Nov 2009

Shurin GV, Aalamian M, Pirtskhalaishvili G et al (2001) Human prostate cancer blocks the generation of dendritic cells from CD34+ hematopoietic progenitors. Eur Urol 39(Suppl 4):37–40. doi:52584 [pii]

Shurin MR, Yurkovetsky ZR, Tourkova IL, Balkir L, Shurin GV (2002) Inhibition of CD40 expression and CD40-mediated dendritic cell function by tumor-derived IL-10. Int J Cancer 101(1):61–68. doi:10.1002/ijc.10576

Shurin MR, Shurin GV, Chatta GS (2007) Aging and the dendritic cell system: implications for cancer. Crit Rev Oncol Hematol 64(2):90–105. doi:S1040-8428(07)00055-8 [pii]

Sigel K, Dubrow R, Silverberg M, Crothers K, Braithwaite S, Justice A (2011) Cancer screening in patients infected with HIV. Curr HIV/AIDS Rep 8(3):142–152. doi:10.1007/s11904-011-0085-5

Smith HA, Kang Y (2013) Acute infection induces a metastatic niche: a double menace for cancer patients. Clin Cancer Res 19(17):4547–4549. doi:10.1158/1078-0432.CCR-13-1524

Stagg J, Divisekera U, Duret H et al (2011) CD73-deficient mice have increased antitumor immunity and are resistant to experimental metastasis. Cancer Res 71(8):2892–2900. doi:10.1158/0008-5472.CAN-10-4246

Steinman RM (1991) The dendritic cell system and its role in immunogenicity. Annu Rev Immunol 9:271–296. doi:10.1146/annurev.iy.09.040191.001415

Steinman RM, Hawiger D, Nussenzweig MC (2003) Tolerogenic dendritic cells. Annu Rev Immunol 21:685–711

Stosor V, Zembower T (eds) (2014) Infectious complications in cancer patients, vol 161. Springer, Chicago

Suso EM, Dueland S, Rasmussen AM et al (2011) hTERT mRNA dendritic cell vaccination: complete response in a pancreatic cancer patient associated with response against several hTERT epitopes. Cancer Immunol Immunother 60(6):809–818. doi:10.1007/s00262-011-0991-9

Sutton SH (2014) Infections associated with solid malignancies. Cancer Treat Res 161:371–411. doi:10.1007/978-3-319-04220-6_13

Tourkova IL, Yamabe K, Foster B et al (2004) Murine prostate cancer inhibits both in vivo and in vitro generation of dendritic cells from bone marrow precursors. Prostate 59(2):203–213. doi:10.1002/pros.10369

Tuting T, Zorina T, Ma DI et al (1997) Development of dendritic cell-based genetic vaccines for cancer. Adv Exp Med Biol 417:511–518

van de Peppel RJ, Dekkers OM, von dem Borne PA, de Boer MG (2014) Relapsed and secondary disease drive the risk profile for invasive aspergillosis prior to stem cell transplantation in patients with acute myeloid leukemia or myelodysplastic syndrome. Med Mycol 52(7):699–705. doi:10.1093/mmy/myu036

Van Tendeloo VF, Van de Velde A, Van Driessche A et al (2010) Induction of complete and molecular remissions in acute myeloid leukemia by wilms' tumor 1 antigen-targeted dendritic cell vaccination. Proc Natl Acad Sci U S A 107(31):13824–13829. doi:10.1073/pnas.1008051107

Wald O, Izhar U, Amir G et al (2006) CD4+ CXCR4highCD69+ T cells accumulate in lung adenocarcinoma. J Immunol 177(10):6983–6990. Accessed 20 Nov 2009

Weiss VL, Lee TH, Song H et al (2012) Trafficking of high avidity HER-2/neu-specific T cells into HER-2/neu-expressing tumors after depletion of effector/memory-like regulatory T cells. PLoS One 7(2):e31962. doi:10.1371/journal.pone.0031962

Wenneras C, Hagberg L, Andersson R et al (2014) Distinct inflammatory mediator patterns characterize infectious and sterile systemic inflammation in febrile neutropenic hematology patients. PLoS One 9(3):e92319. doi:10.1371/journal.pone.0092319

Weycker D, Barron R, Edelsberg J, Kartashov A, Legg J, Glass AG (2014) Risk and consequences of chemotherapy-induced neutropenic complications in patients receiving daily filgrastim: the

importance of duration of prophylaxis. BMC Health Serv Res 14:189-6963-14-189. doi:10.1186/1472-6963-14-189
Whelan MC, Casey G, Larkin JO, Guinn BA, O'Sullivan GC, Tangney M (2014) Oral tolerance to cancer can be abrogated by T regulatory cell inhibition. PLoS One 9(5):e97602. doi:10.1371/journal.pone.0097602
Wimmers F, Schreibelt G, Skold AE, Figdor CG, De Vries IJ (2014) Paradigm shift in dendritic cell-based immunotherapy: from in vitro generated monocyte-derived DCs to naturally circulating DC subsets. Front Immunol 5:165. doi:10.3389/fimmu.2014.00165
Wu P, Wu D, Ni C et al (2014) gammadeltaT17 cells promote the accumulation and expansion of myeloid-derived suppressor cells in human colorectal cancer. Immunity 40(5):785–800. doi:10.1016/j.immuni.2014.03.013
Yan L, Cai Q, Xu Y (2013) The ubiquitin-CXCR4 axis plays an important role in acute lung infection-enhanced lung tumor metastasis. Clin Cancer Res 19(17):4706–4716. doi:10.1158/1078-0432.CCR-13-0011
Yazici O, Sendur MA, Aksoy S (2014) Hepatitis C virus reactivation in cancer patients in the era of targeted therapies. World J Gastroenterol 20(22):6716–6724. doi:10.3748/wjg.v20.i22.6716
Yuan J, Kashiwagi S, Reeves P et al (2014) A novel mycobacterial Hsp70-containing fusion protein targeting mesothelin augments antitumor immunity and prolongs survival in murine models of ovarian cancer and mesothelioma. J Hematol Oncol 7:15-8722-7-15. doi:10.1186/1756-8722-7-15
Zhao T, Ren H, Jia L et al (2014) Inhibition of HIF-1alpha by PX-478 enhances the anti-tumor effect of gemcitabine by inducing immunogenic cell death in pancreatic ductal adenocarcinoma. Oncotarget. doi:2948 [pii]
Zhong H, Gutkin DW, Han B et al (2014) Origin and pharmacological modulation of tumor-associated regulatory dendritic cells. Int J Cancer 134(11):2633–2645. doi:10.1002/ijc.28590

Part IV
Infection and Cancer: Comorbid Development

Chapter 15
Comorbid Development of Infection and Cancer

Nahed Ismail, Michael R. Shurin, and Abdulrahman Zaghloul

Abstract Infectious diseases remain the leading cause of death around the world, causing more than 12 million deaths each year. The long-term effects of these illnesses, as a major public health problem, have raised particular concerns since some infectious agents have been associated with different types of chronic diseases, including cancer. Although cancer development can be multi-factorial in origin, several types of solid organ or hematologic cancers are caused by infectious agents (Shurin MR, Immunol Targets Ther 1:1–6, 2012). Approximately 15 % of all cancers occurring worldwide could be attributed to infections, a global total of 1.2 million cases per year (Kuper H, Adami HO, Trichopoulos D et al., J Intern Med 248:171–183, 2000). A substantial body of evidence also support the notion that infection itself or associated production of pro-inflammatory cytokines such as tumor necrosis factor (TNF), which can cause genetic mutations, thus promoting cancer development or enhancing epithelial-mesynchymal transition (EMT), an important mechanism that enable cancer metastasis (Voronov E et al., Proc Natl Acad Sci USA 100:2645–2650, 2003; Grivennikov SI and Karin M, Curr Opin Genet Dev 20:65–71, 2010). On the other hand, immunosuppression caused by the cancer itself or immunosuppressive drugs used for cancer treatment increase the risk and severity of infections, which in turn provide positive feedback mechanism that further enhance cancer metastasis (Khayr W, Haddad RY, Noor SA, Dis Mon 58:239–249, 2012). For instance, ~50–80 % of patients suffering from various hematological malignancies develop infections, which contribute to a higher incidence of mortality in these patients (Yadegarynia D, Tarrand J, Raad I, Rolston K, Clin Infect Dis 37:1144–1145, 2003).

The occurrence of infection and cancer as unrelated comorbidities are clinically common scenarios, however, the effect of infection or its therapy on cancer progression or regression remains elusive. Immune-surveillance and immunoediting are a major mechanism that control cancer development and progression as

---

N. Ismail, M.D., Ph.D. (✉) • M.R. Shurin • A. Zaghloul
Departments of Pathology, School of Medicine, University of Pittsburgh,
Scaife Hall S735, 3550 Terrace Street, Pittsburgh, PA 15261, USA
e-mail: ismailn@upmc.edu

M.R. Shurin et al. (eds.), *Infection and Cancer: Bi-Directorial Interactions*, DOI 10.1007/978-3-319-20669-1_15

explained in the following sections. However, information on the effect of acute or chronic systemic bacterial infections on tumor immune-surveillance or editing is lacking. Future research are warranted to answer several questions and addressed gaps in knowledge such as: (1) How does the host respond to systemic acute or persistent infections with bacterial agents during tumor development and therapy?; (2) How can independent acute or chronic infections affect cancer progression?; (3) How bacterial infection modulates the effector functions of myeloid and lymphoid regulatory cells in the tumor microenvironment? and finally (4) Does the immune responses against cancer and co-morbid infection are independent?

In this review article we will address these questions in the context of what we know about cancer immunosurveillance, tumor microenvironment, and the effect of infection as co-morbidity on cancer development, progression, and metastasis. We will also discuss how infectious agents can be used as therapeutic vaccine candidates in the management of cancer patients.

Keywords Cancer • Infection • *Toxoplasm* • *Mycobacterium tuberculosis* • Malaria • *Ehrlichia* • Co-morbidity • Independent response • Immunoediting • Tumor microenvironment • Regulatory cells • T cells • Myeloid cells

History of Infection-Tumor Association

The association between cancer and infection goes back to 1772 when scientists made the first observations about the development of bronchogenic lung carcinoma at sites of Mycobacterium Tuberculosis pulmonary scars (Herrera et al. 2005). However, it was not until the year 1810 when coexistence of TB and Lung carcinoma was first described, and histologically demonstrated (Dacosta and Kinare 1991). In 1863, a German physician Rudolf Virchow made the first observation linking Parasitic infestation with Schistosoma haematobium in endemic regions of North Africa with the frequent development of bladder cancer in those endemic areas (Parsonnet 1999), which suggested that cancer development and progression tends to occur at sites of chronic infections (Balkwill and Mantovani 2001). In 1890, Russell W reported first evidence for the microbial cause of cancer and that followed by studies showing that infection promote cancer development and progression. In early 1900s, Paul Ehrlich was the first to propose the protective effect of the immune system against cancer; a concept that formulated the "cancer immunosurveillance hypothesis". However, it was not until the end of nineteenth century, where several evidences suggested a protective role of infection against cancer progression. For example, injection of extracts from Gram positive or Gram negative bacteria, *S. Pyogenes* and *Serratia Marcescens*, respectively, into the skin of patients with advanced stages of cancer resulted in cancer regression. This observation formulated the basis of the discovery of Coley's toxin, which has paved the way for the use of Mycobacterium Bacillus Calmette-Guerin (BCG) in the treatment of early stages of bladder carcinoma in humans. This therapy has been shown to be highly effective, and is still used till now as a treatment for bladder cancer (Herr et al. 1995). These initial studies paved the way for future discoveries relating infection to cancer progression or regression (Herrera et al. 2005).

Cancer Immunosurveillance/Immunoediting

Immunoediting or cancer immunosurveillance is a mechanism by which the tumor is modified by the immune system. The hypothesis of immunoediting was derived from comparative analysis of carcinogen-induced tumors harvested from immuno-competent and from immuno-deficient mice. The results of these analyses showed that tumors from the immuno-deficient mice were more immunogenic than the ones from immuno-competent mice. Hence, the tumors derived from immuno-deficient mice were classified as "unedited", while the ones that were derived from immuno-competent mice were classified as "edited". Further studies classified the unedited tumors into "unedited progressors" and "unedited regressors" depending on the tumor stages and development (Shankaran et al. 2001). Three stages of cancer immunoediting have been defined (Vesely et al. 2011)**:** first stage consists of immunosurveillance where innate and adaptive immunity develop against early transformed tumor cells, thus preventing cancer development. Failure of this process led to stage of immune evasion or stage of equilibrium or dormant stage where transformed tumor cells evade the immune system and grow, although the growth of tumor is still partially contained by the adaptive immune system. Failure of containing tumor by adaptive immunity will lead to tumor-escape of immune surveillance and subsequence relapse of primary tumor or metastatic spread. Several mechanisms have been proposed to explain tumor-escape or evasion of immune system that are reviewed in (Dunn et al. 2002, 2004a, b; Vesely et al. 2011; Zitvogel et al. 2006; Khong and Restifo 2002). Examples of the mechanisms that contribute to tumor escaping the immune surveillance are: loss of tumor antigen expression, down regulation of MHC class I antigens, defective antigen processing function of the antigen presenting cells (APCs), thus preventing activation of tumor-specific T cells. Another escape mechanism is through the induction of immunosuppressive state within the tumor microenvironment. Interestingly, these immune evasion mechanisms have been attributed to concomitant infection and/or chronic inflammation. Before discussing the effect of acute or chronic infections on tumor microenvironment, we will highlight some of the studies that illustrate the importance of tumor microenvironment in cancer development and metastasis.

Tumor Microenvironment

The mechanisms that govern the pro-tumorogenic or anti-tumorogenic outcomes of interactions between host, microbe and the tumor environment are multifactorial, complex and poorly understood. The tumor microenvironment is a critical factor in tumorigenesis, and especially in progression, as the pathogenesis of cancer critically depends on the complex interactions between various components within this micro-environment. It is well established that tumor-associated myeloid regulatory cells (MRC), including myeloid derived suppressor cells (MDSC), macrophages type 2 (M2), and regulatory neutrophils type 2 (N2) are critical components of pro-tumorigenic environment (Shurin et al. 2012).

Myeloid APC play major roles in the induction and differentiation of T cells and their anti-tumor effector functions. Mature conventional myeloid cells can promote the activation, expansion and differentiation of $CD4^+$ T cells and cytotoxic $CD8^+$T cells into type 1 cells with anti-tumor effector functions. In contrast, regulatory myeloid cells promote the differentiation of T cells into Th2 or T_{reg} cells with immunosuppressive functions and pro-tumorogenic effect. Thus, tumor-associated myeloid cells may promote stimulatory and inhibitory forces on the proliferative, angiogenic, and immuno-modulating properties of the tumor, as well as its potential to spread and metastasize. These suppressive or stimulatory functions of myeloid cells are mediated in part by cytokines they produce. For example, TGF-β produced by MDSCs, along with other inflammatory cytokines (e.g. TNF-α, IL-1, IL-4, and IL-13) have the ability to suppress anti-tumor immunity mediated by cytotoxic $CD8^+$T cells, $CD4^+$Th1 cells, and cytotoxic IFN-γ producing NK cells (Gabrilovich and Nagaraj 2009; Laouar et al. 2005; Rook et al. 1986). TGF-β also support the induction and expansion of immune-suppressor T_{reg} cells, which are further maintained within tumor microenvironment by intrinsic production of TGF-β and adenosine expression by the tumor cells (Zarek et al. 2008). TGF-β along with IL-10 promotes the differentiation of tumor- associated dendritic cells and macrophages into MDSC and M2 macrophages, which in turn induce T_{reg} cells (Luo et al. 2007; Levings et al. 2002; Jarnicki et al. 2006). Paradoxically, some studies have shown that TGF-β has anti-tumor function as it induces production of IL-17 by $CD8^+$ T-cells or inhibit IL-6 production and chronic inflammation (Nam et al. 2008; Hinrichs et al. 2009; Muranski et al. 2008; Martin-Orozco et al. 2009). The initially molecular and cellular events that determine whether TGF-β exert pro-or anti-tumorgenic effect are not completely understood. However, Studies suggested that production of high concentration of TGF-β, not only by myeloid cells, but also by tumor cells, especially at late stages of cancer development, are more likely to induce pro-tumorigenic, rather than, anti-tumorigenic effect. On the other hand, TGF-β is a potent stimulator of EMT (Kalluri and Weinberg 2009; Bates and Mercurio 2003; Sullivan et al. 2009), which is another mechanism that promote tumor metastasis. Together, these studies suggest that TGF-β, MDSCs, macrophages, and Tregs are all immunosuppressive mechanisms that inhibit anti-tumor immunity, and promote cancer invasiveness, and metastasis (Sica et al. 2008; Arteaga 2006). Thus, on a conceptual level, an independent microbial infection that induce production of TGF-β within tumor microenvironment can promote tumor progression via the above immunosuppressive mechanisms.

Microbiome Influence Tumorigenesis

The effect of human microbiome in mucosal surfaces such as skin, oral cavity, female genital tract, lung, and gut on human health and disease has been a major discovery recently. For example, human gastrointestinal (GI) tract is colonized by a wide range of microbiota, approximately 10^{13} bacteria composed of over 500 microbial species. The commensal intestinal microbiota genes outnumber that of their human host by approximately 100 to one. What allows the symbiotic coexistence between host and microbiota is the anatomical separation of microbiota from the host compartment by well-maintained, multi-level barriers. Failure of these barriers due to several reasons such as injury, infection, or other disorders promotes inflammation and diseases, including cancer (O'Hara and Shanahan 2006). The same perplexing concept of cause and effect also exists between barrier failure and inflammation/carcinogenesis. The most prominent and widely known example for this barrier failure leading to tumorigenesis is the increased risk of intestinal cancer in patients with ulcerative colitis. Recently, these facts have driven the attention of scientists towards in depth studies of the bacterial microbiota and their relation and effects on tumorigenesis. There are accumulating evidences that support the notion that bacterial microbiota plays a key role in tumorigenesis and, more specifically, that dysbiosis (imbalance of the host microbiota) can influence tumorigenesis. This might explain, for example, the higher rate of cancer in the large intestine—where microbial densities are much higher than in the small intestine (Reddy et al. 1975). In contrast to gastric carcinogenesis that is caused by a specific bacterium, the tumorigenic effects of the microbiota in colorectal cancer (CRC) seem to be caused by dysbiosis (imbalance of the host microbiota). Accordingly, germ-free status and treatment with wide-spectrum antibiotics led to a significant reduction of the numbers of tumors in experimental models of colorectal carcinogenesis (Reddy et al. 1974; Klimesova et al. 2013; Wen et al. 1995).

Interestingly, several studies have shown that TLR4 signaling by lipopolysaccharide of Gram negative bacteria has direct correlation with increased rate of tumorigenesis (Tang et al. 2012; O'Hara and Shanahan 2006). The mechanism by which TLR4 signaling enhances tumor progression is not completely understood. However, studies have linked TLR4 signaling to activation of activation of NF_kB- and STAT3-dependent pathway (O'Hara and Shanahan 2006), which mediate production of pro-tumorigenic pro-inflammatory cytokines. In addition, activation of NF_kB can exert an anti-apoptotic effect on tumor cells, which enhance tumor cell survival and proliferation. These findings are further supported by studies in which abrogation of TLR4 signaling reduced intestinal tumorigenesis, while overexpression of TLR4 in intestinal epithelium promoted tumorigenesis (Tang et al. 2012). The role of microbiota in carcinogenesis in other organs such as the lungs, skin, and female genital tract, that have vast numbers of bacterial microbiome, is still in need to be further investigated to be determined. Thus, for its future plausible implications in preventive and in therapeutic approaches, further investigations need to be done to better understand the mechanism of Microbiome in tumorigenesis.

Effect of Chronic Inflammation on Cancer Development and Progression

Several studies have provided strong evidence that link inflammation to cancer progression. Chronic inflammation mediated by NF-κB and STAT3 signaling is known to promote proliferation of malignant cells, have anti-apoptotic effect on tumor cells, enhance EMT changes, and induce genetic mutation, which collectively lead to progression of certain tumors (Pikarsky et al. 2004). Murine and human studies have shown that STAT3 control the production of reactive oxygen species (ROS) that correlates with suppressive function of MDSCs in murine and human studies (Ohtsu et al. 2010). Activation of NF-κB and STAT3 also enhances EMT, which in turn increase the production of matrix metallo-proteinases (MMP) (Voronov et al. 2003; Grivennikov and Karin 2010), and subsequent local invasion of tumor (Pollard 2004; Coussens et al. 1999). As discussed above, NF-κB signaling and production of high level of pro-inflammatory cytokines such as TNF can cause destruction of tumor vasculature, prevents tumor invasiveness, and thus promotes cancer regression (Balkwill 2002).

The next sections of this chapter will discuss the effect of acute or chronic infections on tumor microenvironment, which is a major factor influencing not only tumor development, but also resistance to anticancer therapies.

How Can Independent Acute or Chronic Infections Affect Cancer Progression?

Although a crucial role of certain types of chronic infections in supporting or causing tumor the development by maintaining the pro-tumorogenic microenvironment is well proven (infection-related cancer), the phenomenon of independent development of cancer and infection in the same host, i.e. comorbid cancer-infection progression, has not been well studied. The increased risk of acute or chronic in patients with solid tumor or hematologic malignancy due to tumor itself or associated chemotherapy is common clinical phenomenon. For examples, *P aeruginosa* bacteremia, which is a severe disease with high mortality in the general population, is commonly encountered in patients with cancer or a hematologic malignancy.

Several risk factors that promote sepsis, high mortality rate in *P aeruginosa* bacteremia, and emergence of antimicrobial resistant *P aeruginosa* in patients with hematologic malignancies exist and include hospitalization, invasive procedures, nosocomial infection, as well as selective pressure as a result of continuous usage of antibiotics against which the bacteria was initially sensitive. Other risk factors include the development of neutropenia or leukopenia in cancer patients who are on chemotherapy. In addition, various factors predispose patients with hematologic malignancies to invasive *P aeruginosa* infections, including the disruption of mucosal barriers as a result of mucosal inflammation, use of steroids, impaired humoral immunity, and neutropenia. This increased risk of infection in cancer patients raises the question of whether development of independent acute or chronic infection affects cancer progression.

We provide several examples where infection with intracellular pathogen alters the tumor microenvironment and thus influence tumor development and progression.

Effect of **Toxoplasma gondii** *Infection on Cancer Development and Metastasis*

Toxoplasma gondii is an obligate intracellular protozoan parasite that preferentially invade myeloid cells; DCs, Macrophages, and monocytes. Several studies have demonstrated an anti-tumor effect of T. gondii infection as it enhances anti-tumor immunity and abrogate the tumor-associated immunosuppression. These studies suggested that antigens from *T. gondii* can be used as a candidate therapeutic vaccine as a novel cancer immunotherapeutic strategy. The immunotherapeutic effects of non-replicating *T. gondii* strain (cps strain) completely depends on their ability to induce production of immunostimulatory Th1 cytokines such as IL-12 and IFNγ. IL-12 enhances the activation, differentiation and expansion of $CD4^+$Th1 cells, type 1 cytotoxic $CD8^+$T cells and NK cells that exhibit anti-tumor effector functions (Curtsinger et al. 2003; Kalinski et al. 1999) and inhibits angiogenesis (Airoldi et al. 2007). IL-12 induced by *Toxoplasma* infection enhances the T cell priming functions of antigen presenting cells (macrophages and dendritic cells) as it promotes differentiation of macrophages into T cells- stimulatory M1 phenotype and enhances the expression of co-stimulatory molecules (e.g. CD80, CD86) on dendritic cells, which in turn promote the induction and expansion of $CD4^+$ and $CD8^+$T cells (Biswas and Mantovani 2010; Hagemann et al. 2008).

The effect of *T. gondii* infection on tumor-associated APCs appears to be not restricted to infected cells but also influence other uninfected host cells via bystander effect. The latter could be explained by production of inflammatory and Th1 promoting cytokines such as IL-12 by infected cells (Fox et al. 2013). Interestingly, it has been shown that intralesional injection of lung cancer-bearing mice with formalin-fixed *T. gondii* significantly prevented cancer metastasis and enhanced mice survival when compared to uninfected tumor-bearing mice (Suzuki and Kobayashi 1985; Kim et al. 2007). The effect of infection with *T. gondii* CPS strain on cancer progression in these mice was dependent of NK cells and effector $CD8^+$ T cells. Although the exact mechanism by which intra-lesional injection of CPS strain improved CD8 T cells priming, it was suggested that infection of tumor-associated dendritic cells enhance their ability to cross-present exogenous tumor antigens to naïve $CD8^+T$ cells, and thus activate tumor antigen-specific $CD8^+T$ cells. Thus, it appears that *Toxoplasma* infection converts the immunosuppressive function of tumor-associated dendritic cells within tumor microenvironment into immune-stimulatory functions (Baird et al. 2013).

Effect of Mycobacterium Tuberculosis Infection on Cancer Development and Metastasis

Earlier studies have shown that *M. Tuberculosis*- Heat Shock Protein (HSP) enhances T-cell priming functions of DCs and induction of Th1 response via ligation of TLR4 and MYD88-dependent signaling (Baird et al. 2013). Similar to the immunotherapeutic effect of *T. gondii* cps strain, the anti-tumor effect of M. TB-HSP was mediated in part by IFNγ. IFNγ-mediated type 1 response induced upon immunization with HSP also correlated with decreased frequency of immunosuppressive T_{reg} cells (Jung et al. 2014).

Interestingly, studies have shown that the ant-tumor effect of infectious agents can be specific to certain types of tumor and not universal against all cancers. For example, intra-vesicle instillation of live Mycobacterium bovis bacillus Calmette-Guérin (BCG) into bladder cancer improved the outcome and prevented cancer progression. In contrast, co-infection of patients with cervical cancer with BCG worsen cancer progression as it inhibits the maturation and antigen presenting functions of DCs, which in turn cause immunosuppression and favor cervical cancer growth (Manickam and Sivanandham 2011). Although it is not clear why and how these opposing effects of BCG on cancer development occur in different types of cancers, these studies suggest that alteration of myeloid cells phenotype and function within tumor microenvironment may not be the only mechanism by which infectious agent mediate anti-tumor effect. Nevertheless, these conflicting results in different types of cancer are not surprising if we consider the fact that cancer development is multi-factorial. However, these studies raises several questions such as: (1) whether infection of tumor-associated myeloid cells, inflammation or both are indispensable for induction of protective anti-tumor immunity; (2) whether certain infection alter only local, but not systemic or distant, effect on tumor microenvironment; (3) How infection influence interaction of parynchymal cells within tumor microenvironment such as endothelial cells with tumor cells and the effect of this interaction on ensuing angiogenesis, which is an important mechanism that mediate cancer metastasis; and (4) what the effect of infection on multiple components of tumor microenvironment that are known to affect immune response against tumor? Further studies are needed to address these gaps in knowledge.

Effect of Malaria Infection on Cancer Development and Metastasis

Malaria is a protozoal infectious disease caused by plasmodium species that infect red blood cells. Certain plasmodium species also infect parenchymal liver cells. Systemic plasmodium infection induce strong innate and adaptive immune responses marked by excessive production of pro-inflammatory cytokines (i.e. TNF-alpha, IL-1α, IL-1β), Th1-promoting cytokine (e.g.IFNγ). Plasmodium infection also increases the expression of the co-stimulatory molecules on APCs and their maturation, which together with secreted cytokines induce activation and proliferation of T cells (Ing et al. 2006; Roetynck et al. 2006). Recent studies have evaluated the effect of malaria infection on lung cancer progression in mice. These studies have shown that malaria inhibits cancer progression and metastasis, and that effect that was due to enhanced innate and adaptive immunity mediated by IFNγ, IFN-α, NK, and T cells.

Effect of Ehrlichia *Infection on Cancer Development and Metastasis*

Ehrlichiamuris (EM) is an obligate tick-transmitted Gram negative intracellular bacterium that target and replicates within myeloid-derived cells (macrophages, monocytes, dendritic cells, neutrophils) and non-hematopoietic cells (hepatocytes and endothelial cells). *Ehrlichia muris* exhibit several features that make it as optimal model to examine the effect of independent acute or chronic bacterial infection on tumor development and be used in development of effective therapeutic vaccine approach for cancer metastasis or hematologic malignancies. These include: (1) Live *E. muris* is avirulent/attenuated strain of *Ehrlichia* species that cause acute mild/self-limited disease in humans referred to as human monocytic ehrlichiosis (HME) (Ismail et al. 2004, 2006; Shibata et al. 2000; Olano et al. 2004; McBride and Walker 2011); (2) murine models of HME are available where *E. muris* cause either acute infection and mild disease or chronic and persistent infection. The ability of this bacteria to causes controllable acute and chronic infections is ideal for inducing alterations in the tumor environment; (3) the bacterium targets and replicates within myeloid-derived cells and parynchymal cells as mentioned above can preferentially cause infection of tumor-associated myeloid cells or endothelial cells where these cells are predominant; (4) the bacteria utilize type I and type IV secretion systems to secrete several effector proteins into the host cell cytoplasm and outside the cells, which was shown to influence expression of several host genes known to be involved in activation of innate and acquired immune responses, signaling, actin polymerization and phagocytosis (Wakeel 2009; Rikihisa 2010); (5) *E. muris* induces the generation of NKT and NK, both of which are critical early innate cells that shown to mediate anti-tumor immunity (Mattner et al., 2005; Ismail 2006, 2007; Stevenson 2006); (6) Unlike other intracellular bacteria, *E. muris* stimulate strong cytotoxic responses mediated by $CD4^+$ and $CD8^+$T cells. Cytotoxic $CD4^+$ T cells has been shown to be an alternative effector immune mechanism in conditions when $CD8^+$T cell response against cancer is absent due to the down-regulation of MHC class I molecules; (7) *E. muris* cause disseminated infection, particularly in the organs of reticuloendothelial system and thus can be used as a vehicle to deliver tumor specific antigen as a therapeutic vaccine for cancer metastasis; (8) *Ehrlichiae* exits the infected host cells via filopodium (Stevenson et al. 2006, 2010; Ismail et al. 2007; Thomas et al. 2010; Ghose et al. 2011; Nandi et al. 2009; Overwijk and Restifo 2001), which could potentially stimulate cell-cell communications within the tumor microenvironment.

We have utilized these immunestimulatory feature of *E. muris* and examine whether *Ehrlichia* infection of tumor-bearing mice alter the tumor microenvironment from a suppressive to stimulatory phenotype and thus enhance anti-tumor immune responses. To this end, we developed a murine model of infection-cancer comorbidity using *Ehrlichia muris* and B16 melanoma cells that form metastases in the lungs. As a tumor model system, we have selected the spontaneous C57BL/6-derived B16 melanoma, a well-established and widely used tumor model in which treatment is notoriously difficult (Shurin et al. 2012). In this mode, B16 melanoma cells are inoculated intravenously and that cause primary tumor nodule in the lung as well as multiple nodules in other lung lobes (i.e. pulmonary metastasis) and other organs. This B16 pulmonary metastasis model has been used in our studies to provide the proof-of-principle of the effect of acute infection of tumor progression.

We have developed two models of cancer and infection comorbidities: acute and chronic *E. muris* infection in tumor-bearing mice. In the <u>'acute' model</u>, C57BL/6 (B6) mice received i.v. injection of 3×10^6 B16 cells. On day 14 or 21 post-injection, mice were infected i.p. with a high dose of *E. muris.* We refer to this group as tumor+*E. muris* group. In the <u>'chronic' model</u>, mice were initially infected with a high dose of *E. muris* that cause persistent infection and then these infected mice were re-challenged via i.v. route, on day 30 post-infection with B16 melanoma cells. Control mice included tumor-bearing mice injected with PBS ("tumor+PBS" group) and mice infected only with *E. muris.*

Interestingly, our results showed that a*cute Ehrlichia infection prolonged survival of tumor-bearing mice.* Thus, while 100 % of control group (tumor+PBS) mice died on day 20 after inoculation of tumor cells, tumor-*E. muris* mice group survived till day 40 after tumor inoculation. Prolonged survival of tumor-*E. muris* mice group was associated with significant decrease in tumor growth macroscopically and microscopically (Fig. 15.1). On the other hand, unlike acute infection, chronic infection with *E. muris* resulted in enhanced tumor dissemination and growth (data not shown).

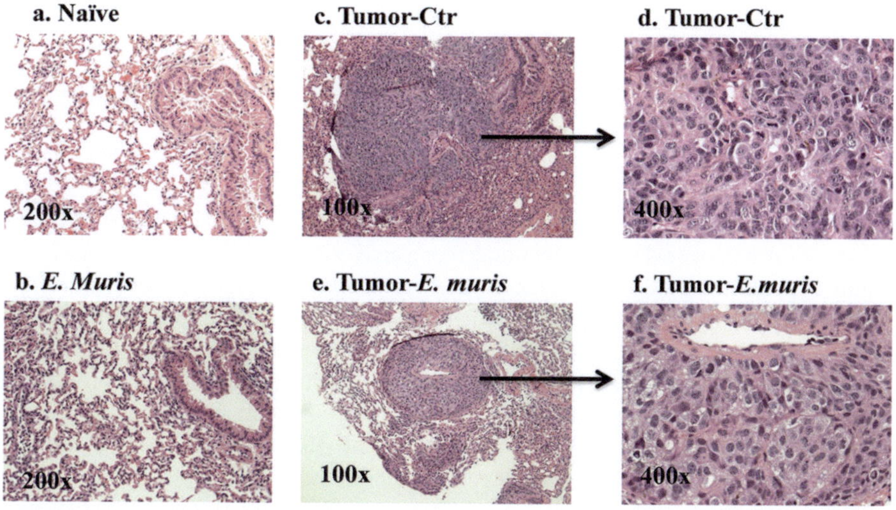

Fig. 15.1 Histopathology of lung sections. **(a)** Lung of naive mice with normal histology. **(b)** Lung from *E.* muris-infected mice with mild inflammatory infiltrate. **(c)** and **(d)** Lung from "tumor+PBS" group with abundant foci of malignant neoplasm (X100 and X400). **(e)** and **(f)**. Lung sections from " tumor+E. muris" with mild interstitial inflammation and microfoci of malignant neoplasm (X100 and X400)

Analysis of the host immune response to tumor in all mice groups demonstrated that acute *Ehrlichia* infection induced Th1-type responses as evidenced by increased frequency of IFN-γ producing type 1 cells as well as total number of $CD4^+T$ and $CD8^+T$ cells in the liver and spleen of tumor+*E. muris* mice group compared to control tumor+PBS group. The number of granzyme B expressing-cytotoxic $CD4^+$ and $CD8^+T$ cells in the spleen and lung of tumor+EM mice group was higher than that detected in control tumor+PBS group. Intriguingly, enhanced anti-tumor immunity in the "tumor+*E. muris*" group correlated with significant decline in the percentage of $CD4^+CD25^+Foxp3^+$ Treg cells compared to the "tumor+PBS" group. Further, *E. muris* infection significantly decreased the levels of $CD11b^+Gr\text{-}1^+MDSC$ and $CD11c^{low}CD11b^{high}$ regulatory DC in the lungs of "tumor+*E. muris*" mice compared to "tumor+PBS" animals. Together, these data suggest that *E. muris* infection abrogated immunosuppressive regulatory T cells while enhancing the magnitude and effector functions of anti-tumor $CD4^+$ and $CD8^+$ effector T cells, possibly via enhancing the T cell priming and antigen-presenting and/or co-stimulatory functions of DC in the "tumor+*E. muris*" group. *E. muris* induced enhanced anti-tumor immune response are more likely account for tumor regression, and prolonged survival of tumor-bearing mice. Although these results suggest the beneficial anti-tumor effect of mildly virulent *E. muris* infection, further mechanistic analysis of the molecular pathways that contribute to the down-regulation of tumor-associated MDSC and T regulatory cells by *E. muris and how this effect lead to* inhibition of tumor growth. Information gained from this novel model system can be valuable for therapeutic purposes and translational application such as development of therapeutic vaccine against cancer using *Ehrlichia*-derived antigens.

Summary

The above studies suggested that emergence/survival of immune regulatory cell populations might be changed during tumor progression in response to infection and thus establishes that immune regulator depleting/blocking approaches should be changed accordingly.

Understanding the role of infection-induced inflammatory and immune responses in controlling cancer development and metastasis may lead to new immunotherapeutic approaches to control cancer. Further mechanistic studies will foster development of rationally designed therapeutics aimed to revert the immunosuppressive circuits that undermine a meaningful anti-tumor immune response. For example, controlling MRC accumulation and function in cancer represents is a promising strategy to overcome disease-induced immune defects, which might be a key step in enhancing the effectiveness of immune-based therapies. Although several approaches have been offered to decrease MDSC and regulatory T cells (T_{reg}) in cancer (Gabrilovich and Nagaraj 2009; Suzuki et al. 2005; Ugel et al. 2009; Corthay 2014), there are few or no data on how to control and regulate regDC, M2, and neutrophils accumulation or function in the tumor environment. Based on the analysis of cellular and molecular events controlling pro-tumorigenic MRC accumulation by infection including Toxoplasma, BCG, and E. muris in the tumor environment, we propose a clinically relevant approach to block accumulation of MRC in cancer using infectious agents that target myeloid cells, which opens new opportunity to conquer tumor-induced immune unresponsiveness. In conclusion, there is critical need to understand the effect of infection on tumor using comorbid infection-tumor models that allow differentiating the role of specific subsets of effector and regulatory immune cells in the tumor-bearing host. This is highly significant for several reasons:

1. Understanding of mechanisms underlying tumor immune escape.
2. Development of therapeutic approaches targeting specific cell subsets with genetically engineered bacteria.
3. Evaluating a new concept suggesting that the presence of comorbid bacterial infection in cancer not only controls appearance of specific immune regulators, but also changes this path during tumor progression and spreading.

Understanding the cellular interplay between MRC subsets in the tumor environment during acute bacterial infection is also a new direction, which will gain new insight into how cancers progress and ultimately how cancer might be treated with bacterium-based approaches that selectively target MRC. Data obtained with *E. muris*-cancer co-morbidity model will allow the analysis of inflammatory-associated and tumor-associated MRC at different points of bacterium- and tumor-induced immune responses. The development of infection-cancer co-morbidity models will not only increase our knowledge of the mechanisms of comorbid development of immune-mediated diseases, but would also expand an opportunity to use antimicrobial and anti-cancer agents in a more controllable fashion. The results generated from infection-tumor models demonstrate the importance of cross-talk between cancer- and infection-induced immune responses and provide new insights in understanding the role of local and systemic immune environments in disease development and progression. Such information will allow better manipulation of the adaptive immune system at the molecular level to develop effective and optimal immunotherapy against cancer. Finally, information gained from the protective model of cancer associated with acute *E. muris* infection will be applied towards designing a new cancer therapeutic vaccine using *Ehrlichia* as a delivery system for tumor antigens. As the majority of current approaches focus on depleting T_{reg} and MDSC, the proposed opportunity to reprogram and harness suppressor MRC in the tumor environment is also a new therapeutic direction.

Time for the "Wounds That Never Heal" to Heal!

As we started our review with the German physician Rudolf Virchow -who made the 1st observation linking parasitic infestation with the development of bladder cancer- we end our review with his description of tumors as "wounds that never heal". So, is this description going to change in the future? From where we stand right now, one can conclude that future studies could pave the way in the near future for new immune-based therapeutic strategies and/or vaccination against cancer. We need a better understanding of the tumor micro-environment, and a better understanding of the actual mechanisms that direct the interrelationship between co-infection and associated inflammation on cancer development, progression, and metastasis.

References

Airoldi I et al (2007) Endogenous IL-12 triggers an anti-tumorogenic program in melanoma cells. Proc Natl Acad Sci U S A 104:3996–4001

Arteaga CL (2006) Inhibition of TGFbeta signaling in cancer therapy. Curr Opin Genet Dev 16:30–37

Baird JR, Fox BA, Sanders KL, Lizotte PH, CubillosRuiz JR, Scarlett UK, Rutkowski MR, ConejoGarcia JR, Fiering S, Bzik DJ (2013) Avirulent Toxoplasma gondii generates therapeutic antitumor immunity by reversing immunosuppression in the ovarian cancer microenvironment. Cancer Res 73(13):3842–3851

Balkwill F (2002) Tumor necrosis factor or tumor promoting factor? Cytokine Growth Factor Rev 13:135–141

Balkwill F, Mantovani A (2001) Inflammation and cancer: back to Virchow? Lancet 357:539–545

Bates RC, Mercurio AM (2003) Tumor necrosis factor-α stimulates the epithelial-to-mesenchymal transition of human colonic organoids. Mol Biol Cell 14:1790–1800

Becker C, Fantini MC, Schramm C et al (2004) TGF-h suppresses tumor progression in colon cancer by inhibition of IL-6 trans-signaling. Immunity 21:491–501

Biswas SK, Mantovani A (2010) Macrophage plasticity and interaction with lymphocyte subsets: cancer as a paradigm. Nat Immunol 11:889–896

Chen W et al (2003a) Conversion of peripheral CD4+ CD25–naive T cells to CD4+ CD25+ regulatory T cells by TGF-β induction of transcription factor Foxp3. J Exp Med 198:1875–1886

Chen LW, Egan L, Li ZW, Greten FR, Kagnoff MF, Karin M (2003b) The two faces of IKK and NF-e^B inhibition: prevention of systematic inflammation but increased local injury following intestinal ischemia-reperfusion. Nat Med 9:575–581

Chen L, He Z, Qin L, Li Q, Shi X et al (2011a) Antitumor effect of malaria parasite infection in a murine lewis lung cancer model through induction of innate and adaptive immunity. PLoS One 6(9):e24407

Chen L, He Z, Qin L, Li Q, Shi X et al (2011b) Antitumor effect of malaria parasite infection in a murine lewis lung cancer model through induction of innate and adaptive immunity. PLoS One 6(9):e24407. doi:10.1371/journal.pone.0024407

Corthay A (2014) Does the immune system naturally protect against cancer? Front Immunol 5:197.

Coussens LM, Raymond WW, Bergers G et al (1999) Inflammatory mast cells upregulate angiogenesis during squamous epithelial carcinogenesis. Genes Dev 13:1382–1397

Curotto de Lafaille MA, Lafaille JJ (2009) Natural and adaptive foxp3+ regulatory T cells: more of the same or a division of labor? Immunity 30:626–635

Curtsinger JM et al (2003) Signal 3 determines tolerance versus full activation of naïve CD8 T cells: dissociating proliferation and development of effector function. J Exp Med 197:1141–1151

Dacosta NA, Kinare SG (1991) Association of lung carcinoma and tuberculosis. J Postgrad Med 37(4):185–189. PMid:1841965

Dunn GP, Bruce AT, Ikeda H, Old LJ, Schreiber RD (2002) Cancer immunoediting: from immunosurveillance to tumor escape. Nat Immunol 3:991

Dunn GP, Old LJ, Schreiber RD (2004a) The immunobiology of cancer immunosurveillance and immunoediting. Immunity 21:137

Dunn GP, Old LJ, Schreiber RD (2004b) The three Es of cancer immunoediting. Annu Rev Immunol 22:329

Endo Y et al (2008) Activation-induced cytidine deaminase links between inflammation and the development of colitis-associated colorectal cancers. Gastroenterology 135:889–898

Engle SJ et al (2002) Elimination of colon cancer in germ-free transforming growth factor β1-deficient mice. Cancer Res 62:6362–6366

Fantini MC et al (2004) Cutting edge: TGF-β induces a regulatory phenotype in CD4+ CD25–T cells through Foxp3 induction and down-regulation of Smad7. J Immunol 172:5149–5153

Fox BA, Sanders KL, Bzik DJ (2013) Non-replicating Toxoplasma gondii reverses tumor-associated immunosuppression. Oncoimmunology 2(11):e26296

Gabrilovich DI, Nagaraj S (2009) Myeloid-derived suppressor cells as regulators of the immune system. Nat Rev Immunol 9(3):162–174

Ghose P, Ali AQ, Fang R, Forbes D, Ballard B, Ismail N (2011) The interaction between IL-18 and IL-18 receptor limits the magnitude of protective immunity and enhances pathogenic responses following infection with intracellular bacteria. J Immunol 187(3):1333–1346

Greten FR et al (2004) IKKβlinks inflammation and tumorigenesis in a mouse model of colitis-associated cancer. Cell 118:285–296

Grivennikov SI, Karin M (2010) Inflammation and oncogenesis: a vicious connection. Curr Opin Genet Dev 20:65–71

Grivennikov S et al (2009) IL-6 and Stat3 are required for survival of intestinal epithelial cells and development of colitis-associated cancer. Cancer Cell 15:103–113

Gunn L et al (2012) Opposing roles for complement component microenvironment C5a in tumor progression and the tumor. J Immunol 189:2985–94

Hagemann T et al (2008) "Re-educating" tumor-associated Macrophages by targeting NF-kB. J Exp Med 205:1261–1268

Herr H et al (1995) Intravesical bacilli Calmette-Guerin therapy prevents tumor progression and death from superficial bladder cancer: ten-year follow up of prospective randomized trial. J Clin Oncol 13:1404–1408

Herrera LA, Benitez-Bribiesca L, Mohar A, Ostrosky-Wegman P (2005) Role of infectious diseases in human carcinogenesis. Environ Mol Mutagen 45:284–303

Hinrichs CS et al (2009) Type 17 CD8$^+$T cells display enhanced antitumor immunity. Blood 114:596–599

Ing R, Segura M, Thawani N, Tam M, Stevenson MM (2006) Interaction of mouse dendritic cells and malaria-infected erythrocytes: uptake, maturation, and antigen presentation. J Immunol 176:441–450

Inokuma H, Brouqui P, Drancourt M, Raoult D (2001) Citrate synthase gene sequence: a new tool for phylogenetic analysis and identification of Ehrlichia. J Clin Microbiol 39:3031–3039

Ismail N, Shurin MR (2012) Cancer and infection: friends or foes? Future Oncol 8(9):1061–1064. doi:10.2217/fon.12.110

Ismail N, Soong L, McBride JW, Valbuena G, Olano JP, Feng HM, Walker DH (2004) Overproduction of TNF-alpha by CD8$^+$ type 1 cells and down-regulation of IFN-gamma production by CD4$^+$ Th1 cells contribute to toxic shock-like syndrome in an animal model of fatal monocytotropic ehrlichiosis. J Immunol 172(3):1786–1800

Ismail N, Stevenson HL, Walker DH (2006) Role of tumor necrosis factor alpha (TNF-alpha) and interleukin-10 in the pathogenesis of severe murine monocytotropic ehrlichiosis: increased resistance of TNF receptor p55- and p75-deficient mice to fatal ehrlichial infection. Infect Immun 74:1846–1856

Ismail N, Crossley EC, Stevenson HL, Walker DH (2007) Relative importance of T-cell subsets in monocytotropic ehrlichiosis: a novel effector mechanism involved in Ehrlichia-induced immunopathology in murine ehrlichiosis. Infect Immun 75:4608–4620

Jarnicki AG, Lysaght J, Todryk S, Mills KH (2006) Suppression of antitumor immunity by IL-10 and TGF-beta-producing T cells infiltrating the growing tumor: influence of tumor environment on the induction of CD41 and CD81 regulatory T cells. J Immunol 177:896–904

Jung D et al (2014) Enhancement of tumor-specific T cell based – mediated immunity in dendritic cell heat mycobacterium tuberculosis vaccines by shock protein X. J Immunol 193(3):1233–1245.

Kalinski P et al (1999) T-cell priming by type-2 polarized dendritic cells: the concept of a third signal. Immunol Today 20:561–567

Kalluri R, Weinberg RA (2009) The basics of epithelial mesenchymal transition. J Clin Invest 119:1420–1428

Karin M, Lin A (2002) NF-nB at the crossroads of life and death. Nat Immunol 3:221–227

Karin M, Cao Y, Greten FR, Li ZW (2002) NF-nB in cancer: from innocent bystander to major culprit. Nat Rev Cancer 2:301–310

Khayr W, Haddad RY, Noor SA (2012) Infections in hematological malignancies. Dis Mon 58:239–249

Khong HT, Restifo NP (2002) Natural selection of tumor variants in the generation of "tumor escape" phenotypes. Nat Immunol 3:999

Kim JO et al (2007) Inhibition of Lewis lung carcinoma growth by Toxoplasma gondii through induction of Th1 immune responses and inhibition of angiogenesis. J Korean Med Sci 22:S38–S46

Klimesova K et al (2013) Altered gut microbiota promotes colitis-associated cancer in IL-1 receptor associated kinase M-deficient mice. Inflamm Bowel Dis 19:1266–1277

Komori J et al (2008) Activation-induced cytidine deaminase links bile duct inflammation to human cholangiocarcinoma. Hepatology 47:888–896

Kuper H, Adami HO, Trichopoulos D (2000) Infections as a major preventable cause of human cancer. J Intern Med 248:171–183.

Laouar Y, Sutterwala FS, Gorelik L, Flavell RA (2005) Transforming growth factor-β controls T helper type 1 cell development through regulation of natural killer cell interferon-γ. Nat Immunol 6:600–607

Levings MK, Bacchetta R, Schulz U, Roncarolo MG (2002) The role of IL-10 and TGF-β in the differentiation and effector function of T regulatory cells. Int Arch Allergy Immunol 129:263–276
Li Q, Verma IM (2002a) NF-nB regulation in the immune system. Nat Rev Immunol 2:725–734. 131
Li MO, Wan YY, Sanjabi S, Robertson AK, Flavell RA (2006) Transforming growth factor-beta regulation of immune responses. Annu Rev Immunol 24:99–146
Li Z, Chen L, Qin Z (2009) Paradoxical roles of IL-4 in tumor immunity. Cell Mol Immunol 6:415–422
Lin A, Karin M (2003) NF-nB in cancer: a marked target. Semin Cancer Biol 13:107–114
Liu X et al (2014) Cytokines as effectors and predictors of responses in the treatment of bladder cancer by bacillus Calmette-Guérin. Future Oncol 10(8):1443–1456
Luo X et al (2007) Dendritic cells with TGF-β1 differentiate naive $CD4^+$ CD25–T cells into islet-rotective $Foxp3^+$ regulatory T cells. Proc Natl Acad Sci U S A 104:2821–2826
Manickam A, Sivanandham M (2011) Mycobacterium bovis BCG and purified protein derivative-induced reduction in the CD80 expression and the antigen up-take function of dendritic cells from patients with cervical cancer. Eur J Obstet Gynecol Reprod Biol 159:413–417.
Martin-Orozco N et al (2009) T helper 17 cells promote cytotoxic T cell activation in tumor immunity. Immunity 31:787–798
Mattner J, Debord KL, Ismail N, Goff RD, Cantu C 3rd, Zhou D, Saint-Mezard P, Wang V, Gao Y, Yin N, Hoebe K, Schneewind O, Walker D, Beutler B, Teyton L, Savage PB, Bendelac A (2005) Exogenous and endogenous glycolipid antigens activate NKT cells during microbial infections. Nature 434:525–529
McBride JW, Walker DH (2011) Molecular and cellular pathobiology of Ehrlichia infection: targets for new therapeutics and immunomodulation strategies. Expert Rev Mol Med 13:e3
McBride JW, Ndip LM, Popov VL, Walker DH (2002) Identification and functional analysis of an immunoreactive DsbA-like thio-disulfide oxidoreductase of Ehrlichia spp. Infect Immun 70:2700–2703
Moossavi S, Rezaei N (2013) Toll-like receptor signalling and their therapeutic targeting in colorectal cancer. Int Immunopharmacol 16:199–209
Muranski P et al (2008) Tumor-specific Th17-polarized cells eradicate large established melanoma. Blood 112:362–373
Nam JS et al (2008) Transforming growth factor βsubverts the immune system into directly promoting tumor growth through interleukin-17. Cancer Res 68:3915–3923
Nandi B, Chatterjee M, Hogle K, McLaughlin M, MacNamara K, Racine R, Winslow GM (2009) Antigen display, T-cell activation, and immune evasion during acute and chronic ehrlichiosis. Infect Immun 77(10):4643–4653
O'Hara AM, Shanahan F (2006) The gut flora as a forgotten organ. EMBO Rep 7:688–693
Ohtsu N, Takaoka K, Segawa E, Hashitani S, Noguchi K, Kishimoto H, Urade M (2010) Antitumor effects of inhibitors of nitric oxide synthase or cyclooxygenase-2 on human KB carcinoma cells overexpressing COX-2. Oncol Rep 24:31–36
Okazaki IM, Kotani A, Honjo T (2007) Role of AID in tumorigenesis. Adv Immunol 94:245–273
Olano JP, Wen G, Feng H-M, McBride JW, Walker DH (2004) Histologic, serologic, and molecular analysis of persistent ehrlichiosis in a Murine model. Am J Pathol 165(3):997–1006
Old LJ, Boyse EA (1964) Immunology of experimental tumors. Annu Rev Med 15:167
Olver S, Apte S, Baz A, Kienzle N (2007) The duplicitous effects of interleukin 4 on tumour immunity: how can the same cytokine improve or impair control of tumour growth? Tissue Antigens 69:293–298
Overwijk WW, Restifo NP (2001) B16 as a mouse model for human melanoma. Curr Protoc Immunol Chapter 20:Unit 20.1
Paddock CD, Childs JE (2003) Ehrlichia chaffeensis: a prototypical emerging pathogen. Clin Microbiol Rev 16:37–64

Parsonnet J (1999) Introduction. In: Parsonnet J (ed) Microbes and malignancy. Oxford University Press, New York, pp 3–15

Pikarsky E et al (2004) NF-κB functions as a tumour promoter in inflammation-associated cancer. Nature 431:461–466

Pollard JW (2004) Tumour-educated macrophages promote tumour progression and metastasis. Nat Rev Cancer 4:71–78

Pritt BS, Sloan LM, Johnson DK, Munderloh UG, Paskewitz SM, McElroy KM, McFadden D, Binnicker MJ, Neitzel DF, Liu G, Nicholson WL, Nelson CM, Franson JJ, Martin SA, Cunningham SA, Steward CR, Bogumill K, Bjorgaard ME, Davis JP, McQuiston JH, Warshauer DM, Wilhelm MP, Patel R, Trivedi VA, Eremeeva ME (2011) Emergence of a new pathogenic Ehrlichia species, Wisconsin and Minnesota, 2009. N Engl J Med 365:422–429

Reddy BS, Weisburger JH, Narisawa T, Wynder EL (1974) Colon carcinogenesis in germ-free rats with 1,2-dimethylhydrazine and N-methyl-N′-nitro- N-nitrosoguanidine. Cancer Res 34:2368–2372

Reddy BS et al (1975) Colon carcinogenesis with azoxymethane and dimethylhydrazine in germ-free rats. Cancer Res 35:287–290

Rikihisa Y (2010) Anaplasma phagocytophilum and Ehrlichia chaffeensis: subversive manipulators of host cells. Nat Rev Microbiol 8:328–339

Roetynck S, Baratin M, Johansson S, Lemmers C, Vivier E et al (2006) Natural killer cells and malaria. Immunol Rev 214:251–263

Rook AH et al (1986) Effects of transforming growth factor β on the functions of natural killer cells: depressed cytolytic activity and blunting of interferon responsiveness. J Immunol 136:3916–3920

Seung L, Rowley D, Dubeym P, Schreiber H (1995) Synergy between T-cell immunity and inhibition of paracrine stimulation causes tumor rejection. Proc Natl Acad Sci U S A 92:6254–6258

Shankaran V et al (2001) IFNgamma and lymphocytes prevent primary tumour development and shape tumour immunogenicity. Nature 410:1107

Shibata S, Kawahara M, Rikihisa Y, Fujita H, Watanabe Y, Suto C, Ito T (2000) New Ehrlichia species closely related to Ehrlichia chaffeensis isolated from Ixodes ovatus ticks in Japan. J Clin Microbiol 38(4):1331–1338

Shurin MR (2012) Cancer as an immune-mediated disease. Immunol Targets Ther 1:1–6

Shurin GV, Ouellette CE, Shurin MR (2012) Regulatory dendritic cells in the tumor immunoenvironment. Cancer Immunol Immunother 61:223–230

Sica A et al (2008) Macrophage polarization in tumour progression. Semin Cancer Biol 18:349–355

Stevenson HL, Jordan JM, Peerwani Z, Wang HQ, Walker DH, Ismail N (2006) An intradermal environment promotes a protective type-1 response against lethal systemic monocytotropic ehrlichial infection. Infect Immun 74(8):4856–4864

Stevenson HL, Estes MD, Thirumalapura NR, Walker DH, Ismail N (2010) Natural killer cells promote tissue injury and systemic inflammatory responses during fatal Ehrlichia -induced toxic shock-like syndrome. Am J Pathol 177(2):766–776

Stutman O (1975) Immunodepression and malignancy. Adv Cancer Res 22:261

Sullivan NJ et al (2009) Interleukin-6 induces an epithelial-mesenchymal transition phenotype in human breast cancer cells. Oncogene 28:2940–2947

Suzuki Y, Kobayashi A (1985) Antitumor effect of intralesional injection with formalin-fixed Toxoplasma gondii organisms on Lewis lung carcinoma in Toxoplasma-infected mice. Cancer Lett 25(3):247–254

Suzuki E, Kapoor V, Jassar AS, Kaiser LR, Albelda SM (2005) Gemcitabine selectively eliminates splenic $Gr\text{-}1^+/CD11b^+$ myeloid suppressor cells in tumor-bearing animals and enhances antitumor immune activity. Clin Cancer Res 11:6713–6721

Takai A et al (2009) A novel mouse model of hepatocarcinogenesis triggered by AID causing deleterious p53 mutations. Oncogene 28:469–478

Tang D, Kang R, Coyne CB, Zeh HJ, Lotze MT (2012) PAMPs and DAMPs: signal 0s that spur autophagy and immunity. Immunol Rev 249(1):158–175

Thomas S, Popov VL, Walker DH (2010) Exit mechanisms of the intracellular bacterium Ehrlichia. PLoS One 5:e15775

Ugel S, Delpozzo F, Desantis G, Papalini F, Simonato F, Sonda N, Zilio S, Bronte V (2009) Therapeutic targeting of myeloid-derived suppressor cells. Curr Opin Pharmacol 9:470–481

Vesely MD, Kershaw MH, Schreiber RD, Smyth MJ (2011) Natural innate and adaptive immunity to cancer. Annu Rev Immunol 29:235–271.

Voronov E et al (2003) IL-1 is required for tumor invasiveness and angiogenesis. Proc Natl Acad Sci U S A 100:2645–2650

Wakeel A, Kuriakose JA, McBride JW (2009) An Ehrlichia chaffeensis tandem repeat protein interacts with multiple host targets involved in cell signaling, transcriptional regulation, and vesicle trafficking. Infect Immun 77:1734–1745

Wen B, Rikihisa Y, Mott J, Fuerst PA, Kawahara M, Suto C (1995) Ehrlichia muris sp. nov., identified on the basis of 16S rRNA base sequences and serological, morphological, and biological characteristics. Int J Syst Bacteriol 45:250–254

Yadegarynia D, Tarrand J, Raad I, Rolston K (2003) Current spectrum of bacterial infections in patients with cancer. Clin Infect Dis 37:1144–1145

Young MRI, Newby M, Wepsic TH (1987) Hematopoiesis and suppressor bone marrow cells in mice bearing large metastatic Lewis lung carcinoma tumors. Cancer Res 47:100–106. [PubMed: 2947676]

Zarek PE et al (2008) A2A receptor signaling promotes peripheral tolerance by inducing T-cell anergy and the generation of adaptive regulatory T cells. Blood 111:251–259

Zitvogel L, Tesniere A, Kroemer G (2006) Cancer despite immunosurveillance: immunoselection and immunosubversion. Nat Rev Immunol 6:715

Part V
Infection Agent-Based Vectors for Cancer Therapy

Chapter 16
Bacterial Cancer Therapy: How Patients Might Benefit from *Salmonella* Infections

Paulina Chorobik, Dominik Czaplicki, Karolina Ossysek, and Joanna Bereta

Abstract A growing tumor undergoes processes of heterogenisation and selection resulting in a complex system that facilitates further growth and progression. Immunosuppressive conditions belong to the hallmarks of this microenvironment, as they prevent efficient anti-cancer immunity. Eligible danger signal seems to be the missing link in the chain of events that would lead to successful elimination of cancer. The concept of bacterial cancer therapy is based on the ability of some microbial species to target tumor site and activate the cancer-specific response via pathogen-associated immunostimulatory signals. To date, a number of bacterial species have been shown to colonize tumor tissue, but strains of *Salmonella* are particularly interesting, since they meet all the requirements for an ideal tumor-targeting agent: a motile, facultatively anaerobic, intracellular microorganism that is prone to genetic manipulations. The most promising, *S.* Typhimurium, has multiple adaptations that are therapeutically relevant, including broad host specificity, specialized secretion systems and virulence factors with proapoptotic and immunomodulatory properties. Attenuation of wild-type strains has rendered them safe for preclinical and clinical use, while additional genetic modifications can add to their capacity to kill tumor cells and stimulate anti-cancer immunity. Given the recent developments in the field and the spectrum of possibilities offered by *S.* Typhimurium and its derivatives, it has a good chance of becoming a novel tool in the anticancer toolbox.

Keywords Bacterial cancer therapy • Immunotherapy • Danger signal • Cancervaccine • Tumor targeting • *Salmonella* Typhimurium • Attenuation

---

Dedication *In memory of Michał Bereta, who 'infected' us with the idea of therapeutic Salmonella*

P. Chorobik • D. Czaplicki • K. Ossysek • J. Bereta (✉)
Department of Cell Biochemistry, Faculty of Biochemistry, Biophysics and Biotechnology, Jagiellonian University, ul. Gronostajowa 7, 30-387 Kraków, Poland
e-mail: joanna.bereta@uj.edu.pl

M.R. Shurin et al. (eds.), *Infection and Cancer: Bi-Directorial Interactions*, DOI 10.1007/978-3-319-20669-1_16

Abbreviations

APC	Antigen-Presenting Cell
CEA	CarcinoEmbryonic Antigen
DAMP	Damage-Associated Molecular Pattern
DC	Dendritic Cell
IDO	Indoleamine 2,3-DiOxygenase
IFN-I	type I InterFeroN
IFNγ	InterFeroN gamma
IL	InterLeukin
LPS	LipoPolySaccharide
MDSC	Myeloid-Derived Suppressor Cells
MHC	Major Histocompatibility Complex
MMP	Matrix MetalloProtease
MTD	Maximum Tolerated Dose
PAMP	Pathogen-Associated Molecular Pattern
scFv	single chain variable fragment antibody
SCV	*Salmonella-Containing Vacuole*
shRNA	short hairpin RNA
SPI	*Salmonella Pathogenicity Island*
STAT3	Signal Transducer and Activator of Transcription 3
TTSS	Type III Secretion System
TAA	Tumor-Associated Antigen
TAM	Tumor-Associated Macrophage
TLR	Toll-Like Receptor
TNF	Tumor Necrosis Factor
TRAIL	TNF-Related Apoptosis-Inducing Ligand
Treg	regulatory T cell
VEGF	Vascular Endothelial Growth Factor

How Cancer Avoids Immunity

Cancer Cells May Elicit Immune Response – But Too Little and Too Late

When discussing the relationship between cancer and the immune system, we should bear in mind that the immune system has not evolved particularly to recognize and eliminate cancer cells. The roles of the immune system are to provide defense against pathogens, remove dying cells of the organism and mediate tissue healing. However, due to certain common elements of pathogen invasion and cancer growth, immune system is able to mount a specific, although usually ineffective, anti-tumor response (Goldszmid et al. 2014) and this fact is the basis for the belief that the development of active immunotherapies is possible.

Despite the nature of the trigger, carcinogenesis is considered a slow process, the first stages of which remain frequently ignored by the immune system because the initially mutated cells do not usually express immunogenic tumor-associated antigens (TAAs) and more importantly no danger signal accompanies the early stage of tumor development (Whiteside 2010; Chow et al. 2012). Dendritic cells (DCs) are necessary for the initiation of the adaptive immune response. Unlike immature DCs which phagocytose tissue antigens but are indolent in their processing and presentation, mature DCs are the most potent antigen presenting cells and have the ability to efficiently promote differentiation of effector T cells. In order to mature fully, DCs must encounter a danger signal: pathogen- or damage-associated molecules that interact with and stimulate Toll-like receptors and/or NOD-like receptors (Chow et al. 2012; Whiteside 2010).

The mechanism of early recognition of transformed cells is not known. It seems reasonable to expect that at the moment of recognition cancer cells already produce mutated proteins or proteins with modified epitopes or express carcinoembrional proteins that may play a role of tumor-specific antigens. It is hypothesized that the danger signal might be provided by dying tumor cells. According to the immunosurveillance hypothesis, the early recognition of transformed cells may lead to eradicating of developing tumors (Chow et al. 2012). However, the extent of this phenomenon is difficult to estimate.

Tumor Growth and Selection Result in a Highly Immunosuppressive Microenvironment

It is believed that frequently, due to weak immunogenicity of TAAs and delayed appearance of a danger signal, the anti-tumor adaptive immune response arises too late and is of too low magnitude to efficiently eradicate tumor cells. The induction of the specific but inefficient anti-tumor immune response results in the so-called cancer immunoediting, the evolutionary process that leads to the selection of the genetically fittest tumor cells. These may further progress into a highly aggressive, malignant population (Dunn et al. 2002; Chow et al. 2012).

The genetic and epigenetic changes that enable tumor escape from the immune system are twofold – they make tumor cells less sensitive to the effects of the immune system (e.g. through increased expression of anti-apoptotic proteins, decreased expression of MHC class I molecules) but they also disarm the immune system by inducing and supporting immunoinhibitory tumor microenvironment. Numerous tumors gain the ability to secrete molecules that limit their infiltration by the immune cells as well as disturb the immune cells maturation and activation.

It is worth noting that the tumor growth itself continuously modulates tumor microenvironment. Hypoxia, a drop in pH due to increased levels of lactate (the glycolytic end-product) (Choi et al. 2013), as well as the accumulation of extracellular adenosine are all hallmarks of tumor growth and all exert immunosuppressive effects by influencing the activities of the immune cells present in tumor millieu (Gabrilovich et al. 2012).

Hypoxia in the tumor tissue leads to necrosis, which induces the recruitment of monocytes and DCs to the site and triggers inflammatory response. Inflammation has paradoxical roles in tumor development (Balkwill and Mantovani 2012). Almost all proinflammatory cytokines and factors (TNF, IL-1, IL-6, nitric oxide) have been shown by numerous studies to have antitumor as well as tumor promoting activities. The net outcome of tumor-associated inflammation depends on the stage of the mutual relationship between tumor cells and the immune cells present in tumor microenvironment.

Cancer-Associated Immune Cells Facilitate Tumor Progression and Afflict Immunity

Similarly to the differences in phenotypic subsets of the immune cells which participate in the triggering and resolution of inflammation or during the conversion of acute to chronic inflammation, also diverse immune cell populations, often with opposite activities, are seen in tumor and tumor inflammatory microenvironment at different stages of its development.

However, in established tumors, most of the tumor-infiltrating immune cells will demonstrate immunosuppressive and tumor-promoting phenotype. Among them tumor-associated macrophages (TAMs), myeloid-derived suppressor cells (MDSCs) and T regulatory cells (Tregs) seem to play a crucial role in tumor progression.

Several lines of evidence indicate that in general TAMs have M2-like phenotype, characterized by a high expression of IL-10 and low expression of IL-12. Immunosuppressive activities of TAMs are also attained by the synthesis of TGF-β, iNOS-derived nitric oxide, indoleamine 2,3-dioxygenase (IDO) – a tryptophan-degrading enzyme, and B7-H1, B7-H3 and B7-H4 molecules – negative regulators of T cell immunity. TAMs may aid tumor cell proliferation by expression of growth factors such as EGF, bFGF, VEGFs and PDGF. The growth factors, including VEGF-A, C and D, together with TAM-derived matrix metalloproteases (MMP2, MMP9), and IL-8 (CXCL8) are also responsible for tumor angiogenesis and lymphangiogenesis, which enable tumor growth and spreading. MMPs and other proteases produced by TAMs such as uPA, plasmin and cathepsin B play a double role; they activate and/or release growth factors entrapped in extracellular matrix (ECM) and are responsible for the degradation of the basement membrane and ECM remodeling, thus facilitating tumor invasion and metastasis (Gabrilovich et al. 2012; Whiteside 2010; Galdiero et al. 2013). TAMs also produce PGE2, a prostaglandin with a broad immunoregulatory activity (Kalinski 2012). PGE2 suppresses Th1 cell-mediated immunity and modulates chemokine production, inhibiting recruitment of $CD8^+$, NK and Th1 cells while enhancing attraction of Tregs and MDSCs (Kalinski 2012).

MDSCs are a highly heterogeneous population of cells that can be subdivided into two major groups: monocytic (M-MDSCs) and granulocytic (G-MDSCs) (Youn and Gabrilovich 2010). MDSCs are defined as cells of myeloid origin characterized by the lack or reduced expression of markers of mature myeloid cells with a strong capacity to suppress cytotoxic T-cell responses (Gabrilovich et al. 2007). Mouse MDSCs are $CD11b^+Gr1^+$ and human MDSCs are $CD11b^+CD33^+$ and some subsets are $CD15^+$, $CD34^+$ or $CD31^+$. MDSCs, recruited from the bone marrow by tumor-derived attractants, migrate to tumor tissue and to lymph nodes, where they impede activation of T cells by DCs. MDSCs share certain immunosuppressive instruments with other cells; e.g. they produce TGF-β and IL-10. What is unique, M-MDSCs express simultaneously high levels of iNOS and arginase 1 (ARG1), which degrade L-arginine, the iNOS substrate (Gabrilovich et al. 2012). Upon L-arginine depletion iNOS produce both NO and superoxide radicals that interact to form peroxynitrite (Xia and Zweier 1997). This potent oxidant affects proliferation, migration and effector functions of T cells. For G-MDSCs, prolonged synthesis of reactive oxygen species (ROS), due to high expression of the NOX2 NADPH oxidase, is the major mechanism of immunosuppression. MDSCs also exert their effects through supporting the development of Tregs populations. Mouse MDSCs have been shown to promote clonal expansion of antigen-specific Tregs and conversion of $CD4^+$ lymphocytes into induced Tregs (Gabrilovich et al. 2012).

Tregs ($CD4^+CD25^+FoxP3^+$) represent a population of immune cells whose physiological role is to prevent autoimmunity and to maintain peripheral tolerance. Tregs support tumor growth by inhibiting activity of tumor-infiltrating effector T cells, DCs, and NK cells. They exert their effects both by contact-dependent suppression through FasL – Fas and PD1 – B7-H1 interactions and by secretion of soluble factors such as TGF-β and IL-10 (Biragyn and Longo 2012).

Also the development and activation of DCs are frequently impaired by tumor microenvironment. DCs in tumors and in tumor-draining lymph nodes show decreased ability to process antigens and to stimulate tumor-specific T-cell responses and what is more, they gain immunosuppressive phenotype (Ma et al. 2013).

It seems that in established tumors, tumor cells and various tumor-infiltrating immune cells have an intense cross-talk, instruct each other and fully cooperate to promote tumor growth. However, the existence of TAAs and the fact that at some stage of tumor development the immune system mounts both humoral and cellular anti-tumor response makes researchers believe that immunotherapies, which destroy the unfavorable balance between a high suppressive and a low activatory response may lead to tumor eradication and prevent its recurrence. This belief may be emphasized by the results of studies involving over 600 patients suffering from colorectal cancer that demonstrated an overwhelming correlation between long term survival of patients after curative surgery and the so called immunoscore reflecting the number of $CD8^+$ and CD45RO activated T cells in the tumor sections. In 5 years after surgery >86 % of patients with immunoscore 4 (the highest number of analyzed T cells), and only 27.5 % of patients with immunoscore 0 or 1 were still alive (Pages et al. 2009). Thus, it seems that the removal of tumor enabled existing effector cells to control residual disease. There are more examples of a correlation between strong

T cell infiltration of different cancers of epithelial origin and good clinical outcomes (Fridman et al. 2012).

The design and desired activities of a growing number of immunotherapeutic drugs and procedures already approved for cancer treatments and those under development include: (*i*) inhibition of tumor cell growth and induction of tumor cell death by antibody-dependent cell-mediated cytotoxicity (ADCC), complement-dependent cytotoxicity (CDC) and antibody-dependent cellular phagocytosis (ADCP) using TAA-specific monoclonal antibodies (mAb); (*ii*) expansion and activation of tumor-specific T cells through an adoptive transfer of autologous tumor-infiltrating lymphocytes (TILs) or genetically-engineered T cells as well as by application of anti-CTLA-4 or anti-PD1 mAb; (*iii*) improvement of TAA presentation and a strength of costimulatory signals utilizing DC-based vaccines; and (*iv*) affecting immunosuppressive tumor environment by small-molecule inhibitors of iNOS, ARG1 and IDO.

Another approach to tumor immunotherapy employs viral or bacterial infection. This more holistic strategy aims to simultaneously act on diverse components of the immune response and may have a number of therapeutic advantages, which are highlighted below.

Rationale for Bacterial Cancer Therapy

Bacteria Can Deliver the Danger Signal Needed for Efficient Cancer-Specific Immune Response

Since the discoveries of alpha-fetoprotein and carcinoembryonic antigen in the mid-twentieth century, a still-increasing number of TAAs has been identified as potential targets for vaccine strategies. However, due to poor antigenicity and self-tolerance, the antigen-based approach suffered major setbacks in the clinic. A growing body of evidence shows that the immune response against tumor antigens requires additional stimulation which would be able to break the tolerance towards the tumor and activate a robust, clinically meaningful antitumor immunity. Hence, the danger signal seems to be the missing link in the chain of events that would lead to successful cancerimmunotherapy (Matzinger 2012). Efficient immunostimulation, including antigen-presenting cells (APCs) activation that is a starting point for primary and secondary immune responses, requires the presence of a danger signal – a molecular trigger that can be of either endo- or exogenous nature. Cells and tissues undergoing stress, damage or necrotic death release a number of factors that can alarm the immune system, including heat-shock proteins, nucleotides, mitochondria-derived molecules, cytokines and products of extracellular matrix degradation; collectively, these factors are referred to as damage-associated molecular patterns (DAMPs). However, the same pattern-recognition receptors that detect DAMP-mediated danger signals can be also triggered by exogenous molecules derived from pathogens, known as pathogen-associated molecular patterns (PAMPs). This term covers a diverse group of compounds, including microbial and viral DNA with non-methylated CpG sites, as well as mannans, flagellin and bacterial cell wall components like lipopolysaccharide (LPS) and peptidoglycan. The pathogen-derived mediators have a potent immunostimulatory effect – upon interaction with pattern-recognition receptors including Toll-like receptors (TLRs), they strongly activate innate- and initiate adaptive immune responses. Not surprisingly, TLR ligands have been identified as potential cancer therapeutics with a few agents from this group already approved for human use. Up to date, those include: bacillus Calmette-Guérin (BCG), an attenuated strain of *Mycobacterium bovis*; picibanil (OK-432), a lyophilized preparation of *Streptococcus pyogenes*; monophosphoryl lipid A (MPL), a detoxified LPS derivative of *Salmonellaminnesota*; and imiquimod (R-837), a synthetic imidazoquinoline. The drugs are followed by a growing number of novel candidates (Vacchelli et al. 2013), however, the efficacy of TLR agonists as single agents for cancer therapy has so far been limited. Moreover, the activation of TLRs seems to be a double-edged sword, as – depending on the tissue context and tumor type – it can promote tumor growth rather than the antitumor response (Lu 2014). For example, prolonged activation of TLR4 via signaling by both PAMPs and DAMPs creates tumor-promoting microenvironment that includes immunosuppressive cytokines, as well as cell populations like Tregs and MDSCs (Mai et al. 2013). However, TLR ligands delivered with an infection have been proved to be clinically effective – intravesical treatment with BCG, which operates as a mixed TLR2/TLR4 agonist, is currently the gold standard of care in non-muscle invasive bladder cancer (Sylvester 2011). Bacterial infection within the tumor tissue can evoke both the damage- and the pathogen-related danger signals; it has been shown that microorganisms within the tumor tissue can indeed promote an inflammatory reaction and potentiate the anti-tumor host response (Avogadri et al. 2005). Notably, bacterial cancer therapies offer many more mechanisms of action than solely the immunostimulatory effect of TLR activation – the most important being the ability to directly influence the suppressive microenvironment.

Many Features of Bacteria Provide Benefits for Cancer Treatment

Successful delivery of the danger signal into the tumor tissue is possible due to a number of bacterial mechanisms that can be therapeutically relevant. Bacteria-based treatments can benefit from microbial metabolism, motility and sensitivity to address the key issues in cancer therapy: low specificity towards cancer tissue and insufficient penetration of the tumor, both of which are limiting to currently used treatment modalities. Cancer cells form a complex and heterogeneous system that is poorly accessible to chemotherapeutic drugs (Saunders et al. 2012), while motile microbial organisms are able to cross biological barriers, act against hemodynamic gradients and preferentially accumulate in the tumor tissue. In contrast to passively-diffusing therapeutics, which produce relatively large drug concentrations in the bloodstream and relatively low drug concentrations in the tumor, bacteria offer unique mechanisms that can facilitate site-specific treatment, highly focused on the tumor and safe to other tissues. The natural ability of bacteria to receive signals via chemoreceptors can be used as a tool to effectively target the unique microenvironment formed by the tumor tissue. Anaerobic bacterial species are able to thrive in the hypoxic areas of tumors, while auxotrophic strains can recognize the tumor microenvironment as a source of nutrients. Moreover, live microorganisms are metabolically active and able to perform specific actions at the tumor site, e.g. produce a prodrug-converting enzyme or a cytotoxic molecule. Strains derived from intracellular pathogens can infect tumor cells and deliver specific proteins or genes into the tumor cells (St Jean et al. 2008). Importantly, bacteria are susceptible to antibiotic treatment and therefore fully manageable in the clinical setting – therapy can be stopped at the onset of adverse effects or when is no longer needed. Taken altogether, the use of bacteria as anticancer agents might have multiple advantages over other therapeutic approaches. Strains of *Salmonella* are particularly apt to this task, as they can readily address all requirements for an ideal tumor-targeting agent: a motile, facultatively anaerobic, intracellular microorganism that is prone to genetic manipulations (Forbes 2010).

Salmonella – Portrait of a Conqueror

Salmonella *Species Are Broad-Host, Perfectly Adapted Intracellular Pathogens*

Bacteria from genus *Salmonella* are enteric pathogens which exploit the common basic strategy to colonize vertebrates, and are able to survive in the host digestive tract as well as in the intracellular niche. *Enterobacteriacae* family are commensal or pathogenic bacteria, with *Escherichia* being the closest known genus to *Salmonella*, since the two branched 120–160 million years ago from the common ancestor (Desai et al. 2013; McClelland et al. 2001; Ochman and Wilson 1987). Two subgroups within the genus are recognized as species: *Salmonella bongori* and *Salmonella enterica. S. bongori* strains are typically although not exclusively isolated from reptiles. Isolates more significant for human pathogenesis belong to *S. enterica* species, divided into six subspecies. *S. enterica* subspecies *enterica* are isolated from warm-blooded vertebrates including humans. Gene inactivation and lateral gene transfer are the main driving forces shaping the genome through *Salmonella* evolution. If we define the core genome as genes which perform the household functions and compare these genes in *S. enterica* and *Escherichia coli*, we find a mere 10 % difference, while there is only a 1 % difference among *S. enterica* serovars (Baker and Dougan 2007). Lateral gene transfer is a source of variability within *S. enterica* species and 11 % of serovar *S.* Typhimurium strain LT2 genes are missing in *S.* Typhi strain CT18 (McClelland et al. 2001). The genetic differences reflect adaptation driven by the interaction with invaded host organisms. Not only physical barriers are defeated by the pathogen, but the resident immune cells of the host, which are utilized in favor of the invader and for its successful propagation.

S. enterica serovars differ in their host preferences and pathogenesis. Most of the strains which are of interest for preventive and therapeutic vaccine development belong to serovar Typhimurium or Typhi. *S. enterica* subspecies *enterica* serovar Typhimurium, for the sake of brevity further referred to as *S.* Typhimurium, is a broad-host range serovar infecting both humans and other animals, and giving rise to enterocolitis or asymptomatic infection. In humans *S.* Typhimurium is associated with gastroenteritis manifested as short-term, acute inflammation limited to the intestine. On the contrary, host-restricted *S. enterica* subspecies *enterica* serovar Typhi (*S.* Typhi) cause systemic infection known as typhoid fever and humans are the only known host.

Systemic S. *Typhimurium Disease in Humans Is Associated with Immune Deficiency*

Some aspects of *Salmonella* interaction with the host immune system were recognized as the source of the host range diversity. Host immune status and the responses to bacteria are crucial for the outcome of infection. Systemic *Salmonella* infection is usually host-dependent (Ruby et al. 2012) and *S. enterica* serovars that are not host specific, such as *S.* Typhimurium, are associated primarily with disease in young animals, suggesting their non-optimal adaptation to a mature immune system, while host-specific serovars (*S.* Typhi) tend to be more virulent (Baumler et al. 1998).

S. Typhimurium infection in immune competent humans leads to gastroenteritis since the infection does not spread outside the intestine lamina propria (Ruby et al. 2012). However, the infection can lead to bacteremia and severe invasive disease in immune compromised individuals (Gordon 2008). In some laboratory mouse strains *S.* Typhimurium infection results in systemic inflammation congenial to *S.* Typhi typhoid fever in humans. Therefore *S.* Typhimurium infection in mice has become a widely accepted model of typhoid fever and immunity to acute *Salmonella* infection. In both mice and humans the susceptibility to intracellular pathogens is associated with polymorphism of *Nramp1* gene (natural resistance associated macrophage protein 1, or Slc11a1). Nramp1 protein is present in the phagosome membranes in macrophages and neutrophils. It impairs the growth of intracellular pathogens such as *Salmonella*, which rely on survival and replication inside the phagosome (Forbes and Gros 2003). In mice resistant to lethal *Salmonella* infection, wild-type Nramp1 removes divalent metal cations essential for bacterial growth from bacteria-containing phagosome. Oral infection with wild-type *S.* Typhimurium in mice with wild type *Nramp1* results in the development of a long-lasting chronic carrier state. The mice showed no apparent signs of illness during the acute phase of infection and oral lethal dose is a few thousand times higher than for mice with mutated *Nramp1* (Monack et al. 2004). On the contrary, mice which are susceptible to acute systemic *S.* Typhimurium infection carry mutated allele of *Nramp1*. Commonly used laboratory mice such as Balb/c and C57Bl/6 have mutated *Nramp1*, and wild-type *S.* Typhimurium cause lethal disease soon after infection, while attenuated strains produce a chronic, persistent infection (Ruby et al. 2012; Voedisch et al. 2009).

Intracellular Lifestyle Predisposes Salmonella *to Therapeutic Vaccine Development*

The intracellular pathogen invasion scenario comprises of seemingly contradictory events decisive for the balance between survival, replication, spread of the pathogen and the host welfare. Regulated expression of virulence factors allows *Salmonella* to reach and conquer an intracellular niche and use it for replication and propagation of invasive phenotype. The molecular interactions of bacterial and host factors along with the aforementioned host susceptibility traits indicate the vast plasticity of interactions which can be exploited for the benefit of therapeutic applications. The superb ability of some attenuated *S.* Typhimurium strains for preferential colonization of solid tumors and the intrinsic immune stimulatory properties prompted the development of anticancer biotherapeutics, with the hope to boost the suppressed immune responses. We will briefly present the various strategies of virulence attenuation, their consequences on the immunity and the results of pre-clinical efficacy of *Salmonella*-based cancer therapeutics.

Salmonella *Virulence Factors Stimulate Immunity and Mediate Intracellular Propagation*

The shortest possible description of *Salmonella* – a Gram-negative, facultatively intracellular, motile enteropathogen – well emphasizes the major factors, which contribute to its virulence.

The molecules common to *Salmonella* and other bacterial pathogens, that fall into the category of PAMPs, are: LPS (endotoxin), flagellin, which constructs flagellae, outer-membrane proteins belonging to Omp family, fimbrial- and non-fimbrial adhesins and non-methylated CpG sequences characteristic for bacterial DNA (Fig. 16.1).

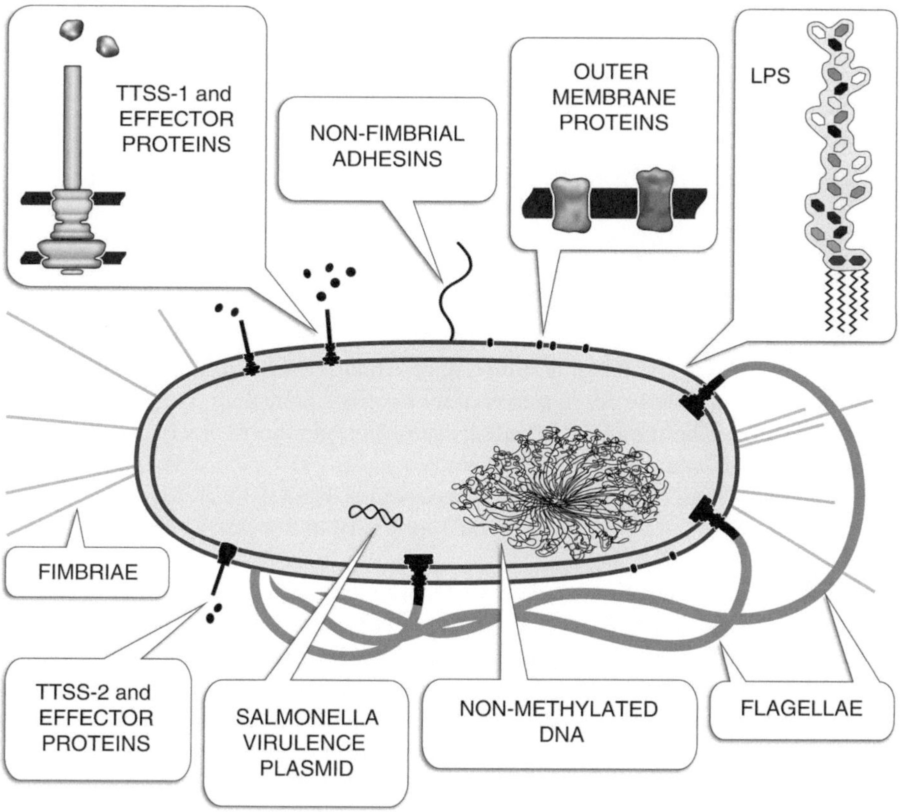

Fig. 16.1 *Salmonella* Typhimurium virulence factors and immunomodulators

The virulence determinants play a bimodal role, as they are important both for a successful bacterial invasion and the initiation of host immunity. *Lipopolysaccharide* confers bacterial resistance to the serum components and antimicrobial peptides Additionally, LPS on bacterial surface as well as released from bacteria during cell division and upon death binds to TLR4 and activates the expression of pro-inflammatory cytokines, chemokines and co-stimulatory molecules. *Fimbriae* and *non-fimbrial adhesins* facilitate colonization of intestinal mucosa and adhesion to epithelial cells (van der Velden et al. 1998; Wagner et al. 2011). *Flagellum*-mediated motility improves the invasion of epithelial cells and the flagellar structural protein, flagellin, binds to TLR5 and activates the expression of pro-inflammatory mediators in epithelial cells (Steiner 2007). Furthermore, the adaptive immune responses are stimulated by *S.* Typhimurium*outer membrane proteins*, which activate DCs maturation (Lee et al. 2010). Binding of *non-methylated bacterial DNA* to intracellular TLR9 stimulates antigen presentation by DCs, increases the expression of surface TLR9 on epithelial cells and stimulates them to secrete chemokines (Lahiri et al. 2010; Ewaschuk et al. 2007). Invasion of epithelial and phagocytic cells is mediated primarily by type three secretion system 1 (*TTSS-1*) effector proteins which harness the host cell cytoskeleton to trigger the uptake and formation of *Salmonella* containing vacuole (SCV). Following the internalization the maturation and trafficking of SCV is guided by *TTSS-2* effector proteins to set up a niche permissive for bacterial replication and to delay killing. TTSS-1 and -2 effector proteins regulate the onset of immune response owing to both activatory and inhibitory actions that stimulate cytokine secretion, induce or manipulate the migration of infected cells (Schleker et al. 2012). *Plasmid*-encoded factors contribute to virulence by interfering with cytoskeletal proteins, mediating resistance to complement-mediated killing or encoding fimbrial proteins.

Elements in the bacterial cell diagram were scaled up relative to the bacterial cell length, e.g. TTSS-1 and flagellar motor, approximately threefold, the widths of flagellar filament and fimbriae, approximately twofold. Schematic representations of TTSS-1 and flagellar motor are based on cryo-electron microscopy data and models presented in (Kawamoto et al. 2013; Radics et al. 2014; Schraidt and Marlovits 2011; Thomas et al. 2006). Similar reconstruction of TTSS-2 is not yet available. Structure of LPS from *S. enterica* serovar Typhimurium was described in (Olsthoorn et al. 2000).

These molecules constitute a danger signal to leukocytes when bound to pattern recognition receptors (PRR) either in their plasma membrane, as Toll-like receptors (TLRs), or to intracellular receptors (NOD-like receptors and intracellular TLRs). Moreover, *Salmonella* has extraordinary genes clustered in pathogenicity islands. These genes encode genus-specific virulence factors, which are strictly related to invasion, intracellular survival and replication. Two crucial for pathogenesis and best known *Salmonella* pathogenicity islands are island 1 and 2 (SPI1 and SPI2), encoding proteins of two distinct type three secretion systems, TTSS-1 and TTSS-2 respectively (Fig. 16.1). Each TTSS consists of proteins that constitute the secretion machinery (structural proteins), chaperones and effector proteins. Type three secretion system structural proteins form a needle-like molecular syringe which delivers the effector proteins into the cytoplasm of eukaryotic cell. SPI effectors harness the host cell signaling pathways and manipulate vesicular trafficking to establish the intracellular niche permissive for bacterial replication.

Type Three Secretion Systems Mediate Invasion and Intra-Phagocyte Survival

Salmonella pathogenicity island 1-encoded proteins facilitate invasion of phagocytes and force bacterial uptake by non-phagocytic epithelial cells of intestinal lining by so called "trigger"-mediated invasion. SPI1 effectors injected by TTSS-1 into the cytoplasm of a eukaryotic cell interfere with actin remodeling and manipulate the cytoskeleton leading to membrane ruffling and formation of a vacuole. After being engulfed by a phagocyte or an epithelial cell, *Salmonella* resides inside the phagosome, which is modified due to the activities of *Salmonella* proteins and is therefore termed *Salmonella* containing vacuole (SCV). Maturation and maintenance of SCV is driven by TTSS-2 effector proteins that manipulate SCV trafficking to endocytic compartment. *Salmonella*-induced filaments (SIFs) are formed, which are the membrane tubules expanding from the surface of SCV and enriched in late endosomal markers (Garcia-del Portillo et al. 1993; Mota et al. 2009). Further steps of SCV development involve dynein-mediated transport along microtubules triggered by SPI1 and SPI2 TTSS effectors resulting in the juxtanuclear positioning of the SCV. *Salmonella*-induced tubules interact with the post-Golgi trafficking in epithelial cells, thus the role for acquiring nutrients and membranes or to control host immune response was suggested (Mota et al. 2009). Intra-phagocyte survival of *S.* Typhimurium may serve to modulate the onset of adaptive immune response, due to the activity of TTSS-2-delivered effector protein SseI, which inhibits the adhesion of infected phagocyte, affecting its migration efficiency and communication with other immune cells (McLaughlin et al. 2009). In some infected epithelial cells not all *Salmonella* cells are enclosed within SCV, but some replicate in the cytoplasm. These cytoplasmic bacteria are primed for infection of adjacent cells and ready to be shed to the environment, due to the induction of TTSS-1 and flagellar motility (Knodler et al. 2010).

The spatial and temporal regulation of virulence genes is governed by a few two-component systems and regulatory feedbacks. For example, the two-component regulatory system of the membrane sensor PhoQ and the response cytoplasmic kinase PhoP senses the transition from an extra- to intracellular environment and converts it into a transcriptional activation signal for many genes including SPI2. The LPS biosynthesis pathway is modulated as well and less negatively charged LPS molecules are produced, with lower affinity to cationic antimicrobial peptides (Shafer et al. 1984).

S. *Typhimurium Infection Resolves in the Intestine or Disseminates into Systemic Disease*

Over the natural course of acute typhoid-like infection in humans or mice, the initial step of invasion starts in the small intestine where non-phagocytic epithelial cells and phagocytic M cells of Peyer's patches, resident phagocytes within lamina propria and tissue-associated DCs, are infected. M cells are the main route of entry for invasive, i.e. SPI1-proficient, *Salmonella* (Wick 2011). Once *Salmonella* enters the subepithelial dome of Peyer's patches, it faces resident cells of intestinal lymphoid tissue, mainly the DCs, which transport bacteria to mesenteric lymph nodes (MLN). The outcome of the infection is significantly affected by the ability of MLN to control the bacterial load (Voedisch et al. 2009; Wick 2011). At this stage the infection can spread to extraintestinal tissues leading to systemic disease, if the bacterial growth is not sufficiently restrained. $CD18^{+}$macrophages and DCs play a major role in transporting *S.* Typhimurium through lymphatics and blood stream to the spleen, liver and bone marrow (Ruby et al. 2012). After reaching the spleen and liver

Salmonella establishes the infection foci, where it multiplies mainly inside the phagocytes. Cell death of infected phagocytes enables bacteria release and the formation of new infection foci in the liver and spleen (Mastroeni et al. 2009; Salcedo et al. 2001).

Salmonella *Attenuation Improves Safety and Tumor Colonization*

S. Typhimurium is a relatively mild pathogen in immune competent organisms, but systemic infection in immune compromised individuals comprises a serious health risk. Therefore the virulence of therapeutic strains is attenuated in order to achieve the proper safety profile, restrain the natural ability for spread between hosts and to enable the termination of the treatment. Moreover, attenuation provides an opportunity to redirecting the bacterial colonization, as attenuated *S.* Typhimurium strains colonize solid tumors preferentially over internal organs.

The balance between safety and immunogenicity is critical to achieve eligible immune responses. Over-attenuating may lead to compromised invasion, colonization or insufficient immune stimulation. For instance, in the experimental protective *S.* Typhi vaccine, simultaneous disruption of aromatic amino acid synthesis pathway and either purine synthesis pathway or virulence regulatory factors, by genetic disruption of *aroA*, *purA* or *phoP/phoQ* loci, respectively, resulted in the restriction of side effects, but at the same time decreased immunogenicity preventing the induction of vaccine-specific immunity (Galen and Curtiss 2013).

Attenuation is achieved either by chemical mutagenesis or by genetic engineering techniques. Regardless of the methodology, attenuation results from metabolic and/or virulence deficiencies. Disruption of genes coding for the proteins essential for bacterial fitness, as the enzymes for biosynthesis of amino acids, e.g. *aroA* mutant strains, leucine and arginine-dependent mutant strain A1-R (described in detail later in this chapter); or purines – *purI*, *purD* mutant strains, results in decreased virulence. Likewise, pathogenicity is directly impaired by the inactivation of virulence genes, such as *msbB* in mutant strains which produce modified LPS, or regulatory genes (catabolite repression regulators *crp* and *cya* or two component regulatory system for virulence genes, *phoP/phoQ*).

Frequently the candidate therapeutic *S.* Typhimurium strains are equipped with heterologous effector therapeutic protein produced by the bacterial translation apparatus. The expression of heterologous proteins inevitably raises a metabolic burden and forced production of foreign protein often results in restrained growth or impaired invasiveness, further aggravating the attenuation. In order to balance the safety and immunogenicity and avoid the risk of over-attenuation, precisely regulated or delayed expression systems were engineered.

S. Typhimurium lipopolysaccharide synthesis pathway involves more than 30 enzymes and since it is the important structural and virulence factor, altering its structure results in attenuation. LPS structural diversity among microorganisms and its differential recognition by the receptor complex has been denoted as the factor which determines the outcome of the infection (Miller et al. 2005). The conservative core of LPS molecule is built of lipid A and oligosaccharides (core sugars), while *Salmonella* serotype-specific portion lies within O-antigen polysaccharide. Antimicrobial host responses are initiated through lipid A binding to membrane receptor complex TLR4-MD2-CD14 present on many cell types, among them DCs and macrophages, which secrete proinflammatory cytokines including TNF.

Modification of LPS structure in order to decrease the endotoxic shock induction has already been applied with satisfactory results, as *S.* Typhimurium VNP20009 strain was safely administered to cancer patients (Toso et al. 2002). The strain is attenuated and auxotrophic due to the disruption of lipid A myristoylation enzyme (inactivation of *msbB* gene) and purine synthesis pathway gene deletion (*purI*), respectively. Intravenous administration of VN20009 to tumor bearing mice results in tumor colonization, with the ratio of intratumoral to intrasplenic accumulation exceeding 1000. However, in humans the clinical benefit was not observed, presumably due to insufficient tumor colonization (Toso et al. 2002). The clinical trial will be discussed in detail later in this chapter.

In some studies, intratumoral hemorrhage induced by TNF was attributed to intratumoral *Salmonella* accumulation (Leschner et al. 2009). The ability of mouse and human to recognize the same lipid A molecule differs. Furthermore, the length and number of acyl side chains is important for TLR4 signaling in humans. Additional diversity in TLR4 responsiveness to LPS comes from the natural gene polymorphism which alters TLR4 signaling outcome in humans (Miller et al. 2005). Recently it has been proposed that the administration of live VNP20009 mixed with molecular wild-type lipid A could overcome the poor clinical tumor targeting (Zhang et al. 2013a). *Salmonella* accumulation in subcutaneous 4 T1 tumors was lipid A-dose-dependent and bacteria distribution was more homogenic after co-administration. However, the study was completed 2 days after bacteria inoculation, therefore a long-term therapeutic effect was not revealed.

Salmonella ***Is Able to Induce Several Different Death Modalities in Infected Cells***

During natural infection *Salmonella* enters two types of cells: intestine epithelium including specialized M cells, and immune cells: DCs and macrophages. *Salmonella*, as an intracellular pathogen, does not immediately kill the cell it has invaded, but rather delays an onset of cell death to earn time to replicate, escape and infect new cells. However, the interplay between infected host cells and bacteria eventually leads to the death of the eukaryotic cells including macrophages. The ability of engulfed *Salmonella* not only to resist the antimicrobial activity of macrophages, but also to proliferate inside and induce the death of these professional phagocytes is the key strategy of the pathogen to circumvent innate immune defense (Lindgren et al. 1996).

Different types of cell death are induced in *Salmonella*-containing epithelial cells and macrophages (Ramos-Morales 2012). In cultured human epithelial cells *Salmonella* triggers an *apoptotic cell death program* with all the hallmarks of this process including exposure of phosphatidylserine on the cell surface, activation of effector caspase-3, fragmentation of DNA, depolarization of mitochondrial membrane and degradation of cytokeratin 18. However, apoptosis is turned on after a lag period of between 12 and 24 h following bacterial entry (Kim et al. 1998; Paesold et al. 2002). A number of virulence factors stimulate apoptosis by affecting general protein synthesis, inducing actin depolimerization, and tipping the balance between expression of pro- and antiapoptotic proteins. Importantly, some bacterial proteins counteract a rapid induction of apoptosis: SopB activates pro-survival kinase Akt and AvrA inhibits c-Jun-N terminal kinase, which, when activated by stress stimuli, is involved in apoptosis (Ramos-Morales 2012).

In macrophages *Salmonella* Typhimurium induces a proinflammatory form of programmed cell death – so-called *pyroptosis* or caspase-1-dependent cell death. This process is induced by TTSS-1-expressing bacteria and occurs in 1–2 h postinfection. Activation of caspase-1 is responsible for maturation of two cytokines: IL-1β and IL-18 and results in cell lysis accompanied by the release of the active proinflammatory mediators (Cookson and Brennan 2001). Pyroptosis requires activation of NLRC4 inflammasome by flagellin and a rod component PrgJ (Zhao et al. 2011). However, the expression of these genes is repressed during systemic infection and in such situation the death of macrophages occurs not earlier than 12–16 h postinfection (Wynosky-Dolfi et al. 2014). This delay is advantageous for *Salmonella* and is in fact caused by bacteria. It has recently been demonstrated that bacterial tricarboxylic acid (TCA) cycle enzymes impair, by a yet unknown mechanism, an expected rapid canonical activation of NLRP3 inflammasome (Wynosky-Dolfi et al. 2014). Instead, delayed caspase-11- and type I IFN (IFN-I)-dependent noncanonical activation of NLRP3 inflammasome takes place. In the proposed scenario *Salmonella* through TLR4 stimulates the expression of both IFN-I and caspase-11. IFN-I induces the synthesis of a group of GTPases known as guanylate binding proteins (GBP), which mediate the lysis of *Salmonella*-containing vacuole. Cytosolic LPS of released bacteria indirectly activates caspase-11/NLRP3 inflammasome followed by activation of caspase-1 and macrophage death (Meunier et al. 2014). It has also been proposed that IFN-I mediates another type of *Salmonella*-induced cell death in macrophages – so-called *necroptosis* or programmed necrosis. This type of cell death is triggered by the formation of a necrosis-inducing complex comprised of receptor-interacting protein kinase 1 (RIPK1) and RIPK3 (Lu and Walsh 2012), which is followed by mitochondrial fragmentation (Wang et al. 2012). In contrast to apoptosis, both pyroptosis and necroptosis trigger inflammatory responses (Hu and Zhao 2013).

Autophagy Is an Important Cellular Process Upon Salmonella *Infection*

Until recently, autophagy was viewed as a process whose primary function in the cell is to degrade faulty organelles and cytosolic macromolecular aggregates as well as to prevent cell death by providing amino acids and other basic compounds during nutritional deprivation (Deretic 2011). On the other hand, due to the fact that autophagosomes and biochemical autophagy markers have often been observed in dying cells, autophagy has been regarded as, distinct from apoptosis, programmed cell death, termed "type II programmed cell death" or "autophagic cell death" (Ryter et al. 2014). However, a growing body of research indicates that autophagy – although often accompanies cell death – usually is not its cause. Therefore, the term "death with autophagy" seems to be more proper in the vast majority of cases than "death by autophagy" (Kroemer and Levine 2008).

Apart from its importance for the cell maintenance, autophagy is presently recognized as a process which plays important roles in innate and adaptive immunity. It has been proposed that autophagy acts as an evolutionary ancient system specialized to eliminate intracellular bacteria and viruses. It also contributes to the presentation of endogenously expressed antigens via major histocompatibility complex class II (MHC-II) (Deretic 2011; Espert et al. 2007). Thus, it is not surprising that autophagy has been observed in macrophages as well as in epithelial cells, fibroblasts and melanoma cells subjected to *Salmonella* infection (Birmingham et al. 2006; Hernandez et al. 2003; Lee et al. 2014; Thurston et al. 2009).

Autophagy involves three stages: (*i*) formation of an initiator crescent membrane (phagophore), (*ii*) growth of the phagophore to enclose cargo in the completed autophagosome and (*iii*) conversion to the autolysosome by fusion with late endosomes or lysosomes. A phagophore membrane contains phosphatidylethanolamine-anchored LC3-II protein, which interacts with autophagic adapter proteins that recognize targets marked for autophagy by ubiquitination (Deretic 2011).

In epithelial cells a subset of invading *Salmonella* is released to the cytosol from TTSS-1-damaged SCV shortly after infection (Birmingham et al. 2006). It has been shown that in HeLa cells bacteria that escaped SCV have become ubiquitin-coated and targeted to the autophagy pathway by p62 and nuclear domain 10 protein 52 (NDP52) adapters (Zheng et al. 2009; Thurston et al. 2009). Alternatively, diacylglycerol present in SCV membrane may constitute a signal for antibacterial autophagy (Shahnazari et al. 2010). A number of reports indicate that although autophagy is a transient process peaking in 1–2 h after *Salmonella* infection, it restricts intracellular growth of bacteria (Birmingham et al. 2006; Thurston et al. 2009; Zheng et al. 2009). Intriguingly and in contrast to the previous reports, Yu et al. (2014) have recently demonstrated, using a live-cell imaging method, that cytosolic *Salmonella* associated with autophagy components p62 and LC3 replicated efficiently in HeLa cells and, what is more, the replication has been diminished when p62 or LC3 availability has been decreased (Yu et al. 2014). This discrepancy may result from the different time schedules of the experimental design and may be explained, at least partially, by the fact that autophagy regulation is associated with the metabolic status of the cell where amino acid starvation seems to be the key trigger of the program. The transient character of cytosolic amino acid depletion induced by *S.* Typhimurium infection may explain the transient execution of autophagy. How the bacteria take advantage of certain autophagy elements at later times postinfection remains to be established.

Although autophagy has often been indicated as a possible cause of death of *Salmonella*-infected cells, in light of current knowledge it should rather be regarded as a natural element of the cell response to the infection that may help in pathogen eradication.

Cell Death Induced by Salmonella *Can Promote Cross-Presentation of Tumor Antigens*

The studies on the types of death caused by *Salmonella* were performed mostly using wild type bacteria or bacteria with knockouts in specific genes or gene clusters whose role in *Salmonella's* deadly potential was under investigation. Therefore, caution is required if the results are extrapolated to the potential effects of attenuated strains, which are candidates for clinical applications. However, the understanding of the biochemical consequences of particular modes of attenuation allows presuming that at least some of the attenuated strains will execute the death programs in a similar way to the wild type strains.

The primary asset of *Salmonella*-based therapies could be an enhancement of tumor antigen cross-presentation by DCs that have taken up cancer cells dying due to infection. Much effort has been made to understand the mechanisms of optimal cross presentation. Some researchers indicate specialized subset of DCs, namely $CD8^+$ DCs in mice and thrombomodulin-expressing DC in humans, as possessing superior capability to cross-present antigens (Villadangos and Shortman 2010; Joffre et al. 2012), whereas others postulate that all DCs may be efficient cross-presenters when appropriately activated by CD40L, inflammatory cytokines, and TLR ligands (Dresch et al. 2012). DCs can ingest dying cancer cells and cross-present a panel of tumor antigens; since necrotic cells are a rich source of danger signals, necrosis would be expected as the type of tumor cell death with the highest potential to activate DCs to efficient cross-presentation. Surprisingly, numerous reports from *in vitro* and *in vivo* studies showed that tumor apoptotic cells are better at facilitating cross-presentation of tumor antigens to $CD8^+$ T cells than their necrotic counterparts (Spel et al. 2013). The higher immunogenicity of apoptotic cells may result from their CLEC9A-mediated targeting to storing phagosomes, in which mild pH and reduced rate of proteolysis constitute proper conditions for optimal cross-presentation of cell-associated antigens (Spel et al. 2013). Interestingly, the exposure of DCs to IFN-I further enhances this process. IFN-I seems to prolong the presence of apoptotic cells inside DCs, supporting localization of MHC class I molecules to antigen-storage compartment as well as promoting DCs survival

(Lorenzi et al. 2011). *Salmonella* may strongly promote cross-presentation of tumor antigens since it: (*i*) induces apoptosis in cells of epithelial origin, (*ii*) stimulates expression of IFN-I both in macrophages and in non-phagocytic cells, (*iii*) provides ligands for TLRs, and (*iv*) stimulates release of proinflammatory cytokines from infected macrophages. Additionally, DAMPs released from necrotic tumor cells, which appear independently of *Salmonella* infection or originate from non-phagocytosed bacteria-infected apoptotic cells may further increase maturation and activation of DCs. It is noteworthy that also autophagy occurring in antigen-donor cells (e.g. tumor cells or virus-containing cells) may support antigen cross-presentation by DCs immunized with these cells. Uhl et al. demonstrated that virally infected fibroblasts undergoing autophagy facilitated antigen-specific $CD8^{+}$ T cell cross-priming more efficiently than their apoptotic counterparts (Uhl et al. 2009).

It is also tempting to speculate that *Salmonella*-induced death of TAMs resulting in the reduction of the number of these immunosuppressive cells in tumors may help to tip the balance of signals in favor of anti-tumor responses. The observation that the killing of TAMs by natural killer T cells in primary human neuroblastoma suppressed tumor growth may support this concept (Song et al. 2009). Currently, the idea of simultaneous targeting of TAMs and tumor cells is gaining attention as a promising approach for tumor therapy (Germano et al. 2013; Xin et al. 2009).

Salmonella-*Mediated Tumor Growth Inhibition Involves Immune Regulatory Mechanisms*

Immune therapy rationale relies on the pre-existence of tumor specific immunity which is not effective in the absence of therapeutic intervention due to a tumor immunosuppressive environment. Therefore the primary value of bacteriotherapy is its potential to stimulate the immune responses to combat suppression in tumor microenvironment and break immune tolerance to tumor. While the elimination of major tumor burden is in the scope of classical treatment, the long-term benefit of immunotherapy would be the induction of tumor-specific memory responses to protect patients from tumor recurrence and metastatic disease. Various approaches have been proposed in the pre-clinical studies to eradicate tumor with *Salmonella*. Attenuated and auxotrophic strains are either used solely to boost the immunity or as vectors for the delivery of tumor-associated antigens to elicit specific T cell response, or other heterologous therapeutic molecules.

Immunoregulatory cells impede tumor immune surveillance, trigger chronic inflammation and tumor-promoting environment. Modification of phenotype and activity of immune regulatory cells, such as M2 macrophages and MDSCs was shown to account for *Salmonella*-induced tumor growth inhibition. Moreover, DC maturation, effective tumor antigen presentation and cytotoxic T cells responses after *Salmonella* treatment prove the concept of danger signal-mediated shift form immunosuppressive, tumor fostering environment to antitumor immune responses (Hong et al. 2013; Jarosz et al. 2013; Kaimala et al. 2014).

Auxotrophic *S.* Typhimurium BRD509E administered intraperitoneally to C57Bl/6 mice bearing subcutaneous B16.F1 melanoma tumors, accumulated in tumors and significantly retarded tumor growth (Kaimala et al. 2014). Tumor colonization was correlated with increased accumulation of $CD11b^{+}Gr1^{+}$ myeloid cells in the tumors. These cells had increased surface expression of maturation and activation markers – MHC class II, co-stimulatory molecule CD80 and Sca-1 (Ly-6A, Stem cell antigen-1) in *Salmonella*-treated, compared to non-treated mice. Moreover, expression of IFNγ was up-regulated in spleens and tumors after *Salmonella* treatment.

Importantly, intratumoral myeloid cells from bacteria-treated mice had lower expression of ARG1, suggesting a reversion of suppressive phenotype. Indeed, intratumoral myeloid cells from *Salmonella*-treated mice were less suppressive towards $CD4^{+}$ T cells than those from control mice (Kaimala et al. 2014).

Natural course of *Salmonella* oral infection results predominantly in mounting the local protective immunity in the mesenteric lymph nodes (Voedisch et al. 2009). Since the exact mechanism of solid tumor colonization is yet to be clarified and bacterial tumor colonization is critical for therapeutic benefit, the optimal route of administration is crucial for the outcome of treatment. Various mutant strains defective in metabolic or virulence factors show diverse extent of tissue colonization in mice. Both orogastric and intravenous administration of different *S.* Typhimurium strains were shown to inhibit transplantable tumor growth in mouse models. The oral route would be technically easier to conduct in clinical settings and would be of reduced health risk, but some strains do not target distant solid tumors after oral administration.

Recently the group led by Siegfried Weiss compared the efficacy of systemic and oral delivery of different *S.* Typhimurium strains. *S.* Typhimurium SL1344Δ*aroA* strain was administered by either of the three routes: intravenous (i.v.), intraperitoneal (i.p.) or oral, to Balb/c mice with subcutaneous CT26 solid tumors. Viable bacteria were recovered from tumor, spleen and liver at all tested time points (1–17 days) after i.v. or i.p. injection, but only on 11th day (not earlier and not later) after oral infection. However, all tested routes resulted in preferential tumor colonization, with about 100 times more bacteria in tumors than in the spleen and liver. Intravenous and intraperitoneal infection led to complete clearance of tumors, while oral infection transiently inhibited tumor growth when bacterial tumor colonization was detectable, but finally the tumors grew as in non-treated control mice (Crull et al. 2011). The same authors provided a comparison of organ colonization by wild type *S.* Typhimurium strain SL1344 and its two derivatives – deletion mutants of *aroA* or *purA*, auxotrophic for aromatic amino acids or purines, respectively. All three strains efficiently colonized tumor and spleen after intravenous injection, but wild-type bacteria caused severe side effects. Colonization with *purA*-deleted strain was delayed and less efficient than that with *aroA*-deleted strain (Crull et al. 2011). Finally, the effects of two strains with the same attenuation type but on different genetic background were compared. SL1344Δ*aroA* colonized tumors more efficiently, leading to their clearance in all mice, while treatment with SL7207, which is also *aroA*-deficient, eliminated tumors in three out of five mice.

These results point out the dependence of attenuation outcome on genetic background (Crull et al. 2011). In line with the aforementioned results, the importance of bacterial tumor colonization, as well as colocalization of tumor antigens and bacterial immune mediators at tumor site was highlighted by the recent study. Attenuated *S.* Typhimurium X9241 strain was injected intratumorally or given orally to Balb/c mice bearing subcutaneous CT26 tumors expressing heterologous tumor antigen, human Her-2/Neu. Intratumoral injection significantly inhibited tumor growth while the oral route was not effective (Hong et al. 2013). Tumor growth inhibition was partially dependent on cytotoxic lymphocyte response, since $CD8^+$ cells depletion reduced the effect. The treatment increased the frequency of splenic M- and G-MDSCs, and intratumoral G-MDSCs. But yet the phenotype of myeloid cells was distinct from suppressive MDSCs, as a significantly higher percentage produced TNF. The therapeutic effect relied also on the decreased percentage of splenic and tumor-infiltrating $CD4^+CD25^+Foxp3^+$ regulatory T cells (Hong et al. 2013). These experimental data support the concept that tumor targeting with attenuated *Salmonella* not only inhibits its growth, but induces the break of the immune suppressive tumor microenvironment.

Transforming the Perfect Pathogen into a Perfect Drug Candidate

Attenuated S. *Typhimurium Inhibits Tumor Growth in Animal Models*

While designing *S.* Typhimurium-based anti-tumor therapies there is a prevalent tendency to improve or enhance therapeutic potential of different attenuated strains by various genetic modifications. These will be described later. Yet, there are also approaches to utilize natural cytotoxic and immunomodulatory capabilities of attenuated bacteria strains without any further modifications. A few examples of the strains that showed promising results when used in animal models are worth mentioning.

The group led by Robert Hoffman used a sophisticated way of attenuation to develop the A1-R strain. At first, *S.* Typhimurium (ATCC 14028) was subjected to chemical mutagenesis with nitrosoguanidine (NTG) that resulted in obtaining an auxotrophic A1 strain dependent on an external source of leucine and arginine (Zhao et al. 2005). In the next step, A1 bacteria were passaged through HT-29 tumor-bearing nude mice and recovered from the excised tumor tissue. As a consequence of this selection step, the strain isolated from the tumor, called A1-R, demonstrated improved targeting towards cancer cells *in vitro* and *in vivo* in comparison to A1 (Zhao et al. 2006). A1-R without any additional modifications was successfully used in a model therapy of numerous human tumor xenografts in nude mice, significantly suppressing tumor growth in all tested models and causing complete tumor eradication in several of them (Hayashi et al. 2009a; Hayashi et al. 2009b; Kimura et al. 2010; Nagakura et al. 2009; Zhang et al. 2012; Zhao et al. 2006, 2007).

The A1-R strain has also been tested in metastatic tumor models, e.g. in the spontaneous popliteal lymph node metastasis model of human HT-1080 fibrosarcoma in nude mice. In five out of six mice the lymph node metastases were completely eradicated within 7–21 days after the administration of A1-R in the footpad, which was the site of the tumor injection. In contrast, metastases in lymph nodes of all mice from the control group, injected with the solvent instead of bacteria, continued to grow (Hayashi et al. 2009a). In two other nude mice models, the lung metastasis model of human osteosarcoma and human HT-1080 fibrosarcoma, intravenous administration of A1-R bacteria resulted in a strong decrease of the number and size of metastases (Hayashi et al. 2009a, b). Recently, the A1-R strain was demonstrated to have increased efficacy compared to standard chemotherapy in treating patient-derived orthotopic xenograft (PDOX) of pancreatic cancer in nude mice (Hiroshima et al. 2014). The results suggest the clinical potential of A1-R bacterial therapy against this highly lethal type of cancer. All the above examples come from the studies performed on immunodeficient animals and thus the results indicate that the A1-R strain may exert strong cytotoxic effects toward tumors independently of the adaptive immune responses. Presently, the effects of various experimental settings for the evaluation of A1-R applicability in immunocompetent mice are under investigation (Tome et al. 2013; Zhang et al. 2013b).

Another interesting approach was proposed by Yu et al. (2012), who developed a novel strain of *S.* Typhimurium, YB1, unable to survive in normal tissue. They placed the *asd* gene, crucial for the synthesis of DAP, an important bacterial cell wall component, under the control of a hypoxia-induced promoter. Limiting the expression of the *asd* gene to hypoxic conditions makes the bacterium an obligate anaerobe. Such a modification renders YB1 strain non-toxic for normal tissues while maintaining its tumor targeting. The safety and anti-tumor efficacy of three *Salmonella* strains: SL7207, YB1 and another attenuated *Salmonella* strain VNP20009 were compared in the MDA-MB-231 breast tumor-bearing nude mice model. Bacteria of all the strains were able to infiltrate and destruct tumors. However, in contrast to deadly SL7207 parental strain, YB1 bacteria were effectively cleared from normal tissues and were barely detectable in the liver 3 days postinfection. They also showed better therapeutic parameters than VNP20009 in MDA-MB-231 tumor model (Yu et al. 2012).

VNP20009 Is Safe But Does Not Colonize Tumors in Humans

VNP20009, *msbB*, *purI*-attenuated strain of *S.* Typhimurium, has been shown to accumulate preferentially at the tumor sites after intravenous administration to the mice bearing spontaneous, syngeneic or human xenograft tumors. The tumor-to-normal tissue ratio ranged from 300:1 to 1000:1 (Clairmont et al. 2000). Also in dogs suffering from different spontaneous tumors and subjected to *S.* Typhimurium intravenous injections, an accumulation of bacteria was demonstrated in more than 40 % of cases (Thamm et al. 2005). The maximum tolerated dose (MTD) of VNP20009 for mice was estimated to be as high as 0.5×10^8 colony forming units (CFU) per kg of body weight (Lee et al. 2000). Similar values of MTD were established for other mammals: dogs – 3.0×10^7, pigs – 1.9×10^8 and monkeys – 2.5×10^8 CFU/kg (Lee et al. 2000; Thamm et al. 2005). Based on the preclinical data, in November 1999, VNP20009 entered Phase I human clinical trial conducted by Vion Pharmaceuticals Inc. Twenty five patients – 24 with metastatic melanoma and one with metastatic renal cancer – were initially enrolled for the trial. VNP20009 have been administered intravenously for 30 min in doses ranging from 10^6 to 10^9 CFU/m^2. MTD has been established at 3×10^8 CFU/m^2, as higher doses were accompanied by substantial adverse effects including fever, hypotension, anemia and thrombocytopenia. In the majority of cases the bacteria were rapidly cleared from the bloodstream. The serum levels of cytokines: TNF, IL-1β, IL-6, IL-12 were transiently increased and correlated with *Salmonella* dose. Unfortunately, only 3 out of 18 tumor biopsies contained viable bacteria and even in those cases tumor growth regression was not observed (Toso et al. 2002). Increasing the infusion time to 4 h, which was applied for the next four patients, did not improve the outcome (Heimann and Rosenberg 2003).

Another pilot clinical trial was performed on three patients in order to evaluate the possible effects of intratumoral application of genetically modified VNP20009 strain called TAPET-CD (Tumor Amplified Protein Expression Therapy – Cytosine Deaminase), expressing *E. coli* cytosine deaminase (Nemunaitis et al. 2003). In principle, intratumoral localization of the enzyme-expressing bacteria would result in the accumulation of cytotoxic 5-fluorouracil (5-FU) at the tumor site following systemic application of non-toxic 5′-fluorocytosine (5-FC). Indeed, some colonization by TAPET-CD bacteria in tumors of two patients was accompanied by the increased conversion of 5-FC to 5-FU. However, intratumoral route of *Salmonella* administration did not significantly improve bacterial load in tumor tissue (Nemunaitis et al. 2003).

The clinical trials revealed that VNP20009 do not colonize tumors in patients to the same degree as observed in murine models, which may be a major reason for the lack of its therapeutic effects. Hence attempts were made to increase the tumor targeting capability of *Salmonella*. Bereta et al. (2007) constructed a VNP20009 strain expressing a fusion protein of OmpA and a single chain antibody fragment (scFv) recognizing carcinoembryonic antigen (CEA), a widespread TAA of human carcinomas. The modified VNP20009 strain showed increased localization in CEA-expressing MC38 tumors compared to wild type-MC38 and inhibited CEA-MC38 tumor growth more efficiently than non-modified, parental VNP20009 (Bereta et al. 2007). A similar approach was also proposed by Massa et al. (2013) who genetically engineered *aroA*-deficient SL3261 strain to express a fusion protein consisting of OmpA and a camelid single-domain (VHH) antibody against human CD20. They observed that the presence of the antibody at the surface of bacteria improves *Salmonella* targeting to the mouse and human tumors expressing CD20 and limits bacterial invasion of the liver and spleen (Massa et al. 2013).

Also *Salmonella* Typhi is taken into account in the development of anti-cancer therapy. Up to date various attenuated *S.* Typhi strains have been tested mainly as live attenuated oral paratyphoid or typhoid fever vaccines, reviewed in (Roland and Brenneman 2013). When used as a vector in humans *S.* Typhi turned out to be inferior to *S.* Typhimurium in eliciting the immune response to a delivered heterologous antigen (*H. pylori* urease), however this may result from the oral route of bacteria administration and longer persistence of *S.* Typhimurium in the intestine (Angelakopoulos and Hohmann 2000). *S.* Typhi Ty21a strain was generated as an anti-angiogenic cancer therapeutic vaccine. It carries a plasmid encoding full length vascular endothelial growth factor receptor-2 (VEGFR-2). *S.* Typhi Ty21a entered Phase I clinical trial in 2012 (Niethammer et al. 2012), but the results of the study are not yet available.

S. *Typhimurium Can Be Optimized Using Genetic Engineering*

The failure of the first clinical trials highlights the need for improvement of *Salmonella* targeting to tumor tissues in humans. This is certainly the biggest challenge facing researchers involved in the development of *Salmonella*-based therapies. However, even in mouse models attenuated *Salmonella* strains, although they strongly inhibited the growth of various tumors, rarely led to their complete eradication. Therefore, numerous attempts are being made to increase efficacy of *Salmonella*-based therapies by introducing genetic modifications into attenuated strains. The modified bacteria, thanks to their intracellular lifestyle, serve as a vector for the delivery of therapeutic molecules into the cells. *Salmonella* usually gets equipped in a plasmid coding for effector molecules whose expression is controlled by eukaryotic or bacterial promoters. The potential therapeutic advantage of various effector molecules including TAAs, cytokines, apoptosis-inducing factors, prodrug-converting enzymes or short hairpin RNAs (shRNAs) able to silence the expression of a protein of choice have already been tested. The intended effects of these molecules can be divided into three groups: (*i*) increasing cytotoxic effects of *Salmonella* infection; (*ii*) enhancement and navigation of the immunomodulatory properties of *Salmonella*; (*iii*) delivery of TAAs. The therapeutic approaches are summarized in Table 16.1 and several examples of interesting and promising ideas are presented below.

Apoptosis-Inducing Factors Can Increase Cytotoxic Potential of Salmonella

In order to enhance the capability of *S.* Typhimurium to kill infected cancer cells, several research groups took advantage of apoptosis-inducing factors and equipped bacteria with appropriate expression vectors.

Table 16.1 Examples of therapeutic approaches based on genetically modified *S.*Typhimurium strains

Cargo	*Salmonella* strain	Tumor model	Therapeutic scheme/outcome	References
TRAIL cDNA under the control of the bacterial anaerobically inducible *nirB* promoter	VNP20009	s.c.[a], B16F10 melanoma and RM-1 prostate cancer in C57Bl/6 mice	i.p., 10^5 CFU/mouse on day 7 for B16F10 or day 9 for RM-1 after tumor inoculation	Chen et al. (2012)
			Increased levels of TRAIL in B16F10 tumor (but not in liver or spleen). Tumor growth inhibition. VNP-TRAIL was not more effective than VNP control strain in tumor growth suppression of TRAIL-resistant RM-1 tumors	
Apoptin gene under the control of the cytomegalovirus (CMV) promoter	*phoP/phoQ* null strain LH430	s.c., Hep-2 cancer[b] in Balb/c nude mice	i.v., 10^7 CFU/mouse on day 21 after tumor inoculation (mice with tumors >75 mm^3), one group of mice was re-treated after 7 days	Guan et al. (2013)
			Significant tumor growth delay especially in mice, which obtained two doses of bacteria. Obliteration of the tumor vasculature	
TRAIL cDNA and apoptin gene under the control of eukaryotic promoters	SL7207	s.c., SGC-7901 human gastric cancer in Balb/c nude mice	i.t., 2×10^6 CFU/mouse on day 7 after tumor implantation repeated every 7 days	Cao et al. (2010)
			Cancer cell apoptosis and tumor regression	
MTD cDNA of human Noxa fused to N-terminal cell penetrating peptide under the control of pBAD promoter of the *E. coli* arabinose operon	SKS002	s.c., CT26 colon carcinoma in Balb/c mice	i.v., 10^7 CFU/mouse when the tumor size reached about 100–150 mm^3. Three days later L-arabinose was administered i.p.	Jeong et al. (2014)
			Necrosis in tumor tissue and tumor regression	

(continued)

Table 16.1 (continued)

Cargo	*Salmonella* strain	Tumor model	Therapeutic scheme/outcome	References
TNF cDNA fused to 160 aa N-terminal fragment of SipB under the control of the bacterial *lac* promoter	BRD509	s.c., B16F10 melanoma and TC-1 cervical in C57Bl/6 mice	Two s.c. injections, 10^8 CFU/mouse on day 7 and day 14 after B16F10 tumor inoculation	Yoon et al. (2011)
			Complete inhibition of tumor cell growth in 90 % of animals. Suggested involvement of NK cells. Lack of protection against second B16F10 challenge	
		s.c., 4T1 breast, CT26 colon, RENCA kidney cancers in Balb/c mice	Reduced growth of all listed tumors, although with diverse efficacy	
IL-18 cDNA under the control of the bacterial *ompC* promoter	VNP20009	s.c., CT26 colon carcinoma in Balb/c mice	i.v., 5×10^6 CFU/mouse on day 7 day after tumor inoculation	Loeffler et al. (2008)
			Reduced tumor growth. Increased number of NK cells, $CD4^+$ but not $CD8^+$ T cells in CT26 tumors, increased intratumoral levels of IL-1β, TNF, IFNγ	
		s.c., D2F2 breast carcinoma in Balb/c mice	i.v., 5×10^6 CFU/mouse on day 9, 14 and 19 after tumor inoculation	
			Significant tumor growth inhibition	
CCL21 cDNA under the control of the *ompC* promoter	VNP20009	s.c., CT26 colon carcinoma in Balb/c mice	i.v., 5×10^6 CFU/mouse on day 9, 14 and 19 after tumor inoculation	Loeffler et al. (2009)
			Significant inhibition of tumor growth	
		i.v., D2F2 breast carcinoma in Balb/c mice (pulmonary retention model)	i.v., 5×10^6 CFU/mouse on day 6, 13 and 20 after tumor inoculation	
			Decreased number of tumor foci in lungs	

LIGHT cDNA under the control of the *ompC* promoter	VNP20009	s.c., D2F2 breast carcinoma or CT26 colon carcinoma in Balb/c mice	i.v., 5×10^6 CFU/mouse on day 9, 14, and 19 after tumor inoculation	Loeffler et al. (2007)
		s.c., Lewis lung carcinoma (LLC) in C57Bl/6 mice	Significant inhibition of tumor growth	
		i.v., D2F2 breast carcinoma in Balb/c mice (pulmonary retention model)	i.v., 5×10^6 CFU/mouse on day 7 day after tumor inoculation	
			Significant inhibition of tumor growth. Massive infiltration of tumors by inflammatory cell. Increased levels of B cells ($CD19^+$) and both $CD4^+$ and $CD8^+$T cells	
			i.v., 5×10^6 CFU/mouse on days 6, 13 and 20 after tumor inoculation	
			Decreased number of tumor foci in lungs	
IL-2 cDNA under the control of the anaerobically inducible *nirB* promoter	BRD509	s.c., B16.F1 murine melanoma in C57BL/6 mice	i.p., 5×10^5 CFU/mouse on day 13 after tumor inoculation	al-Ramadi et al. (2009)
			Tumor growth inhibition and prolonged mouse survival. Decreased angiogenesis and increased cancer cells necrosis within the tumor tissue. A treatment regimen involving multiple low doses of *Salmonella* was more effective than a single high dose regimen.	
IDO shRNA under the control of the human U6 promoter	VNP20009	s.c., B16F10 murine melanoma in C57BL/6 mice	Two i.v. injections, 2.5×10^6 CFU/mouse 4 days apart, into mice when tumor diameter ≥7 mm	Blache et al. (2012)
			Synergistic effect of IDO silencing and *S.*Typhimurium on tumor growth suppression. Increased tumor influx of polymorphonuclear leukocytes ($CD11^+Gr1^+$) and intratumoral cell death. Tumor growth suppression occurred also in the absence of functional adaptive immunity.	

[a] *s.c.* subcutaneously, *i.v.* intravenously, *i.t.* intratumorally, *i.p.* intraperitoneally
[b] Probably wrongly described by the authors as laryngeal cancer

Apoptin (VP3) is a chicken anemia virus (CAV) protein which exhibits p53-independent, tumor cell-specific proapoptotic effects (Zhuang et al. 1995). It does not affect non-malignant cells and this selectivity of action makes apoptin an interesting potential anti-tumor agent. *S.* Typhimurium LH430 carrying apoptin gene under the control of eukaryotic cytomegalovirus (CMV) early promoter (*ST-rC-Apoptin*) increased the delay of the growth of human Hep-2 xenografts in nude mice as compared to the effects of *ST*-rC-EGFP control. The lack of side effects may be explained by an almost thousand times higher accumulation of *ST*-rC-Apoptin in tumor tissue than in the liver, equally after one or two i.v. injections of *Salmonella*. The expression of apoptin in tumor tissue was followed by an increased activity of caspases (Guan et al. 2013).

Another *Salmonella* strain has been equipped in a double proapoptotic weapon: apoptin and TNF-related apoptosis-inducing ligand (TRAIL). Similarly to apoptin, TRAIL induces apoptosis in a wide variety of cancer cells, but hardly in normal cells (Walczak et al. 1999; Yagita et al. 2004). *S.* Typhimurium SL7207 carrying apoptin gene and *TRAIL* cDNA under the control of CMV promoter injected intratumorally to human gastric tumor xenografts in nude mice induced higher apoptosis rate than unmodified SL7207 and strongly suppressed tumor growth with its complete eradication in some animals (Cao et al. 2010).

The idea of combining TRAIL expression with *Salmonella*-tumor targeting was also explored by Ganai et al. (2009). They placed *TRAIL* cDNA under the bacterial RecA promoter activated during SOS response to DNA damage (Anderson and Kowalczykowski 1998) and thus created a radiation-inducible system for temporal and spatial control of TRAIL expression. Intravenous administration of TRAIL-expressing VNP20009 into mice bearing 4T1 mammary carcinoma followed by 2Gy whole body γ-irradiation 2 days later led to a significant inhibition of tumor growth resulting from the combined effects of *Salmonella* infection, irradiation and TRAIL expression (Ganai et al. 2009).

It has been shown that second mitochondria-derived activator of caspases (Smac) sensitizes various tumor cells for TRAIL-induced apoptosis (Deng et al. 2002; Zhang et al. 2001). Therefore, Fu et al. (2008) engineered a modified *S.* Typhimurium SL3261 strain carrying a vector coding for Smac and TRAIL (S.L./SNhTS). Expression of both cDNAs was controlled by the promoter of human telomerase reverse transcriptase (hTERT), highly active in many human cancers and inactive in non-proliferating normal cells (Kim et al. 1994). Indeed, in contrast to normal cells, S.L./SNhTS-infected tumor cells of different origin (LL/2 Lewis lung carcinoma, B16F10 melanoma and 4T1 mammary carcinoma) expressed high levels of exogenous Smac and TRAIL, resulting in an elevated apoptosis rate. *In vivo* studies demonstrated that orally delivered S.L./SNhTS markedly suppressed tumor growth in all tested mice models, without any detectable side-effects (Fu et al. 2008).

Another approach aiming at the enhancement of proapoptotic properties of *Salmonella* was proposed by Joeng et al. (2014), who thought not only of the synthesis of a proapoptotic factor, mitochondrial-targeting domain of Noxa (MTD), at the tumor site, but also carefully designed a system of its transport from bacteria to tumor cells (Jeong et al. 2014). The researchers used ΔppGpp *Salmonella* strain unable to produce a key regulator of vital bacterial processes – guanosine-3′, 5′-bis-diphospate which guarantees very high accumulation of bacteria in the tumor tissue over the liver or spleen. Noxa is a transcriptional target of p53 which contributes to the induction of intrinsic apoptosis pathway via the activation of mitochondrial damage (Seo et al. 2009; Zhang et al. 2011). MTD, its prodeath domain causes extensive necrosis of cells *in vitro* through an increase of the cytosolic calcium levels (Seo et al. 2009). In the designed system, MTD was expressed as a fusion protein with DS4.3, a cell-penetrating peptide facilitating eukaryotic cell entry after bacteria lysis. To release DS4.3-MTD from bacteria phage lysis genes of a newly characterized *Salmonella* phage were employed. Both DS4.3-MTD cDNA and phage lysis genes were placed under the control of pBAD, a promoter activated by L-arabinose. CT26 colon carcinoma-bearing mice were intravenously injected with *Salmonella* carrying $pLYSP_{BAD}$::DS4.3-MTD. Three days later, when the bacteria accumulated in the tumors over the livers at a ratio of 48,000:1 and over the spleens at a ratio of 65,000:1, daily administration of L-arabinose was started. Massive necrosis of tumor tissue and suppression of tumor growth was observed (Jeong et al. 2014).

Immunomodulatory Properties of Salmonella *Can Be Strengthened*

Numerous modifications have been introduced to *Salmonella* to influence immune cells and reinforce anti-tumor immune response by tipping the balance between pro- and anti-tumor activities. The strains of *Salmonella* producing various cytokines and chemokines such as IL-2 (al-Ramadi et al. 2009; Ha et al. 2012), IL-21 (Wang et al. 2013), TNF (Yoon et al. 2011), IL-18 (Loeffler et al. 2008), CCL21 (Loeffler et al. 2009) or LIGHT (Loeffler et al. 2007) were generated and tested in mouse models. Interestingly, *Salmonella* expressing specific shRNA may switch off the synthesis of a host immunosuppressing protein (Blache et al. 2012). Two examples of promising approaches are described below.

John C. Reed's group demonstrated a superiority of CCL21 chemokine-expressing *Salmonella* strain over a parental strain in inhibiting the growth of CT26 colon, D2F2 breast, and B16 melanoma tumors as well as in limiting CT26-lung colonization in immunocompetent mice (Loeffler et al. 2009). The idea of equipping *Salmonella* in CCL21 came from the known activities of this chemokine and was supported by the results of experiments in which CCL21 was directly injected into tumors. As CCL21 is a chemoattractant for T lymphocytes and DCs it may therefore stimulate the colocalization of naive T cells and tumor antigen-presenting DCs which may help to mount effective immune response and lead to subsequent tumor eradication. Indeed, observed inhibition of tumor growth by CCL21-expressing *Salmonella* was accompanied by the increased intratumoral levels of IFNγ, CXCL9, and CXCL10, cytokines known to be induced by CCL21, as well as by enhanced infiltration of tumors by immune cells including $CD4^+$ and $CD8^+$ T lymphocytes. Immunodepletion of different cell populations revealed that both

$CD4^+$ and $CD8^+$ T cells were indispensable for significant inhibition of tumor development by CCL21-expressing *Salmonella*.

An unconventional therapeutic approach was proposed by Blache et al. (2012), who decided to use *Salmonella* vector to modulate immunosuppressive tumor microenvironment via silencing of IDO expression. VNP20009 strain of *S.* Typhimurium carried a plasmid coding for shRNA able to silence IDO expression (shIDO-ST). Bacteria injected intravenously significantly inhibited the growth of subcutaneous B16F10 melanoma as well as considerably diminished a number of lung melanoma foci after intravenous application of the tumor cells and their effects were more pronounced than those of VNP20009 expressing scrambled shRNA. The treatment of mice with shIDO-ST resulted in a significant intratumoral influx of $CD11b^+Gr1^+$ which were mostly $Ly6G^+$, accompanied by markedly elevated ROS levels and massive intratumoral cell death. Unexpectedly, the levels of $CD4^+$ and $CD8^+$ T cells remained unaffected and what is more, shIDO-ST was equally active in normal mice as in mice depleted of $CD4^+$, $CD8^+$, or NK subsets. In contrast, depletion of $Gr1^+$ cells resulted in abrogation of shIDO-ST inhibitory effects. The results indicate that in this tumor model polymorphonuclear leukocytes were obligatory for the shIDO-ST therapeutic efficacy. Interestingly, the silencing of IDO potentiated the colonization of tumor by shIDO-ST (Blache et al. 2012).

Salmonella *May Play the Role of a TAA-Expressing Vector*

The rationale for using *Salmonella* as a vector delivering TAAs is that the presentation of antigens to immune cells will be enhanced by the strong danger signals. Up to date, several natural (mAFP, survivin, endoglin) or artificial (β-galactosidase) tumor antigens have been placed under the control of CMV promoter in the plasmids introduced to attenuated *Salmonella*, which were then used as prophylactic or therapeutic vaccines. This approach assumes that the plasmid carried by the bacteria becomes available to eukaryotic transcriptional machinery which recognizes CMV promoter. Although the mechanism by which the plasmid enclosed in *Salmonella* is delivered to the cell nucleus is not clear, it is well documented that the TAAs whose expression was controlled by CMV promoter were produced in the cytoplasm of infected cells or in the cytoplasm of phagocytes that engulfed bacteria or infected, apoptotic cells (Paglia et al. 1998; Yrlid and Wick 2000). TAAs produced in this way elicited efficient cell-mediated or both cell-mediated and humoral immune responses (Chou et al. 2006; Fest et al. 2009; Jarosz et al. 2013; Paglia et al. 1998). However, the intracellular location of *Salmonella* within the SCV may limit the CMV-driven TAAs expression. Another approach has been developed to circumvent this limitation. It utilizes *Salmonella's* TTSS to deliver TAAs produced by bacteria inside the SCV to the cytoplasm of the host cell. The transgene that is placed in a plasmid under the control of a bacterial promoter encodes a fusion protein consisting of TAA preceded by a short sequence derived from the N-terminus of a bacterial protein, e.g. SopE, SseF, YopE, SipB exported via TTSS. This sequence contains the so-called secretion and translocation signal which enables an efficient delivery of a fusion protein to the host cell cytoplasm and therefore makes it available for MHC-I presentation. The usefulness of this approach in both prophylactic as well as therapeutic vaccine settings was demonstrated by Russmann et al. and followers (Roider et al. 2011; Russmann et al. 1998). The activation of specific $CD8^+$ T cells, epitope spreading, increased intratumoral proliferation of $CD4^+$ and $CD8^+$ T lymphocytes, elevated intratumoral levels of granzyme B, increased serum levels of TNF and IFNγ are among the effects that were demonstrated as resulting from oral or intravenous administration of *Salmonella*-based tumor vaccines. The natural TAAs delivered by *Salmonella*'s TTSS include NY-ESO-1 (Nishikawa et al. 2006), survivin (Manuel et al. 2011), tyrosinase-related protein 2 (Trp2) (Zhu et al. 2010) and human papilloma virus HPV E7 (Yoon et al. 2014). The chimeric transgenes are usually placed under bacterial promoters activated after the invasion of the host cell. Some researchers introduced additional modifications to the therapeutic settings to further improve TAA-directed immune responses. For example Zhu et al. (2010) designed the genetic construct coding for a fusion protein consisting of a melanoma-specific TRP2 antigen and Hsp70 immunochaperone, which should facilitate the proper presentation of antigenic peptides to cytotoxic T cells (Zhu et al. 2010). Manuel et al. (2011) proposed a combined *Salmonella*-based therapy in which subsequent intravenous injections of two different *Salmonella* strains: one – coding for STAT3 shRNA and the second coding for TAA – survivin were applied to mice bearing melanoma tumors (Manuel et al. 2011). The silencing of the tolerogenic transcription factor STAT3 before TAA delivery enabled mounting the strong anti-tumor immune response. More examples of prophylactic and therapeutic approaches utilizing *Salmonella* can be found in the extensive review by Chorobik et al. (Chorobik et al. 2013).

Other Microbial Species Might Also Be Useful in Cancer Treatment

Species of *Salmonella* posses many features of a perfect bacterial cancer therapeutic. However, a number of other microorganisms proved to be advantageous anticancer agents. The most promising so far have been strains of *Clostridium*, *Escherichia* and *Listeria* (St Jean et al. 2008), as well as some lactic acid bacteria from *Lactococcus*, *Lactobacillus* and *Bifidobacterium* genera (Tangney 2010).

Clostridium is a genus of Gram-positive bacteria, obligatory anaerobic and capable of producing endospores. Since *Clostridium* spores can only germinate in oxygen-free conditions, they are particularly suited to target hypoxic or anoxic tumor regions characterized by poor or no perfusion resulting in quiescence and necrosis. This notion sparked interest as early as the 1970s (Heppner and Mose 1978), but the spore treatment alone had limited efficacy against better-vascularized tumor regions, small tumors or metastases. A novel treatment approach, known as combined bacteriolytic therapy, was developed to utilize hypoxia-specific accumulation of bacteria; one of the most effective examples was an attenuated *C. novyi-based* treatment resulting in lysis of experimental tumors formed by human colon carcinoma and murine melanoma (Dang et al. 2001). Various other

Clostridium- based strategies are currently being investigated, including prodrug-, cytokine- and antibody-combined approaches (Umer et al. 2012).

Bacteria from the well-known genus of *Escherichia* also have a therapeutic potential against cancer. Probiotic *E. coli* Nissle 1917 was shown to accumulate in tumors and replicate at the border of live and necrotic tumor tissue, while the colonization levels in the spleen and liver *versus* tumor tissue were very low (Stritzker et al. 2007). The administration of K-12, another strain of *E. coli*, to mice bearing murine breast carcinomas effectively stimulated an anti-tumor immune response and resulted in major reduction of lung metastases (Weibel et al. 2008). Due to its facultatively anaerobic metabolism, the mechanisms of *E. coli* tumor targeting are probably similar to those of *Salmonella*. However, since *Escherichia* does not invade cells, its possible applications are limited to site-specific expression of proteins within the tumor tissue.

A number of tumor therapies using *Listeria* have been under development in recent years. *L. monocytogenes* is a Gram-positive, intracellular pathogen that causes foodborne infections. Most *Listeria*-based treatments use a strategy different than other bacterial cancer therapies – instead of tumor targeting to perform an intratumoral action, the bacteria are used as live vaccine vectors that can deliver tumor-related antigens to non-tumor cells and thereby stimulate systemic anticancer immune responses. An example of this approach is *L. monocytogenes* which is able to express E7 antigen of human papilloma virus (HPV)-16 directly within APCs; human trials with patients bearing HPV-induced tumors provided promising clinical data (Le et al. 2012). Another interesting concept utilizing *Listeria* is targeting tumors via *L. monocytogenes*-infected MDSCs; the bacteria, labeled with ^{188}Rhenium, successfully delivered the radioactive cargo into the tumor tissue (Chandra and Gravekamp 2013).

Another group of bacteria used for tumor targeting are endosymbiotic strains of *Lactococcus*, *Lactobacillus* and *Bifidobacterium* that share ability to produce lactic acid and are commonly utilized in food and dairy fermentations. Having a long track record of safety in humans, or even health-promoting or probiotic benefits, these bacteria are also able to colonize tumor tissue and can be potentially useful for gene-based treatment and/or detection of cancer. For example, oral administration of obligatory anaerobic *B. breve* can result in its translocation from the gastrointestinal tract via bloodstream into the tumor, where it was shown to express a reporter gene (Cronin et al. 2010).

Future Perspectives

In order to be successful, any bacterial cancer therapy will need to address a number of issues that are limiting to current treatment options. One of the most important ones, relevant to virtually all chemotherapeutic and biological agents, is the limited accessibility of the tumor tissue to passively-distributed therapeutics. Bacterial motility and environmental sensing can be particularly useful for tumor localization; however, the effective targeting is required not solely in the animal models, but also in the human context, which to date remains challenging.

Therapeutic bacteria are expected to be safe, but also be able to completely switch the tumor microenvironment from immunosuppression into immunoactivation. While many studies proved the feasibility of this approach, it is important to note that preventing toxic effects by attenuation seems to be a double-edged sword, as it may compromise other therapeutically-relevant features such as invasion, colonization or immunostimulation. The balance between safety and immunogenicity is vital for a clinically-meaningful anticancer effect.

The concept of using bacteria against cancer is not necessary related to a single-agent therapy. A more feasible regimen would include initial treatment with therapeutic strains (administered orally, intravenously or intratumorally, depending on the disease), followed by surgical removal of the tumor mass. Additional follow-up with bacteria to treat the minimal residual disease (MRD) is likely to improve clinical outcomes. Combining bacterial therapy with other treatment modalities may result in stronger impact on the tumor microenvironment, which would improve cancer-specific responses.

Another question is the importance of recurring contact with bacteria to the therapeutic benefits. The role of pre-existing immunity, e.g. against pathogenic *Salmonella*, is unclear – it may increase clearance of the bacteria on one hand, but also potentiate anti-cancer effects on the other. This issue might be particularly relevant to multiple dosing regimens, in which the bacteria-based therapeutic is applied repeatedly to the same patient – for a prolonged, perhaps even life-long therapy.

Acknowledgements The authors would like to thank Dr. Grzegorz Bereta for his help with manuscript preparation, especially for the scientific illustration, as well as fruitful discussions. The authors gratefully acknowledge financial support from the Polish National Centre for Research and Development through INNOTECH grant no. 152553 and the funding from the Jagiellonian University within the SET project co-financed by the European Union. Faculty of Biochemistry, Biophysics and Biotechnology of the Jagiellonian University in Kraków is a partner of the Leading National Research Center (KNOW) supported by the Ministry of Science and Higher Education.

References

al-Ramadi BK, Fernandez-Cabezudo MJ, El-Hasasna H, Al-Salam S, Bashir G, Chouaib S (2009) Potent anti-tumor activity of systemically-administered IL2-expressing Salmonella correlates with decreased angiogenesis and enhanced tumor apoptosis. Clin Immunol 130(1):89–97. doi:10.1016/j.clim.2008.08.021

Anderson DG, Kowalczykowski SC (1998) Reconstitution of an SOS response pathway: derepression of transcription in response to DNA breaks. Cell 95(7):975–979

Angelakopoulos H, Hohmann EL (2000) Pilot study of phoP/phoQ-deleted Salmonella enterica serovar typhimurium expressing Helicobacter pylori urease in adult volunteers. Infect Immun 68(4):2135–2141

Avogadri F, Martinoli C, Petrovska L, Chiodoni C, Transidico P, Bronte V, Longhi R, Colombo MP, Dougan G, Rescigno M (2005) Cancer immunotherapy based on killing of Salmonella-infected tumor cells. Cancer Res 65(9):3920–3927. doi:10.1158/0008-5472.CAN-04-3002

Baker S, Dougan G (2007) The genome of Salmonella enterica serovar Typhi. Clin Infect Dis 45(Suppl 1):S29–S33. doi:10.1086/518143

Balkwill FR, Mantovani A (2012) Cancer-related inflammation: common themes and therapeutic opportunities. Semin Cancer Biol 22(1):33–40. doi:10.1016/j.semcancer.2011.12.005

Baumler AJ, Tsolis RM, Ficht TA, Adams LG (1998) Evolution of host adaptation in Salmonella enterica. Infect Immun 66(10):4579–4587

Bereta M, Hayhurst A, Gajda M, Chorobik P, Targosz M, Marcinkiewicz J, Kaufman HL (2007) Improving tumor targeting and therapeutic potential of Salmonella VNP20009 by displaying cell surface CEA-specific antibodies. Vaccine 25(21):4183–4192. doi:10.1016/j.vaccine.2007.03.008

Biragyn A, Longo DL (2012) Neoplastic “Black Ops”: cancer’s subversive tactics in overcoming host defenses. Semin Cancer Biol 22(1):50–59. doi:10.1016/j.semcancer.2012.01.005

Birmingham CL, Smith AC, Bakowski MA, Yoshimori T, Brumell JH (2006) Autophagy controls Salmonella infection in response to damage to the Salmonella-containing vacuole. J Biol Chem 281(16):11374–11383. doi:10.1074/jbc.M509157200

Blache CA, Manuel ER, Kaltcheva TI, Wong AN, Ellenhorn JD, Blazar BR, Diamond DJ (2012) Systemic delivery of Salmonella typhimurium transformed with IDO shRNA enhances intratumoral vector colonization and suppresses tumor growth. Cancer Res 72(24):6447–6456. doi:10.1158/0008-5472.CAN-12-0193

Cao HD, Yang YX, Lu L, Liu SN, Wang PL, Tao XH, Wang LJ, Xiang TX (2010) Attenuated Salmonella typhimurium carrying TRAIL and VP3 genes inhibits the growth of gastric cancer cells in vitro and in vivo. Tumori 96(2):296–303

Chandra D, Gravekamp C (2013) Myeloid-derived suppressor cells: cellular missiles to target tumors. Oncoimmunology 2(11):e26967. doi:10.4161/onci.26967

Chen J, Yang B, Cheng X, Qiao Y, Tang B, Chen G, Wei J, Liu X, Cheng W, Du P, Huang X, Jiang W, Hu Q, Hu Y, Li J, Hua ZC (2012) Salmonella-mediated tumor-targeting TRAIL gene therapy significantly suppresses melanoma growth in mouse model. Cancer Sci 103(2):325–333. doi:10.1111/j.1349-7006.2011.02147.x

Choi SY, Collins CC, Gout PW, Wang Y (2013) Cancer-generated lactic acid: a regulatory, immunosuppressive metabolite? J Pathol 230(4):350–355. doi:10.1002/path.4218

Chorobik P, Czaplicki D, Ossysek K, Bereta J (2013) Salmonella and cancer: from pathogens to therapeutics. Acta Biochim Pol 60(3):285–297

Chou CK, Hung JY, Liu JC, Chen CT, Hung MC (2006) An attenuated Salmonella oral DNA vaccine prevents the growth of hepatocellular carcinoma and colon cancer that express alpha-fetoprotein. Cancer Gene Ther 13(8):746–752. doi:10.1038/sj.cgt.7700927

Chow MT, Moller A, Smyth MJ (2012) Inflammation and immune surveillance in cancer. Semin Cancer Biol 22(1):23–32. doi:10.1016/j.semcancer.2011.12.004

Clairmont C, Lee KC, Pike J, Ittensohn M, Low KB, Pawelek J, Bermudes D, Brecher SM, Margitich D, Turnier J, Li Z, Luo X, King I, Zheng LM (2000) Biodistribution and genetic stability of the novel antitumor agent VNP20009, a genetically modified strain of Salmonella typhimurium. J Infect Dis 181(6):1996–2002. doi:10.1086/315497

Cookson BT, Brennan MA (2001) Pro-inflammatory programmed cell death. Trends Microbiol 9(3):113–114

Cronin M, Morrissey D, Rajendran S, El Mashad SM, van Sinderen D, O’Sullivan GC, Tangney M (2010) Orally administered bifidobacteria as vehicles for delivery of agents to systemic tumors. Mol Ther 18(7):1397–1407. doi:10.1038/mt.2010.59

Crull K, Bumann D, Weiss S (2011) Influence of infection route and virulence factors on colonization of solid tumors by Salmonella enterica serovar Typhimurium. FEMS Immunol Med Microbiol 62(1):75–83. doi:10.1111/j.1574-695X.2011.00790.x

Dang LH, Bettegowda C, Huso DL, Kinzler KW, Vogelstein B (2001) Combination bacteriolytic therapy for the treatment of experimental tumors. Proc Natl Acad Sci U S A 98(26):15155–15160. doi:10.1073/pnas.251543698

Deng Y, Lin Y, Wu X (2002) TRAIL-induced apoptosis requires Bax-dependent mitochondrial release of Smac/DIABLO. Genes Dev 16(1):33–45. doi:10.1101/gad.949602

Deretic V (2011) Autophagy in immunity and cell-autonomous defense against intracellular microbes. Immunol Rev 240(1):92–104. doi:10.1111/j.1600-065X.2010.00995.x

Desai PT, Porwollik S, Long F, Cheng P, Wollam A, Clifton S, Weinstock GM, McClelland M (2013) Evolutionary genomics of the Salmonella enterica subspecies. mBio 4(2):e00579–12. doi:10.1128/mBio.00579-12

Dresch C, Leverrier Y, Marvel J, Shortman K (2012) Development of antigen cross-presentation capacity in dendritic cells. Trends Immunol 33(8):381–388. doi:10.1016/j.it.2012.04.009

Dunn GP, Bruce AT, Ikeda H, Old LJ, Schreiber RD (2002) Cancer immunoediting: from immunosurveillance to tumor escape. Nat Immunol 3(11):991–998. doi:10.1038/ni1102-991

Espert L, Codogno P, Biard-Piechaczyk M (2007) Involvement of autophagy in viral infections: antiviral function and subversion by viruses. J Mol Med (Berl) 85(8):811–823. doi:10.1007/s00109-007-0173-6

Ewaschuk JB, Backer JL, Churchill TA, Obermeier F, Krause DO, Madsen KL (2007) Surface expression of Toll-like receptor 9 is upregulated on intestinal epithelial cells in response to pathogenic bacterial DNA. Infect Immun 75(5):2572–2579. doi:10.1128/IAI.01662-06

Fest S, Huebener N, Bleeke M, Durmus T, Stermann A, Woehler A, Baykan B, Zenclussen AC, Michalsky E, Jaeger IS, Preissner R, Hohn O, Weixler S, Gaedicke G, Lode HN (2009) Survivin minigene DNA vaccination is effective against neuroblastoma. Int J Cancer 125(1):104–114. doi:10.1002/ijc.24291

Forbes NS (2010) Engineering the perfect (bacterial) cancer therapy. Nat Rev Cancer 10(11):785–794. doi:10.1038/nrc2934

Forbes JR, Gros P (2003) Iron, manganese, and cobalt transport by Nramp1 (Slc11a1) and Nramp2 (Slc11a2) expressed at the plasma membrane. Blood 102(5):1884–1892. doi:10.1182/blood-2003-02-0425

Fridman WH, Pages F, Sautes-Fridman C, Galon J (2012) The immune contexture in human tumours: impact on clinical outcome. Nat Rev Cancer 12(4):298–306. doi:10.1038/nrc3245

Fu W, Chu L, Han X, Liu X, Ren D (2008) Synergistic antitumoral effects of human telomerase reverse transcriptase-mediated dual-apoptosis-related gene vector delivered by orally attenuated Salmonella enterica Serovar Typhimurium in murine tumor models. J Gene Med 10(6):690–701. doi:10.1002/jgm.1191

Gabrilovich DI, Bronte V, Chen SH, Colombo MP, Ochoa A, Ostrand-Rosenberg S, Schreiber H (2007) The terminology issue for myeloid-derived suppressor cells. Cancer Res 67(1):425. doi:10.1158/0008-5472.CAN-06-3037; author reply 426

Gabrilovich DI, Ostrand-Rosenberg S, Bronte V (2012) Coordinated regulation of myeloid cells by tumours. Nat Rev Immunol 12(4):253–268. doi:10.1038/nri3175

Galdiero MR, Bonavita E, Barajon I, Garlanda C, Mantovani A, Jaillon S (2013) Tumor associated macrophages and neutrophils in cancer. Immunobiology 218(11):1402–1410. doi:10.1016/j.imbio.2013.06.003

Galen JE, Curtiss R 3rd (2013) The delicate balance in genetically engineering live vaccines. Vaccine 32(35):4376-4385. doi:10.1016/j.vaccine.2013.12.026

Ganai S, Arenas RB, Forbes NS (2009) Tumour-targeted delivery of TRAIL using Salmonella typhimurium enhances breast cancer survival in mice. Br J Cancer 101(10):1683–1691. doi:10.1038/sj.bjc.6605403

Garcia-del Portillo F, Zwick MB, Leung KY, Finlay BB (1993) Salmonella induces the formation of filamentous structures containing lysosomal membrane glycoproteins in epithelial cells. Proc Natl Acad Sci U S A 90(22):10544–10548

Germano G, Frapolli R, Belgiovine C, Anselmo A, Pesce S, Liguori M, Erba E, Uboldi S, Zucchetti M, Pasqualini F, Nebuloni M, van Rooijen N, Mortarini R, Beltrame L, Marchini S, Fuso Nerini I, Sanfilippo R, Casali PG, Pilotti S, Galmarini CM, Anichini A, Mantovani A, D'Incalci M, Allavena P (2013) Role of macrophage targeting in the antitumor activity of trabectedin. Cancer Cell 23(2):249–262. doi:10.1016/j.ccr.2013.01.008

Goldszmid RS, Dzutsev A, Trinchieri G (2014) Host immune response to infection and cancer: unexpected commonalities. Cell Host Microbe 15(3):295–305. doi:10.1016/j.chom.2014.02.003

Gordon MA (2008) Salmonella infections in immunocompromised adults. J Infect 56(6):413–422. doi:10.1016/j.jinf.2008.03.012
Guan GF, Zhao M, Liu LM, Jin CS, Sun K, Zhang DJ, Yu DJ, Cao HW, Lu YQ, Wen LJ (2013) Salmonella typhimurium mediated delivery of Apoptin in human laryngeal cancer. Int J Med Sci 10(12):1639–1648. doi:10.7150/ijms.6960
Ha XQ, Yin Q, Zhao HB, Hui L, Wang ML, Peng JH, Dong JZ, Deng ZY, Zhao Y, Zhang YY (2012) Inhibitory effects of the attenuated Salmonella typhimurium containing the IL-2 gene on hepatic tumors in mice. J Biomed Biotechnol 2012:946139. doi:10.1155/2012/946139
Hayashi K, Zhao M, Yamauchi K, Yamamoto N, Tsuchiya H, Tomita K, Hoffman RM (2009a) Cancer metastasis directly eradicated by targeted therapy with a modified Salmonella typhimurium. J Cell Biochem 106(6):992–998. doi:10.1002/jcb.22078
Hayashi K, Zhao M, Yamauchi K, Yamamoto N, Tsuchiya H, Tomita K, Kishimoto H, Bouvet M, Hoffman RM (2009b) Systemic targeting of primary bone tumor and lung metastasis of high-grade osteosarcoma in nude mice with a tumor-selective strain of Salmonella typhimurium. Cell Cycle 8(6):870–875
Heimann DM, Rosenberg SA (2003) Continuous intravenous administration of live genetically modified salmonella typhimurium in patients with metastatic melanoma. J Immunother 26(2):179–180
Heppner F, Mose JR (1978) The liquefaction (oncolysis) of malignant gliomas by a non pathogenic Clostridium. Acta Neurochir (Wien) 42(1–2):123–125
Hernandez LD, Pypaert M, Flavell RA, Galan JE (2003) A Salmonella protein causes macrophage cell death by inducing autophagy. J Cell Biol 163(5):1123–1131. doi:10.1083/jcb.200309161
Hiroshima Y, Zhao M, Maawy A, Zhang Y, Katz MH, Fleming JB, Uehara F, Miwa S, Yano S, Momiyama M, Suetsugu A, Chishima T, Tanaka K, Bouvet M, Endo I, Hoffman RM (2014) Efficacy of Salmonella typhimurium A1-R Versus chemotherapy on a pancreatic cancer Patient-derived orthotopic xenograft (PDOX). J Cell Biochem 115(7):1254–1261. doi:10.1002/jcb.24769
Hong EH, Chang SY, Lee BR, Pyun AR, Kim JW, Kweon MN, Ko HJ (2013) Intratumoral injection of attenuated Salmonella vaccine can induce tumor microenvironmental shift from immune suppressive to immunogenic. Vaccine 31(10):1377–1384. doi:10.1016/j.vaccine.2013.01.006
Hu ZQ, Zhao WH (2013) Type 1 interferon-associated necroptosis: a novel mechanism for Salmonella enterica Typhimurium to induce macrophage death. Cell Mol Immunol 10(1):10–12. doi:10.1038/cmi.2012.54
Jarosz M, Jazowiecka-Rakus J, Cichon T, Glowala-Kosinska M, Smolarczyk R, Smagur A, Malina S, Sochanik A, Szala S (2013) Therapeutic antitumor potential of endoglin-based DNA vaccine combined with immunomodulatory agents. Gene Ther 20(3):262–273. doi:10.1038/gt.2012.28
Jeong JH, Kim K, Lim D, Jeong K, Hong Y, Nguyen VH, Kim TH, Ryu S, Lim JA, Kim JI, Kim GJ, Kim SC, Min JJ, Choy HE (2014) Anti-tumoral effect of the mitochondrial target domain of Noxa delivered by an engineered Salmonella typhimurium. PLoS One 9(1):e80050. doi:10.1371/journal.pone.0080050
Joffre OP, Segura E, Savina A, Amigorena S (2012) Cross-presentation by dendritic cells. Nat Rev Immunol 12(8):557–569. doi:10.1038/nri3254
Kaimala S, Mohamed YA, Nader N, Issac J, Elkord E, Chouaib S, Fernandez-Cabezudo MJ, Al-Ramadi BK (2014) Salmonella-mediated tumor regression involves targeting of tumor myeloid suppressor cells causing a shift to M1-like phenotype and reduction in suppressive capacity. Cancer Immunol Immunother 63(6):587–599. doi:10.1007/s00262-014-1543-x
Kalinski P (2012) Regulation of immune responses by prostaglandin E2. J Immunol 188(1):21–28. doi:10.4049/jimmunol.1101029
Kawamoto A, Morimoto YV, Miyata T, Minamino T, Hughes KT, Kato T, Namba K (2013) Common and distinct structural features of Salmonella injectisome and flagellar basal body. Sci Rep 3:3369. doi:10.1038/srep03369
Kim NW, Piatyszek MA, Prowse KR, Harley CB, West MD, Ho PL, Coviello GM, Wright WE, Weinrich SL, Shay JW (1994) Specific association of human telomerase activity with immortal cells and cancer. Science 266(5193):2011–2015

Kim JM, Eckmann L, Savidge TC, Lowe DC, Witthoft T, Kagnoff MF (1998) Apoptosis of human intestinal epithelial cells after bacterial invasion. J Clin Invest 102(10):1815–1823. doi:10.1172/JCI2466

Kimura H, Zhang L, Zhao M, Hayashi K, Tsuchiya H, Tomita K, Bouvet M, Wessels J, Hoffman RM (2010) Targeted therapy of spinal cord glioma with a genetically modified Salmonella typhimurium. Cell Prolif 43(1):41–48. doi:10.1111/j.1365-2184.2009.00652.x

Knodler LA, Vallance BA, Celli J, Winfree S, Hansen B, Montero M, Steele-Mortimer O (2010) Dissemination of invasive Salmonella via bacterial-induced extrusion of mucosal epithelia. Proc Natl Acad Sci U S A 107(41):17733–17738. doi:10.1073/pnas.1006098107

Kroemer G, Levine B (2008) Autophagic cell death: the story of a misnomer. Nat Rev Mol Cell Biol 9(12):1004–1010. doi:10.1038/nrm2529

Lahiri A, Das P, Vani J, Shaila MS, Chakravortty D (2010) TLR 9 activation in dendritic cells enhances salmonella killing and antigen presentation via involvement of the reactive oxygen species. PLoS One 5(10):e13772. doi:10.1371/journal.pone.0013772

Le DT, Dubenksy TW Jr, Brockstedt DG (2012) Clinical development of Listeria monocytogenes-based immunotherapies. Semin Oncol 39(3):311–322. doi:10.1053/j.seminoncol.2012.02.008

Lee KC, Zheng L-M, Luo X, Clairmont C, Fischer J, Margitich D, Turnier J, Almassian B, Bermudes D, King I (2000) Comparative evaluation of the acute toxic effects in monkeys, pigs and mice of a genetically engineered salmonella strain (VNP20009) being developed as an antitumor agent. Int J Toxicol 19(1):19–25

Lee JS, Jung ID, Lee CM, Park JW, Chun SH, Jeong SK, Ha TK, Shin YK, Kim DJ, Park YM (2010) Outer membrane protein a of Salmonella enterica serovar Typhimurium activates dendritic cells and enhances Th1 polarization. BMC Microbiol 10:263. doi:10.1186/1471-2180-10-263

Lee CH, Lin ST, Liu JJ, Chang WW, Hsieh JL, Wang WK (2014) Salmonella induce autophagy in melanoma by the downregulation of AKT/mTOR pathway. Gene Ther 21(3):309–316. doi:10.1038/gt.2013.86

Leschner S, Westphal K, Dietrich N, Viegas N, Jablonska J, Lyszkiewicz M, Lienenklaus S, Falk W, Gekara N, Loessner H, Weiss S (2009) Tumor invasion of Salmonella enterica serovar Typhimurium is accompanied by strong hemorrhage promoted by TNF-alpha. PLoS One 4(8):e6692. doi:10.1371/journal.pone.0006692

Lindgren SW, Stojiljkovic I, Heffron F (1996) Macrophage killing is an essential virulence mechanism of Salmonella typhimurium. Proc Natl Acad Sci U S A 93(9):4197–4201

Loeffler M, Le'Negrate G, Krajewska M, Reed JC (2007) Attenuated Salmonella engineered to produce human cytokine LIGHT inhibit tumor growth. Proc Natl Acad Sci U S A 104(31):12879–12883. doi:10.1073/pnas.0701959104

Loeffler M, Le'Negrate G, Krajewska M, Reed JC (2008) IL-18-producing Salmonella inhibit tumor growth. Cancer Gene Ther 15(12):787–794. doi:10.1038/cgt.2008.48

Loeffler M, Le'Negrate G, Krajewska M, Reed JC (2009) Salmonella typhimurium engineered to produce CCL21 inhibit tumor growth. Cancer Immunol Immunother 58(5):769–775. doi:10.1007/s00262-008-0555-9

Lorenzi S, Mattei F, Sistigu A, Bracci L, Spadaro F, Sanchez M, Spada M, Belardelli F, Gabriele L, Schiavoni G (2011) Type I IFNs control antigen retention and survival of CD8alpha(+) dendritic cells after uptake of tumor apoptotic cells leading to cross-priming. J Immunol 186(9):5142–5150. doi:10.4049/jimmunol.1004163

Lu H (2014) TLR agonists for cancer immunotherapy: tipping the balance between the immune stimulatory and inhibitory effects. Front Immunol 5:83. doi:10.3389/fimmu.2014.00083

Lu JV, Walsh CM (2012) Programmed necrosis and autophagy in immune function. Immunol Rev 249(1):205–217. doi:10.1111/j.1600-065X.2012.01147.x

Ma Y, Shurin GV, Peiyuan Z, Shurin MR (2013) Dendritic cells in the cancer microenvironment. J Cancer 4(1):36–44. doi:10.7150/jca.5046

Mai CW, Kang YB, Pichika MR (2013) Should a Toll-like receptor 4 (TLR-4) agonist or antagonist be designed to treat cancer? TLR-4: its expression and effects in the ten most common cancers. Onco Targets Ther 6:1573–1587. doi:10.2147/OTT.S50838

Manuel ER, Blache CA, Paquette R, Kaltcheva TI, Ishizaki H, Ellenhorn JD, Hensel M, Metelitsa L, Diamond DJ (2011) Enhancement of cancer vaccine therapy by systemic delivery of a tumor-targeting Salmonella-based STAT3 shRNA suppresses the growth of established melanoma tumors. Cancer Res 71(12):4183–4191. doi:10.1158/0008-5472.CAN-10-4676

Massa PE, Paniccia A, Monegal A, de Marco A, Rescigno M (2013) Salmonella engineered to express CD20-targeting antibodies and a drug-converting enzyme can eradicate human lymphomas. Blood 122(5):705–714. doi:10.1182/blood-2012-12-474098

Mastroeni P, Grant A, Restif O, Maskell D (2009) A dynamic view of the spread and intracellular distribution of Salmonella enterica. Nat Rev Microbiol 7(1):73–80. doi:10.1038/nrmicro2034

Matzinger P (2012) The evolution of the danger theory. Interview by Lauren Constable, Commissioning Editor. Expert Rev Clin Immunol 8(4):311–317. doi:10.1586/eci.12.21

McClelland M, Sanderson KE, Spieth J, Clifton SW, Latreille P, Courtney L, Porwollik S, Ali J, Dante M, Du F, Hou S, Layman D, Leonard S, Nguyen C, Scott K, Holmes A, Grewal N, Mulvaney E, Ryan E, Sun H, Florea L, Miller W, Stoneking T, Nhan M, Waterston R, Wilson RK (2001) Complete genome sequence of Salmonella enterica serovar Typhimurium LT2. Nature 413(6858):852–856. doi:10.1038/35101614

McLaughlin LM, Govoni GR, Gerke C, Gopinath S, Peng K, Laidlaw G, Chien YH, Jeong HW, Li Z, Brown MD, Sacks DB, Monack D (2009) The Salmonella SPI2 effector SseI mediates long-term systemic infection by modulating host cell migration. PLoS Pathog 5(11):e1000671. doi:10.1371/journal.ppat.1000671

Meunier E, Dick MS, Dreier RF, Schurmann N, Broz DK, Warming S, Roose-Girma M, Bumann D, Kayagaki N, Takeda K, Yamamoto M, Broz P (2014) Caspase-11 activation requires lysis of pathogen-containing vacuoles by IFN-induced GTPases. Nature 509(7500):366–370. doi:10.1038/nature13157

Miller SI, Ernst RK, Bader MW (2005) LPS, TLR4 and infectious disease diversity. Nat Rev Microbiol 3(1):36–46. doi:10.1038/nrmicro1068

Monack DM, Bouley DM, Falkow S (2004) Salmonella typhimurium persists within macrophages in the mesenteric lymph nodes of chronically infected Nramp1+/+ mice and can be reactivated by IFNgamma neutralization. J Exp Med 199(2):231–241. doi:10.1084/jem.20031319

Mota LJ, Ramsden AE, Liu M, Castle JD, Holden DW (2009) SCAMP3 is a component of the Salmonella-induced tubular network and reveals an interaction between bacterial effectors and post-Golgi trafficking. Cell Microbiol 11(8):1236–1253. doi:10.1111/j.1462-5822.2009.01329.x

Nagakura C, Hayashi K, Zhao M, Yamauchi K, Yamamoto N, Tsuchiya H, Tomita K, Bouvet M, Hoffman RM (2009) Efficacy of a genetically-modified Salmonella typhimurium in an orthotopic human pancreatic cancer in nude mice. Anticancer Res 29(6):1873–1878

Nemunaitis J, Cunningham C, Senzer N, Kuhn J, Cramm J, Litz C, Cavagnolo R, Cahill A, Clairmont C, Sznol M (2003) Pilot trial of genetically modified, attenuated Salmonella expressing the E. coli cytosine deaminase gene in refractory cancer patients. Cancer Gene Ther 10(10):737–744. doi:10.1038/sj.cgt.7700634

Niethammer AG, Lubenau H, Mikus G, Knebel P, Hohmann N, Leowardi C, Beckhove P, Akhisaroglu M, Ge Y, Springer M, Grenacher L, Buchler MW, Koch M, Weitz J, Haefeli WE, Schmitz-Winnenthal FH (2012) Double-blind, placebo-controlled first in human study to investigate an oral vaccine aimed to elicit an immune reaction against the VEGF-Receptor 2 in patients with stage IV and locally advanced pancreatic cancer. BMC Cancer 12:361. doi:10.1186/1471-2407-12-361

Nishikawa H, Sato E, Briones G, Chen LM, Matsuo M, Nagata Y, Ritter G, Jager E, Nomura H, Kondo S, Tawara I, Kato T, Shiku H, Old LJ, Galan JE, Gnjatic S (2006) In vivo antigen delivery by a Salmonella typhimurium type III secretion system for therapeutic cancer vaccines. J Clin Invest 116(7):1946–1954. doi:10.1172/JCI28045

Ochman H, Wilson AC (1987) Evolution in bacteria: evidence for a universal substitution rate in cellular genomes. J Mol Evol 26(1–2):74–86

Olsthoorn MM, Petersen BO, Duus J, Haverkamp J, Thomas-Oates JE, Bock K, Holst O (2000) The structure of the linkage between the O-specific polysaccharide and the core region of the lipopolysaccharide from Salmonella enterica serovar Typhimurium revisited. Eur J Biochem 267(7):2014–2027

Paesold G, Guiney DG, Eckmann L, Kagnoff MF (2002) Genes in the Salmonella pathogenicity island 2 and the Salmonella virulence plasmid are essential for Salmonella-induced apoptosis in intestinal epithelial cells. Cell Microbiol 4(11):771–781

Pages F, Kirilovsky A, Mlecnik B, Asslaber M, Tosolini M, Bindea G, Lagorce C, Wind P, Marliot F, Bruneval P, Zatloukal K, Trajanoski Z, Berger A, Fridman WH, Galon J (2009) In situ cytotoxic and memory T cells predict outcome in patients with early-stage colorectal cancer. J Clin Oncol 27(35):5944–5951. doi:10.1200/JCO.2008.19.6147

Paglia P, Medina E, Arioli I, Guzman CA, Colombo MP (1998) Gene transfer in dendritic cells, induced by oral DNA vaccination with Salmonella typhimurium, results in protective immunity against a murine fibrosarcoma. Blood 92(9):3172–3176

Radics J, Konigsmaier L, Marlovits TC (2014) Structure of a pathogenic type 3 secretion system in action. Nat Struct Mol Biol 21(1):82–87. doi:10.1038/nsmb.2722

Ramos-Morales F (2012) Impact of Salmonella enterica type III secretion system effectors on the eukaryotic host cell. ISRN Cell Biol 2012:36. doi:10.5402/2012/787934

Roider E, Jellbauer S, Kohn B, Berchtold C, Partilla M, Busch DH, Russmann H, Panthel K (2011) Invasion and destruction of a murine fibrosarcoma by Salmonella-induced effector CD8 T cells as a therapeutic intervention against cancer. Cancer Immunol Immunother 60(3):371–380. doi:10.1007/s00262-010-0950-x

Roland KL, Brenneman KE (2013) Salmonella as a vaccine delivery vehicle. Expert Rev Vaccines 12(9):1033–1045. doi:10.1586/14760584.2013.825454

Ruby T, McLaughlin L, Gopinath S, Monack D (2012) Salmonella's long-term relationship with its host. FEMS Microbiol Rev 36(3):600–615. doi:10.1111/j.1574-6976.2012.00332.x

Russmann H, Shams H, Poblete F, Fu Y, Galan JE, Donis RO (1998) Delivery of epitopes by the Salmonella type III secretion system for vaccine development. Science 281(5376):565–568

Ryter SW, Mizumura K, Choi AM (2014) The impact of autophagy on cell death modalities. Int J Cell Biol 2014:502676. doi:10.1155/2014/502676

Salcedo SP, Noursadeghi M, Cohen J, Holden DW (2001) Intracellular replication of Salmonella typhimurium strains in specific subsets of splenic macrophages in vivo. Cell Microbiol 3(9):587–597

Saunders NA, Simpson F, Thompson EW, Hill MM, Endo-Munoz L, Leggatt G, Minchin RF, Guminski A (2012) Role of intratumoural heterogeneity in cancer drug resistance: molecular and clinical perspectives. EMBO Mol Med 4(8):675–684. doi:10.1002/emmm.201101131

Schleker S, Sun J, Raghavan B, Srnec M, Muller N, Koepfinger M, Murthy L, Zhao Z, Klein-Seetharaman J (2012) The current Salmonella-host interactome. Proteomics Clin Appl 6(1–2):117–133. doi:10.1002/prca.201100083

Schraidt O, Marlovits TC (2011) Three-dimensional model of Salmonella's needle complex at subnanometer resolution. Science 331(6021):1192–1195. doi:10.1126/science.1199358

Seo YW, Woo HN, Piya S, Moon AR, Oh JW, Yun CW, Kim KK, Min JY, Jeong SY, Chung S, Song PI, Choi EK, Seol DW, Kim TH (2009) The cell death-inducing activity of the peptide containing Noxa mitochondrial-targeting domain is associated with calcium release. Cancer Res 69(21):8356–8365. doi:10.1158/0008-5472.CAN-09-0349

Shafer WM, Casey SG, Spitznagel JK (1984) Lipid A and resistance of Salmonella typhimurium to antimicrobial granule proteins of human neutrophil granulocytes. Infect Immun 43(3):834–838

Shahnazari S, Yen WL, Birmingham CL, Shiu J, Namolovan A, Zheng YT, Nakayama K, Klionsky DJ, Brumell JH (2010) A diacylglycerol-dependent signaling pathway contributes to regulation of antibacterial autophagy. Cell Host Microbe 8(2):137–146. doi:10.1016/j.chom.2010.07.002

Song L, Asgharzadeh S, Salo J, Engell K, Wu HW, Sposto R, Ara T, Silverman AM, DeClerck YA, Seeger RC, Metelitsa LS (2009) Valpha24-invariant NKT cells mediate antitumor activity via killing of tumor-associated macrophages. J Clin Invest 119(6):1524–1536. doi:10.1172/JCI37869

Spel L, Boelens JJ, Nierkens S, Boes M (2013) Antitumor immune responses mediated by dendritic cells: how signals derived from dying cancer cells drive antigen cross-presentation. Oncoimmunology 2(11):e26403. doi:10.4161/onci.26403

St Jean AT, Zhang M, Forbes NS (2008) Bacterial therapies: completing the cancer treatment toolbox. Curr Opin Biotechnol 19(5):511–517. doi:10.1016/j.copbio.2008.08.004

Steiner TS (2007) How flagellin and toll-like receptor 5 contribute to enteric infection. Infect Immun 75(2):545–552. doi:10.1128/IAI.01506-06

Stritzker J, Weibel S, Hill PJ, Oelschlaeger TA, Goebel W, Szalay AA (2007) Tumor-specific colonization, tissue distribution, and gene induction by probiotic Escherichia coli Nissle 1917 in live mice. Int J Med Microbiol 297(3):151–162. doi:10.1016/j.ijmm.2007.01.008

Sylvester RJ (2011) Bacillus Calmette-Guerin treatment of non-muscle invasive bladder cancer. Int J Urol 18(2):113–120. doi:10.1111/j.1442-2042.2010.02678.x

Tangney M (2010) Gene therapy for cancer: dairy bacteria as delivery vectors. Discov Med 10(52):195–200

Thamm DH, Kurzman ID, King I, Li Z, Sznol M, Dubielzig RR, Vail DM, MacEwen EG (2005) Systemic administration of an attenuated, tumor-targeting Salmonella typhimurium to dogs with spontaneous neoplasia: phase I evaluation. Clinical Cancer Res 11(13):4827–4834. doi:10.1158/1078-0432.CCR-04-2510

Thomas DR, Francis NR, Xu C, DeRosier DJ (2006) The three-dimensional structure of the flagellar rotor from a clockwise-locked mutant of Salmonella enterica serovar Typhimurium. J Bacteriol 188(20):7039–7048. doi:10.1128/JB.00552-06

Thurston TL, Ryzhakov G, Bloor S, von Muhlinen N, Randow F (2009) The TBK1 adaptor and autophagy receptor NDP52 restricts the proliferation of ubiquitin-coated bacteria. Nat Immunol 10(11):1215–1221. doi:10.1038/ni.1800

Tome Y, Zhang Y, Momiyama M, Maehara H, Kanaya F, Tomita K, Tsuchiya H, Bouvet M, Hoffman RM, Zhao M (2013) Primer dosing of S. typhimurium A1-R potentiates tumor-targeting and efficacy in immunocompetent mice. Anticancer Res 33(1):97–102

Toso JF, Gill VJ, Hwu P, Marincola FM, Restifo NP, Schwartzentruber DJ, Sherry RM, Topalian SL, Yang JC, Stock F, Freezer LJ, Morton KE, Seipp C, Haworth L, Mavroukakis S, White D, MacDonald S, Mao J, Sznol M, Rosenberg SA (2002) Phase I study of the intravenous administration of attenuated Salmonella typhimurium to patients with metastatic melanoma. J Clin Oncol 20(1):142–152

Uhl M, Kepp O, Jusforgues-Saklani H, Vicencio JM, Kroemer G, Albert ML (2009) Autophagy within the antigen donor cell facilitates efficient antigen cross-priming of virus-specific CD8+ T cells. Cell Death Differ 16(7):991–1005. doi:10.1038/cdd.2009.8

Umer B, Good D, Anne J, Duan W, Wei MQ (2012) Clostridial spores for cancer therapy: targeting solid tumour microenvironment. J Toxicol 2012:862764. doi:10.1155/2012/862764

Vacchelli E, Eggermont A, Sautes-Fridman C, Galon J, Zitvogel L, Kroemer G, Galluzzi L (2013) Trial watch: toll-like receptor agonists for cancer therapy. Oncoimmunology 2(8):e25238. doi:10.4161/onci.25238

van der Velden AW, Baumler AJ, Tsolis RM, Heffron F (1998) Multiple fimbrial adhesins are required for full virulence of Salmonella typhimurium in mice. Infect Immun 66(6):2803–2808

Villadangos JA, Shortman K (2010) Found in translation: the human equivalent of mouse CD8+ dendritic cells. J Exp Med 207(6):1131–1134. doi:10.1084/jem.20100985

Voedisch S, Koenecke C, David S, Herbrand H, Forster R, Rhen M, Pabst O (2009) Mesenteric lymph nodes confine dendritic cell-mediated dissemination of Salmonella enterica serovar Typhimurium and limit systemic disease in mice. Infect Immun 77(8):3170–3180. doi:10.1128/IAI.00272-09

Wagner C, Polke M, Gerlach RG, Linke D, Stierhof YD, Schwarz H, Hensel M (2011) Functional dissection of SiiE, a giant non-fimbrial adhesin of Salmonella enterica. Cell Microbiol 13(8):1286–1301. doi:10.1111/j.1462-5822.2011.01621.x

Walczak H, Miller RE, Ariail K, Gliniak B, Griffith TS, Kubin M, Chin W, Jones J, Woodward A, Le T, Smith C, Smolak P, Goodwin RG, Rauch CT, Schuh JC, Lynch DH (1999) Tumoricidal activity of tumor necrosis factor-related apoptosis-inducing ligand in vivo. Nat Med 5(2):157–163. doi:10.1038/5517

Wang Z, Jiang H, Chen S, Du F, Wang X (2012) The mitochondrial phosphatase PGAM5 functions at the convergence point of multiple necrotic death pathways. Cell 148(1–2):228–243. doi:10.1016/j.cell.2011.11.030

Wang Y, Chen J, Tang B, Zhang X, Hua ZC (2013) Systemic administration of attenuated in combination with interleukin-21 for cancer therapy. Mol Clin Oncol 1(3):461–465. doi:10.3892/mco.2013.90

Weibel S, Stritzker J, Eck M, Goebel W, Szalay AA (2008) Colonization of experimental murine breast tumours by Escherichia coli K-12 significantly alters the tumour microenvironment. Cell Microbiol 10(6):1235–1248. doi:10.1111/j.1462-5822.2008.01122.x

Whiteside TL (2010) Immune responses to malignancies. J Allergy Clin Immunol 125(2 Suppl 2):S272–S283. doi:10.1016/j.jaci.2009.09.045

Wick MJ (2011) Innate immune control of Salmonella enterica serovar Typhimurium: mechanisms contributing to combating systemic Salmonella infection. J Innate Immun 3(6):543–549. doi:10.1159/000330771

Wynosky-Dolfi MA, Snyder AG, Philip NH, Doonan PJ, Poffenberger MC, Avizonis D, Zwack EE, Riblett AM, Hu B, Strowig T, Flavell RA, Jones RG, Freedman BD, Brodsky IE (2014) Oxidative metabolism enables Salmonella evasion of the NLRP3 inflammasome. J Exp Med 211(4):653–668. doi:10.1084/jem.20130627

Xia Y, Zweier JL (1997) Superoxide and peroxynitrite generation from inducible nitric oxide synthase in macrophages. Proc Natl Acad Sci U S A 94(13):6954–6958

Xin H, Zhang C, Herrmann A, Du Y, Figlin R, Yu H (2009) Sunitinib inhibition of Stat3 induces renal cell carcinoma tumor cell apoptosis and reduces immunosuppressive cells. Cancer Res 69(6):2506–2513. doi:10.1158/0008-5472.CAN-08-4323

Yagita H, Takeda K, Hayakawa Y, Smyth MJ, Okumura K (2004) TRAIL and its receptors as targets for cancer therapy. Cancer Sci 95(10):777–783

Yoon WS, Chae YS, Hong J, Park YK (2011) Antitumor therapeutic effects of a genetically engineered Salmonella typhimurium harboring TNF-alpha in mice. Appl Microbiol Biotechnol 89(6):1807–1819. doi:10.1007/s00253-010-3006-4

Yoon W, Choi JH, Kim S, Park YK (2014) Engineered Salmonella typhimurium expressing E7 fusion protein, derived from human papillomavirus, inhibits tumor growth in cervical tumor-bearing mice. Biotechnol Lett 36(2):349–356. doi:10.1007/s10529-013-1370-8

Youn JI, Gabrilovich DI (2010) The biology of myeloid-derived suppressor cells: the blessing and the curse of morphological and functional heterogeneity. Eur J Immunol 40(11):2969–2975. doi:10.1002/eji.201040895

Yrlid U, Wick MJ (2000) Salmonella-induced apoptosis of infected macrophages results in presentation of a bacteria-encoded antigen after uptake by bystander dendritic cells. J Exp Med 191(4):613–624

Yu B, Yang M, Shi L, Yao Y, Jiang Q, Li X, Tang LH, Zheng BJ, Yuen KY, Smith DK, Song E, Huang JD (2012) Explicit hypoxia targeting with tumor suppression by creating an "obligate" anaerobic Salmonella Typhimurium strain. Sci Rep 2:436. doi:10.1038/srep00436

Yu HB, Croxen MA, Marchiando AM, Ferreira RB, Cadwell K, Foster LJ, Finlay BB (2014) Autophagy facilitates Salmonella replication in HeLa cells. mBio 5(2):e00865–e00914. doi:10.1128/mBio.00865-14

Zhang XD, Zhang XY, Gray CP, Nguyen T, Hersey P (2001) Tumor necrosis factor-related apoptosis-inducing ligand-induced apoptosis of human melanoma is regulated by smac/DIABLO release from mitochondria. Cancer Res 61(19):7339–7348

Zhang L, Lopez H, George NM, Liu X, Pang X, Luo X (2011) Selective involvement of BH3-only proteins and differential targets of Noxa in diverse apoptotic pathways. Cell Death Differ 18(5):864–873. doi:10.1038/cdd.2010.152

Zhang Y, Tome Y, Suetsugu A, Zhang L, Zhang N, Hoffman RM, Zhao M (2012) Determination of the optimal route of administration of Salmonella typhimurium A1-R to target breast cancer in nude mice. Anticancer Res 32(7):2501–2508

Zhang M, Swofford CA, Forbes NS (2013a) Lipid A controls the robustness of intratumoral accumulation of attenuated Salmonella in mice. Int J Cancer 135(3):647–657. doi:10.1002/ijc.28700

Zhang Y, Zhang N, Su S, Hoffman RM, Zhao M (2013b) Salmonella typhimurium A1-R tumor targeting in immunocompetent mice is enhanced by a traditional Chinese medicine herbal mixture. Anticancer Res 33(5):1837–1843

Zhao M, Yang M, Li XM, Jiang P, Baranov E, Li S, Xu M, Penman S, Hoffman RM (2005) Tumor-targeting bacterial therapy with amino acid auxotrophs of GFP-expressing Salmonella typhimurium. Proc Natl Acad Sci U S A 102(3):755–760. doi:10.1073/pnas.0408422102

Zhao M, Yang M, Ma H, Li X, Tan X, Li S, Yang Z, Hoffman RM (2006) Targeted therapy with a Salmonella typhimurium leucine-arginine auxotroph cures orthotopic human breast tumors in nude mice. Cancer Res 66(15):7647–7652. doi:10.1158/0008-5472.CAN-06-0716

Zhao M, Geller J, Ma H, Yang M, Penman S, Hoffman RM (2007) Monotherapy with a tumor-targeting mutant of Salmonella typhimurium cures orthotopic metastatic mouse models of human prostate cancer. Proc Natl Acad Sci U S A 104(24):10170–10174. doi:10.1073/pnas.0703867104

Zhao Y, Yang J, Shi J, Gong YN, Lu Q, Xu H, Liu L, Shao F (2011) The NLRC4 inflammasome receptors for bacterial flagellin and type III secretion apparatus. Nature 477(7366):596–600. doi:10.1038/nature10510

Zheng YT, Shahnazari S, Brech A, Lamark T, Johansen T, Brumell JH (2009) The adaptor protein p62/SQSTM1 targets invading bacteria to the autophagy pathway. J Immunol 183(9):5909–5916. doi:10.4049/jimmunol.0900441

Zhu X, Zhou P, Cai J, Yang G, Liang S, Ren D (2010) Tumor antigen delivered by Salmonella III secretion protein fused with heat shock protein 70 induces protection and eradication against murine melanoma. Cancer Sci 101(12):2621–2628. doi:10.1111/j.1349-7006.2010.01722.x

Zhuang SM, Shvarts A, van Ormondt H, Jochemsen AG, van der Eb AJ, Noteborn MH (1995) Apoptin, a protein derived from chicken anemia virus, induces p53-independent apoptosis in human osteosarcoma cells. Cancer Res 55(3):486–489

Chapter 17
Development of *Salmonella*-Based Cancer Vaccines

Xin Xu, Michael Hensel, and Leonid S. Metelitsa

Abstract One of the major limitations of the modern cancer vaccines is poor delivery of tumor-associated antigens (TAAs) to the intact professional antigen-presenting cells (APCs). To address this challenge, attenuated viral or bacterial vectors have been used in experimental cancer vaccines to deliver TAAs to the host APCs *in situ*. From the clinical application prospective, *Salmonella*-based vectors have an advantage because of excellent safety record of an FDA-approved oral vaccine for typhoid fever and potent immunogenicity with low toxicity in humans shown by recently developed attenuated strains. Live attenuated *Salmonella* vectors have been used in experimental cancer vaccines to deliver TAAs in the form of either DNA or protein. Of particular promise are *Salmonella*-based recombinant vaccines in which a TAA of choice is expressed and delivered to the cytosol of professional APC using effector proteins of the *Salmonella* Pathogenicity Island 2-encoded type III secretion system. This chapter reviews strategies of using natural properties of *Salmonella* for construction of effective cancer vaccines and their clinical translation.

Keywords Cancer vaccine • *Salmonella* vectors • Immunotherapy • Antigen presentation

Abbreviations

APC	antigen-presenting cell(s)
DC	dendritic cell(s)
LPS	lipopolysaccharides
PAMP	Pathogen-associated molecular pattern
TAA	tumor-associated antigen(s)
TLR	toll-like receptor(s)

X. Xu • L.S. Metelitsa (✉)
Department of Pediatrics, Texas Children's Cancer Center, Center for Cell and Gene Therapy, Baylor College of Medicine, Houston, TX 77030, USA
e-mail: lsmeteli@txch.org

M. Hensel
Division of Microbiology, University of Osnabrück, Osnabrück, Germany

M.R. Shurin et al. (eds.), *Infection and Cancer: Bi-Directorial Interactions*, DOI 10.1007/978-3-319-20669-1_17

Introduction

Despite the identification of potentially immunogenic tumor-associated antigens (TAAs) in many types of cancer, current therapeutic cancer vaccines remain largely ineffective (Klebanoff et al. 2011). One of the major limitations of the modern cancer vaccines is that, unlike infectious pathogens, they poorly deliver TAAs in an immunogenic form to intact professional antigen-presenting cells (APCs) at their anatomic location. To address this challenge, attenuated viral or bacterial vectors have been used in experimental cancer vaccines to deliver TAAs to the host APCs *in situ* (Vassaux et al. 2006). Since the discovery that attenuated *Shigella* can deliver DNA for expression in mammalian cells (Sizemore et al. 1995), attenuated strains of several types of intracellular bacteria (*Shigella*, *Salmonella*, *Listeria*, and *Yersinia*) have been studied as delivery vehicles for subunit vaccines against a range of infectious diseases and cancer (Vassaux et al. 2006). From the clinical application prospective, *Salmonella*-based vectors might have an advantage because of excellent safety record of an FDA-approved oral *Salmonella* vaccine (S. *typhi* strain Ty21a, Vivotif®) in children and in adults (Ivanoff et al. 1994; Gentschev et al. 2007). Recently developed attenuated strains S. *typhi* also demonstrated low toxicity in clinical trials (Thamm et al. 2005; Toso et al. 2002). Because *Salmonella* naturally migrate from intestine to mesenteric lymph nodes and spleen, *Salmonella*-based vectors induce a systemic immune response to the bacterially expressed antigens (Pertl et al. 2003; Gentschev et al. 2001; Levine 2009). Live attenuated *Salmonella* has been used in experimental cancer vaccines to deliver either TAA DNA (under control of eukaryotic promoters) or TAA protein expressed and secreted by the bacteria themselves. Each of these approaches has advantages and limitations.

Salmonella-Based DNA Vaccines

The methodology relies on the delivery of a plasmid, in which eukaryotic promoter (e.g. pCMV) drives the expression of a target antigen in the host APCs. After proteolysis of the endogenously expressed protein, peptides enter HLA class-I pathway of antigen presentation (Dietrich et al. 1999; Reisfeld et al. 2004). At the same time, various Pathogen-associated molecular patterns (PAMPs) such as flagellin and lipopolysaccharides (LPS) present in *Salmonella* provide strong "danger" signals via activation of toll-like receptors (TLRs) on DCs that leads to effective antigen presentation and co-stimulation of CD8 T cells and initiation of cell-mediated adaptive immune response against the target (Reisfeld et al. 2004). Thus far, several experimental *Salmonella*-based DNA vaccines conferred protection against various viral and other intracellular pathogens in mice (Schoen et al. 2004). The potency of such vaccines has also been demonstrated against tumor antigens and self-antigens that are preferentially expressed in tumor stroma and/or neovasculature (Reisfeld et al. 2004). To achieve therapeutic activity in cancer models, target antigen was often introduced with a Th1-promoting cytokines such as IL-18 (Luo et al. 2003) or chemokines such as CCL21 (Xiang et al. 2005). The anti-tumor efficacy of *Salmonella-delivered* DNA vaccines against melanoma and neuroblastoma TAAs were synergistically enhanced by an antibody-cytokine fusion protein (ch14.18-IL-2) that targets IL-2 to the tumor microenvironment (Niethammer et al. 2001; Pertl et al. 2003). In these models, vaccinated mice that received suboptimal doses of hu14.18/IL-2 were completely protected against neuroblastoma or melanoma metastases. Both T-cell and NK cell-dependent mechanisms were involved in the induction of a systemic tumor-protective immunity. In another example, an attenuated *Salmonella* strain VNP20009 was modified with a DNA plasmid expressing the shRNA specific for indoleamine 2,3-dioxygenase, which as a single therapeutic was successful in extending survival of mice injected with a melanoma or a pancreatic cancer cell line (Blache et al. 2012; Manuel and Diamond 2013). Despite the demonstrated ability of *S. typhimurium* vectors to mediate somatic DNA transfer and transgene expression in mammalian cells *in vitro* (Darji et al. 1997), the efficacy of such transfer *in vivo* remains low that limits the therapeutic potential of DNA vaccines (Vassaux et al. 2006).

There is no intrinsic mechanism known for transfer of episomal DNA from *Salmonella* into target cells, thus the delivery of DNA vaccines by *Salmonella* carrier strains relies on the destruction of the carrier strain by antigen-presenting cell, the release of the expression vector and, occasionally the expression of the transgene. The efficacy of the various steps is difficult to determine, thereby complicating the rational optimization of delivery of DNA vaccines by *Salmonella* carrier strains. More direct forms of delivery of DNA to target cells such as biobalistic transfection or use of viral vectors thus represent attractive options.

Salmonella-Based Protein Vaccines

To overcome the low rate of gene transfer to the host cells inherent to DNA vaccines, several approaches have been developed to use *Salmonella* vectors to transfer whole antigenic proteins or their immunogenic fragments (Gentschev et al. 2001). However, intracellular location of *Salmonella* within *Salmonella*-containing vacuole (SCV), which is linked to the endosomal compartment, routes antigens for the HLA class-II presentation and restricts their access to the cytosol and, therefore, to the HLA class-I presentation pathway (Verjans et al. 1995). Moreover, *Salmonella* evolved *yej* gene, which product interferes with the MHC class I presentation, and *yej* mutants have been used in the design of *Salmonella*-based cancer vaccines as one of the approaches to enhance CTL generation to heterologous antigens (Qimron et al. 2004; Hummel et al. 2005). In another approach researchers used the type I secretory system for hemolysin A (HlyA) of *Escherichia coli* (Su et al. 1992; Gentschev et al. 1996, 2001) for secretion of heterologous antigens by *Salmonella*. When an antigenic protein is fused to HlyA or HlyAs and expressed in *Salmonella*, it can be secreted from the bacterium into the SCV. A portion of the secreted antigen also enters the cytosol of the infected cell via yet poorly defined mechanism

(Gentschev et al. 2001). However, HlyA-based *Salmonella* vaccines still predominantly elicit humoral rather than cellular immune responses (Gentschev et al. 2004), which is more applicable for vaccine development against bacterial pathogens rather than tumors.

Salmonella-Based Vaccines Using the Type III Secretion System

An alternative way to deliver antigenic proteins to the HLA class-I compartment of the antigen presenting cells is to use one of two type III secretion systems (T3SS) of *Salmonella.* T3SS evolved to deliver sets of bacterial effector proteins into the host-cell cytosol (Russmann et al. 1998). The contact-dependent T3SS function depends on a specialized organelle known as the needle complex (Galan and Wolf-Watz 2006) that links the bacterial envelope to the target cell membrane. A group of T3SS proteins can insert in the target cell membrane, forming a channel through which T3SS effector proteins pass to the target cell cytosol (Marlovits et al. 2004). Some of these effector proteins or their secretion signals have been used in experimental vaccines to direct heterologous proteins expressed in *Salmonella* for secretion via T3SS. Salmonella deploys two T3SS independently during distinct phases of pathogenesis. The *Salmonella* Pathogenicity Island 1 (SPI1)-encoded T3SS translocates effector proteins that remodel the host cell cytoskeleton and induce bacterial invasion of non-phagocytic cells. In contrast, the main function of the SPI2-encoded T3SS is during the intracellular phase of *Salmonella* lifestyle and the cognate effector proteins manipulate host cell endocytic transport. This manipulation results in formation of a specific compartment that is permissive for intracellular bacterial replication (Haraga et al. 2008). Such vaccines have been shown to be effective in eliciting both CD8 and CD4 T cell-mediated immune responses in models of infectious diseases, including those of viral infections (Shams et al. 2001; Evans et al. 2003). Recent studies revealed high therapeutic efficacy of T3SS-based *Salmonella* vaccine that delivered NY-ESO-1 tumor antigen (Nishikawa et al. 2006, 2008). The vaccine elicited NY-ESO-1-specific CD8 and CD4 T cells from peripheral blood mononuclear cells of cancer patients *in vitro* and, when orally administrated to tumor-bearing mice, resulted in the regression of established tumors. One disadvantage of the expression system used in these studies was that the antigen was constitutively expressed in *Salmonella* and could be transported inside any cell that was in contact with the bacteria.

Antigen Delivery by the *Salmonella* Pathogenicity Island 2-Encoded Type III Secretion System

The synthesis and translocation of effector proteins of the SPI1-T3SS and SPI2-T3SS is distinct in kinetics and location. The SPI1-T3SS is active in extracellular Salmonella prior to invasion and the host effector proteins are located in the cytosol of host cells after translocation. In contrast, most effector proteins translocated by the SPI2-T3SS are only synthesized when *Salmonella* is inside host cells, such as DCs and macrophages (Hensel et al. 1995; Hensel 2000; Abrahams and Hensel 2006). A subset of the SPI2-T3SS effector proteins shows a strong association with endosomal membranes after translocation into host cells. The distinct localization of these effector proteins influences their half-life and also that of fused antigens. Furthermore, it is likely that the subcellular localization of the fusion proteins with effectors affects the route of antigen processing and presentation by DCs. This enables the use of live attenuated *Salmonella* vectors for delivery of the heterologous antigens of interest into the class-I antigen presentation pathway of the intact professional APCs (Fig. 17.1). The efficacy of SPI2-T3SS-based vaccination approach has been demonstrated in a mouse model of listeriosis (Husseiny et al. 2007). We and colleagues have reported the use of the SPI2-T3SS to construct cancer vaccines in which human survivin (SVN) or its codon-optimized version (coSVN) was expressed under control of a promoter the a SPI2 operon, P_{sseA} and fused to the gene for effector SseF for translocation (Xiong et al. 2010; Manuel et al. 2011). The vaccines induced CD8 T cell-mediated anti-tumor responses in mouse models of CT26 colon carcinoma, orthotopic DBT glioblastoma, and B16F10 melanoma. Among multiple factors that contribute to the efficacy of SPI2-based vaccines, the choice of a promoter for target antigen expression and an effector protein as a fusion partner for the antigen translocation are of particular importance. Recent studies have evaluated the promoter activities of genes of the SsrAB virulon in *S. typhimurium* and various SPI2-T3SS effector proteins for translocation of heterologous antigens (Xu et al. 2010; Hegazy et al. 2012). Results of these studies identified several candidate SPI2 genes such as *sifB* or *sseJ* with

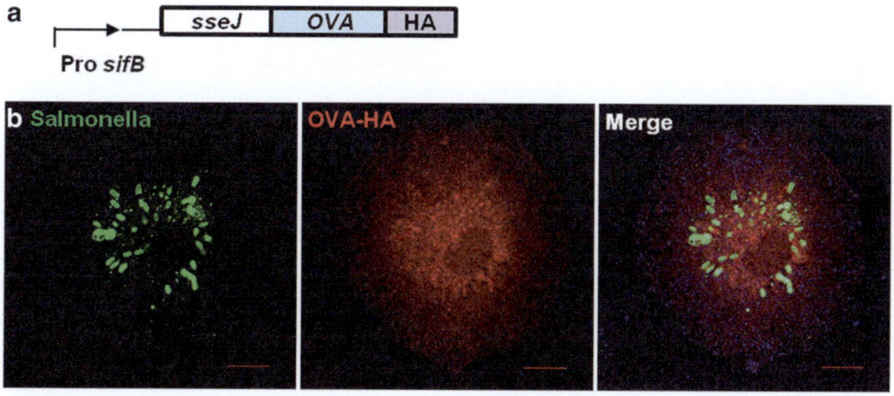

Fig. 17.1 Effective antigen delivery into the cytosol of dendritic cells using the SPI2-encoded T3SS of *Salmonella*. (**a**) A schematic presentation of p3643 expression construct, consisting of *sifB* promoter and *sseJ* encoding an effector protein, fused to OVA and HA tag. (**b**) Murine dendritic cells were infected with attenuated *S. typhimurium,* MvP728 carrying p3643 plasmid. After 16 h, cells were analyzed for *Salmonella* LPS (*green*), HA tag (*red*), and CD11c (*blue*). Shown are representative confocal microscopy images. Scale bar: 10 μm

superior performance that could be used in the next generation of *Salmonella*-based cancer vaccine platform. The anti-tumor potency of SPI2-based vaccines can further be enhanced in combinations with therapies that target the tumor microenvironment. For example, CD8 T-cell infiltration and potent antitumor activity in B16F10 melanoma model was achieved when p3342Max survivin vaccine was combined with a tumor-targeted Stat3 shRNA (Manuel et al. 2011).

The Use of NKT Cell Ligands as Adjutants for Cancer Vaccines

Vaccines are commonly given together with adjuvants (alum, MPL, MF59, AS04, etc.) which stimulate TLRs to enhance immunogenicity (Zepp 2010). Recent advances in the understanding of the mechanisms of adjuvant activity provide a basis for rational combinations of vaccines and adjuvants. In the last decade synthetic ligands of CD1d-reactive Natural Killer T (NKT) cells such as αGalactosylceramide (αGalCer, KRN7000) have been extensively tested as adjuvants for vaccines against microbial pathogens and tumors, including those based on *Salmonella* vectors (Cerundolo et al. 2009; Vasan and Tsuji 2010). The rationale of combining NKT ligands with *Salmonella*-based vectors came from the studies of NKT-cell-mediated response early in the course of *Salmonella* infection. Despite the fact that *Salmonella* does not have glycolipid antigens, which could directly activate NKTs, TLR-mediated signaling in DCs in response to *Salmonella* invasion leads to IL-12 and IL-18 production (Berntman et al. 2005; Nagarajan and Kronenberg 2007) and generation or accumulation of NKT-cell endogenous ligands in DCs (Brigl et al. 2003; Darmoise et al. 2010) that in turn activate NKT-cell-mediated help to DCs, creating a positive amplification loop that plays an important role in the generation of protective immunity against *Salmonella*. Consistent with the described mechanism of NKT-cell help to DCs in *Salmonella* infection, we have demonstrated that NKT ligands strongly enhance IL-12 production induced by an attenuated *Salmonella* vector in human DCs and that *in vivo* co-administration of a NKT ligand with *Salmonella*-based survivin vaccine enhances effector-memory T cell responses that were associated with potent anti-tumor activity in murine cancer models (Xiong et al. 2010). In the search for a glycolipid that can exert more potent stimulatory activity for human NKT cells, M. Tsuji's group synthesized a new glycolipid, 7DW8-5 that has more than 100-fold higher binding affinity for human CD1d and NKT TCR (Li et al. 2010). During *in vivo* testing, 7DW8-5 exhibited a superior Th-1-type responses and adjuvant effect than αGalCer for HIV and malaria vaccines in mice. Moreover, this ligand enhanced CD8 T cell responses induced by an adenovirus-vectored malaria vaccine in non-human primates (Padte et al. 2013), thus representing the primary candidate for entering into clinical testing as a vaccine adjuvant.

Attenuated Strains of *S. typhi* and Clinical Translation

In susceptible mouse strains, infection with *S. typhimurium* results in a system infection that resembles human typhoid fever caused by *S. typhi*. However, human infections with *S. typhimurium* commonly cause gastroenteritis, but no systemic spread of the pathogen. While the vast majority of experimental cancer vaccines have been tested using attenuated strains of *S. typhimurium*, *S. typhi* is uniquely adapted for human host and efficiently traffic to lymphoid tissues in humans (Zhang et al. 2008; Galen et al. 2009). Moreover, there is an excellent safety record of an FDA-approved oral *S. typhi* vaccine (Ivanoff et al. 1994) and the potent immunogenicity with low toxicity in humans shown by the recently developed attenuated strains, such as CVD908, CVD908*htrA* and CVD909 (Tacket et al. 1992; Levine 2009). However, genetic differences between *S. typhimurium* and *S. typhi* need to be considered to ensure optimal functionality of SPI2-T3SS based vectors in the *S. typhi* strains. Yet, alternative *S. typhi* carrier strains are required that harbor attenuating mutation compatible with the function of the SPI2-T3SS. Many attenuating mutations are based on metabolic defects and, in part, these defects diminish the ability to express genes encoding the SPI2-T3SS and to efficiently translocate effector proteins by intracellular *Salmonella*. Combinations of attenuating mutations have been optimized in *S. typhimurium,* but the transfer to *S. typhi* will require a critical assessment of the safety in humans and approval for clinical applications. Another important consideration for the use of bacterial vectors in recombinant vaccines is providing a means of stable expression of recombinant antigens without antibiotic-dependent selection. A recent report described such plasmid stabilization system for *S. typhi* strains in which the single-stranded binding protein (SSB), an essential protein in DNA metabolism, was deleted from the bacterial chromosome and must be provided with the plasmid (Galen et al. 2010). Further development of the vaccination approaches utilizing the molecular machinery of *Salmonella* may provide a foundation for effective oral vaccines against infectious pathogens and cancer. These vaccines should be relatively easy to manufacture, standardize, and administer to patients.

Acknowledgments This work was supported by grants from NIH/NCI (RO1 CA116548), Cancer Prevention and Research Institute of Texas (RP1 121035), Leukemia & Lymphoma Society to LSM; and Deutsche Forschungsgemeinschaft (HE1964) to MH.

References

Abrahams GL, Hensel M (2006) Manipulating cellular transport and immune responses: dynamic interactions between intracellular Salmonella enterica and its host cells. Cell Microbiol 8:728–737

Berntman E, Rolf J, Johansson C, Anderson P, Cardell SL (2005) The role of CD1d-restricted NK T lymphocytes in the immune response to oral infection with Salmonella typhimurium. Eur J Immunol 35:2100–2109

Blache CA, Manuel ER, Kaltcheva TI, Wong AN, Ellenhorn JD, Blazar BR, Diamond DJ (2012) Systemic delivery of Salmonella typhimurium transformed with IDO shRNA enhances intratumoral vector colonization and suppresses tumor growth. Cancer Res 72:6447–6456

Brigl M, Bry L, Kent SC, Gumperz JE, Brenner MB (2003) Mechanism of CD1d-restricted natural killer T cell activation during microbial infection. Nat Immunol 4:1230–1237

Cerundolo V, Silk JD, Masri SH, Salio M (2009) Harnessing invariant NKT cells in vaccination strategies. Nat Rev Immunol 9:28–38

Darji A, Guzman CA, Gerstel B, Wachholz P, Timmis KN, Wehland J, Chakraborty T, Weiss S (1997) Oral somatic transgene vaccination using attenuated S. typhimurium. Cell 91:765–775

Darmoise A, Teneberg S, Bouzonville L, Brady RO, Beck M, Kaufmann SH, Winau F (2010) Lysosomal alpha-galactosidase controls the generation of self lipid antigens for natural killer T cells. Immunity 33:216–228

Dietrich G, Gentschev I, Hess J, Ulmer JB, Kaufmann SH, Goebel W (1999) Delivery of DNA vaccines by attenuated intracellular bacteria. Immunol Today 20:251–253

Evans DT, Chen LM, Gillis J, Lin KC, Harty B, Mazzara GP, Donis RO, Mansfield KG, Lifson JD, Desrosiers RC, Galan JE, Johnson RP (2003) Mucosal priming of simian immunodeficiency virus-specific cytotoxic T-lymphocyte responses in rhesus macaques by the Salmonella type III secretion antigen delivery system. J Virol 77:2400–2409

Galan JE, Wolf-Watz H (2006) Protein delivery into eukaryotic cells by type III secretion machines. Nature 444:567–573

Galen JE, Pasetti MF, Tennant S, Ruiz-Olvera P, Sztein MB, Levine MM (2009) Salmonella enterica serovar Typhi live vector vaccines finally come of age. Immunol Cell Biol 87:400–412

Galen JE, Wang JY, Chinchilla M, Vindurampulle C, Vogel JE, Levy H, Blackwelder WC, Pasetti MF, Levine MM (2010) A new generation of stable, nonantibiotic, low-copy-number plasmids improves immune responses to foreign antigens in Salmonella enterica serovar Typhi live vectors. Infect Immun 78:337–347

Gentschev I, Mollenkopf H, Sokolovic Z, Hess J, Kaufmann SH, Goebel W (1996) Development of antigen-delivery systems, based on the Escherichia coli hemolysin secretion pathway. Gene 179:133–140

Gentschev I, Dietrich G, Spreng S, Kolb-Maurer A, Brinkmann V, Grode L, Hess J, Kaufmann SH, Goebel W (2001) Recombinant attenuated bacteria for the delivery of subunit vaccines. Vaccine 19:2621–2628

Gentschev I, Dietrich G, Spreng S, Neuhaus B, Maier E, Benz R, Goebel W, Fensterle J, Rapp UR (2004) Use of the alpha-hemolysin secretion system of Escherichia coli for antigen delivery in the Salmonella typhi Ty21a vaccine strain. Int J Med Microbiol 294:363–371

Gentschev I, Spreng S, Sieber H, Ures J, Mollet F, Collioud A, Pearman J, Griot-Wenk ME, Fensterle J, Rapp UR, Goebel W, Rothen SA, Dietrich G (2007) Vivotif–a 'magic shield' for protection against typhoid fever and delivery of heterologous antigens. Chemotherapy 53:177–180

Haraga A, Ohlson MB, Miller SI (2008) Salmonellae interplay with host cells. Nat Rev Microbiol 6:53–66

Hegazy WA, Xu X, Metelitsa L, Hensel M (2012) Evaluation of Salmonella enterica type III secretion system effector proteins as carriers for heterologous vaccine antigens. Infect Immun 80:1193–1202

Hensel M (2000) Salmonella pathogenicity island 2. Mol Microbiol 36:1015–1023

Hensel M, Shea JE, Gleeson C, Jones MD, Dalton E, Holden DW (1995) Simultaneous identification of bacterial virulence genes by negative selection. Science 269:400–403

Hummel S, Apte RN, Qimron U, Vitacolonna M, Porgador A, Zoller M (2005) Tumor vaccination by Salmonella typhimurium after transformation with a eukaryotic expression vector in mice: impact of a Salmonella typhimurium gene interfering with MHC class I presentation. J Immunother 28:467–479

Husseiny MI, Wartha F, Hensel M (2007) Recombinant vaccines based on translocated effector proteins of Salmonella Pathogenicity Island 2. Vaccine 25:185–193

Ivanoff B, Levine MM, Lambert PH (1994) Vaccination against typhoid fever: present status. Bull World Health Organ 72:957–971

Klebanoff CA, Acquavella N, Yu Z, Restifo NP (2011) Therapeutic cancer vaccines: are we there yet? Immunol Rev 239:27–44

Levine MM (2009) Typhoid vaccines ready for implementation. N Engl J Med 361:403–405

Li X, Fujio M, Imamura M, Wu D, Vasan S, Wong CH, Ho DD, Tsuji M (2010) Design of a potent CD1d-binding NKT cell ligand as a vaccine adjuvant. Proc Natl Acad Sci U S A 107:13010–13015

Luo Y, Zhou H, Mizutani M, Mizutani N, Reisfeld RA, Xiang R (2003) Transcription factor Fos-related antigen 1 is an effective target for a breast cancer vaccine. Proc Natl Acad Sci U S A 100:8850–8855

Manuel ER, Diamond DJ (2013) A road less traveled paved by IDO silencing: harnessing the antitumor activity of neutrophils. Oncoimmunology 2:e23322

Manuel ER, Blache CA, Paquette R, Kaltcheva TI, Ishizaki H, Ellenhorn JD, Hensel M, Metelitsa L, Diamond DJ (2011) Enhancement of cancer vaccine therapy by systemic delivery of a tumor-targeting Salmonella-based STAT3 shRNA suppresses the growth of established melanoma tumors. Cancer Res 71:4183–4191

Marlovits TC, Kubori T, Sukhan A, Thomas DR, Galan JE, Unger VM (2004) Structural insights into the assembly of the type III secretion needle complex. Science 306:1040–1042

Nagarajan NA, Kronenberg M (2007) Invariant NKT cells amplify the innate immune response to lipopolysaccharide. J Immunol 178:2706–2713

Niethammer AG, Xiang R, Ruehlmann JM, Lode HN, Dolman CS, Gillies SD, Reisfeld RA (2001) Targeted interleukin 2 therapy enhances protective immunity induced by an autologous oral DNA vaccine against murine melanoma. Cancer Res 61:6178–6184

Nishikawa H, Sato E, Briones G, Chen LM, Matsuo M, Nagata Y, Ritter G, Jager E, Nomura H, Kondo S, Tawara I, Kato T, Shiku H, Old LJ, Galan JE, Gnjatic S (2006) In vivo antigen delivery by a Salmonella typhimurium type III secretion system for therapeutic cancer vaccines. J Clin Invest 116:1946–1954

Nishikawa H, Tsuji T, Jager E, Briones G, Ritter G, Old LJ, Galan JE, Shiku H, Gnjatic S (2008) Induction of regulatory T cell-resistant helper CD4+ T cells by bacterial vector. Blood 111:1404–1412

Padte NN, Boente-Carrera M, Andrews CD, McManus J, Grasperge BF, Gettie A, Coelho-dos-Reis JG, Li X, Wu D, Bruder JT, Sedegah M, Patterson N, Richie TL, Wong CH, Ho DD, Vasan S, Tsuji M (2013) A glycolipid adjuvant, 7DW8-5, enhances CD8+ T cell responses induced by an adenovirus-vectored malaria vaccine in non-human primates. PLoS One 8:e78407

Pertl U, Wodrich H, Ruehlmann JM, Gillies SD, Lode HN, Reisfeld RA (2003) Immunotherapy with a posttranscriptionally modified DNA vaccine induces complete protection against metastatic neuroblastoma. Blood 101:649–654

Qimron U, Madar N, Mittrucker HW, Zilka A, Yosef I, Bloushtain N, Kaufmann SH, Rosenshine I, Apte RN, Porgador A (2004) Identification of Salmonella typhimurium genes responsible for interference with peptide presentation on MHC class I molecules: Deltayej Salmonella mutants induce superior CD8+ T-cell responses. Cell Microbiol 6:1057–1070

Reisfeld RA, Niethammer AG, Luo Y, Xiang R (2004) DNA vaccines suppress tumor growth and metastases by the induction of anti-angiogenesis. Immunol Rev 199:181–190

Russmann H, Shams H, Poblete F, Fu Y, Galan JE, Donis RO (1998) Delivery of epitopes by the Salmonella type III secretion system for vaccine development. Science 281:565–568

Schoen C, Stritzker J, Goebel W, Pilgrim S (2004) Bacteria as DNA vaccine carriers for genetic immunization. Int J Med Microbiol 294:319–335

Shams H, Poblete F, Russmann H, Galan JE, Donis RO (2001) Induction of specific CD8+ memory T cells and long lasting protection following immunization with Salmonella typhimurium expressing a lymphocytic choriomeningitis MHC class I-restricted epitope. Vaccine 20:577–585

Sizemore DR, Branstrom AA, Sadoff JC (1995) Attenuated Shigella as a DNA delivery vehicle for DNA-mediated immunization. Science 270:299–302

Su GF, Brahmbhatt HN, de Lorenzo V, Wehland J, Timmis KN (1992) Extracellular export of Shiga toxin B-subunit/haemolysin A (C-terminus) fusion protein expressed in Salmonella typhimurium aroA-mutant and stimulation of B-subunit specific antibody responses in mice. Microb Pathog 13:465–476

Tacket CO, Hone DM, Losonsky GA, Guers L, Edelman R, Levine MM (1992) Clinical acceptability and immunogenicity of CVD 908 Salmonella typhi vaccine strain. Vaccine 10:443–446

Thamm DH, Kurzman ID, King I, Li Z, Sznol M, Dubielzig RR, Vail DM, MacEwen EG (2005) Systemic administration of an attenuated, tumor-targeting Salmonella typhimurium to dogs with spontaneous neoplasia: phase I evaluation. Clin Cancer Res 11:4827–4834

Toso JF, Gill VJ, Hwu P, Marincola FM, Restifo NP, Schwartzentruber DJ, Sherry RM, Topalian SL, Yang JC, Stock F, Freezer LJ, Morton KE, Seipp C, Haworth L, Mavroukakis S, White D, MacDonald S, Mao J, Sznol M, Rosenberg SA (2002) Phase I study of the intravenous administration of attenuated Salmonella typhimurium to patients with metastatic melanoma. J Clin Oncol 20:142–152

Vasan S, Tsuji M (2010) A double-edged sword: the role of NKT cells in malaria and HIV infection and immunity. Semin Immunol 22:87–96

Vassaux G, Nitcheu J, Jezzard S, Lemoine NR (2006) Bacterial gene therapy strategies. J Pathol 208:290–298

Verjans GM, Janssen R, UytdeHaag FG, van Doornik CE, Tommassen J (1995) Intracellular processing and presentation of T cell epitopes, expressed by recombinant Escherichia coli and Salmonella typhimurium, to human T cells. Eur J Immunol 25:405–410

Xiang R, Mizutani N, Luo Y, Chiodoni C, Zhou H, Mizutani M, Ba Y, Becker JC, Reisfeld RA (2005) A DNA vaccine targeting survivin combines apoptosis with suppression of angiogenesis in lung tumor eradication. Cancer Res 65:553–561

Xiong G, Husseiny MI, Song L, Erdreich-Epstein A, Shackleford GM, Seeger RC, Jackel D, Hensel M, Metelitsa LS (2010) Novel cancer vaccine based on genes of Salmonella pathogenicity island 2. Int J Cancer 126:2622–2634

Xu X, Husseiny MI, Goldwich A, Hensel M (2010) Efficacy of intracellular activated promoters for generation of Salmonella-based vaccines. Infect Immun 78:4828–4838

Zepp F (2010) Principles of vaccine design-Lessons from nature. Vaccine 28(Suppl 3):C14–C24

Zhang XL, Jeza VT, Pan Q (2008) Salmonella typhi: from a human pathogen to a vaccine vector. Cell Mol Immunol 5:91–97

Chapter 18
Harnessing the Host Immune Response to Infection – BCG Immunotherapy for Bladder Cancer

Hana Zdimerova, Matthew L. Albert, and Molly A. Ingersoll

Abstract Bacillus Calmette–Guérin, or BCG, an avirulent strain of *Mycobacterium bovis*, was developed as a vaccine for the prevention of tuberculosis. BCG's success in disease prevention resulted in the vaccination of billions of individuals. The observation that bacterial components could induce tumor regression, coupled with reports that BCG-vaccinated individuals demonstrated reduced cancer incidence, led to the development of BCG as an immunotherapeutic agent. The pioneering work of Morales and colleagues in the 1970s, demonstrating that direct instillation of live BCG into the bladder of patients with nonmuscle invasive bladder cancer prevented tumor recurrence, laid the path for what is arguably the most successful immunotherapy to date. Notably, although much work has focused on how BCG mediates tumor immunity, important unknowns regarding the mechanism of action remain. Nonspecific innate pathways, such as neutrophil-mediated killing, and adaptive immunity, such as induction of BCG- and tumor-specific T cells, likely work in concert to exert anti-tumor effects. Finally, as the mechanisms of action are unraveled, questions of pharmaco-equivalency of BCG substrains have arisen with respect to the clinical management of bladder cancer patients. Thus, while BCG immunotherapy is currently the standard of care for nonmuscle invasive bladder cancer, there exists great opportunity to improve upon this treatment through additional research and clinical trials.

Keywords Bacillus Calmette–Guérin • Immunotherapy • Bladder cancer • Neutrophils

---

H. Zdimerova • M.L. Albert • M.A. Ingersoll (✉)
Unité d'Immunobiologie des Cellules Dendritiques, Department of Immunology, Institut Pasteur, 25 Rue du Dr Roux, Batiment Metchnikoff, 75015 Paris, France

INSERM U818, Department of Immunology, Institut Pasteur, 25 Rue du Dr Roux, Batiment Metchnikoff, 75015 Paris, France
e-mail: molly.ingersoll@pasteur.fr

M.R. Shurin et al. (eds.), *Infection and Cancer: Bi-Directorial Interactions*, DOI 10.1007/978-3-319-20669-1_18

Abbreviations

BCG	Bacillus Calmette–Guérin
FAP	fibronectin attachment protein
IL	interleukin
IFN	interferon
NET	neutrophil extracellular traps
TRAIL	tumor necrosis factor-related apoptosis-inducing ligand
TUR	transurethral resection

Introduction

One of the best examples of successful cancerimmunotherapy in use today relies upon the instillation of the bacterial strain BCG, or Bacillus Calmette–Guérin, into the bladder. Used to specifically treat nonmuscle invasive bladder cancer, BCG immunotherapy induces up to a 70 % response rate in patients (Brandau and Suttmann 2007). BCG, an attenuated strain of *Mycobacterium bovis*, is more commonly recognized as childhood vaccine against tuberculosis and its role as a bladder cancer immunotherapeutic is less well known. This chapter highlights early therapeutic attempts to treat multiple malignancies with BCG; and discusses the current knowledge regarding mechanisms of tumor immunity induced by BCG, with a specific focus on the role of neutrophils and adaptive responses mediated by T cells. Finally, the role that variation in BCG substrains plays in clinical response is considered.

The Advent of BCG Immunotherapy

Bacillus Calmette–Guérin arose from the deliberate continuous subculture of a virulent strain of *Mycobacterium bovis*. In an effort to develop a vaccine against tuberculosis, Albert Calmette and Camille Guérin passaged this virulent strain more than 260 times over 11 years, producing an avirulent strain of *M. bovis* incapable of producing disease in experimental animal models (Luca and Mihaescu 2013). The attenuated strain was first tested in humans as a vaccine against tuberculosis by Calmette in 1921 (Luca and Mihaescu 2013). Since this first successful trial, BCG vaccination has been in use for more than 90 years. While vaccine efficacy varies significantly, for reasons that are incompletely understood, BCG vaccination displays a striking safety record, with the numbers of vaccinated individuals numbering in the billions (Mangtani et al. 2014; Gan et al. 2013).

Many consider William Coley as one of the first physicians to treat cancer patients with immunotherapy. At the turn of the twentieth century, Coley deliberately injected live and heat-killed *Streptococcus pyogenes* and *Serratia marcescens* directly into the tumors of cancer patients. Strikingly, he was able to induce tumor regression in a wide variety of tumor types, including sarcomas, lymphomas, and melanomas (Hoption Cann et al. 2003). Importantly, Coley's work laid the foundation for the acceptance of bacteria as a therapy for cancer and the application of "Coley's toxins" as immunotherapy continued after his death (Hoption Cann et al. 2003). The finding, in 1928, that the frequency of cancer was reduced in cadavers harboring tuberculosis granulomas suggested that BCG might also have potential as a therapeutic for cancer (reviewed in (Redelman-Sidi et al. 2014)). Retrospective studies from the early 1970s revealed that school-aged children vaccinated with BCG displayed a reduced incidence of acute leukemia (Davignon et al. 1970; Rosenthal et al. 1972). As the concept of immunotherapy was gaining popularity in the 1960s and 1970s (Rosenberg 1975), these observations prompted testing of BCG as a nonspecific immune stimulant for cancer therapy (McKhann and Gunnarsson 1974). One of the earliest pre-clinical applications of this concept was the direct injection of live BCG into subcutaneous tumors in Guinea pigs that had been sensitized to the bacteria or not, prior to therapy (Zbar and Tanaka 1971). In this study, intratumorally-injected BCG induced tumor regression at the primary site and, significantly, prevented the development of metastases in regional lymph nodes in 70 % of treated animals, whereas mice treated only by tumor excision uniformly developed lymph node metastases (Zbar and Tanaka 1971). Following from this work, Zbar and colleagues tested additional variables and tumor models, leading to a set of requirements necessary to achieve an anti-tumor response with BCG (Zbar and Rapp 1974). The concept that the host, or patient, must have an intact immune system capable of mounting an immune response against antigen, such as BCG, had been put forth concurrently by the immunotherapy community (McKhann and Gunnarsson 1974). Zbar expanded the conditions to include that (i) BCG must be alive, (ii) BCG must be in close proximity to the tumor cells, and (iii) the tumor burden must be small (Zbar and Rapp 1974).

Following from these conditions, BCG was tested as a therapy for several malignancies, primarily as a nonspecific immune stimulant, with varying results. Intratumoral injection of BCG up to 7 days after tumor implantation was effective at limiting spread of tumors in a rat model of mammary adenocarcinoma (Kreider et al. 1979). Of note, this study found no evidence to support that pre-sensitization with BCG conferred superior protection. The authors tested sensitized animals for a delayed type hypersensitivity reaction 15 days after vaccination, but waited only 7 days after vaccination to treat tumor bearing animals by intratumoral injection of BCG (Kreider et al. 1979). It is possible that 7 days was insufficient to permit development of a robust adaptive response following BCG vaccination, as *Mycobacterium* species are known to delay T cell priming (Torrado et al. 2011). In patients, BCG showed considerable promise against acute leukemia in the mid 1970s after several clinical studies demonstrated that BCG scarification, in combination with additional chemotherapies, extended the duration of remission in both children and adult patients (Rosenberg 1975; BCG and cancer 1975; Mathe et al. 1972). BCG, in combination with chemotherapy or surgery, extended disease-free survival in patients with malignant melanoma, although these trials were not randomized (Rosenberg

1975). The use of BCG scarification, with or without the addition of 5-fluorouracil for Dukes' C classification of carcinoma of the large bowel, in which the cancer has spread to at least one lymph node, demonstrated better recurrence-free and disease-free survival as compared to surgery alone (Mavligit et al. 1975).

While many of these attempts did not prove successful in further clinical testing (Tan and Ho 1993; Czarnetzki et al. 1993), BCG for the treatment of nonmuscle invasive bladder cancer showed considerable promise (Brandau and Suttmann 2007; Morales et al. 1976). Preclinical studies in a rat model of bladder cancer demonstrated that BCG intravesical therapy prevented tumor progression in vaccinated animals (Lamm et al. 1977). In a naïve canine bladder model, BCG instillation induced robust immune cell infiltration to the bladder, particularly in animals that had been previously vaccinated with BCG, leading the authors to suggest that BCG might be a viable nonspecific immunostimulant for the treatment of bladder cancer (Bloomberg et al. 1975). The initial clinical trial, published in 1976 by Morales and colleagues, coupling percutaneous BCG injection with instillation of BCG into the bladder of nine patients, demonstrated that BCG induced a marked reduction in the incidence of tumor recurrence after therapy as compared to before therapy (Morales et al. 1976). In addition, Morales observed that there was a small increase in the number of infiltrating lymphocytes after BCG instillation (Morales et al. 1976). The successful reduction in tumor recurrence observed by Morales lead to the initiation of two controlled clinical trials, which showed that the protection provided by BCG extended up to 10 years after therapy, suggesting that BCG induces long term modifications to the immune system (Gandhi et al. 2013; Sarosdy and Lamm 1989; Lamm et al. 1980; Pinsky et al. 1985). These findings laid the foundation for BCG to become the standard of care for nonmuscle invasive bladder cancer.

BCG Immunotherapy for Bladder Cancer

Carcinoma of the bladder is the 4th most common cancer in men, the 8th in women, and due its chronic, recurrent nature, the most costly malignancy to manage (per patient) from diagnosis to death (Sievert et al. 2009; Botteman et al. 2003). Approximately 72,500 people in the US and more than 100,000 individuals in the EU are estimated to be at risk for bladder cancer annually (Ferlay et al. 2007, 2013; Siegel et al. 2013). A majority of urothelial tumors (>75 %) present as superficial lesions that have not infiltrated connective tissue or the bladder wall (Brandau and Suttmann 2007). These non-muscle invasive bladder cancers can be categorized according to their potential to recur and/or progress using the EORTC (European Organisation for Research and Treatment of Cancer) risk tables, generated from observations in seven EORTC monitored clinical trials (Sylvester et al. 2006). Low risk nonmuscle invasive tumors are typically surgically resected and monitored over time for recurrence without additional treatment (Schmitz-Drager et al. 2014). High risk nonmuscle invasive disease includes tumors staged as Ta (urothelial cell layer), T1 (invasion into the mucosa) and/or carcinoma *in situ*, characterized by flat tumors spread over the surface of the bladder urothelium. Patients with high risk bladder cancer typically undergo transurethral resection (TUR) of the tumor, except in the case of carcinoma *in situ*, followed by BCG immunotherapy. During the induction phase of therapy, 1–5 x 10^8 colony forming units of live BCG are administered intravesically once per week over 6 weeks (Brandau and Suttmann 2007). Numerous clinical studies have established that (i) TUR followed by adjuvant BCG therapy results in fewer recurrences as compared to TUR alone; (ii) BCG therapy is superior to intravesical chemotherapy for nonmuscle invasive bladder cancer, and (iii) BCG is the only treatment option with the ability to prevent or delay the progression to muscle invasive disease (Brandau and Suttmann 2007; Gandhi et al. 2013; Herr et al. 1983). Of note, the protocol defined by Morales, *i.e.*, 6 weekly BCG instillations, has changed very little in the last 40 years. The most significant change is the discontinuation of concurrent intradermal injection of BCG during the course of intravesical therapy (Morales et al. 1976). A second significant change – the introduction of maintenance therapy – is based on the fact that while BCG therapy effectively delays or prevents recurrence, there is considerable evidence that BCG therapy requires additional instillation cycles at regular intervals (every 3 months) for its full efficacy to be realized (Gandhi et al. 2013; Sylvester et al. 2002).

Worldwide, approximately 200,000 patients are treated with BCG annually, of which 30–50 % will likely experience tumor recurrence (Botteman et al. 2003). Incidence has increased over the past decades, however, death rates have diminished due to current therapeutic strategies and improved monitoring (Siegel et al. 2013). In the case of progressive disease, treatment includes cystectomy or chemoradiotherapy, which results in a severe decline in quality of life (Singer et al. 2013). Thus, BCG immunotherapy can be considered as a viable bladder-sparing mechanism in eligible patients, thus, leading to better quality of life for these patients as compared to those undergoing radical cystectomy (Cookson et al. 1997).

BCG Induces Nonspecific Tumor Immunity Through the Action of Neutrophils

While it is frequently noted that the mechanisms of BCG-mediated tumor immunity are not well defined, there has been considerable research on the early immune response after BCG intravesical instillation, such as the infiltration of neutrophils and their impact on disease. Neutrophil granulocytes play a critical role in the response to many types of infection or inflammatory disease. Their beneficial role in the immune response to tumors is still debated, as they have the potential to exert both tumor promoting or tumor inhibitory actions (Gutkin and Shurin 2014; Galdiero et al. 2013). In the context of bladder cancer, however, existing evidence predominantly supports neutrophils as tumor inhibitory, even though their specific role is still unclear. Further studies are needed to explore their contribution to bladder cancer immunity in more detail.

As is typical in bacterial infections, neutrophils are the first to arrive to the site of infection and are the most abundant cell type in the bladder in the first hours following BCG instillation (Bisiaux et al. 2009; De Boer et al. 1991). After BCG is instilled into the bladder lumen, the mycobacteria are believed to adhere to urothelial cells via fibronectin attachment protein (FAP), predominantly in areas of bladder injury (Ratliff et al. 1987a; Kavoussi et al. 1990; Zhao et al. 2000). Either the presence or the adherence of bacteria then induces a rapid cytokine response, observed in patients and animal models (Bisiaux et al. 2009; De Boer et al. 1991; Jackson et al. 1995; Biot et al. 2012; Bohle et al. 1990). Several of these cytokines are critical for consequent neutrophil trafficking and shaping of the immune response, such as interleukin (IL)-6, -8, and -18. IL-8, a potent neutrophil chemoattractant, is one of the most abundant cytokines produced after initial and successive BCG instillations and has been observed to be highly stable even after *ex vivo* incubation of urine, making it an attractive candidate for potential biomarker studies (de Boer et al. 1997). Several clinical studies have reported that IL-8 secreted into patient urine in the hours following BCG instillation correlates with outcome after BCG therapy; low cytokine levels have been linked to recurrence, while high levels predict a positive response to the treatment with respect to disease-free survival (de Boer et al. 1997; Thalmann et al. 1997, 2000; Watanabe et al. 2003; Zuiverloon et al. 2012). While less evidence exists, increased levels of IL-6 and IL-18 are also correlated with improved clinical response after BCG therapy (Thalmann et al. 2000; Lima et al. 2012).

Interleukin 17 (IL-17) also increases over the course of BCG therapy, acting to recruit neutrophils to the bladder (Takeuchi et al. 2011). This pro-inflammatory cytokine induces chemotaxis of neutrophils in an indirect manner through other mediators, such as by induction of the chemokine CXCL1 (Kolls and Linden 2004). Takeuchi and colleagues demonstrated that in an orthotopic tumor model, IL-17 deficient mice exhibit a significant decrease in neutrophil infiltration into the bladder, correlating with decreased survival after tumor challenge and BCG therapy (Takeuchi et al. 2011). While $CD4^+$ T_h17 cells produce IL-17, much evidence has shown that γδ T cells also produce high amounts of this cytokine *in vivo* (Chien et al. 2013). It is this subset of T cells, stimulated by BCG, that is the source of IL-17 in the bladder (Takeuchi et al. 2011). In γδ T cell-deficient mice, neutrophil recruitment into the bladder was greatly decreased and similar to the IL-17 knock out mouse model, BCG therapy did not inhibit tumor growth or improve survival, as compared to wild type C57Bl/6 mice treated with BCG following tumor challenge (Takeuchi et al. 2011). Further supporting the hypothesis that γδ T cells are the primary source of IL-17, depletion of $CD4^+$ T cells did not result in a significant decrease in IL-17 production or neutrophil infiltration in the bladder over the course of BCG therapy (Takeuchi et al. 2011).

These data support a role for neutrophils in the initiation of the immune response and suggest that their presence in the bladder is necessary for successful clinical response to BCG. However, these studies have not proposed a mechanism in which neutrophils induce potential tumoricidal activities or how BCG may influence the neutrophil. BCG stimulation, in fact, impacts the phenotype and gene expression of neutrophils, suggesting that their tumoricidal activity is potentiated in the bladder in the presence of mycobacteria (Suttmann et al. 2003). Neutrophils in circulation have an extremely short half-life. However, after tissue injury or inflammation they become activated, extravasating from the blood into the affected tissue and exhibit an increased life span, prolonging the immune response (Coxon et al. 1999). Neutrophils exposed to BCG for only 2 h were observed to upregulate surface integrins and Fc receptors, such as CD11b, CD18, CD16, and CD32, important for adhesion, chemotaxis, and phagocytosis (Suttmann et al. 2003). Exposure to BCG also inhibited apoptosis in *ex vivo*-cultured neutrophils and altered their gene expression, *e.g.*, by inducing the upregulation of cytokines and their receptors (Suttmann et al. 2003). Each of these stimulated neutrophil behaviors favors the host response in BCG bladder cancer therapy. By upregulating crucial surface receptors, neutrophils migrate and extravasate into the bladder to kill mycobacteria. Inhibition of apoptosis allows the prolongation of the inflammatory response in the bladder, while upregulation or *de novo* synthesis of cytokines, chemokines, and their receptors allow for the attraction of monocytes or effector cells into the bladder lumen.

A common hypothesis for the mechanism of action of neutrophils in tumor immunity is that the cells directly kill tumor cells. Studies regarding the direct antitumor action of neutrophils in BCG therapy have pointed to a role for tumor necrosis factor-related apoptosis-inducing ligand (TRAIL). TRAIL induces apoptosis in transformed or tumor cell lines in a ligand-induced manner, without killing non-transformed or healthy cells, by binding to death receptors 4 or 5 (Wiley et al. 1995; Pan et al. 1997a, b; Sheridan et al. 1997). The presence of decoy receptors 1 and 2, not usually expressed by malignant cells, prevents apoptosis in healthy cells (Pan et al. 1997a). Patients responding to BCG therapy exhibit significantly higher levels of functional TRAIL and TRAIL-expressing neutrophils in their urine in the hours following BCG instillation (Ludwig et al. 2004). In addition, urine from these patients induces death in a bladder tumor cell line, while urine from non-responders, or healthy donors, does not impact cell viability. Supporting a role for TRAIL-mediated killing of bladder tumor cells, urine depleted of TRAIL failed to induce cell death (Ludwig et al. 2004). Interferon (IFN)-α stimulation induces TRAIL and has a synergistic effect on the production of TRAIL when combined with BCG stimulation (Kemp et al. 2005; Tecchio et al. 2004). This observation has clinical relevance, as a correlation exists between T^h1-biased inflammatory cytokines, such as IFNs, and TRAIL in the positive responsiveness to BCG therapy for bladder cancer. TRAIL-mediated induction of cell death supports a direct antitumor activity of neutrophils after BCG stimulation. However, neutrophil-mediated direct killing is unlikely to fully account for BCG-mediated tumor immunity. Mathematical modeling of the innate immune response to BCG, in which the hypothesis that only innate cells (*e.g.*, neutrophils, monocytes) are responsible for mechanisms of tumor immunity was tested, found that the contribution of innate cells is insufficient to orchestrate full tumor immunity (Breban et al. 2012). While this model makes several assumptions that remain to be tested, *in vivo* studies support the conclusion that adaptive immunity is required to achieve BCG immunotherapy (Ratliff et al. 1987b, 1993).

Beyond a role for neutrophils in phagocytosis, cytokine release, and direct tumor cell killing, Suttmann and colleagues demonstrated that they provide chemotactic and activating factors to initiate recruitment of the adaptive arm of the immune system to the bladder (Suttmann et al. 2006). Neutrophils stimulated *in vitro* by BCG or activated neutrophils isolated from BCG-treated patient urine express large amounts of IL-8, GRO-α, and MIP-1α (Suttmann et al. 2006). Further, BCG-activated neutrophils induce monocyte chemotaxis, but do not directly induce T cell migration (Suttmann et al. 2006). Only monocytes activated with BCG-stimulated neutrophil supernatants attracted T cells, suggesting the T cells are induced to infiltrate the bladder indirectly via activated monocytes (Suttmann et al. 2006). Indeed, *in vivo*, neutrophil-depleted mice were found to lack significant $CD4^+$ T cell infiltration, supporting the concept that neutrophils are essential for effector cell infiltration (Suttmann et al. 2006). While this study adds to our knowledge regarding the potential role of neutrophils in the recruitment of effector cells, the use of the anti-GR-1 clone RB6-8C5 to deplete neutrophils *in vivo* complicates interpretation of these results. This particular antibody clone also depletes classical, or inflammatory, monocytes and some lymphocyte populations, due to its capacity to recognize the antigens Ly6G and Ly6C (Daley et al. 2008). Therefore, the *in vivo* phenomena observed might be the result of depletion of one or more cell types and additional studies are necessary to resolve this question.

It is likely that neutrophils possess an essential role in the initiation of an immune response against bladder cancer, playing an anti-tumor role. However, recent research has emerged suggesting a role for neutrophils in metastasis and tumor progression. IL-8, produced by tumor cells, may support metastasis by recruiting neutrophils to the tumor, where they release matrix-degrading enzymes, altering the architecture of the extracellular matrix (De Larco et al. 2004). Breakdown of the tumor bed would facilitate metastasis by allowing tumor cells to pass through the endothelium, enter into circulation, and consequently migrate to distal sites (De Larco et al. 2004). IL-8 expressing melanoma cells rapidly recruit neutrophils, in a B2-integrin/ICAM-1 dependent manner (Huh et al. 2010). Subsequently, tumor cells display increased migration through the endothelial layer, leading to proliferation and dissemination (Huh et al. 2010). As neutrophils robustly infiltrate the bladder in the course of BCG immunotherapy, further study into the potential negative role played by these cells is warranted.

More recently, it has been demonstrated that neutrophil extracellular traps (NETs) – extracellular DNA webs coated with granular proteins – play a role in metastasis of tumor cells after post-surgical infection (Cools-Lartigue et al. 2013). NETs, released by neutrophils in response to infection, are a host defense mechanism for the efficient trapping and clearing of invading pathogens (Brinkmann et al. 2004; McDonald et al. 2012; Papayannopoulos and Zychlinsky 2009). In many patients, tumor resection is followed by postsurgical infection and surprisingly, it is severe infections such as pneumonia or sepsis that are associated with poor oncologic outcomes and a higher risk of death from metastatic disease (Schussler et al. 2006; Farid et al. 2010). One possible explanation for this phenomenon is the observation that NETs, released during postsurgical infection, capture remaining circulating tumor cells at sites distant from the original tumor, leading to tumor cell dissemination and consequent metastasis (Cools-Lartigue et al. 2013). Inhibition of NET formation, through depletion of neutrophils or chemical inhibition of NET components, was observed to decrease the number of observed metastases in the presence of systemic sepsis (Cools-Lartigue et al. 2013).

While they have been observed in many tissues and cell lines, the impact of NETs in the bladder has not been investigated. Importantly, BCG immunotherapy is essentially a live mycobacterial infection of a recently tumor-resected bladder. While at first consideration, this might suggest neutrophil recruitment induced by BCG has the potential to promote metastasis, it is important to keep in mind that muscle-invasive stages of bladder cancer (T2-T4) are not treated by BCG therapy. Indeed, superficial bladder tumors (high grade Ta, T1, or CIS) are non-invasive and have not penetrated tissue beyond the lamina propria. We would propose an alternative hypothesis, that after tumor resection, the majority of remaining tumor cells is lost to micturition while neutrophil NETs trap the mycobacteria, extending its exposure to the immune system. Thus, it may be that NET formation, specifically in the context of bladder cancer, does not induce metastasis, but potentially increases the efficacy of BCG therapy. Further investigation is needed to evaluate the role of NETs in the bladder during BCG immunotherapy, including the potential benefit or detriment from increasing NET generation.

Thus, the role of infiltrating neutrophils in tumor immunology is not entirely clear, however when considering general kinetics of an immune response, it seems reasonable that the first cell infiltrating the bladder after BCG instillation would have a crucial role in initiating and orchestrating the immune response and perhaps link the innate and adaptive players of the immune system.

BCG and Tumor Specific Immunity – The Role of the Adaptive Immune System

The success of BCG lies in part with the patient, as immunocompetence is one of the key requirements for the success of BCG immunotherapy (Zbar and Rapp 1974; BCG and cancer 1975). As discussed above, data support a role for the adaptive immune system in the anti-tumor effects of intravesical BCG. Indeed, while BCG-induced immunity to bladder cancers was thought to be mediated by nonspecific immune stimulation (Wolfe et al. 1976), early evidence examining $CD4^+$ and $CD8^+$T cell infiltration after BCG therapy suggested that specific adaptive immune mechanisms might play a role, as well (Boccafoschi et al. 1995; Sarica et al. 1995; Prescott et al. 1992). Supporting this, intravesical instillation of BCG results in the priming of BCG-specific $CD4^+$ and $CD8^+$T cells (Biot et al. 2012; Lattime et al. 1992). Both $CD4^+$ and $CD8^+$ T cells are required for tumor immunity, as athymic mice do not respond to BCG therapy unless they are reconstituted with splenocytes from BCG-vaccinated animals (Ratliff et al. 1987b). Further, depletion of either T cell subset eliminates BCG-mediated antitumor activity (Ratliff et al. 1993). Of note, mice that rejected their tumors after BCG therapy in this study did not exhibit anti-tumor immunity upon tumor challenge, suggesting that T cell mediated tumor immunity induced by BCG is not tumor specific (Ratliff et al. 1993).

In the context of BCG therapy, $CD4^+$, or T helper, T cell polarization (*e.g.*, T_h1 or T_h2,) is an important consideration. T_h1-mediated responses are generally thought to be necessary for response to bacterial infection and tumor immunity. Thus, it is hypothesized that a strong T_h1 bias is necessary for the success of BCG immunotherapy in bladder cancer patients. BCG therapy induces a predominantly T_h1-biased T cell response, characterized by the expression of IL-2, IL-12, and IFN-γ, but has been shown to induce T_h2 related cytokines, such as IL-10, as well (Redelman-Sidi et al. 2014). Supporting a role for T_h1 bias, high levels of IL-2 in patient urine during the course of therapy are predictive of a positive response to BCG therapy (Zuiverloon et al. 2012; Saint et al. 2001, 2002; Schwentner et al. 2012). T_h1 bias is crucial for control of tumors in preclinical models, as immunotherapy is ineffective against tumors implanted in IFN-γ and IL-12 deficient mice, but improved in IL-10 knockout mice (Riemensberger et al. 2002). Efforts to induce a more pronounced T_h1 bias have been moderately successful. In a mouse model of BCG therapy, fewer instillations, *i.e.*, at week one and week six, induced similar expression levels of T_h1-associated cytokines as six weekly instillations (de Boer et al. 2005). Interestingly, T_h2 associated cytokine mRNA expression levels were reduced in mice receiving only two instillations as compared to mice instilled six times (de Boer et al. 2005). These data suggest that a reduction in the number of BCG instillations might be lead to a more pronounced T helper cell bias; however, the authors did not test the efficacy of a reduced number of BCG treatments on the induction of tumor immunity (de Boer et al. 2005).

Bladder cancer occurs most frequently in elderly patients, who may have additional health complications, such as high cholesterol. Statins, used to lower LDL levels in patients with high cholesterol, have also been found to be immunomodulatory, biasing T helper cells towards a T_h2 response (Arnaud et al. 2005). While T_h2-associated cytokines are thought to be protective in cardiovascular disease, this imbalance may potentially negatively impact BCG therapy. Retrospective analysis of patients taking statins at the time of BCG therapy revealed no differences in the number of recurrences, however, uncovered a statistically significant increase in tumor progression, leading to an increased necessity for radical cystectomy compared to patients not taking statins (Hoffmann et al. 2006). These data provide additional support for a role of T_h1-biased T cell responses in anti-tumor response during BCG therapy but do not necessarily demonstrate a direct impact of statin therapy on the outcome of BCG-treated patients.

While early studies demonstrated that the absence of $CD4^+$ or $CD8^+$T cells abrogated BCG-mediated anti-tumor immunity (Ratliff et al. 1987b), more recent work has revealed that BCG-specific $CD8^+$ T cells are critical for tumor immunity in a mouse orthotopic bladder cancer model (Biot et al. 2012). In these studies, while several BCG instillations were observed to be necessary to induce T cell homing to the bladder, immunization of mice with BCG 2 weeks prior to initiating immunotherapy accelerated the kinetics of T cell infiltration into the bladder (Biot et al. 2012).

BCG vaccination was found to improve response to BCG intravesical instillation, with 100 % of vaccinated animals surviving tumor challenge, as compared to ~60 % lethality in unvaccinated mice (Biot et al. 2012). As further evidence that prior immunity to BCG enhances BCG therapy, patients who were PPD+ at the onset of their immunotherapy, indicating previous exposure to mycobacteria, exhibited improved recurrence-free survival after BCG immunotherapy (Biot et al. 2012). Similarly, the earliest clinical studies observed that patients who seroconverted from PPD- to PPD+ during therapy had a better clinical outcome (Lamm et al. 1981; Kelley et al. 1985; Winters and Lamm 1981). As stated above, the original protocol for BCG immunotherapy included concurrent intradermal BCG injections (Morales et al. 1976). This practice was discontinued, as it was not shown to be more effective than BCG intravesical therapy alone (Luftenegger et al. 1996). It is possible that concurrent BCG injection did not improve therapy, as an appropriate time interval to develop an immune response was not incorporated into the treatment protocol. To address the potential for improved clinical outcome following BCG vaccination prior to BCG therapy, controlled clinical trials are needed.

Many of the studies examining T cell response during BCG therapy do not measure antigen specific T cells. While tumor specific lymphocytes arise during the course of bladder cancer (Marits et al. 2006), few tools exist to directly measure the anti-tumor response induced by BCG in the laboratory and the clinic. Indeed, very few bladder cancer tumor antigens have been described (*e.g.*, MAGE-A3, NY-ESO-1) and none are specific only to bladder cancer (Dyrskjot et al. 2012; Sharma et al. 2003). Efforts to identify bladder cancer antigens and develop tools to easily identify antigen specific T cells before and during therapy would be expected to guide therapeutic decisions regarding the success of BCG therapy. Further, the identification of specific tumor antigens arising early in disease, for use in mouse models to evaluate novel therapies, or more importantly, in humans to monitor response to therapy, would advance our understanding of the host response as well as guide therapeutic decision making.

Does BCG Strain Matter in the Treatment of Bladder Cancer?

During the serial passage of virulent *Mycobacterium bovis*, the strain underwent radical genomic changes, resulting in the attenuated BCG Pasteur strain (Luca and Mihaescu 2013; Brosch et al. 2007). Once BCG was found to be both safe and effective at preventing childhood tuberculosis, live cultures were disseminated around the world and maintained by continuous culture due to the lack of preservation methodologies (Brosch et al. 2007). Thus, over time, disseminated BCG strains displayed genetic drift, in some cases losing significant portions of the mycobacterial genome, resulting in highly variant BCG substrains (Brosch et al. 2007). The consequence of these changes was unknown at the time; indeed, the changes themselves were likely underappreciated. Of note, BCG vaccination for the prevention of childhood tuberculosis displays an efficacy that ranges from 0 to 80 % (Andersen and Doherty 2005). One possible explanation for the wide range of efficacy may be substrain differences, a subject that has received a lot consideration in the tuberculosis field (Ritz and Curtis 2009). Currently, more than 8 different BCG substrains are used clinically without strong evidence of pharmaco-equivalence (Gan et al. 2013; Ritz and Curtis 2009). Recent work has demonstrated that clear differences exist among the strains with respect to BCG immunotherapy (Gan et al. 2013; Noon and Kulkarni 2014). In a comparison between evolutionarily early and late BCG substrains, it was observed that BCG Russia and BCG Connaught induced the highest inhibition of cell proliferation and production of IL-6 and IL-8 in cultured cells, while BCG Glaxo, Phipps, and Tice strains were the least efficacious (Secanella-Fandos et al. 2013). Sengiku and colleagues could not identify a significant difference in the complete response, recurrence-free survival, or adverse event rate, in a clinical trial of 178 patients randomized to receive BCG Connaught or BCG Tokyo (Sengiku et al. 2013). Unfortunately, the disruption in production of BCG Connaught forced the trial to end early (Sengiku et al. 2013).

More recently, Swiss clinical teams performed a direct head-to-head comparison of the two commonly used strains, BCG Connaught and BCG Tice (Rentsch et al. 2014). In this trial, patients were stratified to receive either BCG Connaught or BCG Tice. BCG Connaught treated patients exhibited a 74 % 5-year recurrence-free survival, while only 48 % of patients treated with BCG Tice showed the same success in treatment. Pre-clinical studies to understand the underlying reasons for the clinical differences pointed toward the differing capacity of the two strains to induce immune cell infiltration and prime BCG-specific T cells (Rentsch et al. 2014). Sequencing of the two strains identified strain-specific genetic differences, including a point mutation in the superoxide dismutase C (*SodC*) gene of BCG Tice, that may underlie the differential immune activation phenotype, however this remains to be tested (Rentsch et al. 2014). Indeed, direct comparison of additional clinical strains currently in use should be performed in order to determine the most efficacious strain at preventing tumor recurrence (Noon and Kulkarni 2014).

Future Perspectives for Improving BCG Immunotherapy for Bladder Cancer

Current practices for monitoring recurrence and progression in bladder cancer include cystoscopy and voided urine cytology, which are invasive and require specific expertise. The focus of many clinical groups has included identifying predictors of successful response to BCG therapy (Schmitz-Drager et al. 2014; Zuiverloon et al. 2012; Lima et al. 2012). The incorporation of biomarkers to detect early detection of changes in patient status, such as tumor recurrence or disease progression, has not been universally employed due to suboptimal specificity and sensitivity (Schmitz-Drager et al. 2014; Schwentner et al. 2012). Discovery and development of companion diagnostics to monitor response to therapy are needed as patients with recurrent disease require further intervention, such as additional

BCG cycles, alternate chemotherapies, or bladder resection. Despite 50–70 % positive clinical response rates, BCG therapy, in its current form, does not prevent tumor recurrence or progression in a significant number of patients. While innovation into therapeutic approach is needed to treat nonresponsive patients, bladder cancer research does not receive significant funding in proportion to its prevalence (Lotan et al. 2009; Kaplan et al. 2014; Lerner 2005). Despite this challenge, further research is needed to understand variables such as substrain differences, pre-therapy vaccination, or concurrent administration of chemotherapeutics, T_h1-polarizing cytokines, or novel immunotherapeutics. Significant headway has been made in several of these approaches (Ingersoll and Albert 2013), however, much study remains in order to truly understand and exploit mechanisms of BCG immunotherapy.

References

Andersen P, Doherty TM (2005) The success and failure of BCG – implications for a novel tuberculosis vaccine. Nat Rev Microbiol 3:656–662

Arnaud C, Braunersreuther V, Mach F (2005) Toward immunomodulatory and anti-inflammatory properties of statins. Trends Cardiovasc Med 15:202–206

BCG and cancer (1975). Can Med Assoc J 113(8):708–710

Biot C et al (2012) Preexisting BCG-specific T cells improve intravesical immunotherapy for bladder cancer. Sci Transl Med 4:137ra172

Bisiaux A et al (2009) Molecular analyte profiling of the early events and tissue conditioning following intravesical bacillus Calmette-Guerin therapy in patients with superficial bladder cancer. J Urol 181:1571–1580

Bloomberg SD, Brosman SA, Hausman MS, Cohen A, Battenberg JD (1975) The effects of BCG on the dog bladder. Investig Urol 12:423–427

Boccafoschi C, Montefiore F, Pavesi M, Pastormerlo M, Betta PG (1995) Late effects of intravesical bacillus Calmette-Guerin immunotherapy on bladder mucosa infiltrating lymphocytes: an immunohistochemical study. Eur Urol 27:334–338

Bohle A et al (1990) Elevations of cytokines interleukin-1, interleukin-2 and tumor necrosis factor in the urine of patients after intravesical bacillus Calmette-Guerin immunotherapy. J Urol 144:59–64

Botteman MF, Pashos CL, Redaelli A, Laskin B, Hauser R (2003) The health economics of bladder cancer: a comprehensive review of the published literature. Pharmacoecon 21:1315–1330

Brandau S, Suttmann H (2007) Thirty years of BCG immunotherapy for non-muscle invasive bladder cancer: a success story with room for improvement. Biomed Pharmacother 61:299–305

Breban R et al (2012) Mathematical model of tumor immunotherapy for bladder carcinoma identifies the limitations of the innate immune response. Oncoimmunol 1:9–17

Brinkmann V et al (2004) Neutrophil extracellular traps kill bacteria. Science 303:1532–1535

Brosch R et al (2007) Genome plasticity of BCG and impact on vaccine efficacy. Proc Natl Acad Sci U S A 104:5596–5601

Chien YH, Zeng X, Prinz I (2013) The natural and the inducible: interleukin (IL)-17-producing gammadelta T cells. Trends Immunol 34:151–154

Cookson MS et al (1997) The treated natural history of high risk superficial bladder cancer: 15-year outcome. J Urol 158:62–67

Cools-Lartigue J et al (2013) Neutrophil extracellular traps sequester circulating tumor cells and promote metastasis. J Clin Invest. 2013 Aug 1; 123(8): 3446–3458

Coxon A, Tang T, Mayadas TN (1999) Cytokine-activated endothelial cells delay neutrophil apoptosis in vitro and in vivo. A role for granulocyte/macrophage colony-stimulating factor. J Exp Med 190:923–934

Czarnetzki BM et al (1993) Long-term adjuvant immunotherapy in stage I high risk malignant melanoma, comparing two BCG preparations versus non-treatment in a randomised multicentre study (EORTC Protocol 18781). Eur J Cancer 29A:1237–1242

Daley JM, Thomay AA, Connolly MD, Reichner JS, Albina JE (2008) Use of Ly6G-specific monoclonal antibody to deplete neutrophils in mice. J Leukoc Biol 83:64–70

Davignon L, Robillard P, Lemonde P, Frappier A (1970) B.C.G. vaccination and leukemia mortality. Lancet 2:638

De Boer EC et al (1991) Presence of activated lymphocytes in the urine of patients with superficial bladder cancer after intravesical immunotherapy with bacillus Calmette-Guerin. Cancer Immunol Immunother 33:411–416

de Boer EC et al (1997) Role of interleukin-8 in onset of the immune response in intravesical BCG therapy for superficial bladder cancer. Urol Res 25:31–34

de Boer EC, Rooyakkers SJ, Schamhart DH, de Reijke TM, Kurth KH (2005) BCG dose reduction by decreasing the instillation frequency: effects on local Th1/Th2 cytokine responses in a mouse model. Eur Urol 48:333–338

De Larco JE, Wuertz BR, Furcht LT (2004) The potential role of neutrophils in promoting the metastatic phenotype of tumors releasing interleukin-8. Clin Cancer Res 10:4895–4900

Dyrskjot L et al (2012) Expression of MAGE-A3, NY-ESO-1, LAGE-1 and PRAME in urothelial carcinoma. Br J Cancer 107:116–122

Farid SG et al (2010) Correlation between postoperative infective complications and long-term outcomes after hepatic resection for colorectal liver metastasis. Ann Surg 251:91–100

Ferlay J et al (2007) Estimates of the cancer incidence and mortality in Europe in 2006. Ann Oncol 18:581–592

Ferlay J et al (2013) Cancer incidence and mortality patterns in Europe: estimates for 40 countries in 2012. Eur J Cancer 49:1374–1403

Galdiero MR et al (2013) Tumor associated macrophages and neutrophils in cancer. Immunobiology 218:1402–1410

Gan C, Mostafid H, Khan MS, Lewis DJ (2013) BCG immunotherapy for bladder cancer–the effects of substrain differences. Nat Rev Urol 10:580–588

Gandhi NM, Morales A, Lamm DL (2013) Bacillus Calmette-Guerin immunotherapy for genitourinary cancer. BJU Int 112:288–297

Gutkin DW, Shurin MR (2014) Clinical evaluation of systemic and local immune responses in cancer: time for integration. Cancer Immunol Immunother 63:45–57

Herr HW, Pinsky CM, Whitmore WF Jr, Oettgen HF, Melamed MR (1983) Effect of intravesical Bacillus Calmette-Guerin (BCG) on carcinoma in situ of the bladder. Cancer 51:1323–1326

Hoffmann P, Roumeguere T, Schulman C, van Velthoven R (2006) Use of statins and outcome of BCG treatment for bladder cancer. N Engl J Med 355:2705–2707

Hoption Cann SA, van Netten JP, van Netten C (2003) Dr William Coley and tumour regression: a place in history or in the future. Postgrad Med J 79:672–680

Huh SJ, Liang S, Sharma A, Dong C, Robertson GP (2010) Transiently entrapped circulating tumor cells interact with neutrophils to facilitate lung metastasis development. Cancer Res 70:6071–6082

Ingersoll MA, Albert ML (2013) From infection to immunotherapy: host immune responses to bacteria at the bladder mucosa. Mucosal Immunol 6:1041–1053

Jackson AM et al (1995) Changes in urinary cytokines and soluble intercellular adhesion molecule-1 (ICAM-1) in bladder cancer patients after bacillus Calmette-Guerin (BCG) immunotherapy. Clin Exp Immunol 99:369–375

Kaplan AL, Litwin MS, Chamie K (2014) The future of bladder cancer care in the USA. Nat Rev Urol 11:59–62

Kavoussi LR, Brown EJ, Ritchey JK, Ratliff TL (1990) Fibronectin-mediated Calmette-Guerin bacillus attachment to murine bladder mucosa. Requirement for the expression of an antitumor response. J Clin Invest 85:62–67

Kelley DR et al (1985) Intravesical bacillus Calmette-Guerin therapy for superficial bladder cancer: effect of bacillus Calmette-Guerin viability on treatment results. J Urol 134:48–53
Kemp TJ et al (2005) Neutrophil stimulation with Mycobacterium bovis bacillus Calmette-Guerin (BCG) results in the release of functional soluble TRAIL/Apo-2L. Blood 106:3474–3482
Kolls JK, Linden A (2004) Interleukin-17 family members and inflammation. Immunity 21:467–476
Kreider JW, Bartlett GL, Boyer CM, Purnell DM (1979) Conditions for effective Bacillus Calmette-Guerin immunotherapy of postsurgical metastases of 13762A rat mammary adenocarcinoma. Cancer Res 39:987–992
Lamm DL, Harris SC, Gittes RF (1977) Bacillus Calmette-Guerin and dinitrochlorobenzene immunotherapy of chemically induced bladder tumors. Investig Urol 14:369–372
Lamm DL et al (1980) Bacillus Calmette-Guerin immunotherapy of superficial bladder cancer. J Urol 124:38–40
Lamm DL, Thor DE, Winters WD, Stogdill VD, Radwin HM (1981) BCG immunotherapy of bladder cancer: inhibition of tumor recurrence and associated immune responses. Cancer 48:82–88
Lattime EC, Gomella LG, McCue PA (1992) Murine bladder carcinoma cells present antigen to BCG-specific CD4+ T-cells. Cancer Res 52:4286–4290
Lerner SP (2005) Bladder cancer clinical trials. Urol Oncol 23:275–279
Lima L, Dinis-Ribeiro M, Longatto-Filho A, Santos L (2012) Predictive biomarkers of bacillus Calmette-Guerin immunotherapy response in bladder cancer: where are we now? Adv Urol 2012:232609
Lotan Y et al (2009) Key concerns about the current state of bladder cancer: a position paper from the Bladder Cancer Think Tank, the Bladder Cancer Advocacy Network, and the Society of Urologic Oncology. Cancer 115:4096–4103
Luca S, Mihaescu T (2013) History of BCG vaccine. Maedica 8:53–58
Ludwig AT et al (2004) Tumor necrosis factor-related apoptosis-inducing ligand: a novel mechanism for Bacillus Calmette-Guerin-induced antitumor activity. Cancer Res 64:3386–3390
Luftenegger W et al (1996) Intravesical versus intravesical plus intradermal bacillus Calmette-Guerin: a prospective randomized study in patients with recurrent superficial bladder tumors. J Urol 155:483–487
Mangtani P et al (2014) Protection by BCG vaccine against tuberculosis: a systematic review of randomized controlled trials. Clin Infect Dis 58:470–480
Marits P et al (2006) Detection of immune responses against urinary bladder cancer in sentinel lymph nodes. Eur Urol 49:59–70
Mathe G et al (1972) Attempts at stimulation of immune response in acute lymphoid leukaemia. Proc R Soc Med 65:266–270
Mavligit GM et al (1975) Adjuvant immunotherapy and chemoimmunotherapy in colorectal cancer of the Dukes' C classification. Preliminary clinical results. Cancer 36:2421–2427
McDonald B, Urrutia R, Yipp BG, Jenne CN, Kubes P (2012) Intravascular neutrophil extracellular traps capture bacteria from the bloodstream during sepsis. Cell Host Microbe 12:324–333
McKhann CF, Gunnarsson A (1974) Approaches to immunotherapy. Cancer 34(suppl): 1521–1531
Morales A, Eidinger D, Bruce AW (1976) Intracavitary Bacillus Calmette-Guerin in the treatment of superficial bladder tumors. J Urol 116:180–183
Noon AP, Kulkarni GS (2014) All bacillus Calmette-Guerin (BCG) strains are equal, but some BCG strains are more equal than others. Eur Urol 66(4):689–691. doi:10.1016/j.eururo.2014.04.008, Epub 2014 Apr 28
Pan G et al (1997a) An antagonist decoy receptor and a death domain-containing receptor for TRAIL. Science 277:815–818
Pan G et al (1997b) The receptor for the cytotoxic ligand TRAIL. Science 276:111–113
Papayannopoulos V, Zychlinsky A (2009) NETs: a new strategy for using old weapons. Trends Immunol 30:513–521

Pinsky CM et al (1985) Intravesical administration of bacillus Calmette-Guerin in patients with recurrent superficial carcinoma of the urinary bladder: report of a prospective, randomized trial. Cancer Treat Rep 69:47–53
Prescott S, James K, Hargreave TB, Chisholm GD, Smyth JF (1992) Intravesical Evans strain BCG therapy: quantitative immunohistochemical analysis of the immune response within the bladder wall. J Urol 147:1636–1642
Ratliff TL, Palmer JO, McGarr JA, Brown EJ (1987a) Intravesical Bacillus Calmette-Guerin therapy for murine bladder tumors: initiation of the response by fibronectin-mediated attachment of Bacillus Calmette-Guerin. Cancer Res 47:1762–1766
Ratliff TL, Gillen D, Catalona WJ (1987b) Requirement of a thymus dependent immune response for BCG-mediated antitumor activity. J Urol 137:155–158
Ratliff TL, Ritchey JK, Yuan JJ, Andriole GL, Catalona WJ (1993) T-cell subsets required for intravesical BCG immunotherapy for bladder cancer. J Urol 150:1018–1023
Redelman-Sidi G, Glickman MS, Bochner BH (2014) The mechanism of action of BCG therapy for bladder cancer–a current perspective. Nat Rev Urol 11:153–162
Rentsch CA et al (2014) Bacillus Calmette-Guerin strain differences have an impact on clinical outcome in bladder cancer immunotherapy. Eur Urol 66(4):677–688. doi:10.1016/j.eururo.2014.02.061, Epub 2014 Mar 12
Riemensberger J, Bohle A, Brandau S (2002) IFN-gamma and IL-12 but not IL-10 are required for local tumour surveillance in a syngeneic model of orthotopic bladder cancer. Clin Exp Immunol 127:20–26
Ritz N, Curtis N (2009) Mapping the global use of different BCG vaccine strains. Tuberculosis 89:248–251
Rosenberg SA (1975) Future prospects for immunotherapy. Cancer 36:821–824
Rosenthal SR et al (1972) BCG vaccination and leukemia mortality. JAMA 222:1543–1544
Saint F et al (2001) T helper 1/2 lymphocyte urinary cytokine profiles in responding and nonresponding patients after 1 and 2 courses of bacillus Calmette-Guerin for superficial bladder cancer. J Urol 166:2142–2147
Saint F et al (2002) Prognostic value of a T helper 1 urinary cytokine response after intravesical bacillus Calmette-Guerin treatment for superficial bladder cancer. J Urol 167:364–367
Sarica K et al (1995) Evaluation of cellular immunity following bacillus Calmette-Guerin therapy in patients with superficial bladder cancer. Urol Int 54:137–141
Sarosdy MF, Lamm DL (1989) Long-term results of intravesical bacillus Calmette-Guerin therapy for superficial bladder cancer. J Urol 142:719–722
Schmitz-Drager BJ et al (2014 Oct) Considerations on the use of urine markers in the management of patients with low-/intermediate-risk non-muscle invasive bladder cancer. Urol Oncol 32(7):1061–1068. doi:10.1016/j.urolonc.2013.10.010, Epub 2014 Jan 9
Schussler O et al (2006) Postoperative pneumonia after major lung resection. Am J Respir Crit Care Med 173:1161–1169
Schwentner C, Stenzl A, Gakis G (2012) Monitoring high-risk bladder cancer. Curr Opin Urol 22:421–426
Secanella-Fandos S, Luquin M, Julian E (2013) Connaught and Russian strains showed the highest direct antitumor effects of different Bacillus Calmette-Guerin substrains. J Urol 189:711–718
Sengiku A et al (2013) A prospective comparative study of intravesical bacillus Calmette-Guerin therapy with the Tokyo or Connaught strain for nonmuscle invasive bladder cancer. J Urol 190:50–54
Sharma P et al (2003) Frequency of NY-ESO-1 and LAGE-1 expression in bladder cancer and evidence of a new NY-ESO-1 T-cell epitope in a patient with bladder cancer. Cancer Immun 3:19
Sheridan JP et al (1997) Control of TRAIL-induced apoptosis by a family of signaling and decoy receptors. Science 277:818–821
Siegel R, Naishadham D, Jemal A (2013) Cancer statistics, 2013. CA Cancer J Clin 63:11–30
Sievert KD et al (2009) Economic aspects of bladder cancer: what are the benefits and costs? World J Urol 27:295–300

Singer S et al (2013) Quality of life in patients with muscle invasive and non-muscle invasive bladder cancer. Support Care Cancer 21:1383–1393

Suttmann H, Lehan N, Bohle A, Brandau S (2003) Stimulation of neutrophil granulocytes with Mycobacterium bovis bacillus Calmette-Guerin induces changes in phenotype and gene expression and inhibits spontaneous apoptosis. Infect Immun 71:4647–4656

Suttmann H et al (2006) Neutrophil granulocytes are required for effective Bacillus Calmette-Guerin immunotherapy of bladder cancer and orchestrate local immune responses. Cancer Res 66:8250–8257

Sylvester RJ, van der Meijden AP, Lamm DL (2002) Intravesical bacillus Calmette-Guerin reduces the risk of progression in patients with superficial bladder cancer: a meta-analysis of the published results of randomized clinical trials. J Urol 168:1964–1970

Sylvester RJ et al (2006) Predicting recurrence and progression in individual patients with stage Ta T1 bladder cancer using EORTC risk tables: a combined analysis of 2596 patients from seven EORTC trials. Eur Urol 49:466–465; discussion 475–467

Takeuchi A et al (2011) IL-17 production by gammadelta T cells is important for the antitumor effect of Mycobacterium bovis bacillus Calmette-Guerin treatment against bladder cancer. Eur J Immunol 41:246–251

Tan JK, Ho VC (1993) Pooled analysis of the efficacy of bacille Calmette-Guerin (BCG) immunotherapy in malignant melanoma. J Dermatol Surg Oncol 19:985–990

Tecchio C et al (2004) IFNalpha-stimulated neutrophils and monocytes release a soluble form of TNF-related apoptosis-inducing ligand (TRAIL/Apo-2 ligand) displaying apoptotic activity on leukemic cells. Blood 103:3837–3844

Thalmann GN, Dewald B, Baggiolini M, Studer UE (1997) Interleukin-8 expression in the urine after bacillus Calmette-Guerin therapy: a potential prognostic factor of tumor recurrence and progression. J Urol 158:1340–1344

Thalmann GN et al (2000) Urinary Interleukin-8 and 18 predict the response of superficial bladder cancer to intravesical therapy with bacillus Calmette-Guerin. J Urol 164:2129–2133

Torrado E, Robinson RT, Cooper AM (2011) Cellular response to mycobacteria: balancing protection and pathology. Trends Immunol 32:66–72

Watanabe E et al (2003) Urinary interleukin-2 may predict clinical outcome of intravesical bacillus Calmette-Guerin immunotherapy for carcinoma in situ of the bladder. Cancer Immunol Immunother 52:481–486

Wiley SR et al (1995) Identification and characterization of a new member of the TNF family that induces apoptosis. Immunity 3:673–682

Winters WD, Lamm DL (1981) Antibody responses to Bacillus Calmette-Guerin during immunotherapy in bladder cancer patients. Cancer Res 41:2672–2676

Wolfe SA, Tracey DE, Henney CS (1976) Introduction of "natural" killer' cells by BCG. Nature 262:584–586

Zbar B, Rapp HJ (1974) Immunotherapy of guinea pig cancer with BCG. Cancer 34(suppl):1532–1540

Zbar B, Tanaka T (1971) Immunotherapy of cancer: regression of tumors after intralesional injection of living Mycobacterium bovis. Science 172:271–273

Zhao W et al (2000) Role of a bacillus Calmette-Guerin fibronectin attachment protein in BCG-induced antitumor activity. Int J Cancer 86:83–88

Zuiverloon TC et al (2012) Markers predicting response to bacillus Calmette-Guerin immunotherapy in high-risk bladder cancer patients: a systematic review. Eur Urol 61:128–145

Index

M.R. Shurin et al. (eds.), *Infection and Cancer: Bi-Directorial Interactions*, DOI 10.1007/978-3-319-20669-1

MIX
Papier aus verantwortungsvollen Quellen
Paper from responsible sources
FSC® C105338

If you have any concerns about our products,
you can contact us on
ProductSafety@springernature.com

In case Publisher is established outside the EU,
the EU authorized representative is:
Springer Nature Customer Service Center GmbH
Europaplatz 3, 69115 Heidelberg, Germany

Printed by Libri Plureos GmbH
in Hamburg, Germany